**Chiral Ferrocenes in
Asymmetric Catalysis**

*Edited by
Li-Xin Dai and Xue-Long Hou*

Related Titles

Stepnicka, P. (ed.)

Ferrocenes

Ligands, Materials and Biomolecules

2008
ISBN: 978-0-470-03585-6

Carreira, E. M., Kvaerno, L.

Classics in Stereoselective Synthesis

2009
ISBN: 978-3-527-32452-1

Mohr, F. (ed.)

Gold Chemistry

Applications and Future Directions in the Life Sciences

2009
ISBN: 978-3-527-32086-8

Plietker, B. (ed.)

Iron Catalysis in Organic Chemistry

Reactions and Applications

2008
ISBN: 978-3-527-31927-5

Hudlicky, T., Reed, J. W.

The Way of Synthesis

Evolution of Design and Methods for Natural Products

2007
ISBN: 978-3-527-32077-6

Chiral Ferrocenes in Asymmetric Catalysis

Synthesis and Applications

Edited by
Li-Xin Dai and Xue-Long Hou

WILEY-VCH

WILEY-VCH Verlag GmbH & Co. KGaA

The Editors

Prof. Dr. Li-Xin Dai
State Key Laboratory of Organometallic Chemistry
Shanghai Institute of Organic Chemistry
Chinese Academy of Sciences
354 Fenglin Lu
Shanghai 200032
People's Republic of China

Prof. Dr. Xue-Long Hou
State Key Laboratory of Organometallic Chemistry
Shanghai Institute of Organic Chemistry
Chinese Academy of Sciences
354 Fenglin Lu
Shanghai 200032
People's Republic of China

All books published by Wiley-VCH are carefully produced. Nevertheless, authors, editors, and publisher do not warrant the information contained in these books, including this book, to be free of errors. Readers are advised to keep in mind that statements, data, illustrations, procedural details or other items may inadvertently be inaccurate.

Library of Congress Card No.: applied for

British Library Cataloguing-in-Publication Data
A catalogue record for this book is available from the British Library.

Bibliographic information published by the Deutsche Nationalbibliothek
The Deutsche Nationalbibliothek lists this publication in the Deutsche Nationalbibliografie; detailed bibliographic data are available on the Internet at http://dnb.d-nb.de.

© 2010 WILEY-VCH Verlag GmbH & Co. KGaA, Weinheim

All rights reserved (including those of translation into other languages). No part of this book may be reproduced in any form – by photoprinting, microfilm, or any other means – nor transmitted or translated into a machine language without written permission from the publishers. Registered names, trademarks, etc. used in this book, even when not specifically marked as such, are not to be considered unprotected by law.

Typesetting Thomson Digital, Noida, India
Printing and Binding T.J. International Ltd, Padstow, Cornwall
Cover Design Grafik-Design Schulz, Fußgönheim

Printed in Great Britain
Printed on acid-free paper

ISBN: 978-3-527-32280-0

Contents

List of Contributors *XV*

1 **Introduction** *1*
Li-Xin Dai and Xue-Long Hou
1.1 Foreword *1*
1.2 Planar Chirality of Ferrocenyl Ligands *4*
1.3 Derivatization of the Ferrocene Scaffold *6*
1.4 Stability, Rigidity and Bulkiness of the Ferrocene Scaffold *9*
1.5 Outlook *10*
References *11*

2 **Stereoselective Synthesis of Planar Chiral Ferrocenes** *15*
Wei-Ping Deng, Victor Snieckus, and Costa Metallinos
2.1 Introduction *15*
2.2 Diastereoselective Directed ortho-Metalation of Ferrocenes with Chiral Auxiliaries *19*
2.2.1 Ugi's Amine and Other Chiral α-Substituted Ferrocenylmethylamines *20*
2.2.2 Chiral Sulfoxides *23*
2.2.3 Chiral Acetals *26*
2.2.4 Chiral Oxazolines *29*
2.2.5 Other Chiral Auxiliaries *31*
2.3 Enantioselective Directed ortho-Metalation of Ferrocenes with Chiral Bases *36*
2.4 Enzymatic and Nonenzymatic Kinetic Resolutions *40*
2.4.1 Enzymatic Kinetic Resolutions *40*
2.4.2 Nonenzymatic Kinetic Resolutions *41*
2.5 Summary and Perspectives *42*
2.6 Selected Experimental Procedures *44*

2.6.1	Synthesis of (R,S_p)-N,N-dimethyl-1-(2-diphenylphosphino-ferrocenyl) ethylamine (PPFA) *44*	
2.6.1.1	Synthesis of Racemic α-Ferrocenylethyldimethylamine *44*	
2.6.1.2	Resolution of α-Ferrocenylethyldimethylamine *45*	
2.6.1.3	Synthesis of (R,S_p)-N,N-dimethyl-1-(2-diphenylphosphinoferrocenyl) ethylamine (PPFA) *45*	
2.6.2	Synthesis of (R,R_p)-1-(tert-butylsulfinyl)-2-(diphenylphosphino)- ferrocene *45*	
2.6.2.1	Synthesis of (R)-tert-butylsulfinylferrocene (R) *45*	
2.6.2.2	Synthesis of (R,R_p)-1-(tert-Butylsulfinyl)-2-(diphenylphosphino) ferrocene *46*	
2.6.3	Synthesis of FcPHOX *46*	
2.6.3.1	Synthesis of (S)-2-ferrocenyl-4-(1-methylethyl)oxazoline *46*	
2.6.3.2	Synthesis of (S,S_p)-2-[2-(diphenylphosphino)ferrocenyl]-4- (1-methylethyl) oxazoline *47*	
2.6.4	Synthesis of (S_p)-N,N-diisopropyl-2-(diphenylphosphino) ferrocenecarboxamide *48*	
2.6.4.1	(S_p)-N,N-diisopropyl-2-(diphenylphosphino) ferrocenecarboxamide *48*	
	References *49*	
3	**Monodentate Chiral Ferrocenyl Ligands** *55*	
	Ji-Bao Xia, Timothy F. Jamison, and Shu-Li You	
3.1	Introduction *55*	
3.2	Nickel-Catalyzed Asymmetric Reductive Coupling Reactions *55*	
3.2.1	Synthesis of P-Chiral, Monodentate Ferrocenyl Phosphines *56*	
3.2.2	Nickel-Catalyzed Asymmetric Reductive Coupling of Alkynes and Aldehydes *56*	
3.2.3	Nickel-Catalyzed Asymmetric Reductive Coupling of Alkynes and Ketones *60*	
3.2.4	Nickel-Catalyzed Asymmetric Three-Component Coupling of Alkynes, Imines, and Organoboron Reagents *61*	
3.3	Copper(I)-Catalyzed Asymmetric Allylic Alkylation Reactions *62*	
3.4	Asymmetric Suzuki–Miyaura Reactions *64*	
3.5	Addition of Organoaluminum to Aldehydes and Enones *65*	
3.6	Asymmetric Nucleophilic Catalysis *66*	
3.7	Conclusion and Perspectives *70*	
	References *70*	
4	**Bidentate 1,2-Ferrocenyl Diphosphine Ligands** *73*	
	Hans-Ulrich Blaser and Matthias Lotz	
4.1	Introduction *73*	
4.2	Type A Both PR_2 Groups Attached to the Cp Ring *74*	
4.3	Type B One PR_2 Group Attached to the Cp Ring, one PR_2 Group Attached to the α-Position of the Side Chain *75*	

4.3.1	Josiphos	76
4.3.2	Immobilized Josiphos	80
4.3.3	Josiphos Analogs	82
4.4	Type C One PR_2 Group Attached to the Cp Ring, one PR_2 Group Attached to the β-Position of the Side Chain	83
4.4.1	BoPhoz and Analogs	83
4.5	Type D, One PR_2 Group Attached to the Cp Ring, One PR_2 Group Attached to Other Positions of the Side Chain	85
4.5.1	Taniaphos	86
4.6	Type E, Both PR_2 Groups Attached to Side Chains	87
4.6.1	Walphos	88
4.6.2	TRAP	90
	References	91
5	**1,2-*P,N*-Bidentate Ferrocenyl Ligands**	**97**
	Yong Gui Zhou and Xue Long Hou	
5.1	Introduction	97
5.2	Asymmetric Hydrogenation and Asymmetric Transfer Hydrogenation	97
5.2.1	Asymmetric Hydrogenation	97
5.2.1.1	Asymmetric Hydrogenation of Alkenes	98
5.2.1.2	Asymmetric Hydrogenation of Ketones	101
5.2.1.3	Asymmetric Hydrogenation of Heteroaromatic Compounds	103
5.2.1.4	Asymmetric Hydrogenation of Imines	104
5.2.2	Asymmetric Transfer Hydrogenation	106
5.2.3	Asymmetric Hydrosilylation Reaction	109
5.2.3.1	Palladium-Catalyzed Asymmetric Hydrosilylation of C=C Bond	110
5.2.3.2	Asymmetric Hydrosilylation of a C=O Bond	112
5.2.3.3	Asymmetric Hydrosilylation of the C=N Bond	117
5.2.4	Asymmetric Hydroboration	118
5.3	Formation of a C–C Bond	121
5.3.1	Pd-Catalyzed Asymmetric Allylic Substitution Reaction	121
5.3.2	Asymmetric Heck Reaction and Cross-Coupling Reactions	126
5.3.3	Addition of Organometallic Reagents to a C=X Bond	129
5.4	Cycloaddition Reactions	132
5.4.1	[3 + 2] Cycloaddition Reactions	132
5.4.2	Asymmetric Cyclopropanation Reactions	136
5.4.3	Asymmetric Diels-Alder Reactions	136
5.5	Miscellaneous Reactions	137
5.6	Conclusion and Perspectives	140
5.7	Experimental: Selected Procedures	140
5.7.1	Ugi's Amine Synthesis	140
5.7.1.1	Synthesis of α-Ferrocenylethyldimethylamine (2)	141
5.7.1.2	Resolution of 2	141

5.7.1.3	Synthesis of (S)-N, N-Dimethyl-1-[(R)-2-(diphenylphosphino) ferrocenyl]ethylamine *141*	
5.7.2	(S, S$_p$)-2-[(S)-2-(diphenylphosphino)ferrocenyl]-4-(1-methylethyl) oxazoline *142*	
5.7.2.1	Synthesis of (2S)-N-(1-hydroxy-3-methylbutyl)ferrocenamide *142*	
5.7.2.2	Synthesis of Ferrocenyloxazoline *142*	
5.7.2.3	Synthesis of (S, S$_p$)-2-[(S)-2-(diphenylphosphino)ferrocenyl]-4-(1-methylethyl)oxazoline *143*	
	References *144*	
6	**N,O-Bidentate Ferrocenyl Ligands** *149*	
	Anne Nijs, Olga García Mancheño, and Carsten Bolm	
6.1	Introduction *149*	
6.2	Addition of Organozinc Reagents to Aldehydes *149*	
6.2.1	Addition of Dialkylzinc *150*	
6.2.1.1	With Ferrocenyl Amino Alcohols *150*	
6.2.1.2	With Ferrocenyl Oxazolinyl Alcohols *158*	
6.2.1.3	With Azaferrocenes *160*	
6.2.2	Addition of Arylzinc *161*	
6.2.3	Addition of Phenylacetylene *165*	
6.3	Addition to Aldehydes with Boron Reagents *165*	
6.3.1	Aryl Transfers with Boronic Esters *165*	
6.3.2	Aryl Transfer with Triphenylborane [Ph$_3$B] *166*	
6.3.3	Boronic Acids as Aryl Source *167*	
6.3.4	Alkenylboronic Acids in Alkenyl Transfer Reactions *168*	
6.4	Other Transformations: Asymmetric Epoxidation *168*	
6.5	Conclusion and Perspectives *169*	
6.6	Experimental: Selected Procedures *170*	
6.6.1	Addition of Diethylzinc to Benzaldehyde with Ferrocene 13 *170*	
6.6.2	Phenylacetylene Addition to Aldehydes with Ferrocene 24 *170*	
6.6.3	Phenyl Transfer (ZnEt$_2$ + ZnPh$_2$) to Aldehydes with Ferrocene 13 *171*	
6.6.4	Phenyl Transfer to Aldehydes with Triphenylborane on a Gram Scale *171*	
6.6.5	Aryl Transfer to Aldehydes with Boronic Acids on a Multigram Scale *171*	
	References *173*	
7	**Symmetrical 1,1'-Bidentate Ferrocenyl Ligands** *175*	
	Wanbin Zhang and Delong Liu	
7.1	Introduction *175*	
7.2	Symmetrical 1,1'-Disubstituted Ferrocenyl Ligands *177*	
7.2.1	P-Centered Chiral Diphosphine Ligands *177*	
7.2.2	C-Centered Chiral Diphosphine Ligands *180*	
7.2.3	Chiral Nitrogen-Containing Ligands *188*	
7.3	Symmetrical 1,1',2,2'-Tetrasubstituted Ferrocenyl Ligands *193*	

7.3.1	Tetrasubstituted Ferrocenyl Diphosphine Ligands with Multi-Chiralities *193*	
7.3.2	Tetrasubstituted Ferrocenyl Ligands with Only Planar Chirality *198*	
7.3.3	Other Tetrasubstituted Ferrocenyl Ligands *201*	
7.4	Analogs of Ferrocenes: Symmetrical 1,1'-Bidentate Ruthenocenyl Ligands *203*	
7.5	Conclusion and Perspectives *208*	
7.6	Experimental: Selected Procedures *208*	
7.6.1	Typical Procedure for the Preparation of Ferrocenyldiphosphines L-2 *208*	
7.6.2	Typical Procedure for the Preparation of the (*R*,*R*)-ferrocenyl Diol 14 *208*	
7.6.3	Typical Procedure for the Preparation of ($\alpha R,\alpha' R$)-2,2'-Bis(α-*N*,*N*-dimethylaminopropyl)-(*S*,*S*)-1,1'-bis(diphenylphosphino)ferrocene L-17c *209*	
7.6.4	Typical Procedure for the Preparation of C_2-Symmetric 1,1',2,2'-tetrasubstituted Ferrocene Derivatives L-18 *209*	
7.6.5	Typical Procedure for the Preparation of the Ferrocene-Based Ester Amide 59 *209*	
7.6.6	Typical Procedure for the Preparation of (−)-(*S*)-(*S*)-1,1'-Bis(diphenylphosphino)-2,2'-bis-(methoxycarbonyl)ferrocene L-21 *210*	
7.6.7	Typical Procedure for the Preparation of 1,1'-Bis[(*S*)-4-isopropyloxazolin-2-yl]-ruthenocene 82 *210*	
	References *211*	
8	**Unsymmetrical 1,1'-Bidentate Ferrocenyl Ligands** *215*	
	Shu-Li You	
8.1	Introduction *215*	
8.2	Palladium-Catalyzed Asymmetric Allylic Substitution Reaction *215*	
8.2.1	A Model Reaction of Symmetrical 1,3-Disubstituted 2-Propenyl Acetates *216*	
8.2.2	Substrate Variants of Palladium-Catalyzed Asymmetric Allylic Substitution Reaction *221*	
8.2.3	Regioselective Control for Unsymmetrical Allylic Acetates *224*	
8.2.4	Applications of Palladium-Catalyzed Asymmetric Allylic Substitution Reaction *228*	
8.3	Gold or Silver-Catalyzed Asymmetric Aldol Reactions *229*	
8.3.1	Gold-Catalyzed Asymmetric Aldol Reactions *229*	
8.3.2	Silver-Catalyzed Asymmetric Aldol Reactions *232*	
8.3.3	Applications of Gold or Silver-Catalyzed Asymmetric Aldol Reactions *232*	
8.4	Asymmetric Hydrogenation *234*	
8.4.1	Rh-Catalyzed Hydrosilylation *234*	
8.4.2	Rh, Ir, Ru-Catalyzed Hydrogenation *235*	
8.4.3	Rh-Catalyzed Hydroboration *238*	

8.5	Asymmetric Cross-Coupling Reaction 239
8.5.1	Nickel-Catalyzed Kumada Coupling Reaction 239
8.5.2	Palladium-Catalyzed Cross-Coupling Reaction 240
8.5.3	Palladium-Catalyzed Suzuki–Miyaura Reaction and α-Arylation of Amides 241
8.6	Asymmetric Heck Reaction 242
8.6.1	Intramolecular Heck Reaction 242
8.6.2	Intermolecular Heck Reaction 243
8.7	Miscellaneous 244
8.7.1	Addition of Zinc Reagent to Aldehydes 244
8.7.2	Conjugate Addition 247
8.7.3	Rh-Catalyzed Ring Opening Reaction of Oxabenzonorbornadiene 247
8.7.4	Asymmetric Cyclopropanation of Alkenes with Diazo Esters 248
8.7.5	Asymmetric Palladium-Catalyzed Hydroesterification of Styrene 249
8.8	Conclusion and Perspectives 250
8.9	Experimental: Selected Procedures 250
8.9.1	Synthesis of Compound (S,Sp)-L4d from (S)-5a 250
8.9.2	Synthesis of Compound (S,R_{phos},R)-L8d and (S,S_{phos},R)-L8d from (S)-5d 251
	References 253

9	**Sulfur- and Selenium-Containing Ferrocenyl Ligands** **257**
	Juan C. Carretero, Javier Adrio, and Marta Rodríguez Rivero
9.1	Introduction 257
9.2	Asymmetric Allylic Substitution 258
9.2.1	Palladium-Catalyzed Allylic Substitution 258
9.2.1.1	Bidentate N,S-ferrocene Ligands 259
9.2.1.2	Bidentate P,S-ferrocene Ligands 260
9.2.2	Copper-Catalyzed Allylic Substitution 265
9.3	Other Asymmetric Palladium-Catalyzed Reactions 266
9.3.1	Asymmetric Heck Reactions 266
9.3.2	Desymmetrization of Heterobicyclic Alkenes 267
9.4	Gold-Catalyzed Reactions 268
9.5	Asymmetric Reductions 269
9.6	Asymmetric 1,2- and 1,4-Nucleophilic Addition 270
9.6.1	Asymmetric 1,2 Addition to Aldehydes and Imines 270
9.6.2	Asymmetric 1,4-Conjugate Addition 272
9.7	Asymmetric Cycloaddition Reactions 272
9.7.1	[4 + 2] Cycloadditions 272
9.7.2	[3 + 2] Cycloadditions 274
9.8	Asymmetric Nucleophilic Catalysis 276
9.9	Conclusion and Perspectives 277
9.10	Experimental: Selected Procedures 278
9.10.1	Palladium-Catalyzed Allylic Alkylation Reaction with Ligand 3a 278

9.10.2	Fesulphos/Palladium-Catalyzed Asymmetric Ring Opening of Bicyclic Alkenes	278
9.10.3	Rhodium-Catalyzed Hydrogenation α-(N-acetamido)acrylate in the Presence of Ligand 20	278
9.10.4	Cu-Fesulphos Catalyzed Formal Aza Diels–Alder Reactions of N-Sulfonyl Imines	278
	References	280
10	**Biferrocene Ligands**	*283*
	Ryoichi Kuwano	
10.1	Introduction	283
10.2	Trans-Chelating Chiral Bisphosphines: TRAP	284
10.2.1	Synthesis of TRAP	285
10.2.2	Metal Complexes of TRAP	286
10.2.3	Asymmetric Reactions of 2-Cyanocarboxylates	287
10.2.4	Asymmetric Hydrosilylation of Ketones	291
10.2.5	Asymmetric Hydrogenation of Olefins	292
10.2.6	Asymmetric Hydrogenations of Heteroaromatics	296
10.2.7	Miscellaneous Reactions using TRAP	298
10.3	2,2-Bis(diarylphosphino)-1,1-biferrocenes: BIFEP	299
10.4	Miscellaneous Biferrocene-Based Chiral Ligands	302
10.5	Conclusion	303
	References	304
11	**Applications of Aza- and Phosphaferrocenes and Related Compounds in Asymmetric Catalysis**	*307*
	Nicolas Marion and Gregory C. Fu	
11.1	Introduction	307
11.2	Background on Aza- and Phosphaferrocenes	308
11.2.1	Azaferrocenes	308
11.2.2	Phosphaferrocenes	309
11.3	Azaferrocenes in Catalysis	310
11.3.1	Nucleophilic Catalysis	310
11.3.1.1	Acylations	310
11.3.1.2	Halogenations	316
11.3.1.3	Cycloadditions	317
11.3.1.4	Nucleophilic Additions to Carbonyl Groups	319
11.3.2	Transition-Metal Catalysis	319
11.3.2.1	Copper-Catalyzed Cyclopropanations of Olefins	319
11.3.2.2	Copper-Catalyzed Insertions of Diazo Compounds	320
11.3.2.3	Copper-Catalyzed Cycloadditions	322
11.3.2.4	Olefin Polymerizations	322
11.3.2.5	Rhodium-Catalyzed Reductions of Ketones	324
11.4	Phosphaferrocenes in Catalysis	324
11.4.1	Reduction Reactions	325

11.4.2	Palladium-Catalyzed Allylic Alkylations 326
11.4.3	Rhodium-Catalyzed Isomerizations 328
11.4.4	Copper-Catalyzed Conjugate Additions 328
11.4.5	Copper-Catalyzed Cycloadditions 329
11.4.6	Palladium-Catalyzed Cross-Couplings 331
11.5	Conclusions 333
	References 333
12	**Metallocyclic Ferrocenyl Ligands** 337
	Christopher J. Richards
12.1	Introduction 337
12.2	Asymmetric Synthesis of Planar Chiral Metallocyclic Complexes 339
12.2.1	Resolution 341
12.2.2	Diastereotopic C–H Activation 342
12.2.3	Oxidative Addition of Palladium(0) 348
12.2.4	Transmetalation 350
12.2.5	Enantioselective Palladation 352
12.3	Stoichiometric Synthetic Applications of Scalemic Planar Chiral Metallocyclic Complexes 354
12.3.1	Phosphine Recognition 354
12.3.2	Generation of Planar Chiral Ferrocene Derivatives by M–C Bond Cleavage 354
12.4	Asymmetric Catalysis with Scalemic Planar Chiral Palladocyclic Complexes 355
12.4.1	Allylic Imidate Rearrangement 356
12.4.2	Intramolecular Aminopalladation 363
12.5	Conclusion 364
	References 364
A	**Show Case of the Most Effective Chiral Ferrocene Ligands in Various Catalytic Reactions** 369
A.1	Asymmetric Allylic Substitution Reactions 369
A.1.1	Copper(I)-Catalyzed Asymmetric Allylic Alkylation Reactions 369
A.1.2	Palladium-Catalyzed Asymmetric Allylic Substitution Reactions 370
A.1.2.1	Alkylation 370
A.1.2.2	Amination 371
A.1.3	Generating π-Allyl Palladium Complex Through the Reaction of Allene with Iodobenzene 372
A.1.4	Regioselective Control Concerned with Unsymmetrically 1,3-Disubstituted 2-Propenyl Acetate 372
A.1.4.1	Regio- and Enantioselective Palladium-Catalyzed Allylic Substitution 372
A.1.4.2	Allylic Substitution of 2-Cyanopropionate 373
A.1.4.3	Allylic Substitution of the Conjugated Dienyl Acetates 373

A.1.4.4	Allylic Substitution with Acyclic Ketone Enolate as Nucleophile	374
A.1.4.5	Allylic Alkylation with N,N-Disubstituted Amides as Nucleophile	374
A.1.4.6	Allylic Alkylation with Activated Cyclic Ketone as Nucleophile	374
A.1.5	Applications of Palladium-Catalyzed Asymmetric Allylic Substitution Reaction	375
A.2	Asymmetric Aldol Reactions	375
A.2.1	Gold-Catalyzed Asymmetric Aldol Reactions	375
A.2.1.1	Aldehyde with Isocyanoacetate	375
A.2.1.2	Imines with Isocyanoacetate	376
A.2.2	Silver-Catalyzed Asymmetric Aldol Reactions	376
A.2.3	Applications of Gold or Silver-Catalyzed Asymmetric Aldol Reactions	376
A.3	Asymmetric Cycloaddition Reactions	377
A.3.1	[3 + 2] Cycloaddition Reactions	377
A.3.2	[4 + 2] Cycloaddition Reactions	377
A.3.3	[3+2] Cycloaddition Reactions	378
A.3.4	Asymmetric 1,3-Dipolar Cycloaddition of Azomethine Ylides with N-Phenyl Maleimide	378
A.3.5	[3 + 2] Cycloaddition of Alkynes with Azomethine Imines	378
A.4	Asymmetric Hydrogenation	379
A.4.1	Asymmetric Hydrogenation of Alkenes (C=C)	379
A.4.1.1	Asymmetric Hydrogenation of Unfunctionalized Alkenes	379
A.4.1.2	Asymmetric Hydrogenation of α,β-Unsaturated Esters	379
A.4.1.3	Hydrogenation of Dehydro α-Amino Acid	380
A.4.1.4	Asymmetric Hydrogenation of Activated Imine C=N Bond	380
A.4.1.5	Applications: Hydrogenation of α,β-Unsaturated Acid	381
A.4.1.6	Asymmetric Hydrogenation of Alkyl α-Acetamido Acrylates	381
A.4.1.7	Industrial Applications	382
A.4.2	Asymmetric Hydrogenation of Alkenes (or C=N)	383
A.4.2.1	Ir-Catalyzed Asymmetric Hydrogenation of Quinolines	383
A.4.2.2	Industrial Applications	384
A.4.3	Asymmetric Hydrogenation of Ketones (C=O)	384
A.4.3.1	Asymmetric Hydrogenation of Acetophenone	384
A.4.3.2	Asymmetric Hydrogenation of β-Aminoketone	385
A.4.4	Asymmetric Hydroboration	385
A.4.5	Asymmetric Hydrosilylation Reaction	385
A.4.6	Asymmetric Transfer Hydrogenation	386
A.5	Pd-Catalyzed Asymmetric Heck Reaction	386
A.6	Addition of Organozinc Reagents	387
A.6.1	Addition of Dialkylzinc to Aldehydes	387
A.6.2	Addition of Diarylzinc to Aldehydes	388
A.6.3	Addition of Dialkylzinc to C=N Bond	388
A.6.4	Addition of Phenylacetylene to Aldehydes	389
A.6.5	Addition to Aldehydes with Boron Reagents	389
A.6.5.1	Aryl Transfer with Boronic Ester	389

A.6.5.2	Aryl Transfer with Triphenylborane	389
A.6.5.3	Aryl Transfer with Boronic Acids	390
A.7	Asymmetric Rearrangement of Allylic Imidates	390
A.7.1	Rearrangement of N-Arylbenzimidates	390
A.7.2	Rearrangement of N-Aryl-trifluoracetimidate	390
A.7.3	Rearrangement with Added Nucleophiles	391
A.8	Cu-Catalyzed Cyclopropanation	391
A.8.1	With Diphenylethene	391
A.8.2	With Aryl, Alkyl or Silylethene	391
A.9	Coupling Reaction of Vinyl Bromide and 1-Phenylethylzinc Chloride	392
A.10	Enantioselective Intramolecular Aminopalladation	392
A.11	Nickel-Catalyzed Asymmetric Three-Component Coupling of Alkynes, Imines, and Organoboron Reagents	392
A.12	Reactions with Ketenes	393
A.12.1	O-Acylation with Ketenes	393
A.12.2	Kinetic Resolution by O-Acylation	393
A.12.3	N-Acylation with Ketenes	394
A.12.4	Kinetic Resolution by N-Acylation	394
A.12.5	C-Acylation via Rearrangement	394
A.12.6	Cycloadditions	395
A.12.7	Copper-Catalyzed Insertions of Diazo Compounds	395
A.12.8	Copper-Catalyzed Cycloadditions	395
A.13	Ring Opening Reaction	396
A.13.1	Desymmetrization by Halogenations	396
A.13.2	Ring Opening of Bicyclic Alkenes	396

Index 397

List of Contributors

Javier Adrio
Universidad Autónoma de Madrid
Departamento de Química Orgánica
Facultad de Ciencias
Cantoblanco
28049 Madrid
Spain

Hans-Ulrich Blaser
Solvias AG
Klybeckstrasse 191
4057 Basel
Switzerland

Carsten Bolm
RWTH Aachen University
Institut für Organische Chemie
52056 Aachen
Germany

Juan C. Carretero
Universidad Autónoma de Madrid
Departamento de Química Orgánica
Facultad de Ciencias
Cantoblanco
28049 Madrid
Spain

Li-Xin Dai
State Key Laboratory of
Organometallic Chemistry
Shanghai Institute of Organic
Chemistry
Chinese Academy of Sciences
354 Fenglin Road
Shanghai 200032
People's Republic of China

Wei-Ping Deng
East China University
of Science and Technology
School of Pharmacy
130 Meilong Road
Shanghai 200237
People's Republic of China

Gregory C. Fu
Massachusetts Institute of Technology
Department of Chemistry
Cambridge
MA 02139
USA

List of Contributors

Xue-Long Hou
State Key Laboratory
of Organometallic Chemistry
Shanghai Institute of Organic
Chemistry
Chinese Academy of Sciences
354 Fenglin Road
Shanghai 200032
People's Republic of China

Timothy F. Jamison
Massachusetts Institute of Technology
Department of Chemistry
77 Massachusetts Avenue
Room 18-492
Cambridge
MA 02139
USA

Ryoichi Kuwano
Kyushu University
Department of Chemistry
Graduate School of Sciences
6-10-1 Hakozaki
Higashi-ku Fukuoka 812-8581
Japan

Delong Liu
Shanghai Jiao Tong University
School of Chemistry
and Chemical Technology
800 Dongchuan Road
Shanghai 200240
People's Republic of China

Matthias Lotz
Solvias AG
Klybeckstrasse 191
4057 Basel
Switzerland

Olga García Mancheño
RWTH Aachen University
Institut für Organische Chemie
52056 Aachen
Germany

Nicolas Marion
Massachusetts Institute of Technology
Department of Chemistry
Cambridge
MA 02139
USA

Costa Metallinos
Brock University
Department of Chemistry
500 Glenridge Avenue
St. Catharines
ON L2S 3A1
Canada

Anne Nijs
RWTH Aachen University
Institut für Organische Chemie
52056 Aachen
Germany

Christopher J. Richards
University of East Anglia
School of Chemical Sciences
and Pharmacy
Norwich
NR4 7TJ
UK

Marta Rodríguez Rivero
Universidad Autónoma de Madrid
Departamento de Química Orgánica
Facultad de Ciencias
Cantoblanco
28049 Madrid
Spain

Victor Snieckus
Queen's University
Department of Chemistry
90 Bader Lane Chernoff Hall
Kingston
ON K7L 3N6
Canada

Ji-Bao Xia
Shanghai Institute of Organic
Chemistry
State Key Laboratory
of Organometallic Chemistry
Chinese Academy of Sciences
354 Fenglin Lu
Shanghai 200032
People's Republic of China

Shu-Li You
Shanghai Institute of Organic
Chemistry
State Key Laboratory
of Organometallic Chemistry
Chinese Academy of Sciences
354 Fenglin Lu
Shanghai 200032
People's Republic of China

Wanbin Zhang
Shanghai Jiao Tong University
School of Chemistry
and Chemical Technology
800 Dongchuan Road
Shanghai 200240
People's Republic of China

Yong Gui Zhou
Chinese Academy of Sciences
Dalian Institute of Chemical Physics
457 Zhongshan Road
Dalian, Liaoning 116023
People's Republic of China

1
Introduction
Li-Xin Dai and Xue-Long Hou

1.1
Foreword

The unique substance, ferrocene, was discovered in 1951 by Kealy and Pauson, [1] and Miller *et al.* [2] independently. Although the discovery was serendipitous, the impact of the discovery was huge. It opened a new era of organometallic chemistry. The unusual stability and unique properties of this new compound contradicted the expected structure **1**, the target that Pauson and Kealy aimed to synthesize, by reaction of cyclopentadienylmagnesium bromide with $FeCl_3$. This structural mystery soon attracted the attention of many chemists, including three prominent chemists, Geoffrey Wilkinson [3], Robert B. Woodward [4], and Emil O. Fisher [5]. Based on the unusual stability, diamagnetic nature, single C–H stretching in the IR region and nonpolar character of the substance, Wilkinson and Woodward deduced the sandwich structure and named this substance "ferrocene" with the "ene" ending implying aromaticity, following the name of benzene. The sandwich structure was later confirmed by X-ray crystallographic analysis [5–7]. The unexpected structure was disclosed in 1952 as **2**, later simplified to **3**. The carbon–metal bond of the ferrocene structure was a new concept of π-bonding of a carbocyclic ring to a metal atom, which was a breakthrough and a major departure from the classical Wernerian model of ligand coordination that had prevailed until that time [8]. This novel finding soon initiated the preparation of this type of compound with other metals. In just two years a vast array of sandwich compounds with other metals were prepared by Wilkinson [9] and Fischer [10]. They made bis-cyclopentadienyl-metal and cyclopentadienylmetalcarbonyl compounds of most of the transition metals. Following the name of ferrocene came the term "metallocene".

Chiral Ferrocenes in Asymmetric Catalysis. Edited by Li-Xin Dai and Xue-Long Hou
Copyright © 2010 WILEY-VCH Verlag GmbH & Co. KGaA, Weinheim
ISBN: 978-3-527-32280-0

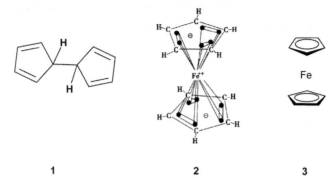

      1   2   3

Metallocene chemistry then became an important sub-branch of organometallic chemistry. Among numerous metallocenes prepared, an outstanding zirconacene 4 [11] has been used as a highly active, single site catalyst for olefin polymerization, which is a milestone in the field of olefin polymerization and a breakthrough in the post Ziegler–Natta era. In addition to the academic study of ferrocene, many applications of ferrocenyl derivatives have also been advanced, such as the control of the burning rate of solid propellant, fuel additives, liquid crystals, and even medicinal applications. In accord with the academic and practical significance of ferrocene, the *Journal of Organometallic Chemistry* published a special issue with 850 pages in 2001 [8], celebrating the fiftieth anniversary of the discovery of ferrocene. In addition, the IUPAC series conference on organometallic chemistry (ICOC) adopted 3 as a logo for the meeting, thus marking the significance of this unique compound, ferrocene.

The most important application to date is the use of ferrocenyl derivatives as catalysts, especially in homogeneous asymmetric catalysis. We should point out that Togni and Hayashi reviewed the pioneering studies in this area in their famous book *Ferrocene* in 1995 [12].

      4   5   6   7

For this monograph, we have concentrated on discussing the use of ferrocene as a ligand scaffold for asymmetric catalysis. The emergence of chiral technology in the past 30 years has led to a surge in the exploration of different ligands. Thousands or tens of thousands of ligands have been designed and synthesized. However, judged by the diversity of the ligands derived, the effectiveness and the stereoselectivity of the catalysts, ligands derived from three types of scaffold, the binaphthalene type 5, the spirobiindane type 6, and the ferrocene type 7 are the most important. In addition, there are many efficient ligands not covered by these three types. A few such ligands are briefly mentioned below. In the 1970s, Kagan et al. [13] applied a tartaric acid derived skeleton for C2-symmetric ligands such as DIOP 8. In addition to DIOP, DIOP* [14] and the well-known TADDOL catalysts, based on this scaffold, have also been developed [15]. Carbohydrate-based dioxiranes [16], developed by Shi, are highly efficient catalysts for the epoxidation reaction of unfunctionalized alkenes. However, only a limited number of efficient ligands derived from the carbohydrate have been developed to date. A simple benzene or cyclohexane skeleton also provides several efficient ligands. The best known are Phox [17] from benzene, and 1,2-cyclohexanediamine and its various derivatives from cyclohexane. The latter is represented by the Salen system which has been developed elegantly by Jacobsen [18] and Katsuki [19] for various catalytic enantioselective processes. A one-carbon system may be the simplest skeleton, ligands derived therefrom include, for example, miniphos [20] and Box [21].

It is true that there are many effective ligands not derived from the three above-mentioned types of skeleton, nevertheless, these three types occupy an important position in chiral ligand design. Various donor atoms such as N, P, O, and S could be introduced onto these skeletons, and a great number of ligand patterns could thus be derived. The binaphthalene 5 derived ligands enjoy a very high reputation and some of them have already found important industrial applications. BINAP (5, G=G′=PPh$_2$) [22], BINOL (5, G=G′=OH) [23], and MOP (5, G=OH, G′=PPh$_2$) [24] are some examples of this class.

In addition, Ferringa catalyst 9 [25], Yang's chiral dioxirane 10 [26], chiral Brønsted acids 11 [27], phase transfer catalyst 12 [28], tetrahydro- and octahydro derivatives [29, 30] are further examples derived from the binaphthalene skeleton. Furthermore, on the skeleton of 5, some supporting groups could be introduced at the 3-, 6-, or other positions to finely tune the steric and electronic properties of the ligands. Thus, it can be envisaged that the ligands derived from this binaphthalene skeleton will have very rich diversity. This type of ligand scaffold is characterized by the presence of an axial chirality and it is also manifested as an atrop-isomer. Similarly, some biphenyl and biheteroaryl ligands, such as Segphos **13** [31], Tunaphos **14** [32], and P-Phos **15** [33], are also very efficient, and this type may also be called bi-aryl ligands.

The second type, spirobiindane **6**, is the skeleton of several new ligands which have appeared only in this millennium [34]. Although no industrial scale application has been documented, a number of very recent reports [35] show that this unique skeleton really deserves to be regarded as one of the three important scaffolds. Almost all the derivatization of the binaphthalene system can be applied to the spirobiindane system, and many of the chiral ligands derived therefrom displayed better performance, either in efficiency or in enantioselectivity. Some examples are SIPHOX **6** (G = oxazoline, G′ = PAr$_2$); SDP **6** (G = G′ = PAr$_2$); SIPIROBOX **6** (G = G′ = oxazoline); SPINOL **6** (G = G′ = OH). The spirobiindane system is also associated with axial chirality. Fortunately, the synthesis of spirobiindanes is quite straightforward.

What makes the third type, the ferrocene skeleton, different from the other two types? The most significant feature is the planar chirality associated with this type in most cases, while the other two types are mainly associated with axial chirality. In a special design, multichiralities, central, planar and axial, could be installed in the catalyst derived from a single ferrocenyl ligand [36]. Furthermore, several ligands of this type have been used in practical applications, from kilogram to pilot plant scales and even to 10 000 ton y^{-1} production. Examples of the ferrocenyl ligands are Josiphos, Taniaphos, Walphos, Bophos, TRAP, Fc-Phox, PPFA, BPPFA, SiocPhox (see Chapter 8), and so on. Therefore, ligands derived from the ferrocene skeleton are really a class of importance, it is the aim of this book to give readers a comprehensive overview of ferrocenyl ligands in asymmetric catalysis.

In the following sections, various characteristics of the ferrocene scaffold that make it different from other types of skeleton will be discussed.

1.2
Planar Chirality of Ferrocenyl Ligands

Since the first preparation of an optically active planar chiral ferrocene 50 years ago [37] and the diastereoselective introduction of planar chirality from Ugi's amine in 1970 [38], planar chirality has become important in ligand design for enhancing or tuning the stereoselectivity of the ligand. Now comes the question: how to introduce carboxylic acid. In addition to the electrophilic reaction, a nucleophilic reaction could also be carried out easily for the derivatization of ferrocene. For example, treatment of bromoferrocene with BuLi could lead to various ferrocenyl compounds upon quenching with different electrophiles. Notably, Nesmeyanov's [50] and Benkeser's groups [51] were pioneers in studying the metalation reaction by using n-butyl lithium.

The lithioferrocene can react further to deliver many useful intermediates. Reaction with CO_2 leads to carboxylic acid and then to oxazolines on treating with amino alcohols. Reaction with PPh_2Cl could install a diphenylphosphino group on the Cp ring. The metalation reaction is even more powerful than the Friedel–Crafts reaction, given the recent reported selective metalation reactions. For example, n-butyl lithium converts ferrocene into 1,1′-dilithioferrocene in the presence of $N,N,N′,N′$-tetramethylethylenediamine (TMEDA) [52] (Scheme 1.2, Eq. 1) while the direct lithiation of ferrocene by *tert*-butyl lithium is a practical method for mono-lithioferrocene (in the presence of excess ferrocene). An improved method using *tert*-butyl lithium and super base (potassium *tert*-butoxide) provides the mono-lithio ferrocene in 90% yield (Scheme 1.2, Eq. 2) [53]. The metalation reaction is the most useful method for substituted ferrocenes. However, compared to the methods for preparing 1,2-disubstituted ferrocene, the preparation of 1,3-disubstituted ferrocene is less studied. Brown [54] developed a method for preparing 1,3-disubstituted ferrocene as shown in Scheme 1.2, Eq. 3. The 1,1′-disubstituted ferrocene could

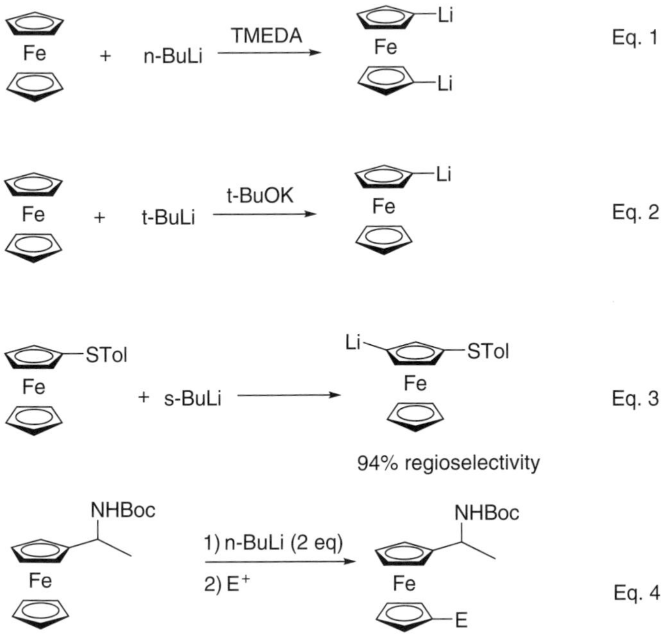

Scheme 1.2 Metalation reaction of ferrocene.

also be prepared by selective lithiation of a Boc-protected ferrocenylalkylamine [55] (Scheme 1.2, Eq. 4) or a ferrocenyl carbaldehyde [56].

With the availability of these efficient methods, partially represented by the above examples, it is very easy to introduce various donor groups or supporting groups onto the skeleton of ferrocene. Ferrocenes with different substitution patterns, such as di-, tri- or multi- substituted, and ferrocenes bearing groups with different electronic or steric properties could thus be synthesized easily in order to study the relationship between the structure and catalytic activity of the catalyst [57].

Hence, thousands of ferrocenyl ligands have been designed and prepared. In this book, the ferrocenyl ligands are discussed according to their substitution patterns. In Chapter 3, Tim Jamison of MIT and You, Xia of SIOC discuss the monodentate chiral ligands, mainly Jamison's work on the P-chiral phosphine ligand. Chapter 4 on bidentate 1,2-ferrocenyl diphosphine ligands is contributed by Blaser and Lotz of Solvias AG. They pay much attention to the industrial application of ferrocenyl ligands. Although the industrial application of asymmetric catalysis has recently been documented in two books [58], the authors of this chapter provide us with many up to date references. We trust that the industrial application of the ferrocenyl ligands will be highly interesting to the readers. Chapter 5, contributed by Zhou and Hou, deals with 1,2-N,P-bidentate ferrocenyl ligands. The ferrocene derivatized ligands are not limited to N- and P- donors, the N,O-bidentate ligands are also an important class. Carsten Bolm of Aachen and his group have contributed greatly to this subject and summarize the progress in this area in Chapter 6. The subsequent two chapters deal with 1,1'-bidentate ligands. The symmetrical 1,1'-bidentate ligands are discussed by Zhang and Liu of Jiaotong University of Shanghai, in Chapter 7, and the unsymmetrical 1,1'-bidentate ligands are discussed by You of SIOC in Chapter 8. In addition, the donor atoms could be further extended to S and Se, as discussed by Carretero, Adrio and Rivero in Chapter 9. Some of their ligands, such as Fesulphos, showed excellent performance. Since the appearance of DIOP **5** in the 1970s C_2-symmetry has become one of major importance in ligand design. The biferrocene ligands discussed in Chapter 10 are the incorporation of this strategy into the ferrocene system. Kuwano describes this type of ligand explicitly and TRAP is a representative of this type. The last two chapters describe special types of derivatization. In addition to changing the groups on the Cp-ring, the Cp-ring itself could be modified. In Chapter 11, Greg Fu of MIT tells us the story of aza- or phospha-ferrocene, in which one CH unit of the Cp ring is replaced by an N-atom or a P-atom. These unique systems and also the pyridino-Cp ring system delivered many very interesting results and showed us how vast the imagination is. In the last chapter, Christopher Richards of UEA (University of East Anglia) discusses metallocyclic ferrocenyl ligands, in which the complexing metal is bonded to an sp^2 carbon atom and becomes a direct component of the element of planar chirality, as illustrated in the example below. These palladacyclic ligands showed excellent selectivity in the asymmetric Overman rearrangement. Finally, an appendix has been added to help the readers access easily most of the significant ligands and reactions.

(R, S_p) [Structure of compound 19: ferrocene with Me, NMe₂, Pd substituents]

19

1.4
Stability, Rigidity and Bulkiness of the Ferrocene Scaffold

Ferrocene is a highly thermal stable compound. It volatilizes without decomposition into a monomeric undissociated vapor which obeys the perfect-gas law, even at 400 °C [59]. Ferrocene is resistant to pyrolysis at 470 °C in a N_2 atmosphere. It also has great chemical stability, which is important for its use as a ligand.

For a chiral ligand, the skeleton should not be too flexible so as to provide an appropriate chiral environment. Therefore, ring compounds with a certain rigidity are usually required for a ligand framework. In Section 1.1, the three types of good ligand skeleton, binaphthalene, spiroindane and ferrocene are all ring compounds with adequate rigidity. On the other hand, the skeleton should not be too rigid so as to allow complexation with a metal and the approach of the reactants. All these three types of skeleton have a certain limited degree of freedom of rotation or tilting. In the case of ferrocene, although the two cyclopentadienyl rings are parallel to each other, they can tilt to some extent if the stereo-environment requires it.

Bulkiness is another factor for consideration in designing chiral ligands. A bulkier group may shield one side from attack thus restricting the attack to only the re or si face. The bulkier groups used are usually *tert*-Bu, TMS, and mesityl. Obviously the volume demanded by a ferrocenyl group is high. Wang *et al.* applied the ferrocenyl group as a bulky substituent [60]. In Chapter 11, contributed by Greg Fu and Nicolas Marion, the elegantly designed catalysts, the azaferrocene and the planar chiral DMAP, contain a pentamethylcyclopentadienyl Cp* moiety. The Cp* moiety apparently enhances the shielding from attack of the lower parts, the catalyst thus becomes a highly selective and efficient one. Fu and Marion give us an insightful discussion of this novel class of catalysts. The bulky shielding effect could be further enhanced by using five phenyl groups or 3,5-dimethylphenyl groups instead of Me groups. In the kinetic resolution of a racemic secondary alcohol, we may perceive the effect of the steric demand of the ligand used. The selectivity factor on using DMAP is 2, while with a more sterically demanding ligand, C_5Ph_5, the selectivity factor increases dramatically to 43 [61]. (For the definition of the selectivity factor, please refer to Chapter 11). A further example is the desymmetrization of a meso-epoxide with $SiCl_4$ [62]. For this ligand, the nucleophilic site is not the N of the DMAP-derivative, but an oxygen atom of the *N*-oxide. As the N−O bond extends further from the pyridine ring, **22b** is not bulky enough to give good shielding while **22c** displays the best enantioselectivity.

[Scheme showing reactions with compounds 20, 21, and 22]

- **20** DMAP, S = 2
- **21** C₅Ph₅DMAP S = 43

a, R = Me, ee = 11% (at 0°C)
b, R = Ph, ee = 60%
c, R = 3,5-Me$_2$-C$_6$H$_3$, ee = 92%

22

1.5 Outlook

From the previous sections, the intrinsic characteristics of ferrocene as a chiral ligand in asymmetric catalysis have fully demonstrated why ferrocene is a truly important and wonderful scaffold for various chiral ligands. Since the preparation and application in asymmetric catalysis of the first chiral ferrocenyl ligands in 1974 by Kumada, Hayashi, and their coworkers, the development of chiral ferrocenyl ligands has been really fast. The use of Josiphos by Togni and Blaser in the catalytic asymmetric transfer hydrogenation of imine for the production of Metachlor still holds the record for the largest scale asymmetric catalysis, which is a really impressive achievement. We may say that the development of chiral ferrocenyl ligands has already reached a high point. When considering practical applications, the cost of the ligand has become a necessary consideration. Presently, ferrocene is a chemical in tonnage production by an electrolytic process. The price is around 65 000 RMB/ton (or 65 RMB/kg), that is around 10 US$/kg, 6–7 €/kg. [63] The price from Sigma-Aldrich (2009) is 1340 RMB/500 g. As described in the Introduction in Chapter 4, we may expect that new industrial applications of chiral ferrocenyl ligands will continue to appear. By checking the SciFinder explorer with "ferrocene" and "chiral" as the two keywords, 50 or more papers have appeared every year in this new millennium, thus we may also expect the development of more important chiral ferrocenyl ligands. We may be proud to say that ferrocene has played and is playing a vital role in both the academic world and in the industrial world serving mankind.

Acknowledgments

We sincerely thank all the authors for their contributions. They are all pioneers or major players in the areas related to the corresponding chapters, which assures the maintenance of a high standard throughout the whole book.

We are indebted to Dr. Tao Tu for compiling the Appendix during his short visit to Shanghai from the University of Bonn. We are grateful to Dr Elke Maase of Wiley–VCH who initiated and motivated our interest in editing this book and to Lesley Belfit for her help during the whole editing process. We also thank Professor Shu-Li You for instructive suggestions and Miss Di Chen for the typing work.

References

1 Kealy, T.J. and Pauson, P.L. (1951) *Nature*, **168**, 1039–1040.
2 Millers, S.A., Tebboth, J.A., and Tremaine, J.F. (1952) *J. Chem. Soc.*, 632–635. The submitting date of this paper was even earlier than that of reference [1].
3 Wilkinson, G., Rosenblum, M., Whiting, M.C., and Woodward, R.B. (1952) *J. Am. Chem. Soc.*, **74**, 2125–2126.
4 Woodward, R.B., Rosenblum, M., and Whiting, M.C. (1952) *J. Am. Chem. Soc.*, **74**, 3458–3459.
5 Fischer, E.O. and Pfab, W. (1952) *Z. Naturforsch.*, **76**, 377–379.
6 Eiland, P.F. and Pepinsky, R. (1952) *J. Am. Chem. Soc.*, **74**, 4971.
7 Dunitz, J.D. and Orgel, L.E. (1953) *Nature*, **171**, 121–122.
8 Adams, R.D.(ed.) (2001) Special Issue: 50th Anniversary of the Discovery of Ferrocene. *J. Organomet. Chem.*, **637–639**, 1.
9 Wilkinson, G., Pauson, P.L., and Cotton, F.A. (1954) *J. Am. Chem. Soc.*, **76**, 1970–1974.
10 Fischer, E.O., Seus, D., and Jira, R. (1953) *Z. Naturforsch.*, **86**, 692.
11 Alt, H.G. and Köppl, A. (2000) *Chem. Rev.*, **100**, 1205–1222.
12 Togni, A. and Hayashi, T. (eds) (1995) *Ferrocene*, VCH Verlagsgesellshaft, Weinheim.
13 Kagan, H.B. and Dang, T.P. (1972) *J. Am. Chem. Soc.*, **94**, 6429–6433.
14 Li, W. and Zhang, X. (2000) *J. Org. Chem.*, **65**, 5871–5874.
15 Seebach, D., Beck, A.K., and Heckel, A. (2001) *Angew. Chem. Int. Ed.*, **40**, 92–138.
16 Shi, Y. (2004) *Acc. Chem. Res.*, **37**, 488–496.
17 Helmchen, G. and Pfaltz, A. (2000) *Acc. Chem. Res.*, **33**, 336–345.
18 Jacobsen, E.N. (2000) *Acc. Chem. Res.*, **33**, 421–431.

19 Katsuki, T. (2002) *Adv. Synth. Catal.*, **344**, 131–147.
20 Imamoto, T. (2001) *Pure Appl. Chem.*, **73**, 373–376.
21 Johnson, J.C. and Evans, D.A. (2000) *Acc. Chem. Res.*, **33**, 325–335.
22 Miyashita, A., Yasuda, A., Takaya, H., Toriumi, K., Ito, T., Souchi, T., and Noyori, R. (1980) *J. Am. Chem. Soc.*, **102**, 7932–7934.
23 Brunel, J.M. (2005) *Chem. Rev.*, **105**, 857–898.
24 Hayashi, T. (2000) *Acc. Chem. Res.*, **33**, 354–362.
25 van den Berg, M., Minnaard, A.J., Schudde, E.P., Van Esch, J., De Vries, A.H.M., De Vries, J.G., and Ferringa, B.L. (2000) *J. Am. Chem. Soc.*, **122**, 11539–11540.
26 Yang, D. (2004) *Acc. Chem. Res.*, **37**, 497–505.
27 (a) Akiyama, T., Itoh, J., Yokota, K., and Fuchibe, K. (2004) *Angew. Chem. Int. Ed.*, **43**, 1566–1568; (b) Uraguchi, D. and Terada, M. (2004) *J. Am. Chem. Soc.*, **126**, 5356–5357.
28 (a) Ooi, T. and Maruoka, K. (2004) *Acc. Chem. Res.*, **37**, 526–533; (b) Hashimoto, T. and Maruoka, K. (2007) *Chem. Rev.*, **107**, 5656–5682.
29 Shen, X., Guo, H., and Ding, K. (2000) *Tetrahedron: Asymmetry*, **11**, 4321–4327.
30 Chan, A.S.C., Zhang, F.Y., and Yip, C.W. (1997) *J. Am. Chem. Soc.*, **119**, 4080–4081.
31 Saito, T., Yokozawa, T., Ishizaki, T., Moroi, T., Soyo, N., Miura, T., and Kumobayashi, H. (2001) *Adv. Synth. Catal.*, **343**, 264–267.
32 Zhang, Z.G., Qian, H., Longmire, J., and Zhang, X. (2000) *J. Org. Chem.*, **65**, 6223–6226.
33 Xu, L.J., Lam, K.H., Ji, J.X., Wu, J., Fan, Q.H., Lo, W.H., and Chan, A.S.C. (2005) *Chem. Comm.*, 1390–1392.
34 Xie, J.-H. and Zhou, Q.-L. (2008) *Acc. Chem. Res.*, **41**, 581–593.
35 (a) Yang, Y., Zhu, S.F., Zhou, C.Y., and Zhou, Q.L. (2008) *J. Am. Chem. Soc.*, **130**, 14052–14053; (b) Duan, H.F., Xie, J.H., Qiao, X.C., Wang, L.X., and Zhou, Q.L. (2008) *Angew. Chem. Int. Ed.*, **47**, 4351–4353; (c) Li, S., Zhu, S.F., Zhang, C.M., Song, S., and Zhou, Q.L. (2008) *J. Am. Chem. Soc.*, **130**, 8584–8585; (d) Zhang, Y.Z., Zhu, S.F., Wang, L.X., and Zhou, Q.L. (2008) *Angew. Chem. Int. Ed.*, **47**, 8496–8498; (e) Hou, G.H., Xie, J.H., Yan, P.C., and Zhou, Q.L. (2009) *J. Am. Chem. Soc.*, **131**, 1366–1367; (f) Xie, J.H., Liu, S., Kong, W.L., Bai, W.J., Wang, X.C., Wang, L.X., and Zhou, Q.L. (2009) *J. Am. Chem. Soc.*, **131**, 4222–4223.
36 Deng, W.-P., You, S.-L., Hou, X.-L., Dai, L.-X., Yu, Y.-H., Xia, W., and Sun, J. (2001) *J. Am. Chem. Soc.*, **123**, 6508–6519.
37 Thomson, J.B. (1959) *Tetrahedron Lett.*, **1**, 26–27.
38 Marquarding, D., Klusacek, H., Gokel, G., Hoffmann, P., and Ugi, I. (1970) *J. Am. Chem. Soc.*, **92**, 5389–5393.
39 Riant, O., Samuel, O., Flessner, T., Taudien, S., and Kagan, H.B. (1997) *J. Org. Chem.*, **62**, 6733–6745.
40 Richards, C.J., Damalidis, T., Hibbs, D.E., and Hursthouse, M.B. (1995) *Synlett*, 74–76.
41 Sammakia, T., Latham, H.A., and Schaad, D.R. (1995) *J. Org. Chem.*, **60**, 10–11.
42 Nishibayashi, Y. and Uemura, S. (1995) *Synlett*, 79–81.
43 Ireland, T., Almena Perea, J.J., and Knochel, P. (1999) *Angew. Chem. Int. Ed.*, **38**, 1457–1460.
44 Enders, D., Peters, R., Lochman, R., and Runsink, J. (1997) *Synlett*, 1462–1464.
45 Bolm, C., Kesselgruber, M., Muñiz, K., and Raabe, G. (2000) *Organometallics*, **19**, 1648–1651.
46 Whisler, M.C., MacNeil, S., Snieckus, V., and Beak, P. (2004) *Angew. Chem. Int. Ed.*, **43**, 2206–2225.
47 Tsukazaki, M., Tinkl, M., Roglans, A., Chapell, B.J., Taylor, N.J., and Snieckus, V. (1996) *J. Am. Chem. Soc.*, **118**, 685–686.
48 Philip, J.P., Kai, R., Robert, A.R., Nancy, N.T., Volante, R.P., and Paul, J.R. (1997) *J. Am. Chem. Soc.*, **119**, 6207–6208.

49 Hou, X.-L., Wu, X.-W., Dai, L.-X., Cao, B.-X., and Sun, J. (2000) *Chem. Comm.*, 1195–1196.

50 Nesmeyanov, A.N., Perevalova, E.G., Golovnya, B.V., and Nesmeyanova, O.A. (1954) *Doklady Akad. Nauk S. S. S. R.*, **97**, 459–461.

51 Benkeser, R.A., Goggin, D., and Schroll, G. (1954) *J. Am. Chem. Soc.*, **76**, 4025–4026.

52 McCulloch, B., Ward, D.L., Woollins, J.D., and Brubaker, Jr., C.H. (1985) *Organometallics*, **4**, 1425–1432.

53 Sanders, R. and Mueller-Westerhoff, U.T. (1996) *J. Organomet. Chem.*, **512**, 219–224.

54 Pichon, C., Odell, B., and Brown, J.M. (2004) *Chem. Comm.*, 598–599.

55 Chong, J.M. and Hegedus, L. (2004) *Organometallics*, **23**, 1010–1014.

56 Iftime, G., Moreau-Bossuet, C., Manoury, E., and Balavoine, G.G.A. (1996) *Chem. Comm.*, 527–528.

57 Tu, T., Deng, W.-P., Hou, X.-L., Dai, L.-X., and Dong, X.-C. (2003) *Chem. Eur. J.*, **9**, 3073–3081.

58 (a) Štěpnička, P. (ed.) (2007) *Ferrocenes: Ligands, Materials and Biomolecules*, John Wiley & Sons, Chichester; (b) Oro, L.A. and Claver, C. (eds) (2009) *Iridium Complexes in Organic Synthesis*, Wiley-VCH Verlag GmbH, Weinheim.

59 Kaplan, L., Kester, W.L., and Katy, J.J. (1952) *J. Am. Chem. Soc.*, **74**, 5531–5532.

60 (a) Wang, M.-C., Wang, D.-K., Zhu, Y., Liu, L.-T., and Guo, Y.-F. (2004) *Tetrahedron: Asymmetry*, **15**, 1289–1294; (b) Wang, M.-C., Liu, L.-T., Zang, J.-S., Shi, Y.-Y., and Wang, D.-K. (2004) *Tetrahedron: Asymmetry*, **15**, 3853–3859.

61 Ruble, J.C., Latham, H.A., and Fu, G.C. (1997) *J. Am. Chem. Soc.*, **119**, 1492–1493.

62 Tao, B., Lo, M.M.-C., and Fu, G.C. (2001) *J. Am. Chem. Soc.*, **123**, 353–354.

63 http://cn.chemnet.com/.

2
Stereoselective Synthesis of Planar Chiral Ferrocenes
Wei-Ping Deng, Victor Snieckus, and Costa Metallinos

2.1
Introduction

This chapter focuses on the stereoselective introduction of planar chirality into the ferrocene backbone by diastereoselective and enantioselective metalation methods and describes methods of kinetic resolution of these compounds. Although it has been nearly half a century since the first optically active planar chiral ferrocene compound (**1**) was synthesized by chemical resolution [1], it was not until the discovery by Ugi that *N,N*-dimethylferrocenylethylamine **2** [2] effects highly diastereoselective ortho-directed lithiation, that ferrocene derivatives containing planar chirality attracted interest [3]. Subsequently, directed ortho-metalation (DoM) methods [4] became the major protocol for the selective preparation of planar chiral ferrocenes.

1 (*R*)-**2**

First optically active planar chiral ferrocene by Thomson   Ugi's amine

The term "planar chirality" was first proposed by Cahn, Ingold, and Prelog (CIP) for cyclophane systems [5]. The concept of planar chirality was then adapted by Schlögl [6] for use in ferrocenes, although this class of molecules does not meet the definition of planar chirality as defined by CIP rules. Eventually, the term "planar chirality" as modified by Schlögl's rules [7] became widely accepted for ferrocene-based chirality derived by breaking the symmetry plane of ferrocene [7, 8]. As Schlögl clearly described, planar chirality appears upon breaking the plane of symmetry by introducing two or more different substituents in one cyclopentadienyl (Cp) ring (Figure 2.1), irrespective of free rotation about the iron–Cp bond.

Chiral Ferrocenes in Asymmetric Catalysis. Edited by Li-Xin Dai and Xue-Long Hou
Copyright © 2010 WILEY-VCH Verlag GmbH & Co. KGaA, Weinheim
ISBN: 978-3-527-32280-0

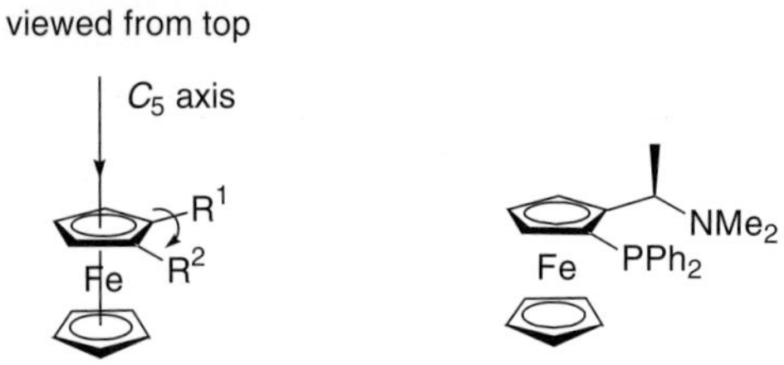

viewed from top

if $R^1 > R^2$, R configuration
if $R^1 < R^2$, S configuration

(R,S_p)-PPFA

Figure 2.1 Assignment of planar chirality in ferrocenes.

The corresponding absolute configuration notation is determined according to Schlögl's rules [2, 7] as follows: The observer looks along the C_5 axis of ferrocene with the more highly substituted Cp ring directed towards the observer; the absolute configuration is then assigned as "(R)" if R^1 and R^2 descend in priority (according to CIP rules) in the shortest clockwise arc. "(S)" stereochemistry requires that the same groups descend in priority in anticlockwise fashion. Since planar chirality differs from central chirality, the symbols denoting "(R)" and "(S)" of planar chirality are often modified with a subscripted italic letter "p" (R_p and S_p) to avoid ambiguity. For ferrocenes with more than one type of chirality, the priority order of chiral elements is *central* > *axial* > *planar* as shown in Figure 2.1.

Since Ugi's seminal work on diastereoselective generation of planar chirality, the preparation of non-racemic planar chiral ferrocenes and their applications in asymmetric synthesis have been important topics in both academia [3] and industry [9]. In principle, the methods of introduction of planar chirality into the ferrocene skeleton can be classified by three types (Scheme 2.1): (a) Diastereoselective directed ortho-metalation (DoM). This method utilizes a chiral auxiliary in the form of a chiral directed metalation group (DMG) that is able to coordinate an alkyllithium or lithium dialkylamide base (complex induced proximity effect) [10] and simultaneously differentiate the two possible prochiral ortho-positions on the same cyclopentadienyl ring in the deprotonation event. The diastereomerically enriched lithioferrocene intermediates generated in this manner may be trapped with electrophiles to afford planar chiral 1,2-disubstituted ferrocenes. The various chiral DMGs characteristically exhibit features of *N*- or *O*-lone pair coordinating sites and appropriate geometrical orientation for promoting the ortho-metalation process. (b) Enantioselective DoM. This method makes use of mono-substituted ferrocenes that bear achiral DMGs and relies on chiral alkyllithium or lithium dialkyl amide bases to differentiate the prochiral ortho-positions on the Cp ring during lithiation, resulting in lithioferrocene intermediates with only planar chirality. (c) Resolution of racemic (chiral, when $R^1 \neq R^2$) planar chiral ferrocenes. This method can be further subdivided into enzymatic kinetic resolution and non-enzymatic kinetic resolution, depending on the reagents employed. Lipases are commonly used for enzymatic resolution via transesterification of alcohols, hydrolysis of esters and similar transformations. The reaction of one *matched* enantiomer

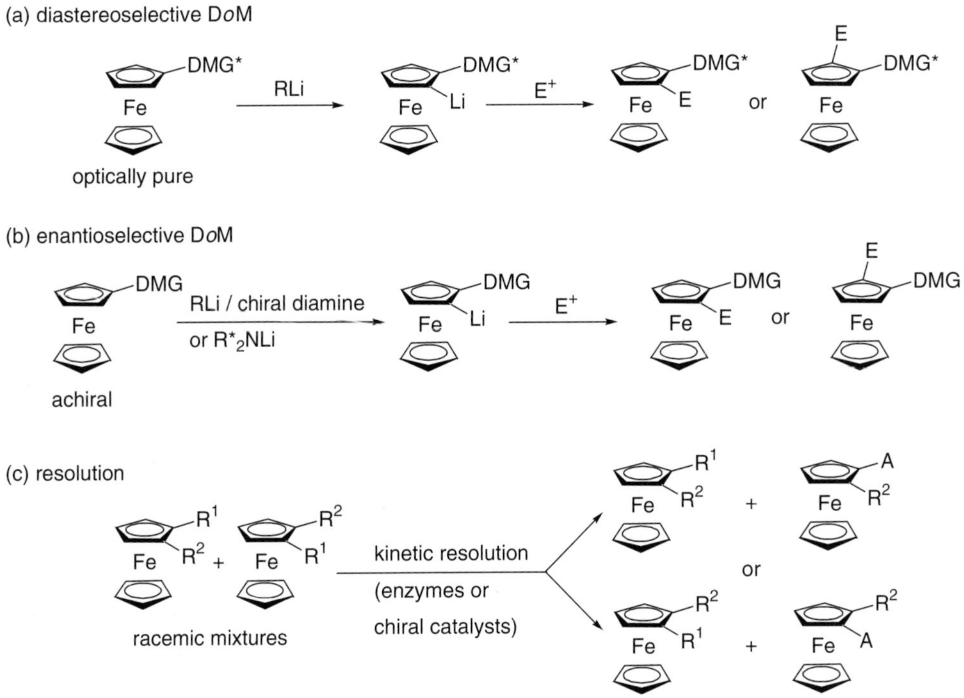

(a) diasteroselective DoM

(b) enantioselective DoM

(c) resolution

Scheme 2.1 Methods for the introduction of planar chirality into the ferrocene skeleton.

of a racemic mixture with an enzyme gives the corresponding enantiomerically enriched product, while its antipode remains untouched and can be isolated, also in optically active form. Non-enzymatic kinetic resolution is the most recent development in this area.

Of the three methods above, type (a) is the most developed and is extensively exploited. Much of the early research on ferrocene-based ligands and reagents for asymmetric catalysis utilized Ugi's chiral amine (N,N-dimethylferrocenylethylamine) almost exclusively. In 1993, chiral sulfoxides were reported as alternate chiral DMGs by Kagan [11]. In the ensuing years, a spate of research evolved concerning the use of DMGs for the preparation of optically active ferrocenes containing planar chirality, which included chiral acetals by Kagan [12], chiral oxazolines (reported independently by Richards [13], Sammakia [14] Uemura [15], and subsequently by Ahn [16]), SAMP-hydrazones by Enders [17] and methoxy analogues of Ugi's amine by Knochel [18].

Owing to the diversity of readily available methods for the introduction of planar chirality into ferrocenes, research into the synthesis of a variety of planar chiral ferrocene ligands for application in various asymmetric catalytic reactions increased dramatically at the turn of the millennium [3b–i, 19]. The availability of diverse planar chiral ferrocene ligands raised questions regarding the processes by which planar chirality, or the combination of planar and central chirality, induce the enantio- or diastereo-selectivity. Most studies aimed to shed light on this subject have shown that planar chirality, either with or without other chiral elements such as central chirality, plays a significant and even a decisive role in determining the degree of enantioselectivity, and also the configuration of the corresponding products [20].

Hou [21] and coworkers have investigated the role played by ferrocene planar chirality in influencing certain catalytic asymmetric reactions. For instance, matching both the planar and central chiral elements of the ligand is essential for obtaining excellent enantiomeric excess using 1-ferrocenyloxazoline-1′-phosphines in Pd-catalyzed Heck and allylic substitution reactions [21a]. In this instance, the planar chirality influenced dramatically not only the *ee* value but also the absolute configuration of the products. Upon complexation of this class of P,N ligand with palladium, a new type of chirality, axial chirality, is generated by two rotamers A and B (Scheme 2.2), the ratio of which is found to be governed by the planar chiral ferrocene moiety.

The modern demands of asymmetric catalysis, as applied to the synthesis of optically pure building blocks, natural products and many drug intermediates, make the development of new planar chiral ferrocene-based ligands increasingly important. This demand has in turn driven the development of new DMGs, such as sulfoximines by Bolm [22], imidazolines by Peters [23, 24], oxazaphospholidine-oxide by Xiao [25] and free alcohols by Ueberbacher [26], which give good to excellent diastereoselectivity in ortho-functionalization of ferrocenes. In addition, advances in enantioselective DoM and the resolution of racemic planar chiral ferrocenes have provided new methods for the synthesis of ferrocenyl ligands, building blocks, and key intermediates containing planar chirality which will be discussed herein. However, a comprehensive coverage of work regarding the stereoselective induction of planar chirality in ferrocenes is beyond the scope of this chapter. Not covered is, for example, the Schmalz [27] report describing the asymmetric insertion of carbenoids into the Cp–H bond of ferrocene to achieve the enantioselective construction of planar chiral ferrocenes with moderate stereoselectivity. In addition, the synthesis of

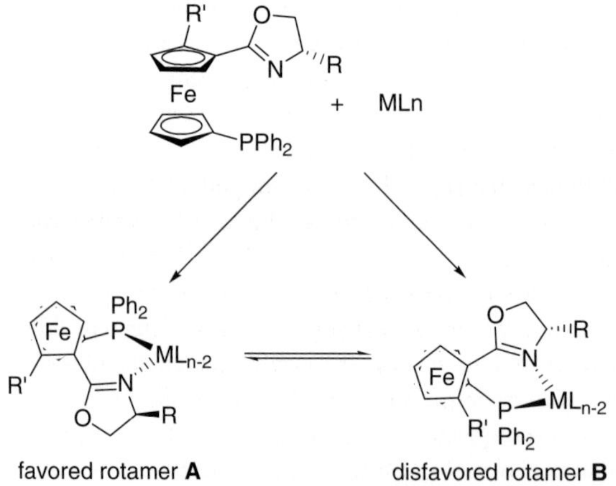

Scheme 2.2 Axial chirality induced upon complexation of a ferrocenyl oxazoline with a metal.

the planar chirality efficiently? Chapter 2 of this book, contributed by Deng, Snieckus and Metallinos gives a thorough answer to this question. In Ugi's work, the amino-group served as a directed metalation group (DMG). Since then many DMGs have been developed, such as Kagan's chiral acetal [39], chiral oxazoline independently developed by Richards [40], Sammakia [41] and Uemura [42], methoxy DMG developed by Knochel [43], Enders' SAMP hydrazone [44], and sulfoximines developed by Bolm [45]. All of these DMGs can give a highly selective directed orthometalation (DoM) reaction, followed by trapping with an E^+ to introduce the E group at a definite position, giving the system an additional planar chirality. The DoM reaction or the complex induced proximate effect (CIPE) [46], makes this protocol highly diastereoselective. Snieckus is one of the major players in this field and his chapter will give readers a good understanding of DoM. He [47] went further by using a directing reagent, sparteine, instead of a DMG. This is unarguably a significant improvement. Planar chirality can then be introduced by using optically pure sparteine or other chiral amines.

There are some other systems in addition to ferrocene that may offer planar chirality. Cyclophane is one such system of interest. Although Pyne has developed a highly selective hydrogenation catalyst with a chiral cyclophane skeleton [48], the introduction of planar chirality with a DMG on a cyclophane system is sometimes problematic. It has been reported [49] that the DoM reaction of **16** (G = oxazoline) gave a mixture of ortho- and benzylic substituted products, which are not easily separable. The difficulty in making a selective and distinct product has somewhat limited the using of cyclophane as a viable skeleton.

π-Arene chromium tricarbonyl is another system that may provide planar chirality, however, the derivatization is also tedious with limited variations. Thus, ferrocene is now the most appropriate scaffold to provide planar chirality for various ligand patterns. Chapter 2 not only describes the preparation of ligands with planar chirality, but also discusses the role of planar chirality in asymmetric catalysis.

How to denote the absolute configuration is clearly described in Chapter 2. There are several symbols or methods used to designate the absolute configuration of planar chirality in the literature. Some authors use a subscript Fc to symbolize the planar chirality of ferrocene. For instance, compound **18** is denoted as (R, S_{Fc}) where the central chirality is R in absolute configuration and the planar chirality is S. Other authors give no additional signs but list the absolute configurations by an order of priority for the elements of chirality, that is: central > axial > planar. Thus **18** is described as (R, S), the R listed first for central chirality and the S listed second for planar chirality. In many cases, a subscript italic letter P (planar) is used with the absolute configuration R or S. Then **18** is described as (R, S_p) where S_p means the planar chirality with S configuration. Similar to this, some authors prefer to put the letter P before the letter indicating the absolute configuration, as (R, $_pS$). For unification of this whole book, (R, S_p) for **18** and this symbol are used throughout. If a phosphorus chiral center exists, *Rphos* or *Sphos* is used for the P-central chirality.

1.3
Derivatization of the Ferrocene Scaffold

During the study of the unusual structure of ferrocene, Woodward *et al.* [4] demonstrated the remarkable similarity between the reactivity of ferrocene and that of a benzene ring. Ferrocene undergoes typical aromatic electrophilic substitution reactions such as the Friedel–Crafts acylation reaction (Scheme 1.1, Eq. 1). For a comparison of the reactivity, when one molecule of ferrocene and ten molecules of anisole were allowed to react with a limited amount of acetyl chloride–aluminum chloride complex in chloroform solution, acetyl-ferrocene was formed exclusively, without any detectable amount of methoxy acetophenone. This is attributed to the fact that the electronic density of the cyclopentadienyl ring is much higher than that of a benzene ring. The π-back donation of the d-orbital of iron may even strengthen the electron negativity of the Cp ring. The reactivity of the Friedel–Crafts reaction of ferrocene is 10^6 times higher than that of a benzene ring. The same trend is evident for other electrophilic substitution reactions, like the Vilsmeier reaction for introducing a formyl group (Scheme 1.1, Eq. 2) and the Mannich reaction for introducing a dimethylaminomethyl group (Scheme 1.1, Eq. 3). These reliable transformations allow ready access to useful intermediates such as Ugi's amine and ferrocenyl

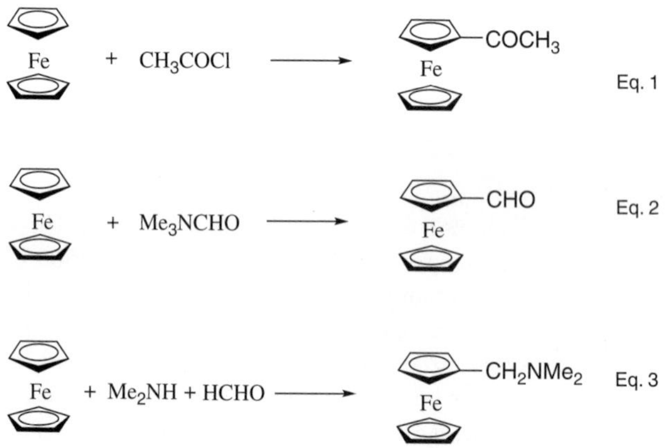

Fe + CH_3COCl ⟶ Fe–$COCH_3$ Eq. 1

Fe + Me_3NCHO ⟶ Fe–CHO   Eq. 2

Fe + Me_2NH + HCHO ⟶ Fe–CH_2NMe_2 Eq. 3

Scheme 1.1 Typical electrophilic reactions of ferrocene in the early days.

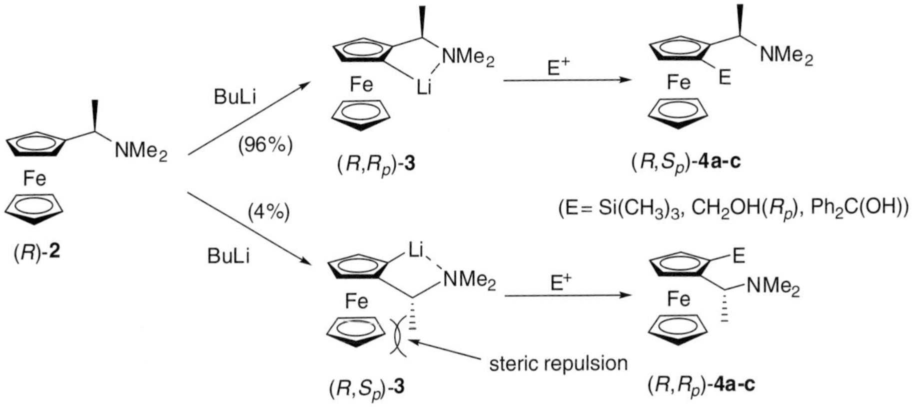

2.2 Diastereoselective Directed ortho-Metalation of Ferrocenes with Chiral Auxiliaries

Scheme 2.4 Diastereoselective ortho-lithiation of Ugi's amine (R)-2.

BPPFA, for use in asymmetric catalytic hydrosilylation reactions [34]. Since this pioneering work, numerous other ligands derived from Ugi's amine have been prepared, either by treatment of ferrocenyllithium species with various electrophiles or by S_N1-like substitution of the N,N-dimethylamino group, which proceeds with retention of configuration. Figure 2.2 shows the representative chiral ligands BPPFOH **6** [35], N,S ligand **7** [36], BoPhoz **8** [37], Josiphos **9** [9, 38], Pigiphos **10** [39], TRAP **11** [40], Walphos **12** [41] and P,N ligand **13** [42] that have been made

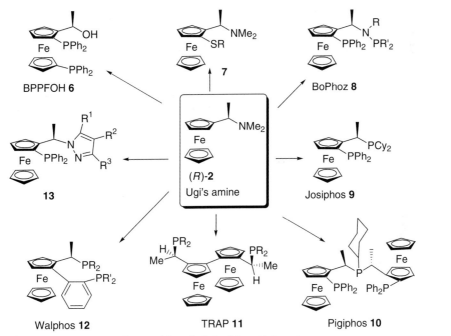

Figure 2.2 Representative ferrocenyl ligands derived from Ugi's amine.

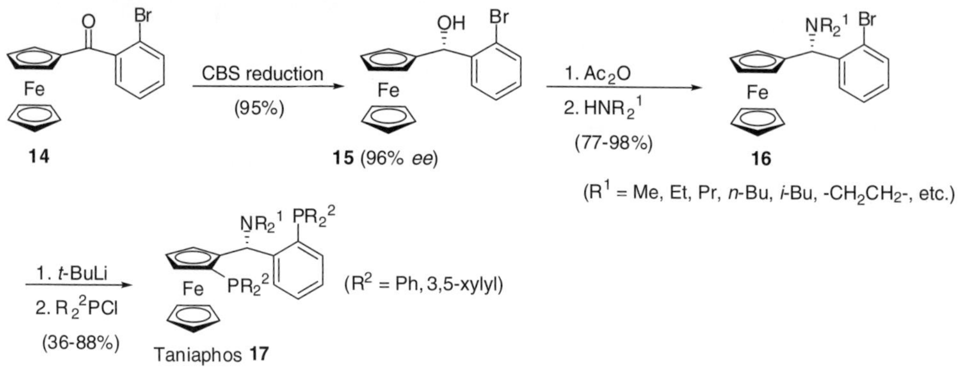

Scheme 2.5 Synthesis of Taniaphos **17**.

using a combination of these strategies. Walphos **12** is prepared by transmetalation of the ferrocenyllithium species **3** with $ZnBr_2$ followed by Negishi cross-coupling with bromoiodobenzene. TRAP ligands such as **11** are formed by copper-catalyzed homocoupling of a 2-iodo ferrocenylphosphine oxide derived from Ugi's amine. Other notable ligands include the PingFer series by Chen [43], which features *P*-chiral phosphines.

Alternatively, chiral DMGs in which the chiral Ugi amine is modified by incorporation of an aryl phosphine at the stereogenic center are also capable of diastereoselectively inducing planar chirality. Thus, treatment of **16** with *t*-BuLi and subsequent quenching with chlorodiphenylphosphine gives Taniaphos **17** [44] in good yield and excellent stereochemical purity (Scheme 2.5).

Chiral amine DMGs with stereogenic centers γ to the Cp ring and with axial chirality have been investigated. These include derivatives with pyrrolidine, piperidine, and *O*-methylephedrine moieties and the unusual chiral binaphthyl azepine (*S*)-**21** (Figure 2.3).

Although Nozaki's γ-chiral (*S*)-**19** was ultimately shown to give poor diastereoselectivity, chiral amines containing an additional methoxy coordination site such as (*S*)-(2-methoxymethylpyrrolidin-1-yl) ferrocene **18a** (FcSMP) [45], do in fact afford the 1,2-disubstituted ferrocenylaminophosphines (*S*,S_p)-**22a** in 86% *de* (Scheme 2.6). Diastereoselectivity can be improved to 98% *de* by using bulkier alkyllithiums (*s*-BuLi) for the deprotonation step. The C_2-symmetric chiral ferrocenylpyrrolidine

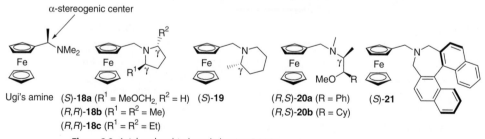

Figure 2.3 Axial and γ-chiral methylamine DMGs.

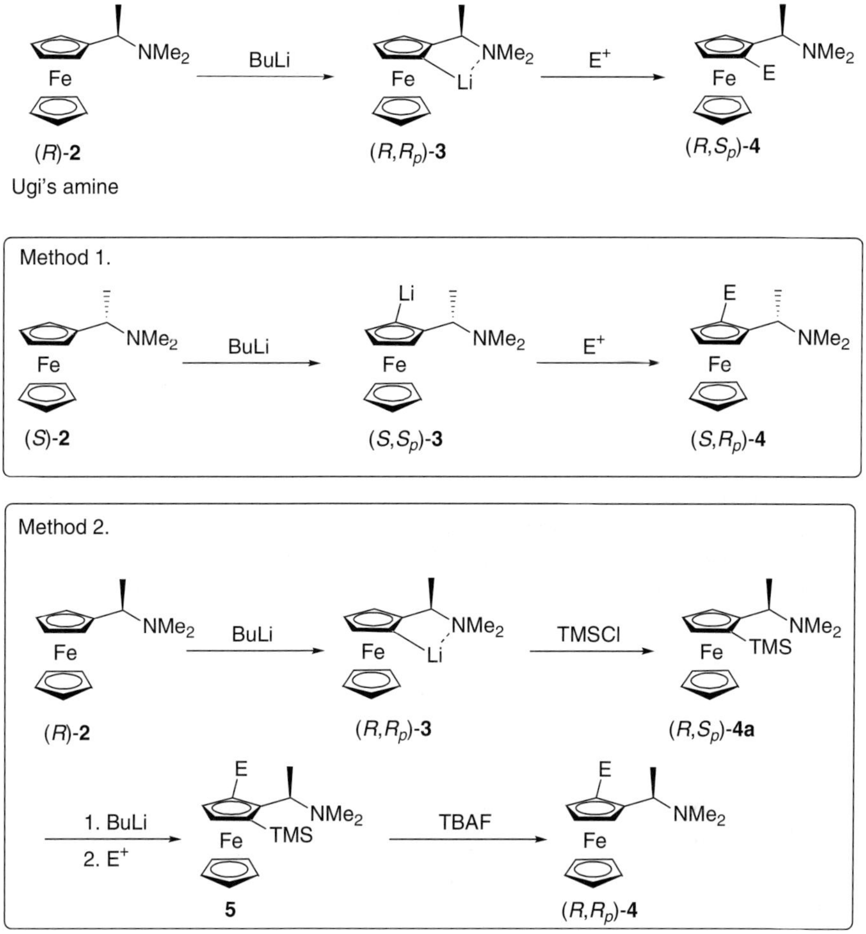

Scheme 2.3 Methods to synthesize ferrocenyl derivatives with opposite relative planar chirality.

2.2.1
Ugi's Amine and Other Chiral α-Substituted Ferrocenylmethylamines

The lithiation of enantiopure chiral amine (R)-2 with butyllithium results in the preferential formation of (R,R_p)-3 with high stereoselectivity (up to 96 : 4 ratio or 92% de) (Scheme 2.4) [2] as demonstrated by treatment with electrophiles such as TMSCl, formaldehyde and benzophenone to give products (R,S_p)-4 and (R,R_p)-4. It was suggested that the high (R,S_p) diastereoselectivity arises from a disfavored steric repulsion between the methyl group at the chiral center and the lower Cp ring.

Although 4% of minor (R,R_p) diastereoisomers are obtained, the major isomers (R,S_p)-4 may be easily isolated in high optical purity by column chromatography in many cases. This method was successfully employed by Kumada and Hayashi for the synthesis of many ferrocenylphosphines, such as the chiral ligands (R,S_p)-PPFA and optically active 1,3-disubstituted [28] and rare multisubstituted planar chiral ferrocenes, are also not presented in this chapter. The work on planar chiral azaferrocenes and heterocyclic ring fused ferrocenes by Fu and planar chiral metallocyclic ferrocenes by Richards will be discussed in Chapters 11 and 12, respectively. As a result of the great activity in this field, monographs and reviews [3, 9, 19, 29, 55] which focus mainly on the preparation of non-racemic ferrocene ligands and their applications in asymmetric catalysis are available. To the best of our knowledge, however, there is no comprehensive account concerning the stereoselective construction of planar chiral ferrocenes.

2.2
Diastereoselective Directed ortho-Metalation of Ferrocenes with Chiral Auxiliaries

The first example of stereoselective synthesis of planar chiral ferrocenes can be traced to the original work of Schlögl [30] who, in 1965, reported the synthesis of an (S,S_p) analogue of **1** bearing an α-phenyl group, which was derived from (S)-α-phenyl-γ-ferrocenylbutyric acid. However, it was not until 1969 that Nozaki [31] reported the very first method for the introduction of planar chirality into the ferrocene skeleton via diastereoselective DoM of a chiral piperidide. Soon after this report, Ugi [32] showed that the diastereoselectivity obtained with Nozaki's piperidide was much lower than claimed. In response, the N,N-dimethylethylaminoferrocene **2**, now known as Ugi's amine [2], showed excellent and reproducible diastereoselectivity in DoM-substitution reactions. Here, the (S,R_p) planar chiral diastereomer **4** was obtained preferentially with high diastereomeric excess (de), despite the fact that the chiral center of the directing group is remote to the reacting site. (Scheme 2.3, Method 1). Since Ugi's seminal observations, many efforts have been devoted towards the development of new chiral auxiliaries to serve as chiral DMGs in the diastereoselective introduction of planar chirality into ferrocenes. These DMGs may be divided into five types: α-substituted methylamines and their congeners, sulfoxides, acetals, oxazolines and miscellaneous chiral auxiliaries. In general, (R) or (S) configured DMGs preferentially afford, after electrophile quenching, one of two possible Cp ring-substituted diastereomers where the major product has (R_p) or (S_p) absolute stereochemistry. Notably, a single enantiomer of a particular starting ferrocene will afford products of the same relative stereochemistry upon quenching of the intermediate ortho-lithiated species, while the antipode of the starting ferrocene can be used to prepare products of the corresponding enantiomeric series, which necessarily have opposite planar and central chirality. To obtain products with opposite relative planar chirality without starting from the antipode of a particular starting material, a latent group such as trimethylsilyl (TMS) may be introduced by normal diastereoselective lithiation and quench with trimethylchlorosilane (TMSCl). An additional ortho-lithiation–quench sequence then affords an intermediate trisubstituted ferrocene that, upon desilylation, gives products with the opposite relative planar chirality. This "silicon trick" [33] is feasible for many but not all DMGs (Scheme 2.3, Method 2).

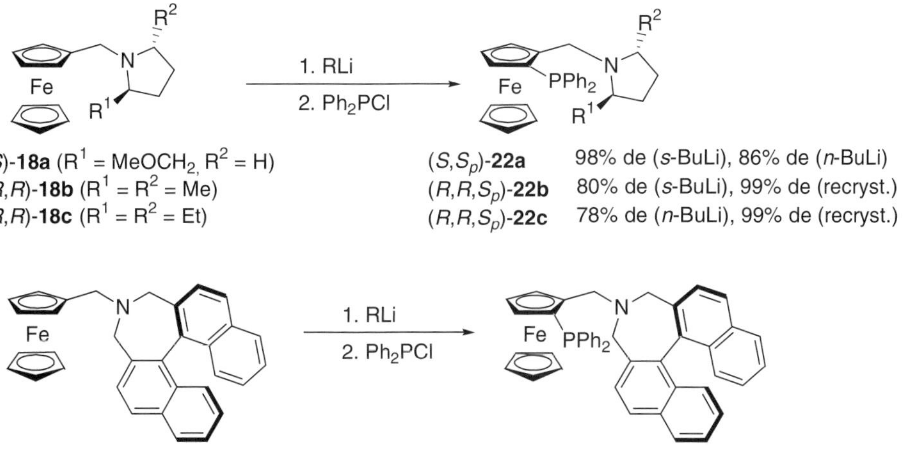

Scheme 2.6 Diastereoselective introduction of the *ortho*-diphenylphosphino group in **18** and **21**.

Compound	R¹, R²	Product	de
(S)-18a	R¹ = MeOCH₂, R² = H	(S,Sp)-22a	98% de (s-BuLi), 86% de (n-BuLi)
(R,R)-18b	R¹ = R² = Me	(R,R,Sp)-22b	80% de (s-BuLi), 99% de (recryst.)
(R,R)-18c	R¹ = R² = Et	(R,R,Sp)-22c	78% de (n-BuLi), 99% de (recryst.)

(S)-21 → (S,Sp)-23 (80% de)

(R,R)-**18b** has been used to prepare ferrocenylaminophosphine ligands (R,R,S$_p$)-**22b** in 80% de through diastereoselective DoM followed by quenching with chlorodiphenylphosphine. As in the previous case, s-BuLi gave superior results compared to n-BuLi. Lithiation of the corresponding diethyl analogue (R,R)-**18c** is more difficult and requires n-BuLi in ether at 25 °C to afford product (R,R,S$_p$)-**22c** in 10% yield and 78% de. A single recrystallization from ethanol delivers products (R,R,S$_p$)-**22b** and (R,R,S$_p$)-**22c** in 99% de. Diastereoselective DoM and Ph₂PCl quenching of ferrocenyl binaphthyl azepine (S)-**21** gives (S,S$_p$)-**23** as the major product in 80% de under many different lithiation conditions [46].

O-Methylephedrine derivatives have also been shown to participate in diastereoselective DoM reactions. Thus, when (R,S)-**20a** is treated sequentially with t-BuLi in pentane and then iodine, (R,S,R$_p$)-**24a** is obtained with 98% de (Scheme 2.7). The selectivity of the transformation decreases to 79% de using s-BuLi as the base [47]. The cyclohexyl analogue (R,S)-**20b** [48] gives (R,S,R$_p$)-**25** with lower diastereoselectivity (78% de) under optimum reaction conditions. The derived planar chiral ferrocenyl O-methylephedrine derivatives may be further transformed into amino, acetoxy, and phosphine derivatives by the S$_N$1 chemistry described previously.

2.2.2
Chiral Sulfoxides

Although chiral sulfoxide methods are of more recent vintage compared to those based on Ugi's amine, they have nonetheless proven to be of comparable utility for the synthesis of chiral ferrocenyl ligands. This utility is based on the ease of adjusting the oxidation state of sulfur and on the ready displacement of the p-tolyl sulfoxide group using t-BuLi [49] or PhLi [50] to give a lithioferrocene species. The enantiomerically pure ferrocenylsulfoxide starting materials may be prepared by

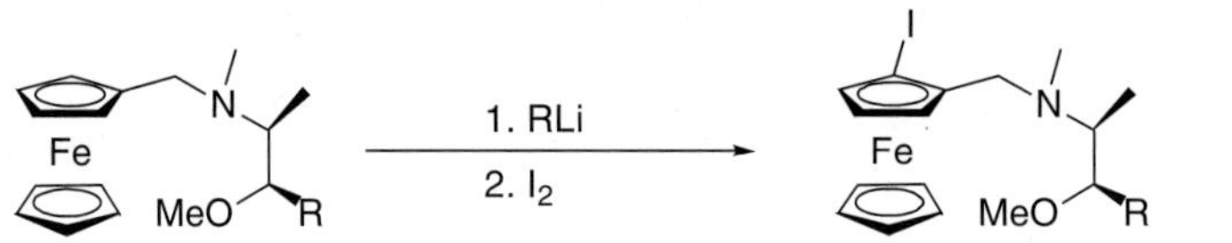

(*R*,*S*)-**20a** R = Ph
(*R*,*S*)-**20b** R = Cy

(*R*,*S*,*R*$_p$)-**24a,** R = Ph, 98% de (*t*-BuLi, pentane)
(*R*,*S*,*R*$_p$)-**25,** R = Cy, 78% de (*t*-BuLi, pentane)

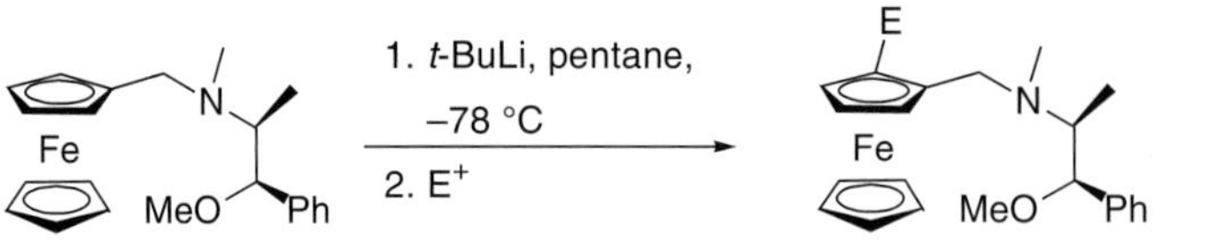

(*R*,*S*)-**20a**

(*R*,*S*,*R*$_p$)-**24a** (E = I, 90%)
(*R*,*S*,*R*$_p$)-**24b** (E = Br, 84%)
(*R*,*S*,*R*$_p$)-**24c** (E = PPh$_2$, 97%)
(*R*,*S*,*R*$_p$)-**24d** (E = C(OH)Ph$_2$, 64%)

Scheme 2.7 Diastereoselective ortho-lithiation of *O*-methylephedrine (*R*,*S*)-**20a,b**.

enantioselective oxidation of ferrocenylsulfides [11] or by nucleophilic attack of monolithioferrocene on optically pure menthyl *p*-toluenesulfinate [51] or other chiral sulfinates [11, 52] such as Ellman's reagent *t*-Bu-S(O)-S*t*-Bu [53].

The first examples of diastereoselective DoM of ferrocenyl sulfoxides were reported by Kagan [11] in 1993. As depicted in Scheme 2.8, enantiopure *t*-Bu ferrocenyl sulfoxide (*S*)-**26a** is smoothly deprotonated with *n*-BuLi and the resulting lithioferrocene species is quenched with iodomethane to afford the 1,2-disubstituted product (*S*,*S*$_p$)-**28** in excellent diastereoselectivity (96% de). The *p*-tolyl sulfoxide congener (*R*)-**26b** may be lithiated with LDA at −78 °C to give species (*R*,*R*$_p$)-**29** which, upon electrophile quench (TMSCl), affords (*R*,*R*$_p$)-**30** in 98% de [49, 54]. The stereochemistry of the products can be predicted based on examination of the

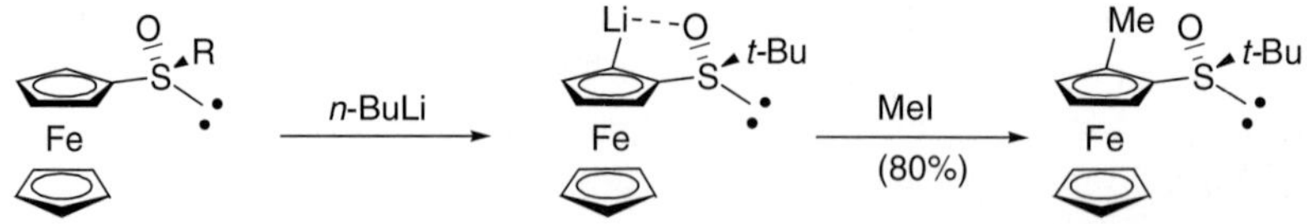

(*S*)-**26a** (R = *t*-Bu)
(*S*,*S*$_p$)-**27**
(*S*,*S*$_p$)-**28** (>96% de)

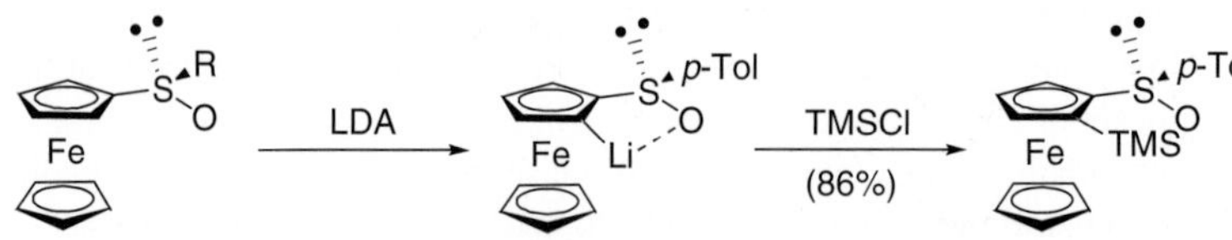

(*R*)-**26b** (R = *p*-Tol)
(*R*,*R*$_p$)-**29**
(*R*,*R*$_p$)-**30** (>98% de)

Scheme 2.8 Diastereoselective ortho-lithiation of ferrocenylsulfoxides **26a,b** with *n*-BuLi and LDA.

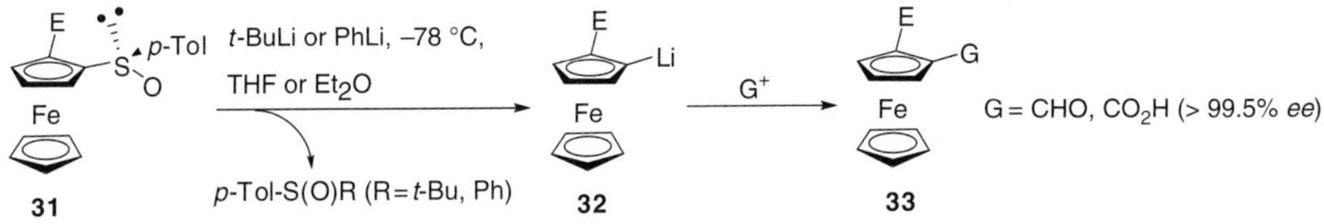

2.2 Diastereoselective Directed ortho-Metalation of Ferrocenes with Chiral Auxiliaries | 25

Scheme 2.9 The synthesis of planar chiral 1,2-disubstituted ferrocenes 33 via sulfoxide-lithium exchange.

preferential conformation of the sulfoxide moiety during lithiation such that the bulky t-Bu or p-Tol groups are positioned anti to the ferrocene core.

The ability to convert the sulfoxide moiety to other functional groups confers a large degree of flexibilty to this method [55], similar to the S_N1 substitution reactions available for Ugi's amine. As depicted in Scheme 2.9, the p-Tol sulfoxide 31 undergoes sulfur-lithium exchange with t-BuLi [49] or PhLi [50] to give the putative intermediate 32 which may be trapped with various electrophiles to produce a series of 1,2-disubstituted ferrocenes 33 featuring, in many cases, exclusively planar chirality. Enantiopure sulfoxides 26a,b are thus of great synthetic value and lead to diverse ligands such as 1,2-diphosphines 34 by Kagan [56], diphosphines 35 and second generation Taniaphos 36 by Knochel [57], biferrocenyl bisphosphine 37 by Weissensteiner [58], P,N ligand 38 [59], aminophosphine 39 [60], Fesulphos 40 by Carretero [61], MOPF 41 [62] and ferrocenyl DMAP 42 [63] by Johannsen (Figure 2.4).

Representative syntheses of the preceding ligands are depicted in Scheme 2.10. Starting from (S)-26b, diastereoselective DoM with LDA is followed by quenching

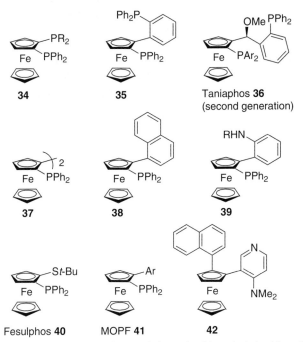

Figure 2.4 Some 1,2-disubstituted planar chiral ligands derived from ferrocenylsulfoxides 26a,b.

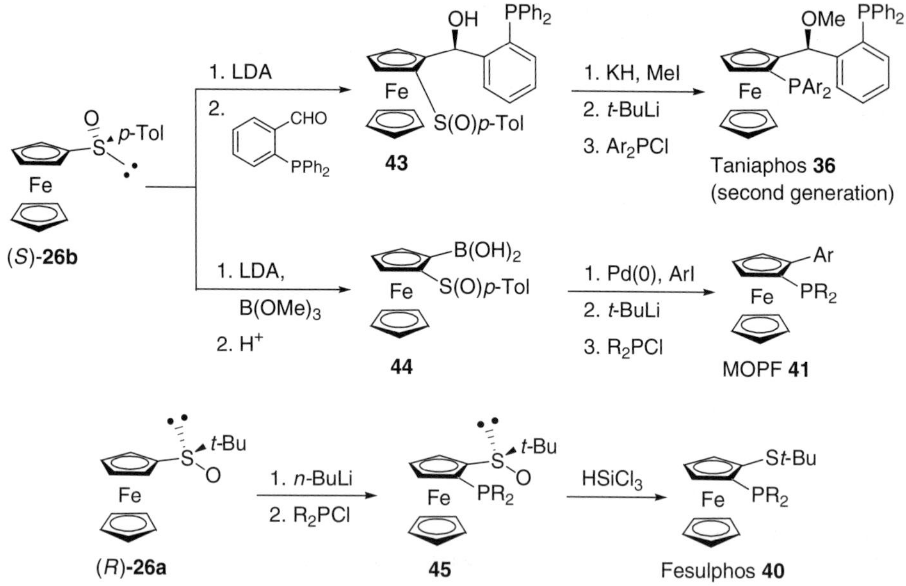

Scheme 2.10 Representative syntheses of planar chiral ligands 36, 40 and 41 based on ferrocenylsulfoxides 26a,b.

with ortho-phosphinylbenzaldehyde to afford diastereomer 43 as the major product. Sequential methylation of the alcohol, sulfoxide-lithium exchange using t-BuLi, and addition of chlorodiarylphosphine produces the second generation Taniaphos ligand 36 in good yield [57]. The same series of transformations using B(OMe)$_3$ as the electrophile yields the ferrocenylboronic acid 44, which can undergo Pd-catalyzed Suzuki coupling with iodoarenes. Standard sulfoxide-lithium exchange and treatment with chlorodiarylphosphine smoothly furnishes MOPF 41 [62]. The Fesulphos ligand 40 [61] is prepared through diastereoselective lithiation of the t-Bu sulfoxide (S)-26a with n-BuLi followed by R$_2$PCl quenching. The resulting intermediate 45 is readily reduced with HSiCl$_3$ to give sulfide 40. Details of the synthesis of 40 and related ferrocene ligands are available in Chapter 9.

The final example (Scheme 2.11) involves DoM of the chiral sulfoxide (S)-26b followed by treatment with N-tosylimine to give a hybrid Ugi–Kagan type ferrocene which, without isolation, is subjected to sulfoxide-lithium exchange and quenching with chlorodiphenylphosphine to produce (S,S$_p$)-46 in 41% yield and 96% de (four steps, one-pot process) [64]. Note that (S,S$_p$)-46 has the same relative stereochemistry as the *minor* diastereomer obtained by Ugi's procedure.

2.2.3
Chiral Acetals

The chiral dioxolane moiety offers an alternative to chiral sulfoxides for the synthesis of planar chiral ferrocene derivatives. As reported by Kagan in 1993 [12], the

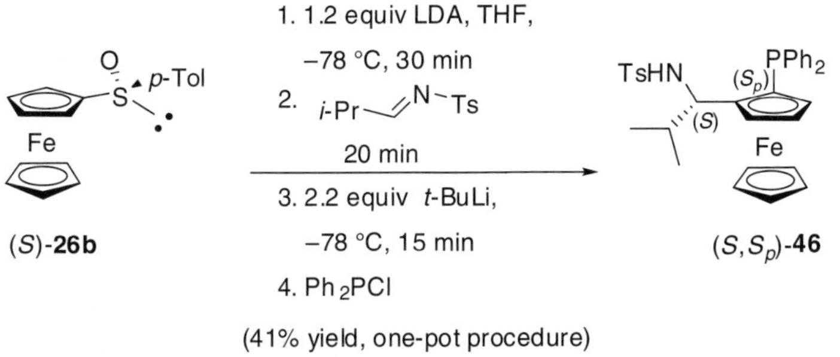

Scheme 2.11 The synthesis of the planar chiral (S,S$_p$)-1,2-disubstituted ferrocene **46**.

enantiopure acetal **48** is conveniently synthesized in three steps and 80% yield from commercially available ferrocenylcarboxaldehyde and (S)-1,2,4-butanetriol (Scheme 2.12).

From acetal **48**, 2-substituted products **49** may be prepared in high yield and excellent de (98%) via selective lithiation of the prochiral R hydrogen of the Cp ring. In this process, the temperature is crucial for the observed diastereoselectivity; lithiation at 0 °C gives products in 80% de. Hydrolysis of the recoverable (S)-1,2,4-butanetriol from **49** gives a range of only planar chiral 1,2-disubstituted ferrocenecarboxaldehydes **51** (Scheme 2.13) [65]. Notably, Manoury and coworkers reported a highly enantioselective synthesis of 1,2,1'-trisubstituted and 1,2,1',2'-tetrasubstituted planar chiral ferrocenedicarbaldehydes. Their synthesis proceeds by

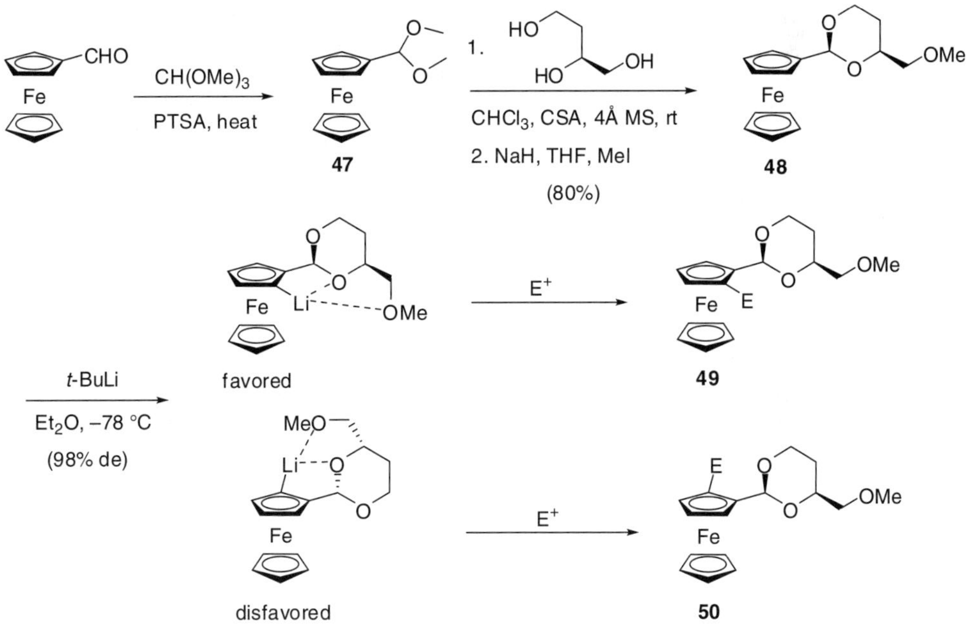

Scheme 2.12 Diastereoselective *ortho*-lithiation-functionalization of acetal **48**.

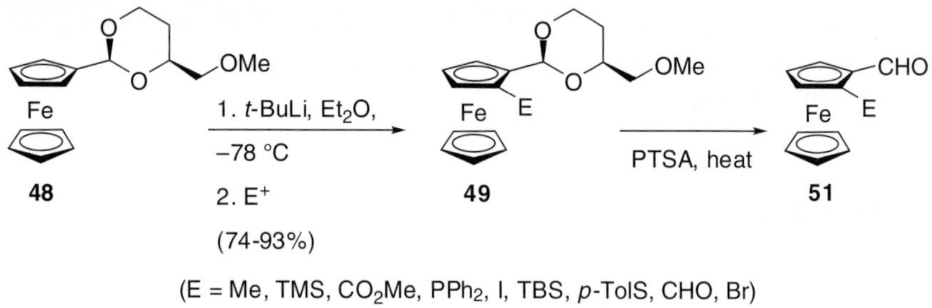

Scheme 2.13 Diastereoselective synthesis of enantiopure 1,2-disubstituted ferrocenecarboxaldehydes **51** from chiral acetal **48**.

(E = Me, TMS, CO_2Me, PPh_2, I, TBS, p-TolS, CHO, Br)

diastereoselective lithiation of a latent aminoalkoxide adduct between (S)-1-(2-pyrrolidinylmethyl) pyrrolidine and the ferrocenyl aldehyde which also effects deprotonation of the C_p' ring [65c].

The chiral 2-lithio acetal has served as a starting point for the synthesis of C_2-symmetric ferrocenyl ligands. These include BMPD **52** (Scheme 2.14) used in yttrium-catalyzed silylcyanation [66], biferrocenyl diol **53** [67], and the N-heterocyclic carbene **54** [68]. The latter is prepared from carboxaldehyde **51** which, after oxidation to the carboxylic acid and acyl azide formation, allows the installation of a

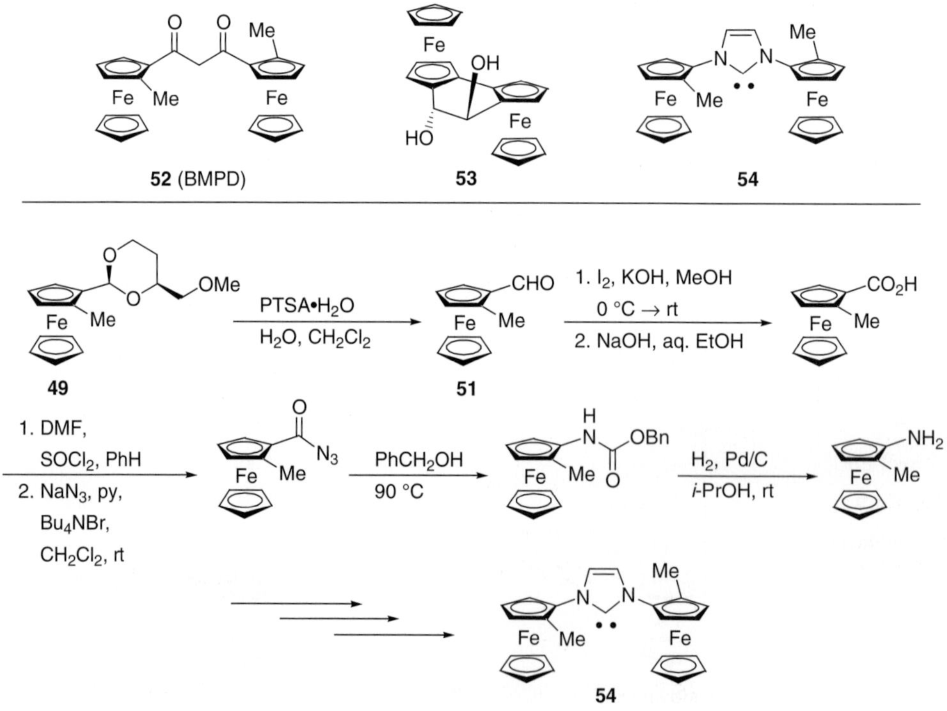

Scheme 2.14 Planar chiral ligands **52**, **53** and **54** prepared from acetal **48**.

CBZ-protected nitrogen on the Cp ring by Curtius rearrangement. Subsequent deprotection gives the key intermediate 2-methyl-1-aminoferrocene which is used to complete the synthesis of NHC **54**.

2.2.4
Chiral Oxazolines

Although several different enantiopure DMGs have been introduced for asymmetric induction of planar chirality into ferrocenes, the directing group often needs to be modified or replaced with suitable donor ligands such as P, S or N (as in Scheme 2.14) to satisfy metal ligation requirements. In the case of chiral oxazolines, however, [29d], [69] the directing group that induces diastereoselective lithiation may double as an excellent N-donor ligand for transition metals. To take advantage of these ligating properties, enantiomerically pure ferrocenyloxazolines (S)-**55** are first synthesized from the commercially available ferrocenecarboxylic acid and optically pure amino alcohols. As reported independently in 1995 by Richards [13], Sammakia [14a] and Uemura [15], treatment of chiral ferrocenyl oxazolines (S)-**55** with n-BuLi or s-BuLi in ether at −78 °C results in a highly diastereoselective DoM process that favors removal of the prochiral R hydrogen in up to 39 : 1 diastereomeric ratio to give (S,R_p)-**56**. Soon after these reports, Sammakia and coworkers [14b] disclosed results which showed a further increase in diastereoselectivity to >500 : 1 by changing the solvent to hexanes and by adding TMEDA. However, many factors may influence the diastereoselectivity during DoM of these substrates, such as the bulk and aggregation of the alkyllithium reagent, the solvent, additives, and the size of the substituents at the stereogenic center of the oxazoline. As depicted in Scheme 2.15, the high diastereoselectivity in the lithiation step may be rationalized by transition states (S,R_p)-**56** (favored) and (S,S_p)-**56** (disfavored), in which the R'

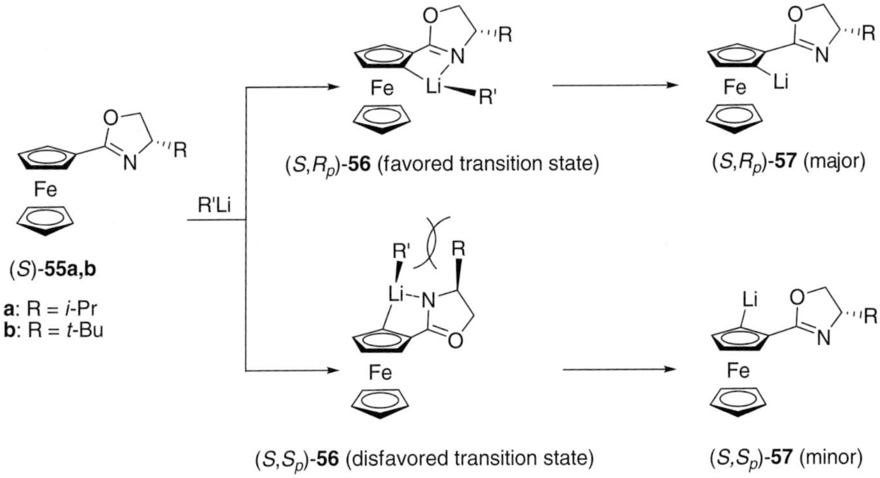

Scheme 2.15 Diastereoselective ortho-lithiation of ferrocenyloxazolines (S)-**55**.

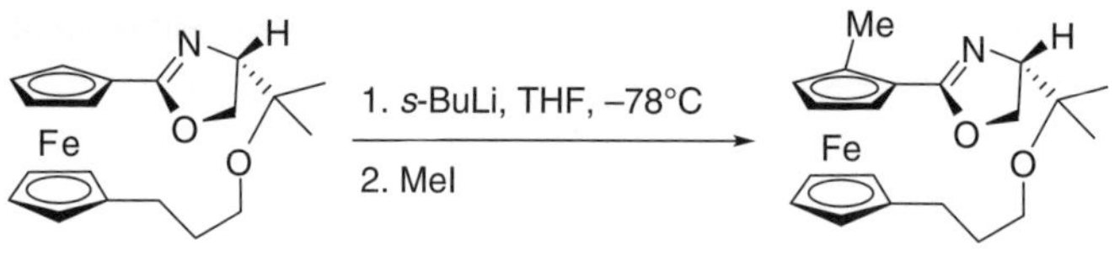

Scheme 2.16 Diastereoselective ortho-lithiation of macrocyclic ferrocenyloxazoline (R)-58.

group of the alkyllithium reagent avoids a steric interaction with the substituent on the oxazoline chiral center. This monomeric model, although inadequate, also accounts for the dramatic increase in the diastereoselectivity of ortho-lithiation to >100:1 by the addition of TMEDA, if one assumes that the diamine increases the effective size of the alkyllithium reagent by chelating to it, thereby increasing steric repulsion between R' and R. In general, ether and hexanes are optimal solvents, depending on the solubility of the corresponding ferrocenyloxazolines, while s-BuLi and n-BuLi are the preferred bases. t-BuLi tends to give lower diastereoselectivities (34:1) for some oxazolines.

Sammakia [70] has expanded the method to the macrocyclic ferrocenyloxazoline (R)-58 to afford the planar chiral derivative 59 in 70% yield and >100:1 diastereomeric ratio (Scheme 2.16). This work shows that nitrogen, rather than oxygen, controls the stereoselectivity of the reaction during ortho-lithiation.

Numerous multidentate chiral ferrocenyl ligands containing the oxazoline DMG such as 1,2-P,N ligands (S_p)-60 [71], (R_p)-60 [16], 61 [72], 62 [73], 1,1'-P,N ligands 63 [74], N,Se(S) ligands 64 [75], 1,2-N,O ligands 65 [76], silyl ethers 66 [77], biferrocenylbisoxazolines 67 [78] and 1,1'-bisoxazolines 68 [79] have been reported (Figure 2.5). The availability of ferrocenyl oxazoline methods has greatly accelerated the applications of ferrocenyl ligands for various asymmetric catalytic reactions [3e, 19c, 29a,d]. Applications have included Pd-catalyzed conjugate addition of Grignard reagents to enones, Pd-catalyzed allylic alkylation and aminations [74a,b], Pd-catalyzed intermolecular asymmetric Heck reactions [74c,d], Ru-catalyzed asymmetric transfer hydrogenation of ketones and Ir-catalyzed asymmetric hydrogenation of 2-substituted quinolines [80].

The synthesis of representative 1,1'-P,N ligands 73 and 74 [74b] is depicted in Scheme 2.17. Lithium-bromide exchange of 1,1'-dibromoferrocene 69 is followed by quenching with carbon dioxide to afford carboxylic acid 70. After oxazoline formation, a second lithium-bromide exchange followed by trapping with Ph₂PCl yields 1,1'-P,N ligand (S)-72, which is then subjected to standard diastereoselective DoM, followed by introduction of electrophiles to produce compounds 73a–c. The description of 73b as having R_p configuration is due to the reversal of priority of the methyl group compared to groups in 73a and 73c. Further DoM of (S,S_p)-73a and subsequent methylation or bromination followed by "silicon trick" desilylation [33] produces (S,S_p)-73b and 74, which have opposite planar chirality compared to the (S,R_p)-73b series. Lithium-bromine exchange of (S,R_p)-74 followed by TMSCl quenching provides (S,R_p)-73c in 73% yield. Diastereomerically pure products are isolated by column chromatography.

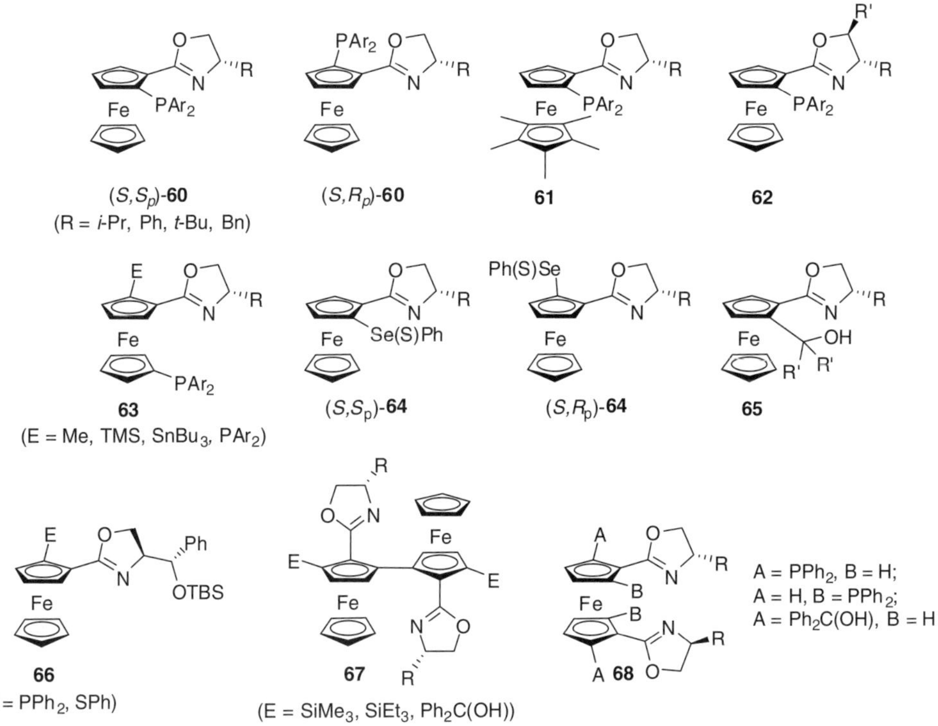

2.2 Diastereoselective Directed ortho-Metalation of Ferrocenes with Chiral Auxiliaries | 31

Figure 2.5 Representative planar chiral ferrocene ligands derived from ferrocenyloxazolines.

2.2.5
Other Chiral Auxiliaries

As early as 1976, Enders introduced (S)-1-amino-2-methoxymethylpyrrolidine, now well known by the SAMP acronym, as a chiral auxiliary for the asymmetric synthesis of optically pure α-alkyl ketones. Twenty years later, the same group reported that ferrocenyl SAMP-protected ketones serve as excellent DMGs [17]. Thus, lithiation of **75** with *n*- or *s*-BuLi in ether at −70 °C followed by electrophile quench affords 1,2-disubstituted products **76** in good yields and excellent diastereoselectivities (87–98%) [81]. Removal of the SAMP auxiliary with $SnCl_2$ or $TiCl_3$ furnishes exclusively planar chiral ketones **77** in good to excellent yields and 70–96% *ee* (Scheme 2.18). As this chapter was being completed, Enders and coworkers extended the SAMP-derived chiral hydrazone strategy to diferrocenyl ketones, which provided planar chiral bisferrocenes in up to 96% *de* [81c].

Successful diastereoselective DoM of FcSAMP hydrazones, acetals and O-methylephedrines indicates that methoxy substituents may play a crucial role in stabilizing transition states during lithiation. It is perhaps not surprising then that other chiral methoxy-containing functional groups can also selectively induce planar chirality [82]. A case in point is the α-methoxybenzyl directing group developed by Knochel [18]

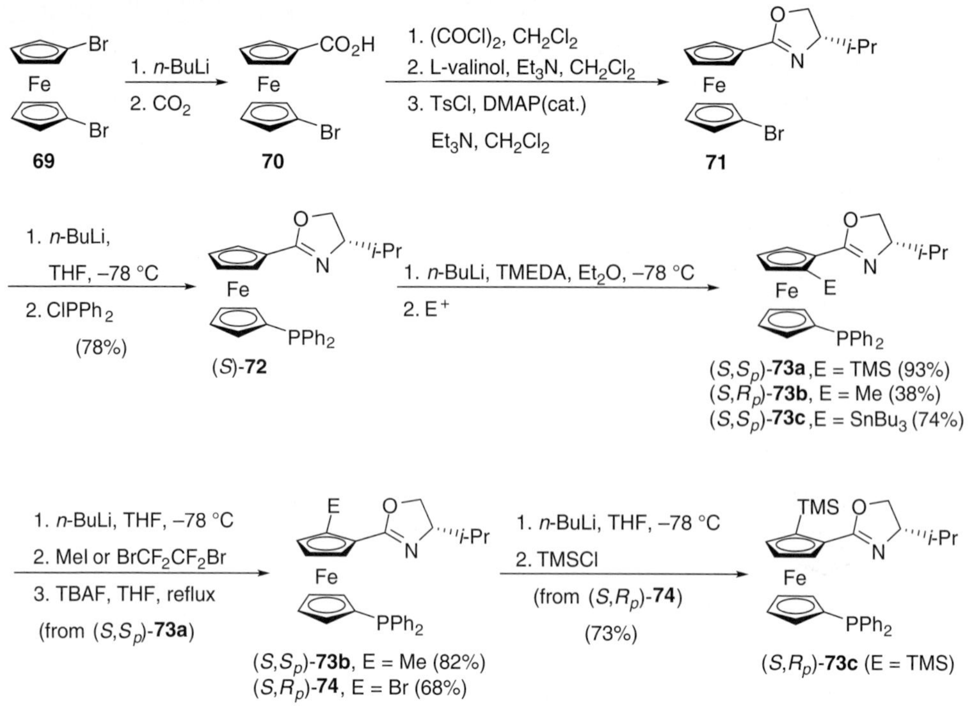

Scheme 2.17 Representative synthesis of 1,1′-P,N ligands **73** and **74**.

which affords planar chiral adducts in 98% de upon DoM with t-BuLi in ether at −78 °C followed by addition of various electrophiles (**78** → **80a–d**) (Scheme 2.19). The lithioferrocene intermediate of **79** also participates in Pd-catalyzed Negishi arylation after transmetalation with $ZnBr_2$ to give **80e**, which upon treatment with $HPPh_2$ and borane protection, gives P,N ligands **81** in variable yields and with retention of configuration due to the α-ferrocenyl carbocation effect.

The efficient conversions to products **80a–e** may not have been expected *a priori*, based on considerations of the relative acidity of 2- and α-ferrocenyl C−H bonds. However, although the preparation α-ferrocenyllithiums from *ortho*-substituted α-ferrocenyl amines, ethers or sulfides **82** (X = NMe_2, OMe or SPh) requires lithium

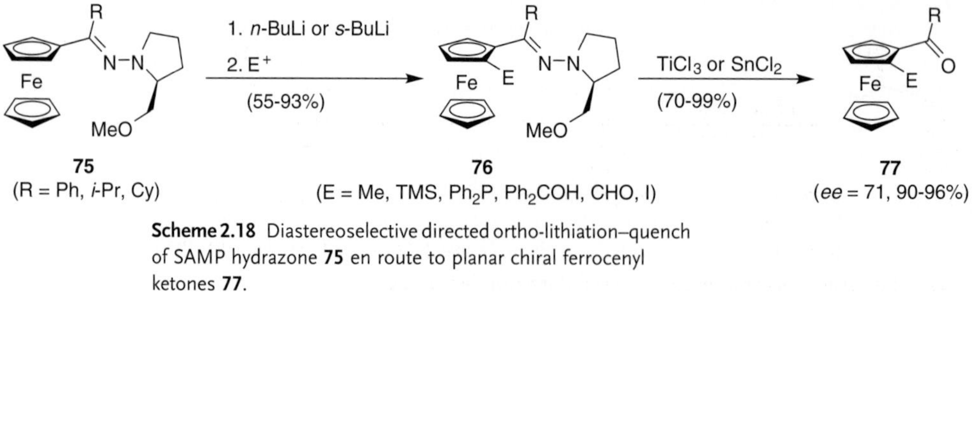

Scheme 2.18 Diastereoselective directed ortho-lithiation–quench of SAMP hydrazone **75** en route to planar chiral ferrocenyl ketones **77**.

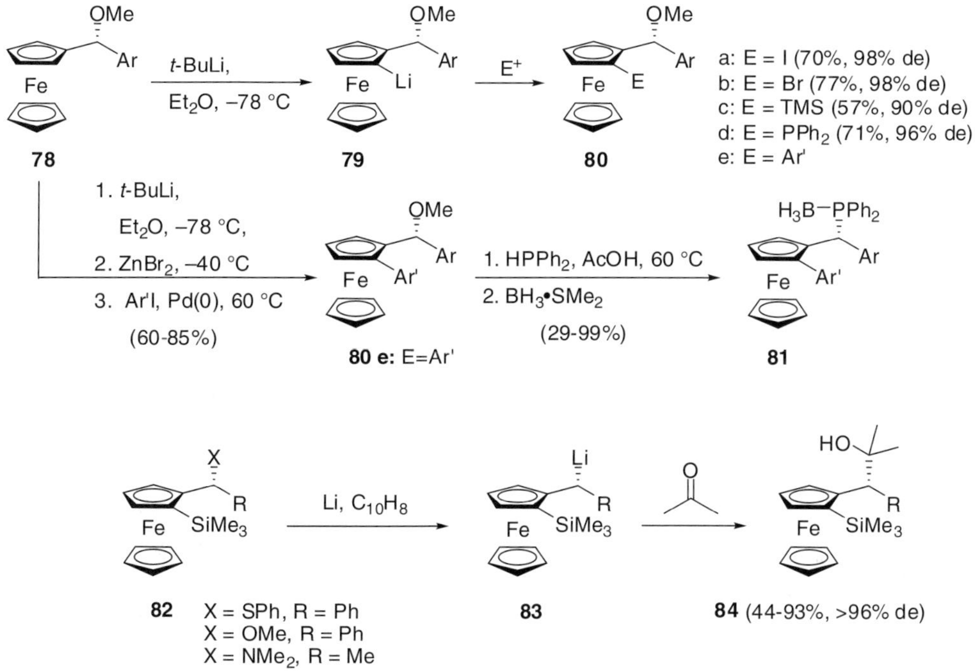

Scheme 2.19 The α-methoxybenzyl DMG in diastereoselective lithiation and reductive lithiation of ortho-substituted α-sulfides, ethers and amines.

naphthalenide, this reaction proceeds without epimerization of **83** at low temperature (Scheme 2.19) to give products such as **84** [83]. This transformation represents an umpolung of the normal reactivity of α-ferrocenyl positions and increases the scope of products derived from **78**. The lack of α-reactivity of α-ferrocenyl ethers **78** towards alkyllithiums contrasts with the α-deprotonation of benzyl ethers whose consequence is the [1,2]-Wittig rearrangement [84].

Recently, Ueberbacher [26] described a similar transformation using the α-ferrocenyl carbinol (S)-**85** (Scheme 2.20). Thus, using excess n-BuLi followed by electrophile quench leads, perhaps via the putative dianion intermediate (S,R_p)-**86**, to products (S,R_p)-**87** in good yield and up to 90% de. The optimal lithiation conditions require 2.2 equiv of n-BuLi in ether at −20 °C. As with the α-methoxybenzyl DMG

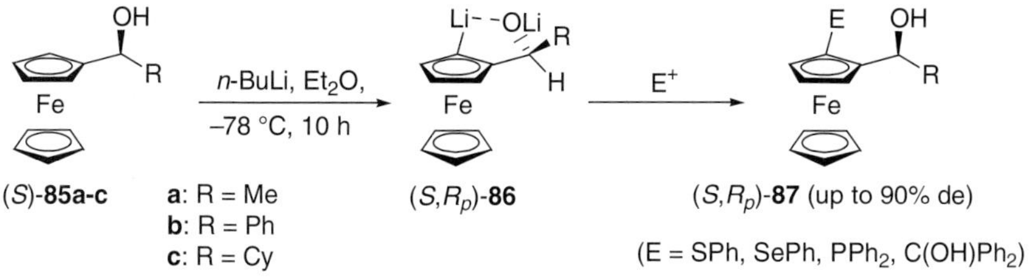

Scheme 2.20 Diastereoselective DoM of chiral α-ferrocenyl carbinols.

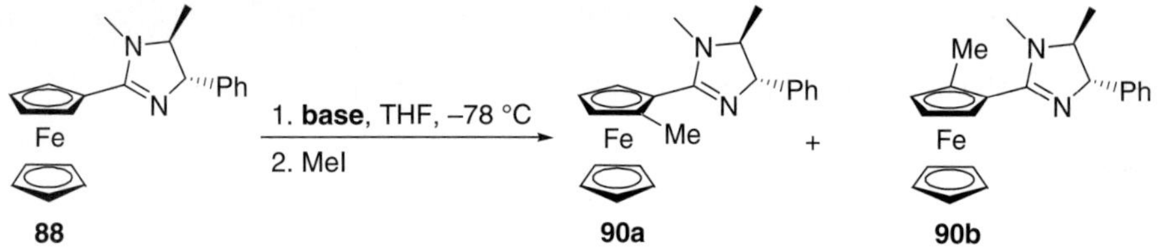

Scheme 2.21 Diastereoselective ortho-lithiation of ferrocenylimidazolines **88** and **89**.

base	90a:90b
1.1 equiv t-BuLi	5:1
0.95 equiv t-BuLi + 1.5 equiv LDA	1:12

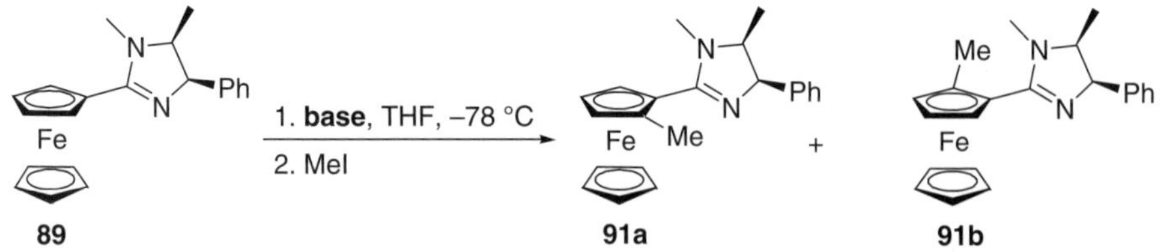

base	91a:91b
1.0 equiv t-BuLi	1:3
1.0 equiv t-BuLi + 1.0 equiv LDA	1:31

above (**78**), a comparison of metalation of **85** (R = Ph) can be made with DoM of benzyl alcohol, which occurs under rather special conditions [85].

Epimeric ferrocenylimidazolines **88** and **89** were recently shown by Peters [23] to participate in diastereoselective DoM reactions. Interestingly, the relative stereochemistry of the lithiation step is highly dependent upon the nature of the base. For example, t-BuLi gives inferior results compared to the combination of t-BuLi and LDA, while LDA alone fails to promote lithiation of these substrates. As depicted in Scheme 2.21 for ferrocenylimidazoline **88**, the ratio of two diastereoisomers **90a**/**90b** was reversed from 5 : 1 to 1 : 12 by using a 1.5 : 0.95 LDA/t-BuLi mixture. Similarly, DoM of ferrocenylimidazoline **89** with a 1 : 1 mixture of LDA/t-BuLi resulted in a dramatic increase in diastereoselectivity from 1 : 3 to 1 : 31. DoM chemistry of the imidazoline DMG is well established but sparsely used [86].

In 2000, Bolm [22] reported that enantiopure ferrocenylsulfoximines **92** [22b] afford diastereomerically pure products **94** (Scheme 2.22) after DoM–electrophile quench for all electrophiles except prochiral p-anisaldehyde, which gives a 1 : 1 mixture of epimeric benzyl alcohols. The origin of diastereoselectivity in DoM of **92** has been ascribed to sulfoximine oxygen rather than nitrogen atom coordination effects (intermediate **93**). Sulfoximine DoM chemistry of non-metallocene aromatic substrates has not been explored.

Also in 2000, Widhalm, van Leeuwen and coworkers [87] reported the (i-Pr$_2$N)MgBr induced diastereoselective ortho-magnesiation of P-chiral phosphine oxide

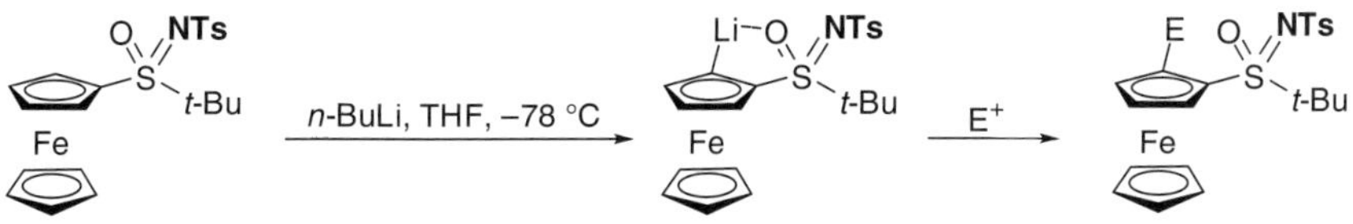

92 **93** **94** (36-78%)

(E = Me, SMe, TMS, $SnBu_3$, I, Me_2COH, p-MeOPhCHOH)

Scheme 2.22 Diastereoselective ortho-lithiation of ferrocenylsulfoximine **92**.

(*R*)-**95** which results, upon iodine quenching, in the formation of the 2-iodoferrocene (*R*,S_p)-**96** in quantitative yield and up to 94% de (Scheme 2.23).

Similarly, *t*-BuLi effects highly diastereoselective ortho-lithiation of oxazaphospholidine-oxide **97** (Scheme 2.24) to give a range of products **98** in variable yields but greater than 99% de [25]. This transformation is closely related to the BH_3-activated diastereoselective lithiation of chiral ferrocenyl phosphorodiamidites **99**, as patented by Pfaltz [88].

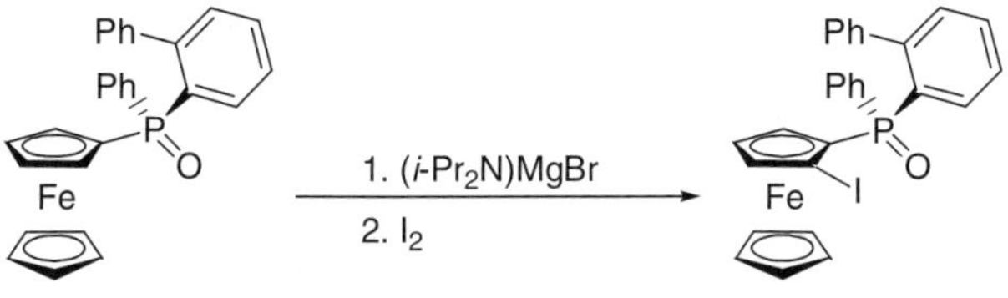

(*R*)-**95** (*R*,S_p)-**96** (94% de)

Scheme 2.23 Diastereoselective DoM of *P*-chiral ferrocenyl phosphine oxide (*R*)-**95**.

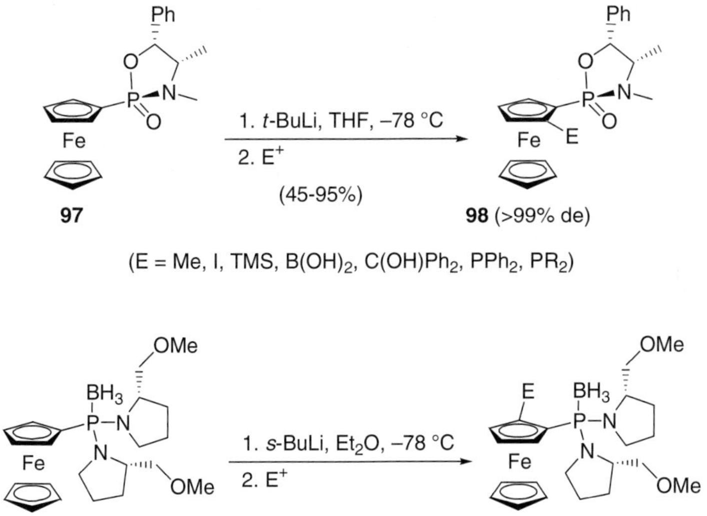

97 **98** (>99% de)

(E = Me, I, TMS, $B(OH)_2$, $C(OH)Ph_2$, PPh_2, PR_2)

99 **100** (99% de)

(E = TMS, PPh_2, Br)

Scheme 2.24 Diastereoselective ortho-lithiation of chiral **97** and **99**.

2.3
Enantioselective Directed ortho-Metalation of Ferrocenes with Chiral Bases

Methods for constructing 1,2-disubstituted ferrocenes via diastereoselective DoM tactics accumulated rapidly after Kagan's seminal contribution describing the ferrocenyl sulfoxide as an excellent DMG for introducing the new stereogenic element of planar chirality. In contrast to these results, the strikingly prescient (−)-sparteine-mediated lithiation of isopropylferrocene **101** by Nozaki in 1970 afforded only mixtures of products **103** in low enantiomeric excess (3%) (Scheme 2.25-1) [89]. This disappointing observation may have discouraged the search for enantioselective methods in the ensuing years. In 1995, Simpkins [90] described a DoM reaction of a diphenylphosphine oxide **104** using a chiral lithium amide base to afford the ortho-substituted trimethylsilyl adduct with moderate enantioinduction (Scheme 2.25-2). Almost concurrently, Snieckus [91] reported that (−)-sparteine, a now widely used alkaloid for asymmetric reactions, as demonstrated especially by the comprehensive work of Hoppe [92], is a particularly effective ligand in the enantioselective n-BuLi mediated DoM reaction of N,N-diisopropyl ferrocenecarboxamide **107** to afford, after quenching the resulting lithium species with various electrophiles, 1,2-disubstituted ferrocene products **108** in excellent enantiomeric excess (up to 99% ee) (Scheme 2.24-3). In the Snieckus work, an interesting and apparently unprecedented observation was the slow racemization of ferrocenes **108** (E = SPh, SePh, PPh$_2$) at room temperature, which is due to a substituted Cp–aryl ligand exchange by an unknown mechanism. The (−)-sparteine mediated DoM reaction is also applicable to 1,1'-N,N,N',N'-tetraisopropyl ferrocene dicarboxamide **109** [93] providing highly enantioenriched singly (**110**) or doubly (dl-**111**) functionalized derivatives.

In subsequent work [94], the N-cumyl-N-ethylferrocenecarboxamide **112** was introduced for (−)-sparteine-mediated ortho-functionalization chemistry in order to overcome the limitation of the diisopropyl amide for conversion to other functionalities. As shown in Scheme 2.26, the N-cumyl-N-ethyl amide behaves equally to the diisopropyl amide [91] in terms of yield and ee values. Product **114f** (E = PPh$_2$), prepared by N-decumylation of **113f** (E = PPh$_2$) in acetic acid at room temperature, underwent rapid racemization in some solvents similar to the diisopropyl analogue **108h** (E = PPh$_2$), but could be crystallized from dichloromethane in enantiomerically pure form. By using even milder decumylation conditions (2,2,2-trifluoroethanol at reflux) and further modifications, a series of secondary N-ethylferrocenecarboxamides possessing only planar chirality was synthesized. The intermediate secondary amides were easily transformed into unprecedented planar chiral esters, other tertiary amides, and a planar chiral isomer of Ugi's amine. An additional interesting sequence of transformations involving a Grubbs metathesis reaction of N-allyl derivative **116** led to a new azepinone-fused planar chiral ferrocene **117**, which was successfully ortho-lithiated to yield the trisubstituted diphenylphosphine **118** after ClPPh$_2$ quenching.

O'Brien has demonstrated that the antipode of (S_p)-**113b** can be prepared in 92% ee by the DoM reaction of **112** using the (+)-sparteine surrogate **119** [95] (Scheme 2.27).

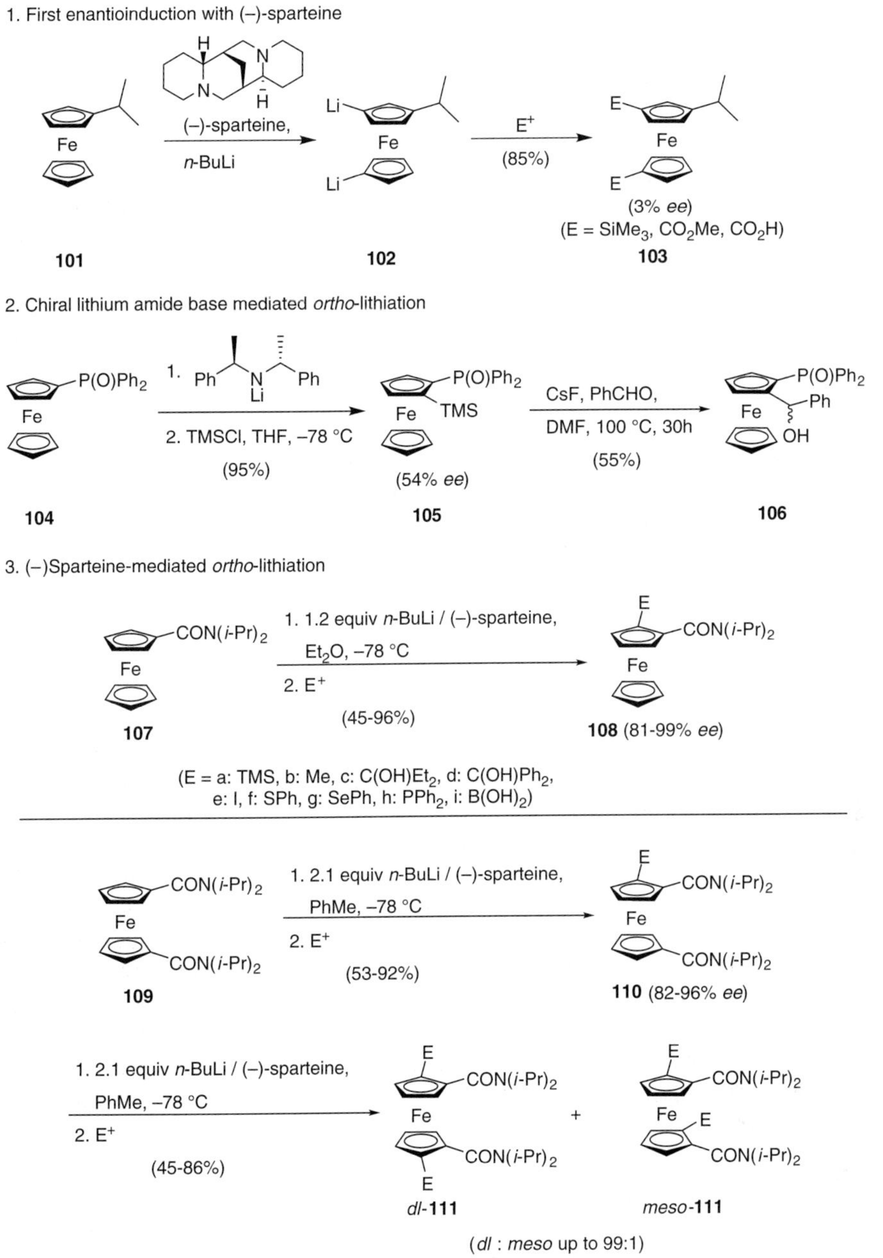

2.3 Enantioselective Directed ortho-Metalation of Ferrocenes with Chiral Bases

1. First enantioinduction with (−)-sparteine

101 → (−)-sparteine, n-BuLi → **102** (85%) → E⁺ → **103** (3% ee) (E = SiMe$_3$, CO$_2$Me, CO$_2$H)

2. Chiral lithium amide base mediated ortho-lithiation

104 (Fc–P(O)Ph$_2$) → 1. Ph–CH(Li)–N–CH(Ph) chiral amide; 2. TMSCl, THF, −78 °C (95%) → **105** (54% ee) → CsF, PhCHO, DMF, 100 °C, 30h (55%) → **106**

3. (−)Sparteine-mediated ortho-lithiation

107 (Fc–CON(i-Pr)$_2$) → 1. 1.2 equiv n-BuLi / (−)-sparteine, Et$_2$O, −78 °C; 2. E⁺ (45–96%) → **108** (81–99% ee)

(E = a: TMS, b: Me, c: C(OH)Et$_2$, d: C(OH)Ph$_2$, e: I, f: SPh, g: SePh, h: PPh$_2$, i: B(OH)$_2$)

109 (1,1′-bis-CON(i-Pr)$_2$ ferrocene) → 1. 2.1 equiv n-BuLi / (−)-sparteine, PhMe, −78 °C; 2. E⁺ (53–92%) → **110** (82–96% ee)

1. 2.1 equiv n-BuLi / (−)-sparteine, PhMe, −78 °C; 2. E⁺ (45–86%) → *dl*-**111** + *meso*-**111**

(*dl* : *meso* up to 99:1)

Scheme 2.25 (−)-Sparteine and chiral lithium amide base-mediated enantioselective ortho-lithiation.

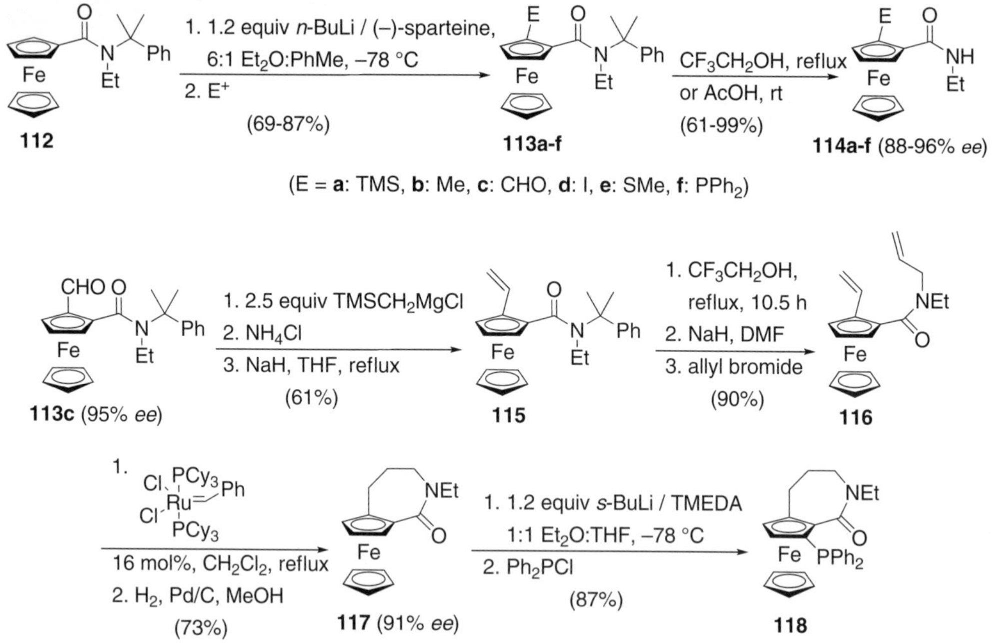

Scheme 2.26 (−)-Sparteine-mediated enantioselective ortho-lithiation of N-cumyl-N-ethylferrocenecarboxamide **112** and the synthesis of an azepinone-fused planar chiral ferrocene.

Uemura reported that N,N,N',N'-tetramethyl-(1R,2R)-cyclohexanediamine-[(R,R)-TMCDA)]-mediated asymmetric lithiation of (dimethylaminomethyl)ferrocene **120** proceeds in up to 80% ee [96] (Scheme 2.28) to give R_p-configured products **121** (E = CHO, PPh$_2$).

Recently, the BF$_3$-complexed dimethylaminoferrocene **122** was found by Metallinos [97] to undergo ortho-lithiation upon treatment with n-BuLi. Exposure of the lithioferrocene to nine different carbon and heteroatom-based electrophiles gave racemic products in 76–94% yield. Among the products were rare derivatives such as 2-phosphino-1-aminoferrocenes, which are of interest as ligands [98] for transition metal catalysis. Although initial attempts to develop an enantioselective variant of this process (Scheme 2.29) by mediating the lithiation with chiral diamines [(−)-sparteine, (S,S)-TMCDA and (R,R)-N,N'-dimethyl-N,N'-di-(3,3-dimethylbutyl)

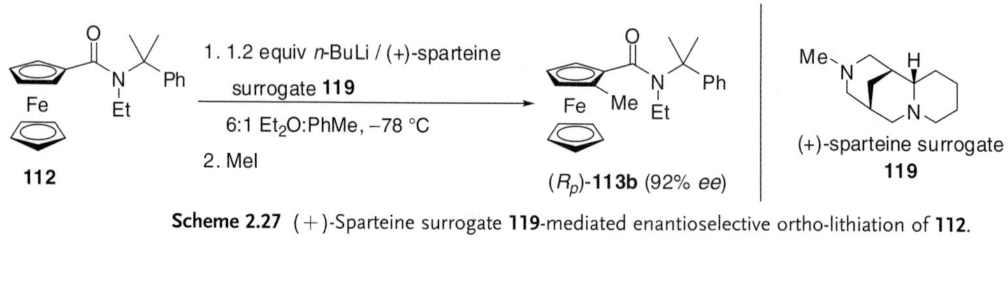

Scheme 2.27 (+)-Sparteine surrogate **119**-mediated enantioselective ortho-lithiation of **112**.

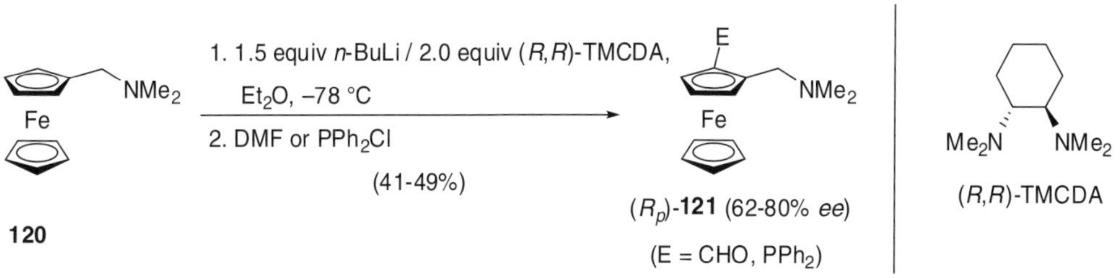

Scheme 2.28 (R,R)-TMCDA-mediated enantioselective ortho-lithiation of (dimethylaminomethyl)ferrocene.

cyclohexanediamine] gave the formyl adduct **123** in low enantiomeric purity (10–22% ee), significant improvements to the stereoselectivity of this transformation have now been realized [99].

As a concluding comment in this section, it has been demonstrated that low diastereoselectivities in the ortho-lithiation of ferrocenes with chiral directing groups may be increased dramatically with (−)-sparteine. A case in point is the lithiation of (−)-menthylferrocenesulfonate **124** with n-BuLi, which provides the formyl derivative **125** in only 20% de after quenching with DMF. However, *matched pair* lithiation of **124** with s-BuLi/(−)-sparteine gives aldehyde **125** in much improved 81% de [100] (Scheme 2.30). This double asymmetric process differs conceptually [101] from the simple diastereoselective or enantioselective induction of ferrocene planar chirality covered in Sections 2.2 and 2.3. Matched–mismatched pair interactions serve as the foundation for kinetic resolutions, which are the topic of the next section.

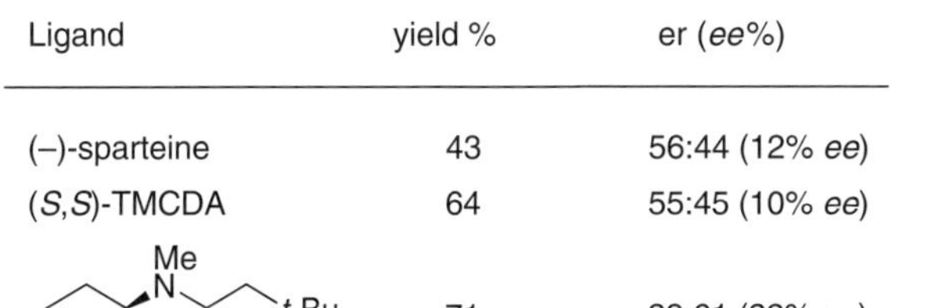

Ligand	yield %	er (ee%)
(−)-sparteine	43	56:44 (12% ee)
(S,S)-TMCDA	64	55:45 (10% ee)
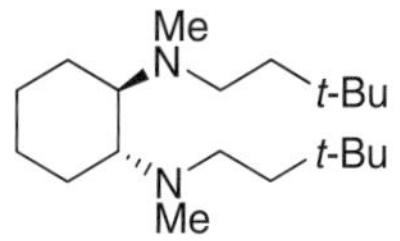 (R,R)-N,N'-dimethyl-N,N'-di-(3,3-dimethylbutyl)cyclohexanediamine	71	39:61 (22% ee)

Scheme 2.29 Enantioselective ortho-lithiation–functionalization of BF$_3$-complexed dimethylaminoferrocene.

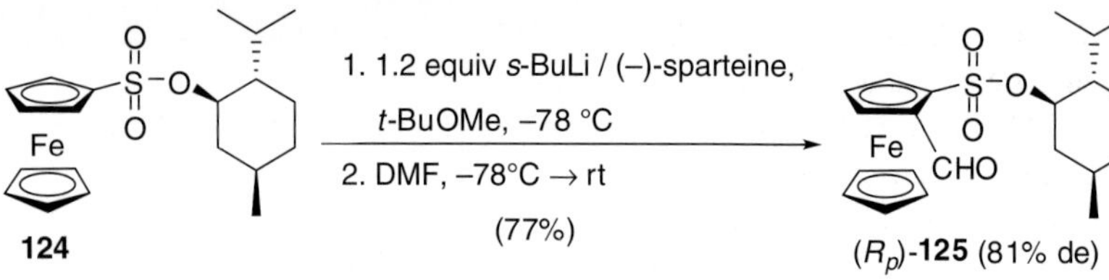

Scheme 2.30 Double asymmetric induction of planar chirality in (−)-menthylferrocenesulfonate **124** with s-BuLi/(−)-sparteine.

2.4
Enzymatic and Nonenzymatic Kinetic Resolutions

In addition to enantioselective and diastereoselective ortho-functionlization strategies, resolution of racemic planar chiral ferrocenes, either enzymatically or nonenzymatically, serves as the third method to prepare optically active 1,2-disubstituted ferrocenes.

2.4.1
Enzymatic Kinetic Resolutions

Enzyme catalyzed asymmetric reactions have a long history of application in the synthesis of optically active compounds in both academic and industrial laboratories [102]. The first examples of the asymmetric preparation of planar chiral ferrocenes via enzyme catalyzed resolution strategies appeared in the late 1980s. In these cases [103], reactions involved mainly enantioselective microbial reduction of both *meso-* and *rac-* 1,2-disubstituted ferrocenyl aldehydes with Baker's yeast to afford the corresponding reduced product and unreacted starting material with moderate to good enantiomeric excess. In terms of potential applications, 1,2-disubstituted ferrocenyl alcohols **126** may be esterified by lipases to afford both enantiomers (S_p)-**127** and (R_p)-**126**.

As depicted in Scheme 2.31, racemic 1,2-disubstituted ferrocenes **126a–e** may be kinetically resolved by lipase-catalyzed esterification using vinyl acetate as the acyl donor to furnish esters **127a–e** favoring the S_p configuration, and unreacted **126** in predominantly R_p configuration. According to Nicolosi [104], *Candida cylindracea*

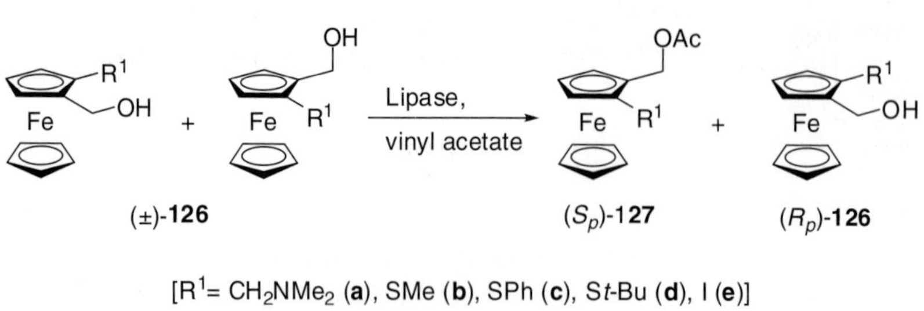

[R^1= CH$_2$NMe$_2$ (**a**), SMe (**b**), SPh (**c**), S*t*-Bu (**d**), I (**e**)]

Scheme 2.31 Lipase-catalyzed resolution of 1,2-disubstituted ferrocenyl carbinols (±)-**126**.

lipase (CCL) in *t*-butylmethyl ether gave (−)-**127a** in 92% *ee* at 42% conversion and (+)-**126a** in 95% *ee* and 32% yield. Interestingly, kinetic resolution by deacylation of racemic **127a** using CCL, gave (−)-**126a** in 95% *ee* and 40% chemical yield. In a subsequent study [105], *Candida antarctica* (Novozyme®) was shown to afford **127b** in 90% *ee* at 32% conversion, although **127c** and **127d** were obtained in lower enantiomeric excess (88% and 76%, respectively). Iodide **127e** was obtained in excellent 96% *ee* at 25% conversion. For substrate **126d**, use of a lipase from *Mucor miehei* (Lipozyme) led to the corresponding dimimg **127d** with better enantiomeric purity (90% *ee* at 35% conversion).

This technique can also be applied to the kinetic resolution of racemic 1-(1-hydroxyethyl)-2-(hydroxymethyl)ferrocene [106] containing both central and planar chirality to furnish unreacted alcohol in up to 97% *ee* and 27% chemical yield. A more complicated case involving the kinetic resolution of a 1,3-disubstituted ferrocene with both central and planar chiralities was reported recently by Nicolosi [107]. Lipase from *Candida antarctica* recognized the central chiral *R*-configured enantiomer, thereby effecting resolution of the racemic mixture.

2.4.2
Nonenzymatic Kinetic Resolutions

Traditional nonenzymatic resolution, also known as stoichiometric chemical resolution [30], has been widely used for the synthesis of chiral ferrocenes possessing either one or mulitple stereogenic centers including the element of planar chirality since the time of the first reported planar chiral ferrocene in 1959 [1]. However, chiral catalyst mediated kinetic resolution, a potential alternative to enzymatic methods, was first applied to the synthesis of planar chiral ferrocenes by Moyano in 2006 [108]. In this work (Scheme 2.32), the Sharpless asymmetric dihydroxylation (AD) reaction was employed on racemic 2-substituted-1-vinylferrocene **128a-d** to give products **129** with stereochemical outcomes in accordance with the Sharpless' mnemonic rule.

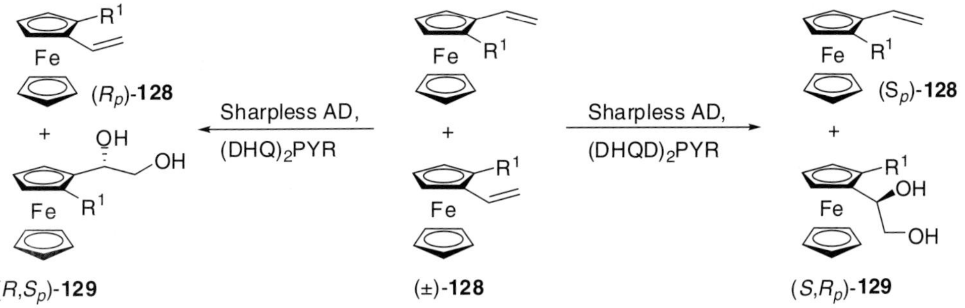

[R^1 = CON(*i*-Pr)$_2$ (**a**), 4,4-dimethyl-1,3-oxazolin-2-yl (**b**), TMS (**c**), I (**d**)]

Scheme 2.32 Kinetic resolution of 2-substituted 1-vinylferrocene **128** using Sharpless AD.

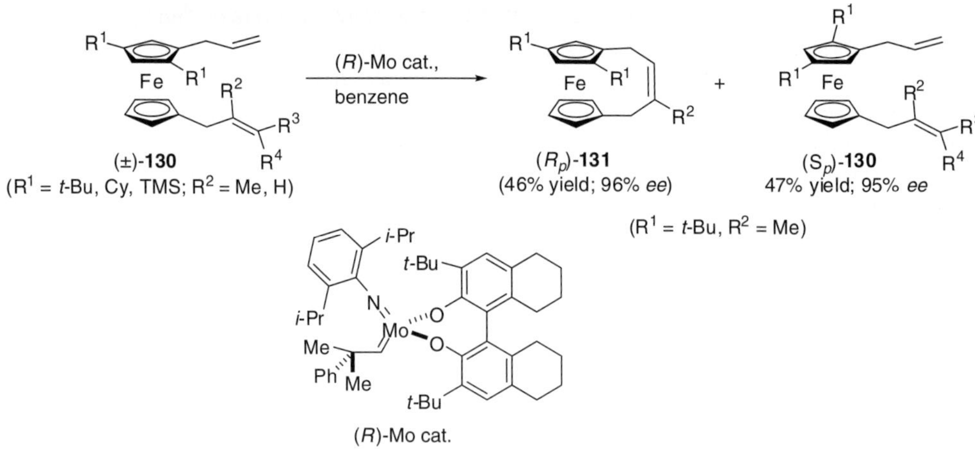

Scheme 2.33 Kinetic resolution of ferrocenylalkene **130** via ARCM.

Thus, as shown in Scheme 2.32, Sharpless AD of (±)-**128** using (DHQD)$_2$PYR as ligand gave (S,R$_p$)-**129** as the major product (90% ee, 44% yield for **129a**; 86% ee, 47% yield for **129b**; 88% ee, 47% yield for **129c**; 80% ee, 43% yield for **129d**). Ligand (DHQ)$_2$PYR produced the enantioisomeric series (R,S$_p$)-**129** as the major products (80% ee, 44% yield for **129a**; 86% ee, 44% yield for **129b**; 64% ee, 48% yield for **129c**; 68% ee, 48% yield for **129d**). The unreacted alkenes, (S$_p$)-**128**, were recovered in somewhat lower yields in most cases but with better enantiomeric purity (e.g., 92% ee for **128a**; 94% ee for **128d**) when the (DHQD)$_2$PYR ligand was employed; the ee values for (R$_p$)-**128** by using (DHQ)$_2$PYR were 64% and 72%, respectively.

Subsequently, Ogasawara [109] described the first example of catalytic kinetic resolution by asymmetric ring-closing metathesis (ARCM) of racemic planar chiral ferrocenylalkene **130** using a chiral molybdenum catalyst. It was found that 0.5 mol% of (R)-Mo* catalyst in benzene at 50 °C gave optimal yields and stereoselectivity. The products (R$_p$)-**131** and untouched (S$_p$)-**130** were obtained in high yield and enantioselectivity (96% ee and 95% ee, respectively) when R^1 = t-Bu and R^2 = Me (Scheme 2.33). Although the ee value of product (R$_p$)-**131** were further increased to 99.5% by lowering the reaction temperature to 23 °C, the yields of both (R$_p$)-**131** (23% yield) and untouched (S$_p$)-**130** (30% yield) were much lower, as was the enantiomeric purity of (S$_p$)-**130** (78% ee) (Scheme 2.33).

2.5
Summary and Perspectives

Following the early bold steps of Schlögl [30] and Nozaki [31], the appearance of Ugi's seminal contribution in 1970 on the highly diastereoselective directed ortho-metalation (DoM) of chiral N,N-dimethylferrocenylethylamine [32] opened the road to the development of stereoselective methods for the synthesis of planar chiral ferrocenes. The immense impact of Ugi's amine on the abundant availability of planar chiral ferrocene ligands (Figure 2.2) is especially evident in their current use in large-scale industrial processes [9, 102]. Beginning with the insightful studies of Kagan on chiral ferrocenyl sulfoxides in 1993 [11] and following the demonstration of chiral base-mediated enantioinduction by Simpkins [91] and Snieckus [90], the DoM tactic has provided a number of enantio- and diastereo-selective routes to 1,2-substituted planar chiral ferrocene derivatives bearing a diversity of directed metalation groups (DMGs) many of which today still require demonstration of DMG modification to establish compound utility. Some notable exceptions are the Ugi amines which undergo ferrocenylmethyl cation-mediated α-displacement (Figure 2.2), chiral sulfoxides which show DMG displacement (Scheme 2.9), and N-cumylcarboxamides which allow modification to a variety of other FGs (Scheme 2.26). Further fundamental studies of sulfoximine (Scheme 2.22), acetal (Scheme 2.12), methoxy analogues of Ugi's amine (Scheme 2.19), oxazoline (Figure 2.5), SAMP (Scheme 2.18), and the inviting P-chiral (Schemes 2.23 and 2.24) DMGs may be pursued to the advantage of accelerating use of ferrocenyl ligands in asymmetric catalytic reactions. The intriguing pyrrolidinyl and related DMG systems (Scheme 2.6) enrich Ugi amine chemistry, furnish many new ferrocenyl P-ligands and provide an opportunity to probe deeper into the question of enantioinduction by chiral sites considerably remote from the deprotonation position. A synthetic method to derive hybrid Ugi–Kagan ferrocenylmethylamine (Scheme 2.11) shows potential for generalization to obtain derivatives unavailable by alternative routes.

Although to date the enantioselective DoM strategies have relied mainly on (−)-sparteine (Schemes 2.25 and 2.26), the recent demonstration of sparteine surrogates [95] (Scheme 2.27) and (R,R) and (S,S)-configured trans-cyclohexane-1,2-diamine derivatives [110] (Scheme 2.28) inducing asymmetric lithiation chemistry will undoubtedly result in the continuing search for new chiral bases and complexes. The TMS "trick" (Schemes 2.3 and 2.17) constitutes a viable and underused alternative for the construction of the antipode of the planar chiral ferrocene resulting from (−)-sparteine-mediated induction. The double asymmetric induction result (Scheme 2.30), different conceptually from either enantio- or diastereo-selection reactivity, may deserve additional scrutiny.

What loci exist in ferrocenyl DMGs for fundamental stereoselective DoM studies? N-, O-, and P-DMG-bearing systems are clearly missing or underdeveloped. Although aminoferrocenes are valuable ligands and optically active materials, the generation of the nitrogen–ferrocene bond has constituted a major synthetic challenge [68]. Hence, the recent work on amino- (Scheme 2.29, see also Scheme 2.14) DMGs is an initial promising foray into this area. Phosphine oxide DMG studies (Schemes 2.23 and 2.24) also constitute early positive results for efforts to develop new P-chiral groups as well as to achieve enantioinduction in prochiral P-DMG systems. Oxygen-bearing DMG systems have not been tested in enantioselective DoM chemistry with the exception of ferrocenyl O-carbamates which have not been substantially investigated [111]. Ferrocenyl boronic acids have been prepared usually in the service of Suzuki–Miyaura cross coupling chemistry (e.g., Scheme 2.10). The test of ferrocenyl boronates as well as boronamides, racemic or with an incorporated chiral auxiliary, in DoM chemistry remains an intriguing undertaking.

Methodology based on enzymatic and nonenzymatic (known for 50 years) catalyzed kinetic resolution of racemic ferrocenes is still in its infancy. Among other valuable observations, the resolution by Nicolosi [107] of a 1,3-disubstituted ferrocene with both central and planar chiralities using lipase from *Candida antarctica* constitutes an advance in this area. The recent chemical Sharpless kinetic resolution technique using catalytic asymmetric reactions (Scheme 2.32) provides a new chemical path to planar chiral ferrocenes containing functional groups which are difficult to obtain by other strategies. The success of the catalytic kinetic resolution process based on asymmetric RCM (Scheme 2.33) appears to be a harbinger for future activity.

In conclusion, the synthesis of planar chiral ferrocenes has evolved to the state of versatility in methodology and, in cases of demand by industry, availability on a larger scale. Nevertheless, the synthesis of optically pure building blocks for organic synthesis demands availability of new ligands that will be met, in part, by emerging (e.g., Scheme 2.14) planar chiral ferrocene derivatives. Although the use of planar chiral ferrocenes as ligands in catalytic asymmetric reactions is now well established in industrial practice, for example, in the 10 000 t/y production of S-metolachlor [9b], further work on ferrocene derivatives providing greater versatility in substitution patterns, subsequent transformations of robust functional groups to useful functionalities, and the application of the derived products in significant catalytic processes is required and may be anticipated.

2.6
Selected Experimental Procedures

2.6.1
Synthesis of (R,S$_p$)-N,N-dimethyl-1-(2-diphenylphosphino-ferrocenyl)ethylamine (PPFA)

2.6.1.1 Synthesis of Racemic α-Ferrocenylethyldimethylamine (2) [2]

At $-20\,°C$, a solution of 23.0 g of 1-ferrocenylethanol in 150 mL of toluene was added dropwise to a stirred solution of 12.5 g of phosgene in 100 mL of toluene. 30 min after the addition was complete, the reaction mixture was allowed to warm to $20\,°C$ and, without isolation of the chloride, added to a $-20\,°C$ solution of 22.5 g of dimethylamine in 200 mL of isopropyl alcohol. The temperature was allowed to rise to $20\,°C$. The reaction mixture was filtered and evaporated to dryness. The residue was taken up in benzene, extracted with 8.5% phosphoric acid, washed with benzene, neutralized with Na_2CO_3, extracted with benzene, dried, and evaporated; crude yield: 24.4 g (95%); yield after distillation (bp $110\,°C$ (0.45 mmHg), with some decomposition): 17.5 g (68%). The undistilled product was sufficiently pure for the resolution of the antipodes of α-ferrocenylethyldimethylamine.

Note: enantiomerically pure 1-ferrocenylethanol can be prepared by enantioselective reduction using ruthenium [112] or CBS [113] catalysts and converted to **2** by the preceding method, in which case the resolution described below is unnecessary.

2.6 Selected Experimental Procedures

2.6.1.2 Resolution of α-Ferrocenylethyldimethylamine 2 [2]

Solutions of 51.4 g of racemic α-ferrocenylethyldimethylamine 2 and 30.0 g of (R)-(+)-tartaric acid, each in 100 mL of methanol, were mixed at 55 °C with stirring. Seeding crystals were added. The temperature was lowered at a rate of 2 °C h^{-1}. After 24 h, 30.0 g (75% of one antipode) of 2-tartrate was collected, and base wash gave 19.0 g of partially optically active 2 ($[\alpha]_D^{25}$ −11.0 (c = 1.5, EtOH)). The solutions of (−)-2 and 11.1 g of (R)-(+)-tartaric acid, each in 50 mL of methanol, were combined at 55 °C and seeded. After slow cooling, 27.5 g of (−)-2-tartrate was obtained, which was converted into 17.0 g (66% overall yield) of (S)-2 (bp 120–121 °C (0.7 mmHg); $[\alpha]_D^{35}$ −14.1 (c = 1.6, EtOH)). The mother liquor from the first crystallization was concentrated to one-quarter of its original volume. Diethyl ether was added until no further precipitate was formed. After standing at 0 °C overnight, 48.6 g of 2-tartrate was collected ($[\alpha]_D^{25}$ 8.0 (c = 1.5, EtOH)) and recrystallized from 800 mL of acetone–water (10 : 1) to yield 34.5 g (85%) of enriched product; $[\alpha]_D^{25}$ +12.0 (c = 1, EtOH). A second recrystallization from 500 mL of aqueous acetone returned 28.0 g (69%) of material ($[\alpha]_D^{25}$ +14.0). By working up the mother liquors, an overall yield of 80–90% of both antipodes can easily be obtained.

2.6.1.3 Synthesis of (R,S$_p$)-N,N-dimethyl-1-(2-diphenylphosphinoferrocenyl)ethylamine (PPFA) [114]

A solution of (R)-2 (12.9 g, 50 mmol) in Et$_2$O (60 mL) was cooled to −78 °C, and a t-BuLi solution (32.4 mL, 55 mmol, 1.1 equiv, 1.7 N in pentane) was added slowly. After addition, the reaction mixture was warmed to room temperature and stirred for 1 h. Chlorodiphenylphosphine (18.5 mL, 100 mmol, 2.0 equiv) was added at 0 °C. After heating to reflux for 2 h, aqueous NaHCO$_3$ was slowly added with cooling in an ice bath. The mixture was extracted with Et$_2$O and the combined organic layers were washed with H$_2$O, dried (MgSO$_4$) and concentrated *in vacuo* to afford a red oil. The oil was chromatographed (silica, hexane/ethyl acetate, 5 : 1) and the product was purified by recrystallization (EtOH). Yield: 15.34 g (34.8 mmol, 69%), orange crystals. TLC (hexane/EtOAc, 5 : 1): R_f = 0.28. ^{1}H NMR (CDCl$_3$): δ 1.27 (d, J_{CHMe} = 6.8 Hz, 3H, CH*Me*), 1.78 (s, 6H, N*Me$_2$*), 3.87 (m, J_{CHCH} = 1.1 Hz, 1H, Cp), 3.95 (s, 5H, Cp′), 4.16 (dq, J_{CHP} = 2.6 Hz, J_{CHMe} = 6.8 Hz, 1H, C*H*Me), 4.25 (m, J_{CHCH} = 2.3 Hz, 1H, Cp), 4.38 (m, J_{CHCH} = 1.1 Hz, 1H, Cp), 7.19 (m, 5H, P*Ph$_2$*), 7.36 (m, 3H, P*Ph$_2$*), 7.00 (m, 2H, P*Ph$_2$*). ^{13}C NMR (CDCl$_3$): δ 9.36 (CH*Me*), 39.03 (N*Me$_2$*), 57.24 (CHMe), 68.38 (CH, Cp), 69.34 (CH, Cp), 69.66 (CH, Cp′), 71.75 (CH, Cp), 76.80 (C, Cp), 77.22 (C, Cp), 127.13, 127.28, 127.38, 127.81, 127.93, 128.69 (C, CH, P*Ph$_2$*), 132.31 (d, J_{CP} = 18.8 Hz, CH, P*Ph$_2$*), 135.14 (d, J_{CP} = 21.4 Hz, CH, P*Ph$_2$*). ^{31}P NMR (CDCl$_3$): δ −21.67 (*PPh$_2$*).

2.6.2 Synthesis of (R,R$_p$)-1-(tert-butylsulfinyl)-2-(diphenylphosphino)-ferrocene 45

2.6.2.1 Synthesis of (R)-tert-butylsulfinylferrocene (R)-26a [52]

To a cold solution (0 °C) of ferrocene (6.17 g, 33.18 mmol) in THF (50 mL) under argon was added t-BuLi (17.0 mL, 1.7 M in pentane, 28.75 mmol). The reaction mixture was stirred at 0 °C for 2 h, and then a solution of (R)-S-tert-butyl-1,1-dimethylethanethiosulfinate (4.30 g, 22.1 mmol, 80% ee) in THF (20 mL) was added. The reaction mixture was warmed to room temperature and then kept at this temperature for 2 h. Brine was added, the organic layer was separated and the aqueous layer was extracted with Et$_2$O. The combined organic layers were dried (MgSO$_4$), filtered and the solvent was evaporated. The resulting mixture was purified by flash column chromatography (ethyl acetate/hexane, 2 : 1) to afford sulfoxide (R)-**26a** (2.52 g, 40%), which was recrystallized from diethyl ether and hexane (1 : 1) to give pure sulfoxide (R)-**26a** (1.25 g, 99% ee, HPLC: Daicel Chiracel OD, i-PrOH/hexane 2/98, flow rate 0.70 mL min^{-1}, t_R: 19.3 min (R)-isomer and 24.3 min (S)-isomer, detection at 254 nm). mp 150–151 °C; $[\alpha]_D^{20}$ −355 (c = 0.5, CHCl$_3$), 99% ee; ^{1}H NMR (200 MHz, CDCl$_3$) δ 4.68 (m, 1H, Cp-H), 4.41 (m, 2H, Cp-H), 4.38 (s, 5H, Cp'-H), 4.35 (m, 1H, Cp-H), 1.12 (s, 9H, t-Bu).

2.6.2.2 Synthesis of (R,R$_p$)-1-(tert-Butylsulfinyl)-2-(diphenylphosphino)ferrocene 45 [61b]

To a solution of sulfoxide (R)-**26a** (0.70 g, 2.41 mmol) in THF (24 mL) was added a solution of t-BuLi in pentane (2.1 mL, 1.7 M, 3.62 mmol). The mixture was stirred at for 1.5 h, and chlorodiphenylphosphine (0.65 mL, 3.62 mmol) was added at −78 °C. The reaction mixture was stirred for 30 min and treated with water (20 mL). The organic layer was separated and the aqueous layer was extracted with Et$_2$O (2 × 20 mL). The combined organic layers were dried (MgSO$_4$), filtered, and the solvents were evaporated under reduced pressure. The residue was purified by flash column chromatography (n-hexanes/EtOAc, 1 : 1) to afford a yellow solid **45** (91%). $[\alpha]_D^{20}$ −437 (c = 0.4, CHCl$_3$); mp 162–163 °C; ^{1}H NMR (300 MHz, CDCl$_3$) δ 7.61–7.52 (m, 2H), 7.35–7.28 (m, 3H), 7.27–7.14 (m, 5H), 4.60–4.56 (m, 1H), 4.53–4.48 (m, 1H), 4.22–4.18 (m, 1H), 4.10 (s, 5H), 0.98 (s, 9H); ^{13}C NMR (75 MHz, CDCl$_3$) δ 140.6, 140.4, 138.8, 138.6, 135.8, 135.5, 132.9, 132.7, 129.2, 128.1, 127.9, 127.8, 90.1, 89.8, 76.5, 76.2, 75.3, 75.2, 74.0, 72.5, 71.5, 55.9, 23.7. EIMS: m/z 474 (M$^+$, 13), 418 (91), 352 (100), 228 (25), 170 (22).

2.6.3
Synthesis of FcPHOX 60 [71]

2.6.3.1 Synthesis of (S)-2-ferrocenyl-4-(1-methylethyl)oxazoline 55a

To a stirred suspension of ferrocenecarboxylic acid (1.03 g, 4.49 mmol) in CH$_2$Cl$_2$ (15 mL) at room temperature under nitrogen was added via syringe oxalyl chloride (0.79 mL, 9 mmol). Gas evolution was accompanied by the formation of a dark red homogeneous solution after 10 min. The reaction mixture was stirred for an additional 10 min, followed by removal of the solvent *in vacuo*. The resultant crude acid chloride, isolated as a dark oil that crystallized on standing, was taken up in CH$_2$Cl$_2$ (10 mL) and added via syringe to a solution of (S)-(+)-valinol (0.554 g, 5.37 mmol) and NEt$_3$ (1.25 mL, 9 mmol) in CH$_2$Cl$_2$ (7 mL) at room temperature under nitrogen. After stirring overnight, the dark reaction mixture was washed with H$_2$O (2 × 20 mL), dried over Na$_2$SO$_4$, filtered and evaporated *in vacuo*. The crude product was purified by column chromatography (3% MeOH/CH$_2$Cl$_2$) to give the amide as a yellow crystalline solid (1.18 g, 84%). mp 109–110 °C; [α]$_D^{21}$ −8 (c = 1.34, EtOH); Found: C, 61.11; H, 6.74; N, 4.21. C$_{16}$H$_{21}$FeNO$_2$ calcd for: C, 60.97; H, 6.72; N, 4.44; ν$_{max}$ (nujol) 3284 (N−H), 3192 (O−H), 1611 (C=O amide I), 1551 (amide II) cm^{-1}; ^{1}H NMR (360 MHz, CDCl$_3$) δ 0.97 (d, J = 6.8 Hz, 3H, −CH$_3$), 0.98 (d, J = 6.8 Hz, 3H, −CH$_3$), 1.93 (octet, J = 6.8 Hz, 1H, −CH(CH$_3$)$_2$), 2.71 (brt, 1H, −OH), 3.65–3.83 (m, 3H, −CH$_2$OH and −NHCH-), 4.16 (s, 5H, C$_5$H$_5$), 4.30 (brs, 2H, Fc), 4.60 (brs, 1H, Fc), 4.62 (brs, 1H, Fc), 5.80 (brd, 1H, −NH−); ^{13}C NMR (90 MHz, CDCl$_3$) δ 19.04 (−CH$_3$), 19.69 (−CH$_3$), 28.99 (−CH(CH$_3$)$_2$), 57.01 (−NHCH−), 63.87 (−OCH$_2$−), 67.91 (Fc), 68.42 (Fc), 69.75 (C$_5$H$_5$), 70.46 (Fc), 70.51 (Fc), 75.95 (Fc-*ipso*), 171.37 (C=O). EIMS: m/z 315 (M$^+$, 100%), 213 (81).

To a light orange solution of (2S)-N-1-hydroxy-3-methylbutyl)ferrocenamide (0.817 g, 2.59 mmol) and PPh$_3$ (2.49 g, 9.5 mmol) in acetonitrile (60 mL) was added NEt$_3$ (1.6 mL, 11.5 mmol) followed by CCl$_4$ (2.2 mL, 22.8 mmol) and the resultant solution was stirred at room temperature under nitrogen overnight. After quenching with H$_2$O (80 mL), the reaction mixture was extracted with petroleum ether (5 × 50 mL), the organics combined, dried over MgSO$_4$, filtered and evaporated. The crude product, which was contaminated by a substantial quantity of triphenylphosphine oxide, was purified by column chromatography (silica, 30% EtOAc/70% 40–60 petroleum ether) to give the pure ferrocenyl oxazoline **55a** as a dark yellow crystalline solid (0.685 g, 89%). Note: removing the solvent from the reaction mixture before work-up results in a substantial reduction in yield of the oxazoline. mp 71.5–72.5 °C; [α]$_D^{22}$ −129 (c = 1.5, EtOH); CD (CHCl$_3$) λ$_{max}$ (Δε) 464 (−0.80), 358 (+1.0) nm; Anal. calcd for C$_{16}$H$_{19}$FeNO: C, 64.67; H, 6.44; N, 4.71; Found: C, 64.46; H, 6.59; N, 4.59.; ν$_{max}$ (nujol) 1657 (C=N) cm^{-1}; ^{1}H NMR (360 MHz, CDCl$_3$) δ 0.87 (d, J = 6.8 Hz, 3H, -CH$_3$), 0.94 (d, J = 6.8 Hz, 3H, −CH$_3$), 1.78 (sextet, J = 6.6 Hz, 1H, −CH(CH$_3$)$_2$), 3.41 (q, J = 7.0 Hz, 1H, −NCH−), 3.89–3.95 (m, 1H, −OCHH−), 4.00 (t, J = 7.7 Hz, 1H, −OCHH−), 4.06 (s, 5H, C$_5$H$_5$), 4.26 (brs, 2H, Fc), 4.66 (brs, 1H, Fc), 4.70 (brs, 1H, Fc); ^{13}C NMR (90 MHz, CDCl$_3$) δ 17.79 (−CH$_3$), 18.81 (−CH$_3$), 32.27 (−CH(CH$_3$)$_2$), 68.92, 68.95, 69.27, 69.51 (C$_5$H$_5$), 70.06, 70.09, 70.60 (Fc-*ipso*), 72.27, 165.56 (C=N); EIMS: m/z 297 (M$^+$, 100), 254 (64), 211 (47), 121 (92).

2.6.3.2 Synthesis of (S,S$_p$)-2-[2-(diphenylphosphino)ferrocenyl]-4-(1-methylethyl)oxazoline 60

A stirred solution of ferrocenyl oxazoline **55a** (0.158 g, 0.53 mmol) and TMEDA (0.10 mL, 0.7 mmol) in Et$_2$O (6 mL) under nitrogen was cooled to −78 °C, resulting in the formation of a yellow precipitate. To this mixture was added n-BuLi (0.38 mL, 0.7 mmol) dropwise, and the reaction mixture darkened to red/brown. After stirring at −78 °C for 2 h, the Schlenk tube containing the orange non-homogeneous reaction mixture was transferred to an ice bath and stirring was maintained for a further 5 min. To the resulting homogeneous orange–red solution was added chlorodiphenylphosphine (0.12 mL, 0.70 mmol) and the reaction mixture allowed to warm to room temperature. After 15 min the reaction mixture was quenched with saturated NaHCO$_3$ (10 mL) and diluted with ether (10 mL). The two layers were separated and the aqueous phase was extracted with Et$_2$O (10 mL). The organics were combined, dried (MgSO$_4$), filtered and evaporated to give an orange crystalline solid. The crude material was purified by column chromatography (10% EtOAc/90% petroleum ether, preadsorbed on silica) to afford 0.163 g (64%) of a yellow–orange crystalline solid. Recrystallization from hexane gave pure (S,S$_p$)-60 as a single diastereomer. mp 157–158 °C; $[\alpha]_D^{24}$ +112 (c = 0.1, EtOH); CD (CHCl$_3$) λ_{max} ($\Delta\varepsilon$) 456 (+2.20), 368 (+0.49), 342 (−1.00), 315 (+2.16) nm; Anal. calcd for C$_{28}$H$_{28}$FeNOP: C, 69.87; H, 5.86; N, 2.91; Found: C, 70.09; H, 6.11; N, 2.85; ν_{max} (nujol) 1652 (C=N) cm^{-1}; ^{1}H NMR(360 MHz, CDCl$_3$) δ 0.68 (d, J = 7 Hz, 3H, −CH$_3$), 0.82 (d, J = 7 Hz, 3H, −CH$_3$), 1.61–1.69 (m, 1H, −CH(CH$_3$)$_2$), 3.61 (brs, 1H, Fc), 3.67 (t, J = 8 Hz, 1H, −OCHH−), 3.83–3.90 (m, 1H, −NCH−), 4.22 (s, 5H, C$_5$H$_5$), 4.22–4.30 (m, 1H, −OCHH−), 4.37 (brs, 1H, Fc), 4.99 (brs, 1H, Fc), 7.18–7.24 (m, 5H, Ph), 7.36–7.37 (m, 3H, Ph), 7.46–7.51 (m, 2H, Ph); ^{13}C NMR (90 MHz, CDCl$_3$) δ 17.52 (−CH$_3$), 18.61 (−CH$_3$), 32.05 (−CH(CH$_3$)$_2$), 69.57 (−OCH$_2$−), 70.72 (C$_5$H$_5$), 72.02, 72.14, 73.81, 73.85, 75.32 (d, J = 16 Hz, Fc), 78.55 (d, J = 15 Hz, Fc), 127.81 (Ph), 127.92 (Ph), 127.99 (Ph), 128.10 (Ph), 128.18 (Ph), 128.89 (Ph), 132.40 (d, J = 20 Hz, Ph), 134.86 (d, J = 22 Hz, Ph), 138.21 (d, J = 13 Hz, Ph-*ipso*), 139.54 (d, J = 12 Hz, Ph-*ipso*); ^{31}P NMR (360 MHz, CDCl$_3$) δ −16.92; EIMS: m/z 481 (M$^+$, 100), 410 (68), 404 (44), 170 (38), 121 (76).

2.6.4
Synthesis of (S$_p$)-N,N-diisopropyl-2-(diphenylphosphino)ferrocenecarboxamide 108h [91]

2.6.4.1 (S$_p$)-N,N-diisopropyl-2-(diphenylphosphino)ferrocenecarboxamide 108h

A solution of N,N-diisopropylferrocenecarboxamide 107 (332 mg, 1.06 mmol) in Et$_2$O (5 mL) was added to a mixture of n-BuLi (1.49 mL, 2.33 mmol, 1.57 M solution) and (−)-sparteine (0.54 mL, 2.33 mmol) in Et$_2$O (10 mL) at −78 °C. After addition of chlorodiphenylphosphine (0.57 mL, 3.18 mmol), the mixture was worked up and purified by flash chromatography using 5:1 hexane/EtOAc to afford 423 mg (82%) of 108h (E = PPh$_2$) as an orange solid. mp 176–177 °C (decomp.); ^{1}H NMR (250 MHz, CDCl$_3$) δ 1.02 (bm, 12H), 3.22 (b, 1H), 3.83 (dd, 1H, J = 1.0 Hz and unresolved J), 3.86 (b, 1H), 4.22 (s, 5H), 4.25 (dd, 1H, J = 2.4 Hz and unresolved J), 4.45 (dd, 1H, J = 1.1 Hz and unresolved J), 7.29 (m, 8H), 7.55 (m, 2H); ^{13}C NMR (62.5 MHz, CDCl$_3$) δ 20.3, 20.8, 68.0, 68.7, 70.9, 79.6, 79.8, 90.8, 91.2, 127.8, 127.9, 128.5, 132.8, 133.1, 134.2, 134.5, 138.1, 138.4, 139.5, 139.7, 166.9; ν_{max} (KBr) 3089, 3061, 3047, 3012, 2970, 2361, 1739, 1623, 1445, 1433, 1371, 1323, 1040, 1028, 1006, 815, 754, 744 cm^{-1}; EIMS: m/z 497 (M$^+$, 30), 457 (7), 412 (100), 346 (32), 222 (8), 201 (78), 183 (8), 170 (10), 121 (14), 98 (3), 56 (6); HRMS calcd for C$_{29}$H$_{32}$FeNOP: 497.1571; found: 497.1569. The enantiomeric excess was found to be 90% using a CHIRALCEL OD chiral HPLC column eluting with 10% Et$_2$O/0.5% Et$_2$NH in hexane as the mobile phase at a flow rate of 1.5 mL min^{-1}.

References

1 Thomson, J.B. (1959) *Tetrahedron Lett.*, **1**, 26–27.
2 Marquarding, D., Klusacek, H., Gokel, G., Hoffmann, P., and Ugi, I. (1970) *J. Am. Chem. Soc.*, **92**, 5389–5393.
3 For reviews, see: (a) Togni, A. and Hayashi, T. (eds) (1995) *Ferrocenes: Homogeneous Catalysis, Organic Synthesis, Materials Science*, VCH, Weinheim; (b) Togni, A. and Halterman, R.L. (eds) (1998) *New Chiral Ferrocenyl Ligands for Asymmetric Catalysis in Metallocenes–Synthesis, Reactivity, Applications*, vol. **2**, Wiley-VCH Verlag GmbH, Weinheim, pp. 685–721; (c) Stepnicka, P. (ed.) (2008) *Ferrocene: Ligands, Materials and Biomolecules*, John Wiley & Sons, Ltd., Chichester; (d) Richards, C.J. and Locke, A.J. (1998) *Tetrahedron: Asymmetry*, **9**, 2377–2407; (e) Colacot, T.J. (2003) *Chem. Rev.*, **103**, 3101–3118; (f) Togni, A. (1996) *Angew. Chem., Int. Ed.*, **35**, 1475–1477; (g) Arrayás, R.G., Adrio, J., and Carretero, J.C. (2006) *Angew. Chem., Int. Ed.*, **45**, 7674–7715; (h) Dai, L.-X., Tu, T., You, S.-L., Deng, W.-P., and Hou, X.-L. (2003) *Acc. Chem. Res.*, **36**, 659–667; (i) Miyake, Y., Nishibayashi, Y., and Uemura, S. (2008) *Synlett*, 1747–1758; (j) Perseghini, M. and Togni, A. (2001) *Science of Synthesis*, vol. **1** (ed. M. Lautens), Georg Thieme-Verlag KG, Stuttgart, pp. 889–929.
4 Snieckus, V. (1990) *Chem. Rev.*, **90**, 879–933.
5 Cahn, R.S., Ingold, C., and Prelog, V. (1966) *Angew. Chem., Int. Ed.*, **5**, 385–415.
6 Schlögl, K. (1967) *Top. Stereochem.*, **1**, 39–91.
7 (a) Schlögl, K. (1984) *Top. Curr. Chem.*, **125**, 27–62; (b) Schlögl, K. (1986) *J. Organomet. Chem.*, **300**, 219–248.
8 Prelog, V. and Helmchen, G. (1982) *Angew. Chem., Int. Ed.*, **21**, 567–583.
9 (a) Blaser, H.-U., Brieden, W., Pugin, B., Spindler, F., Studer, M., and Togni, A. (2002) *Top. Catal.*, **19**, 3; (b) Blaser, H.-U. (2002) *Adv. Synth. Catal.*, **344**, 17–31.
10 Whisler, M.C., MacNeil, S., Snieckus, V., and Beak, P. (2004) *Angew. Chem. Int. Ed.*, **43**, 2206–2225.
11 Rebière, F., Riant, O., Ricard, L., and Kagan, H.B. (1993) *Angew. Chem., Int. Ed.*, **32**, 568–570.
12 Riant, O., Samuel, O., and Kagan, H.B. (1993) *J. Am. Chem. Soc.*, **115**, 5835–5836.
13 (a) Richards, C.J., Damalidis, T., Hibbs, D.E., and Hursthouse, M.B. (1995) *Synlett*, 74–76; (b) Richards, C.J. and Mulvaney, A.W. (1996) *Tetrahedron: Asymmetry*, **7**, 1419–1430.
14 (a) Sammakia, T., Latham, H.A., and Schaad, D.R. (1995) *J. Org. Chem.*, **60**, 10–11; (b) Sammakia, T. and Latham, H.A. (1995) *J. Org. Chem.*, **60**, 6002–6003.
15 (a) Nishibayashi, Y. and Uemura, S. (1995) *Synlett*, 79–81; (b) Nishibayashi, Y., Segawa, K., Ohe, K., and Uemura, S. (1995) *Organometallics*, **14**, 5486–5487.
16 Ahn, K.H., Cho, C.-W., Baek, H.-H., Park, J., and Lee, S. (1996) *J. Org. Chem.*, **61**, 4937–4943.
17 Enders, D., Peters, R., Lochtman, R., and Runsink, J. (1997) *Synlett*, 1462–1464.
18 (a) Lotz, M., Ireland, T., Tappe, K., and Knochel, P. (2000) *Chirality*, **12**, 389; (b) Kloetzing, R.J., Lotz, M., and Knochel, P. (2003) *Tetrahedron: Asymmetry*, **14**, 255–264.
19 (a) Laurenti, D. and Santelli, M. (1999) *Org. Prep. Proced. Int.*, **31**, 245–294; (b) Dai, L.-X., Tu, T., You, S.-L., Deng, W.-P., and Hou, X.-L. (2003) *Acc. Chem. Res.*, **36**, 659–667; (c) Sutcliffe, O.B. and Bryce, M.R. (2003) *Tetrahedron: Asymmetry*, **14**, 2297–2325.
20 (a) You, S.-L., Hou, X.-L., Dai, L.-X., Yu, Y.-H., and Xia, W. (2002) *J. Org. Chem.*, **67**, 4684–4695 and references therein; (b) Stangeland, E.L. and Sammakia, T. (1997) *Tetrahedron*, **53**, 16503–16510; (c) Richards, C.J., Hibbs, D.E., and

Hurthose, M.B. (1995) *Tetrahedron Lett.*, **36**, 3745–3748; (d) Kuwano, R., Uemura, T., Saitoh, M., and Ito, Y. (1999) *Tetrahedron Lett.*, **40**, 1327–1330; (e) Donde, Y. and Overman, L.E. (1999) *J. Am. Chem. Soc.*, **121**, 2933–2934; (f) Zhang, W., Shimanuki, T., Kida, T., Nakatusuji, Y., and Ikeda, I. (1999) *J. Org. Chem.*, **64**, 6247; (g) Shintani, R., Lo, M.M.-L., and Fu, G.C. (2000) *Org. Lett.*, **2**, 3695–3697; (h) Mancheño, O.G., Priego, J., Cabrera, S., Arrayás, R.G., Lluamas, T., and Carretero, C. (2003) *J. Org. Chem.*, **68**, 3679–3686.

21 (a) Deng, W.-P., You, S.-L., Hou, X.-L., Dai, L.-X., Yu, Y.-H., Xia, W., and Sun, J. (2001) *J. Am. Chem. Soc.*, **123**, 6508–6519; (b) Li, M., Yuan, K., Li, Y.-Y., Cao, B.-X., Sun, J., and Hou, X.-L. (2003) *Tetrahedron: Asymmetry*, **14**, 3347–3352.

22 (a) Bolm, C., Kesselgruber, M., Muñiz, K., and Raabe, G. (2000) *Organometallics*, **19**, 1648–1651; (b) Bolm, C., Muñiz, K., Aguilar, N., Kesselgruber, M., and Raabe, G. (1999) *Synthesis*, 1251–1260.

23 Peters, R. and Fischer, D.F. (2005) *Org. Lett.*, **7**, 4137–4140.

24 For non-stereoselective lithiation of ferrocenyl benzimidazoles, see: Hérault, D., Aelvoet, K., Blatch, A.J., Al-Majid, A., Smethurst, C.A., and Whiting, A.J. (2007) *J. Org. Chem.*, **72**, 71–75.

25 Vinci, D., Mateus, N., Wu, X., Hancock, F., Steiner, A., and Xiao, J. (2006) *Org. Lett.*, **8**, 215–218.

26 Ueberbacher, B.J., Griengl, H., and Weber, H. (2008) *Chem. Comm.*, 3287–3289.

27 Siegel, S. and Schmalz, H.-G. (1997) *Angew. Chem., Int. Ed.*, **36**, 2456–2458.

28 (a) Steurer, M., Tiedl, K., Wang, Y., and Weissensteiner, W. (2005) *Chem. Comm.*, 4929–4931; (b) Steurer, M., Wang, Y., Mereiter, K., and Weissensteiner, W. (2007) *Organometallics*, **26**, 3850–3859.

29 (a) Arrayás, R.G., Adrio, J., and Carretero, J.C. (2006) *Angew. Chem., Int. Ed.*, **45**, 7674–7715; (b) Atkinson, R.C.J., Gibson, V.C., and Long, N.J. (2004) *Chem. Soc. Rev.*, **33**, 313–328; (c) Schwink, L. and Knochel, P. (1998) *Chem. Eur. J.*, **4**, 950–968; (d) Miyake, Y., Nishibayashi, Y., and Uemura, S. (2008) *Synlett*, 1747–1758.

30 Falk, H. and Schlögl, K. (1965) *Monatsh. Chem.*, **96**, 1065–1080.

31 Aratani, T., Gonda, T., and Nozaki, H. (1969) *Tetrahedron Lett.*, **10**, 2265–2268.

32 Gokel, G., Hoffmann, P., Kleimann, H., Klusacek, H., Marquarding, D., and Ugi, I. (1970) *Tetrahedron Lett.*, **11**, 1771–1774.

33 Laufer, R., Veith, U., Taylor, N.J., and Snieckus, V. (2006) *Can. J. Chem.*, **84**, 356–369.

34 Hayashi, T., Yamamoto, K., and Kumada, M. (1974) *Tetrahedron Lett.*, **15**, 4405–4408.

35 (a) Hayashi, T. and Yamazaki, A. (1991) *J. Organomet. Chem.*, **413**, 295–302; (b) Hayashi, T., Mise, T., Fukushima, M., Kagotani, M., Nagashima, N., Hamada, Y., Matsumoto, A., Kawakami, S., Konishi, M., Yamamoto, K., and Kumada, M. (1980) *Bull. Chem. Soc. Jpn.*, **53**, 1138–1151.

36 (a) Honeychuck, R.V., Okoroafor, M.O., Shen, L.H., and Brubaker, C.H.J. (1986) *Organometallics*, **5**, 482–490; (b) Okoroafor, M.O., Ward, D.L., and Brubaker, C.H.J. (1988) *Organometallics*, **7**, 1504–1511.

37 (a) Boaz, N.W., Ponasik, J.A., and Large, S.E. (2005) *Tetrahedron: Asymmetry*, **16**, 2063–2066; (b) Boaz, N.W., Ponasik, J.A., and Large, S.E. (2006) *Tetrahedron Lett.*, **47**, 4033–4035.

38 (a) Thommen, M. and Blaser, H.-U. (2002) *Pharm. Chem.*, **7/8**, 33; (b) Lee, D., Kim, D., and Yun, J. (2006) *Angew. Chem., Int. Ed.*, **45**, 2785–2787.

39 Barbaro, P. and Togni, A. (1995) *Organometallics*, **14**, 3570–3573.

40 (a) Sawamura, M., Hamashima, H., and Ito, Y. (1991) *Tetrahedron: Asymmetry*, **2**, 593–596; (b) Sawamura, M., Hamashima, H., Sugawara, M., Kuwano, R., and Ito, Y. (1995) *Organometallics*, **14**, 4549–4558;

(c) Kuwano, R., Sawamura, M., Okuda, S., Asai, T., Ito, Y., Redon, M., and Krief, A. (1997) *Bull. Chem. Soc. Jpn.*, **70**, 2807–2822.

41 (a) Sturm, T., Weissensteiner, W., and Spindler, F. (2003) *Adv. Synth. Catal.*, **345**, 160–164; (b) Kong, J.-R., Ngai, M.-Y., and Krische, M.J. (2006) *J. Am. Chem. Soc.*, **128**, 718–719.

42 Schnyder, A., Hintermann, L., and Togni, A. (1995) *Angew. Chem., Int. Ed.*, **34**, 931–933.

43 (a) Chen, W., Mbafor, W., Roberts, S.M., and Whittall, J. (2006) *J. Am. Chem. Soc.*, **128**, 3922–3923; (b) Chen, W., Roberts, S.M., Whittall, J., and Steiner, A. (2006) *Chem. Comm.*, 2916–2918; (c) Chen, W., McCormack, P.J., and Mohammed, K. et al. (2007) *Angew. Chem., Int. Ed.*, **46**, 4141–4144.

44 (a) Ireland, T., Grossheimann, G., Wieser-Jeunesse, C., and Knochel, P. (1999) *Angew. Chem., Int. Ed.*, **38**, 3212–3215; (b) Ireland, T., Tappe, K., Grossheimann, G., and Knochel, P. (2002) *Chem. Eur. J.*, **8**, 843–852.

45 Ganter, C. and Wagner, T. (1995) *Chem. Ber.*, **128**, 1157–1161.

46 (a) Widhalm, M., Mereiter, K., and Bourghida, M. (1998) *Tetrahedron: Asymmetry*, **9**, 2983–2986; (b) Widhalm, M., Nettekoven, U., and Mereiter, K. (1999) *Tetrahedron: Asymmetry*, **10**, 4369–4391.

47 Kitzler, R., Xiao, L., and Weissensteiner, W. (2000) *Tetrahedron: Asymmetry*, **11**, 3459–3462.

48 Xiao, L., Kitzler, R., and Weissensteiner, W. (2001) *J. Org. Chem.*, **66**, 8912–8919.

49 Riant, O., Argouarch, G., Guillaneux, D., Samuel, O., and Kagan, H.B. (1998) *J. Org. Chem.*, **63**, 3511–3514.

50 Kloetzing, R.J. and Knochel, P. (2006) *Tetrahedron: Asymmetry*, **17**, 116–123.

51 Klunder, J.M. and Sharpless, K.B. (1987) *J. Org. Chem.*, **52**, 2598–2602.

52 Priego, J., Mancheno, O.G., Cabrera, S., and Carretero, J.C. (2002) *J. Org. Chem.*, **67**, 1346–1353.

53 Cogan, D.A., Liu, G., Kim, K., Backes, B.J., and Ellman, J.A. (1998) *J. Am. Chem. Soc.*, **120**, 8011–8019.

54 Lagneau, N.M., Chen, Y., Robben, P.M., Sin, H.S., Takasu, K., Chen, J.S., Robinson, P.D., and Hua, D.H. (1998) *Tetrahedron*, **54**, 7301–7334.

55 Ferber, B. and Kagan, H.B. (2007) *Adv. Synth. Catal.*, **349**, 493–507.

56 Argouarch, G., Samuel, O., Riant, O., Daran, J.-C., and Kagan, H.B. (2000) *Eur. J. Org. Chem.*, 2893–2899.

57 (a) Lotz, M., Kramer, G., and Knochel, P. (2002) *Chem. Comm.*, 2546–2547; (b) Lotz, M., Polborn, K., and Knochel, P. (2002) *Angew. Chem., Int. Ed.*, **41**, 4708–4711.

58 (a) Xiao, L., Mereiter, K., Spindler, F., and Weissensteiner, W. (2001) *Tetrahedon: Asymmetry*, **12**, 1105–1108; (b) Sturm, T., Xiao, L., and Weissensteiner, W. (2001) *Chimia*, **55**, 688–693; (c) Chen, W.P., Roberts, S.M., Whittall, J., and Steiner, A. (2006) *Chem. Comm.*, 2916–2918.

59 Kloetzing, R.J. and Knochel, P. (2006) *Tetrahedron: Asymmetry*, **17**, 116–123.

60 Jensen, J.F., Sotofte, I., Sorensen, H.O., and Johannsen, M. (2003) *J. Org. Chem.*, **68**, 1258–1265.

61 (a) Priego, J., Mancheno, O.G., Cabrera, S., Arrayas, R.G., Llamas, T., and Carretero, J.C. (2002) *Chem. Comm.*, 2512–2513; (b) Mancheno, O.G., Priego, J., Cabrera, S., Arrayas, R.G., Llamas, T., and Carretero, J.C. (2003) *J. Org. Chem.*, **68**, 3679–3686.

62 (a) Pedersen, H.L. and Johannsen, M. (1999) *Chem. Comm.*, 2517–2518; (b) Pedersen, H.L. and Johannsen, M. (2002) *J. Org. Chem.*, **67**, 7982–7994.

63 Seitzberg, J.G., Dissing, C., Sotofte, I., Norrby, P., and Johannsen, M. (2005) *J. Org. Chem.*, **70**, 8332–8337.

64 Grach, G., Lohier, J.-F., Santos, J.S.D.O., Reboul, V., and Metzner, P. (2007) *Chem. Comm.*, 4875–4877.

65 (a) Riant, O., Samuel, O., Flessner, T., Taudien, S., and Kagan, H.B. (1997) *J. Org. Chem.*, **62**, 6733–6745; (b) Wölfle, H.,

Kopacka, H., Wurst, K., Ongania, K.H., Görtz, H.H., Preishuber-Pflügl, P., and Bildstein, B. (2006) *J. Organomet. Chem.*, **691**, 1197–1215; (c) Balavoine, G.G.A., Daran, J.-C., Iftime, G., Manoury, E., and Moreau-Bossuet, C. (1998) *J. Organomet. Chem.*, **567**, 191–198.

66 Abiko, A. and Wang, G. (1996) *J. Org. Chem.*, **61**, 2264–2265.

67 Larsen, A.O., Taylor, R.A., White, P.S., and Gagne, M.R. (1999) *Organometallics*, **18**, 5157–5162.

68 (a) Bertogg, A., Camponovo, F., and Togni, A. (2005) *Eur. J. Inorg. Chem.*, 347–356; (b) Bertogg, A. and Togni, A. (2006) *Organometallics*, **25**, 622–630.

69 McManus, H.A. and Guiry, P.J. (2004) *Chem. Rev.*, **104**, 4151–4202.

70 Sammakia, T. and Latham, H.A. (1996) *J. Org. Chem.*, **61**, 1629–1635.

71 Richards, C.J. and Mulvaney, A.W. (1996) *Tetrahedron: Asymmetry*, **7**, 1419–1430.

72 Geisler, F.M. and Helmchen, G. (2006) *J. Org. Chem.*, **71**, 2486–2492.

73 (a) Nishibayashi, Y., Segawa, K., Ohe, K., and Uemura, S. (1995) *Organometallics*, **14**, 5486–5487; (b) Nishibayashi, Y., Segawa, K., and Arikawa, Y. et al. (1997) *J. Organomet. Chem.*, **545–546**, 381–398; (c) Takei, I., Nishibayashi, Y., and Arikawa, Y. et al. (1999) *Organometallics*, **18**, 2271–2274; (d) Chung, K.-G., Miyake, Y., and Uemura, S. (2000) *Perkin Trans. 1*, 2725–2729; (e) Chung, K.-G., Miyake, Y., and Uemura, S. (2000) *Perkin Trans. 1*, 15–18.

74 (a) Deng, W.-P., Hou, X.-L., Dai, L.-X., Yu, Y.-H., and Xia, W. (2000) *Chem. Comm.*, 285–286; (b) Deng, W.-P., You, S.-L., Hou, X.-L., Dai, L.-X., Yu, Y.-H., Xia, W., and Sun, J. (2001) *J. Am. Chem. Soc.*, **123**, 6508–6519; (c) Tu, T., Deng, W.-P., Hou, X.-L., Dai, L.-X., and Dong, X.-C. (2003) *Chem. Eur. J.*, **9**, 3073–3081; (d) Tu, T., Hou, X.-L., and Dai, L.-X. (2004) *Org. Lett.*, **5**, 3651–3653.

75 You, S.-L., Zhou, Y.-G., Hou, X.-L., and Dai, L.-X. (1998) *Chem. Comm.*, 2765–2766.

76 (a) Fukuzawa, S.-I. and Kato, H. (1998) *Synlett*, 727–728; (b) Bolm, C., Muñiz-Fernández, K., Seger, A., Raabe, G., and Günther, K. (1998) *J. Org. Chem.*, **63**, 7860–7867; (c) Özçubukçu, S., Schmidt, F., and Bolm, C. (2005) *Org. Lett.*, **7**, 1407–1409.

77 Manoury, E., Fossey, J.S., Aït-Haddou, H., Daran, J.C., and Balavoine, G.G.A. (2000) *Organometallics*, **19**, 3736–3739.

78 Kim, S.-G., Cho, C.-W., and Ahn, K.H. (1997) *Tetrahedron: Asymmetry*, **8**, 1023–1026.

79 (a) Ahn, K.H., Cho, C.-W., Park, J., and Lee, S. (1997) *Tetrahedron: Asymmetry*, **8**, 1179–1185; (b) Zhang, W.B., Yoshinaga, H., Imai, Y., Kida, T., Nakatsuji, Y., and Ikeda, I. (2000) *Synlett*, 1512–1514.

80 Lu, S.-M., Han, X.-W., and Zhou, Y.-G. (2004) *Adv. Synth. Catal.*, **346**, 909–912.

81 (a) Enders, D., Peters, R., Lochtman, R., and Raabe, G. (1999) *Angew. Chem., Int. Ed.*, **38**, 2421–2423; (b) Job, A., Janeck, C.F., Bettray, W., Peters, R., and Enders, D. (2002) *Tetrahedron*, **58**, 2253–2329; (c) Enders, D., Joans, E.A., and Klumpen, T. (2009) *Eur. J. Org. Chem.*, 2149–2162.

82 Carrol, M.A., Widdowson, D.A., and Williams, D.J. (1994) *Synlett*, 1025–1026.

83 Ireland, T., Almena Perea, J.J., and Knochel, P. (1999) *Angew. Chem., Int. Ed.*, **38**, 1457–1460.

84 Wittig, G., Davis, P., and Koenig, G. (1951) *Chem. Ber.*, **84**, 627–632.

85 Gilman, H., Brown, G.E., Webb, F.J., and Spatz, S.M. (1940) *J. Am. Chem. Soc.*, **62**, 977–979.

86 Houlihan, W.J. and Parrino, V.A. (1982) *J. Org. Chem.*, **47**, 5177–5180.

87 Nettekoven, U., Widhalm, M., Kamer, P.C.J., van Leeuwen, P., Mereiter, K., Lutz, M., and Spek, A.L. (2000) *Organometallics*, **19**, 2299–2309.

88 Pfaltz, A., Lotz, M., Schoenleber, M., Pugin, B., Kesselgruber, M., and Thommen, M. (2005) WO 2005056566. 98 p.

89 Aratani, T., Gonda, T., and Nozaki, H. (1970) *Tetrahedron*, **26**, 5453–5464.

90 Price, D. and Simpkins, N.S. (1995) *Tetrahedron Lett.*, **36**, 6135–6136.
91 Tsukazaki, M., Tinkl, M., Roglans, A., Chapell, B.J., Taylor, N.J., and Snieckus, V. (1996) *J. Am. Chem. Soc.*, **118**, 685–686.
92 Hoppe, D. and Hense, T. (1997) *Angew. Chem., Int. Ed*, **36**, 2282–2316.
93 Laufer, R.S., Veith, U., Taylor, N.J., and Snieckus, V. (2000) *Org. Lett.*, **2**, 629–631.
94 Metallinos, C., Szillat, H., Taylor, N.J., and Snieckus, V. (2003) *Adv. Synth. Catal.*, **345**, 370–382.
95 Dixon, A.J., McGrath, M.J., and O'Brien, P. (2006) *Org. Synth.*, **83**, 141–154.
96 Nishibayashi, Y., Arikawa, Y., Ohe, K., and Uemura, S. (1996) *J. Org. Chem.*, **61**, 1172–1174.
97 Metallinos, C., Zaifman, J., and Dodge, L. (2008) *Org. Lett.*, **10**, 3527–3530.
98 Wechsler, D., Rankin, M.A., McDonald, R., Ferguson, M.J., Schatte, G., and Stradiotto, M. (2007) *Organometallics*, **26**, 6418–6427.
99 Metallinos, C.Provisional US patent No. 61/117,400 (submitted November 24, 2008).
100 Metallinos, C. and Snieckus, V. (2002) *Org. Lett.*, **4**, 1935–1938.
101 Masamune, S., Choy, W., Petersen, J.S., and Sita, L.R. (1985) *Angew. Chem., Int. Ed.*, **24**, 1–30.
102 Blaser, H.U. and Schmidt, E. (eds) (2003) *Asymmetric Catalysis on Industrial Scale*, Wiley-VCH Verlag GmbH, Weinheim.
103 (a) Yamazaki, Y. and Hosono, K. (1989) *Biotechnol. Lett.*, **11**, 679–684; (b) Izumi, T., Murakami, S., and Kasahara, A. (1990) *Soc. Chem. Ind. (London)*, **3**, 79–80; (c) Izumi, T., Tamura, F., and Sasaki, K. (1992) *Bull. Chem. Soc. Jpn.*, **65**, 2784–2788; (d) Izumi, T., Hino, T., and Ishihara, A. (1993) *J. Chem. Tech. Biotechnol.*, **56**, 45–49.
104 Lambusta, D., Nicolosi, G., Patti, A., and Piattelli, M. (1996) *Tetrahedron Lett.*, **37**, 127–130.
105 Patti, A., Lambusta, D., Piattelli, M., and Nicolosi, G. (1998) *Tetrahedron: Asymmetry*, **9**, 3073–3080.
106 Nicolosi, G., Patti, A., and Piattelli, M. (1994) *J. Org. Chem.*, **59**, 251–254.
107 D'Antona, N., Lambusta, D., Morrone, R., Nicolosi, G., and Secundo, F. (2004) *Tetrahedron: Asymmetry*, **15**, 3835–3840.
108 Bueno, A., Rosol, M., García, J., and Moyano, A. (2006) *Adv. Synth. Catal.*, **348**, 2590–2596.
109 (a) Ogasawara, M., Watanabe, S., Fan, L., Nakajima, K., and Takahashi, T. (2006) *Organometallics*, **25**, 5201–5203; (b) Ogasawara, M., Watanabe, S., Nakajima, K., and Takahashi, T. (2008) *Pure Appl. Chem.*, **60**, 1109–1113.
110 (a) Mealy, M.J., Luderer, M.R., Bailey, W.F., and Sommer, M.B. (2004) *J. Org. Chem.*, **69**, 6042–6049; (b) Stead, D., O'Brien, P., and Sanderson, A. (2008) *Org. Lett.*, **10**, 1409–1412.
111 Metallinos, C. (2001) Ph. D. dissertation, Queen's University.
112 (a) Xie, J.-H., Wang, L.-X., Fu, Y., Zhu, S.-F., Fan, B.-M., Duan, H.-F., and Zhou, Q.-L. (2003) *J. Am. Chem. Soc.*, **125**, 4404–4405; (b) Lam, W.-S., Kok, S.H.L., Au-Yeung, T.T.-L., Wu, J., Cheung, H.-Y., Lam, F.-L., Yeung, C.-H., and Chan, A.S.C. (2006) *Adv. Synth. Catal.*, **348**, 370–374.
113 (a) Wright, J., Frambes, L., and Reeves, P. (1994) *J. Organomet. Chem.*, **476**, 215–217; (b) Review: Corey, E.J. and Helal, C.J. (1998) *Angew. Chem., Int. Ed*, **37**, 1986–2012.
114 Gischig, S. and Togni, A. (2004) *Organometallics*, **23**, 2479–2487.

3
Monodentate Chiral Ferrocenyl Ligands
Ji-Bao Xia, Timothy F. Jamison, and Shu-Li You

3.1
Introduction

This chapter focuses on the synthesis of chiral monodentate ferrocenyl ligands and their applications in asymmetric catalytic reactions. Compared with commonly discussed planar chiral ferrocenyl ligands the chiral monodentate ferrocenyl ligands normally have only one chiral center, they tend to be structurally simple and synthetically straightforward. One would think the monodentate ferrocenyl ligands might not be as efficient chiral ligands as bidentate ligands which commonly bind with metal in a more rigid manner and therefore provide a better chiral environment. However, interestingly, this has been proven to be not the case as many monodentate ferrocenyl ligands display excellent performance in specific enantioselective catalytic processes such as the nickel-catalyzed reductive coupling reactions (*P*-chiral monodentate ferrocenyl phosphines) and copper-catalyzed allylic substitution reactions (primary ferrocenyl amines) (Figure 3.1). The topic of this chapter will be divided into several parts based on the reaction type. Note that the synthesis and application of chiral ferrocenyl oxazolines or ferrocenyl imidazolines (monodentate ferrocenyl ligands) derived palladacycles will be discussed in Chapter 12.

3.2
Nickel-Catalyzed Asymmetric Reductive Coupling Reactions

Nickel-catalyzed asymmetric reductive coupling reactions of alkynes with aldehydes, ketones, or imines are important reactions for the synthesis of enantiopure allylic alcohols and allylic amines, both of which are useful building blocks in the preparation of various organic molecules and exist as key structural units in many natural products [1]. Interestingly, chiral monodentate ferrocenes have been found to be very efficient ligands in this catalytic process.

Chiral Ferrocenes in Asymmetric Catalysis. Edited by Li-Xin Dai and Xue-Long Hou
Copyright © 2010 WILEY-VCH Verlag GmbH & Co. KGaA, Weinheim
ISBN: 978-3-527-32280-0

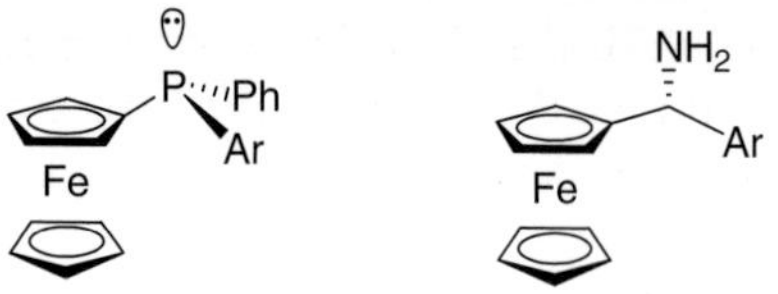

Figure 3.1 Representative monosubstituted chiral ferrocenyl ligands.

3.2.1
Synthesis of P-Chiral, Monodentate Ferrocenyl Phosphines

Following Jugé's ephedrine-based method for the synthesis of *P*-chiral phosphines [2], Jamison and coworkers recently developed the synthesis of a new class of *P*-chiral monodentate ferrocenyl phosphines in high enantiomeric excess (>95% ee in most cases). As shown in Scheme 3.1, methyl phosphinite borane **3** was obtained by the treatment of oxazaphospholidine borane **1** with various organolithium reagents (retention of the stereochemistry) followed by acid-promoted methanolysis of the resulting phosphinamide borane **2** (inversion of the stereochemistry). After installation of the ferrocenyl substituent by addition of ferrocenyl lithium to borane **3**, the BH_3 component in **4** was removed by refluxing in diethylamine, furnishing the free *P*-chiral, monodentate ferrocenyl phosphines **L1a-p** in good to excellent yields [3].

3.2.2
Nickel-Catalyzed Asymmetric Reductive Coupling of Alkynes and Aldehydes

Transition metal-catalyzed reductive coupling of alkynes and aldehydes is a useful method for construction of allylic alcohols [4]. In 2000, Jamison and coworkers found that an achiral ferrocenyl phosphine nickel complex catalyzed intermolecular

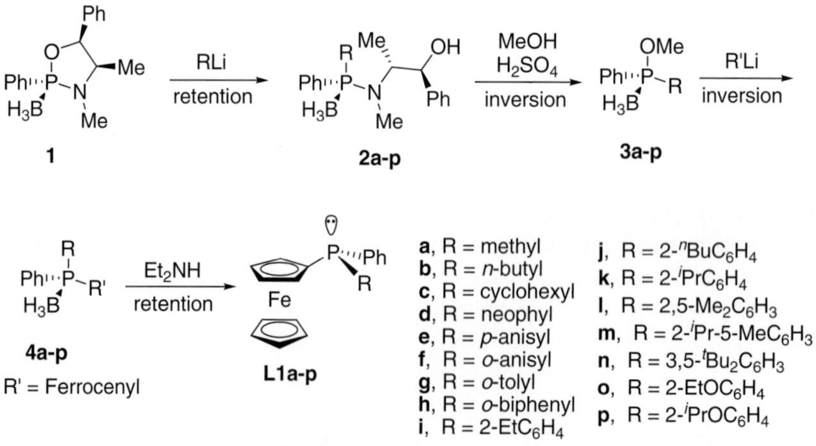

Scheme 3.1 Ligands **L1a–p** and their general synthesis.

reductive coupling reactions of alkynes and aldehydes. In order to develop an asymmetric version of this reaction, enantiopure P-chiral monodentate ferrocenyl phosphines **L1a–h** were synthesized. They were synthesized from oxazaphospholidine borane **1** in 14–55% overall yields (four steps) with the ee ranging from 80% to >98%. In the presence of 10 mol% of Ni(cod)$_2$, 10 mol% of monodentate ferrocenyl phosphine, and 200 mol% of Et$_3$B, asymmetric reductive coupling of alkynes and aldehydes gave optically active allylic alcohols with up to 68% ee (Table 3.1) [3].

As summarized in Table 3.1, phosphines **L1a–h** were examined in the nickel-catalyzed asymmetric reductive cross-coupling of prop-1-ynylcyclohexane (R^1 = c-C$_6$H$_{11}$, R^2 = Me) with isobutyraldehyde (R^3 = i-Pr), and (S)-**L1a** afforded the highest enantioselectivity (46% ee). It should be noted that in all cases the (E)-allylic alcohol (cis-addition across the alkyne) was obtained exclusively (>98/2). With (S)-**L1a** as the ligand, several alkynes and aldehydes were examined in the reductive coupling

Table 3.1 Nickel-catalyzed asymmetric reductive cross-coupling of alkynes and aldehydes.

7a,b: R^1 = c-C$_6$H$_{11}$, R^2 = Me, R^3 = i-Pr
8: R^1 = R^2 = n-Pr, R^3 = Ph
9: R^1 = R^2 = R^3 = n-Pr
10: R^1 = R^2 = n-Pr, R^3 = i-Pr
11a,b: R^1 = c-C$_6$H$_{11}$, R^2 = Me, R^3 = n-Pr

Entry	Ligand	Product	Yield (%)	a : b	ee of a (%)	ee of b (%)
1	(S)-L1a	7a, 7b	65	2.2 : 1	46	45
2	(S)-L1b	7a, 7b	27	1.8 : 1	8	12
3	(S)-L1c	7a, 7b	53	1.6 : 1	−34	−28
4	(S)-L1d	7a, 7b	33	1 : 1	−44	−10
5	(R)-L1e	7a, 7b	60	2.4 : 1	2	4
6	(R)-L1f	7a, 7b	60	3.8 : 1	−28	−17
7	(R)-L1g	7a, 7b	46	5.7 : 1	−55	−19
8	(R)-L1h	7a, 7b	33	1 : 1	−52	−37
9	(R)-L1a	8	85	—	49	—
10	(R)-L1c	8	80	—	−4	—
11	(R)-L1f	8	81	—	12	—
12	(R)-L1g	8	79	—	−28	—
13	(R)-L1h	8	87	—	−36	—
14	(S)-L1a	9	80	—	55	—
15	(S)-L1a	10	80	—	55	—
16	(S)-L1a	11a, 11b	30	2.2 : 1	67	68

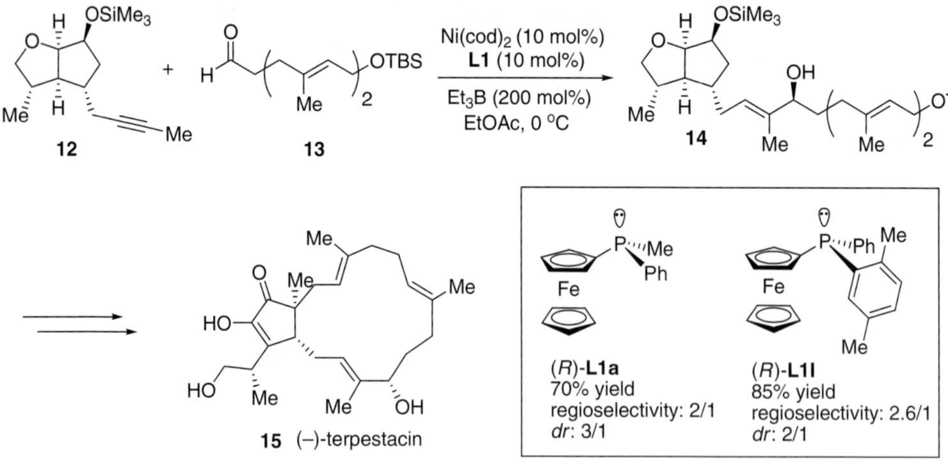

Scheme 3.2 A total synthesis of (−)-terpestacin employing the reductive coupling reaction as a key step.

reaction, the alcohol products were obtained with up to 68% *ee* but moderate regioselectivities in all cases.

The importance of the nickel-catalyzed asymmetric reductive coupling reaction of alkyne and aldehyde is obvious. The subsequent efforts from the same group disclosed the application of this reaction in the total synthesis of natural product, (−)-terpestacin. (−)-Terpestacin was found to inhibit the formation of syncytia by HIV-infected T cells and angiogenesis, and therefore the total synthesis of (−)-terpestacin and structurally related compounds has attracted considerable attention from synthetic communities [5]. As shown in Scheme 3.2, in the total synthesis of (−)-terpestacin, nickel-catalyzed asymmetric reductive coupling of fragment alkyne **12** and aldehyde **13** was employed as the key step. By utilizing (*R*)-**L1l** as ligand, the intermediate chiral allylic alcohol **14** was obtained in an 85% combined yield with 2.6/1 regioselectivity and 2/1 diastereoselectivity, favoring the desired isomer [6].

When 1,3-enynes were used as the coupling partner, chiral conjugated dienols were obtained by nickel-catalyzed asymmetric reductive coupling of alkynes and aldehydes. In the presence of 10 mol% of Ni(cod)$_2$ and monodentate ferrocenyl phosphine (*R*)-**L1k**, coupling of 1,3-enyne with aldehydes gave conjugated dienols **17** with excellent regioselectivities and modest enantioselectivities (Scheme 3.3) [7]. This methodology provides a facile synthesis of 1,3-dienes, which are important and versatile intermediates in organic synthesis, such as the starting materials for the Diels–Alder cycloaddition reaction.

During a mechanistic study of the nickel-catalyzed asymmetric reductive coupling of alkyne and aldehyde, Jamison and coworkers found that high regio- and diastereoselectivity were obtained when chiral 1,6-enyne was used in the absence of phosphine ligands. This can be explained by carbon–carbon bond formation occurring while the olefin tether is coordinated to the nickel. With the phoshine ligands,

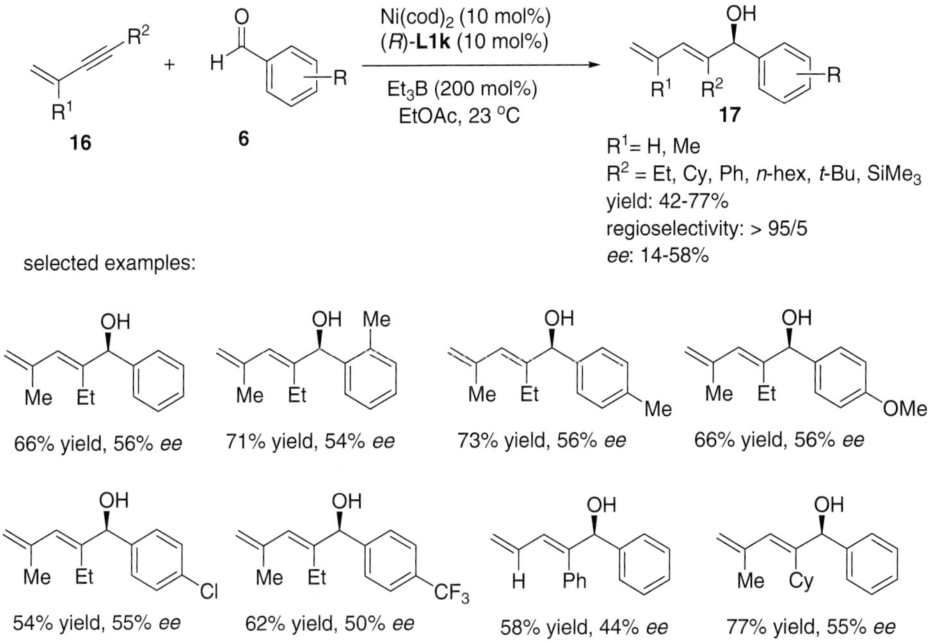

3.2 Nickel-Catalyzed Asymmetric Reductive Coupling Reactions

Scheme 3.3 Nickel-catalyzed asymmetric reductive coupling of enynes and aldehydes.

Reaction of **16** + **6** with Ni(cod)$_2$ (10 mol%), (R)-**L1k** (10 mol%), Et$_3$B (200 mol%), EtOAc, 23 °C → **17**

R^1 = H, Me
R^2 = Et, Cy, Ph, n-hex, t-Bu, SiMe$_3$
yield: 42–77%
regioselectivity: > 95/5
ee: 14–58%

selected examples:

- 66% yield, 56% ee (Me, Et, Ph)
- 71% yield, 54% ee (Me, Et, o-Me-C$_6$H$_4$)
- 73% yield, 56% ee (Me, Et, p-Me-C$_6$H$_4$)
- 66% yield, 56% ee (Me, Et, p-OMe-C$_6$H$_4$)
- 54% yield, 55% ee (Me, Et, p-Cl-C$_6$H$_4$)
- 62% yield, 50% ee (Me, Et, p-CF$_3$-C$_6$H$_4$)
- 58% yield, 44% ee (H, Ph)
- 77% yield, 55% ee (Me, Cy)

the olefin tether was replaced, losing the control during the carbon–carbon bond formation, and resulting in the loss of the selectivity control. Consistent with this hypothesis the observation that the use of a ferrocenyl phosphine results in 1 : 1 regioselectivity suggests that the phosphine displaces the alkene prior to carbon–carbon bond formation (Scheme 3.4) [8].

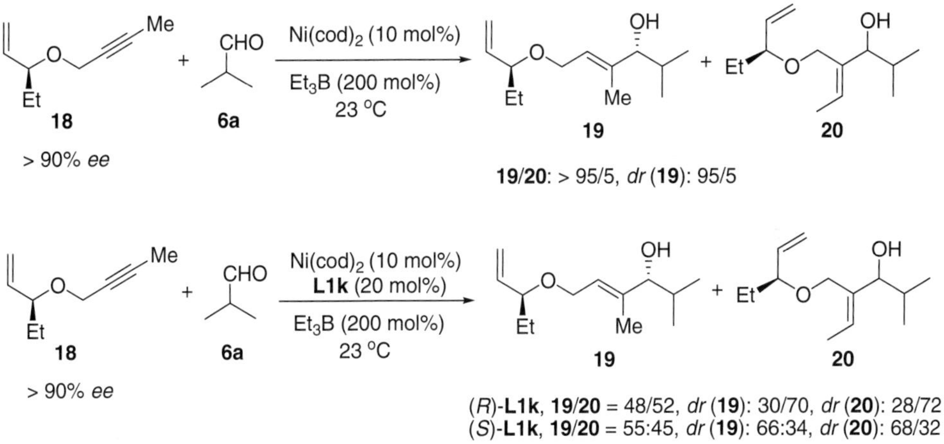

18 (> 90% ee) + **6a** → Ni(cod)$_2$ (10 mol%), Et$_3$B (200 mol%), 23 °C → **19** + **20**

19/20: > 95/5, dr (**19**): 95/5

18 (> 90% ee) + **6a** → Ni(cod)$_2$ (10 mol%), **L1k** (20 mol%), Et$_3$B (200 mol%), 23 °C → **19** + **20**

(R)-**L1k**, 19/20 = 48/52, dr (**19**): 30/70, dr (**20**): 28/72
(S)-**L1k**, 19/20 = 55/45, dr (**19**): 66/34, dr (**20**): 68/32

Scheme 3.4 Nickel-catalyzed asymmetric reductive coupling of chiral alkynes and aldehydes.

3.2.3
Nickel-Catalyzed Asymmetric Reductive Coupling of Alkynes and Ketones

In addition to the nickel-catalyzed asymmetric reductive coupling of alkynes and aldehydes, the Jamison group recently demonstrated that ketones are also suitable electrophilic partners. Utilizing 10 mol% of Ni(cod)$_2$ and 20 mol% of P-chiral, monodentate ferrocenyl phosphine (S)-**L1g**, conjugated dienes with a quaternary carbinol stereogenic center were obtained with excellent regioselectivity (>95/5 in all cases) and up to 70% ee. The ketone substrate scope was quite general and covered aryl with different electronic properties, heteroaryl and 1-cyclohex-1-enyl methyl ketones (Scheme 3.5 Eq. (a)) [9]. As shown in Scheme 3.5 (Eq. (b)), Rh-catalyzed hydrogenation of dienol **22a** led to optically active trisubstituted allylic alcohol **23** that was not accessible via a catalytic method previously. Ozonolysis of allylic alcohol **23** afforded the α-hydroxy ketone **24**.

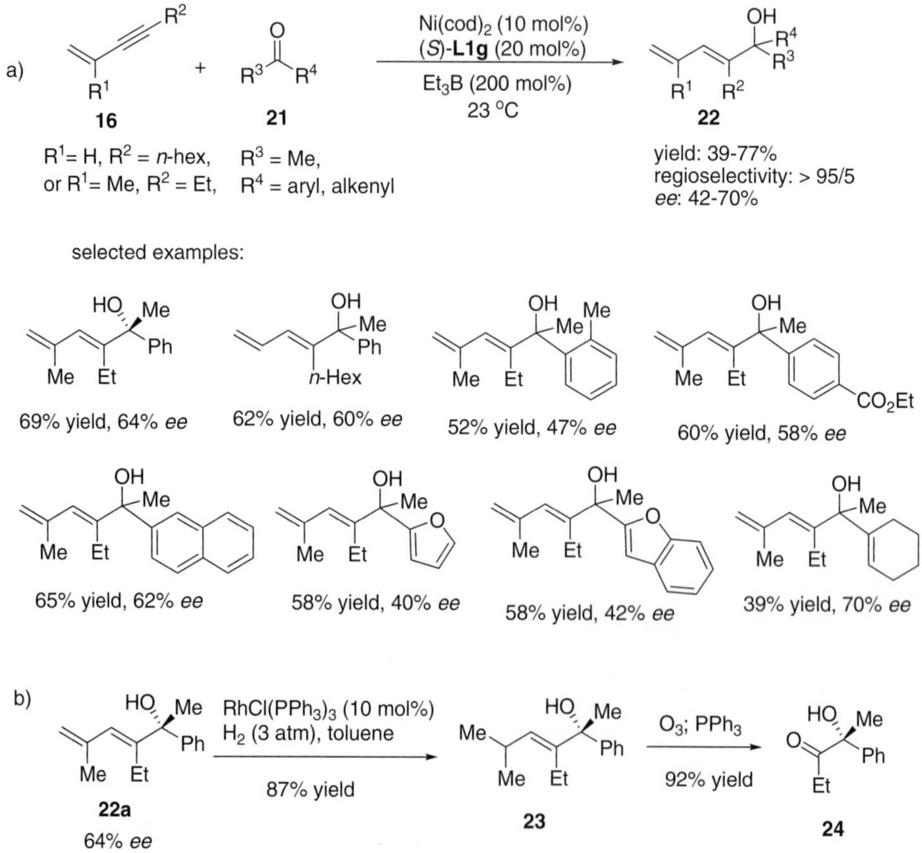

Scheme 3.5 Nickel-catalyzed asymmetric reductive coupling of enynes and ketones.

3.2.4
Nickel-Catalyzed Asymmetric Three-Component Coupling of Alkynes, Imines, and Organoboron Reagents

Given the importance of the enantiomerically enriched allylic amines in organic synthesis, Jamison and coworkers also looked into the nickel-catalyzed reductive coupling reaction of alkynes and imines. In 2003, Patel and Jamison reported a nickel-catalyzed three-component coupling of alkynes, imines, and organoboron reagents (boronic acids and organoboranes) to afford tetrasubstituted allylic amines [10]. Surprisingly, although the nickel-catalyzed coupling reactions of alkynes and aldehydes in the presence of neomenthyldiphenylphosphine afforded the allylic alcohols in excellent yields and *ee*s, this ligand was not effective for the nickel-catalyzed three-component coupling of alkynes, imines, and organoboron reagents [11].

A further extensive evaluation of monophosphines by Patel and Jamison demonstrated that *P*-chiral, monodentate ferrocenyl phosphines were effective ligands, affording the desired tetrasubstituted allylic amines in good yields and enantioselectivites (up to 89% *ee*) (Table 3.2). In addition to the symmetrical dialkyl acetylenes, the unsymmetrical alkynes were also effective in giving enantiomerically enriched tetrasubstituted allylic amines with four different substituents on the alkene with good to complete olefin geometry and regioselectivity. The (*tert*-butyldimethylsilyloxy)ethyl protecting group in the products was removed by deprotection of the TBS

Table 3.2 Ligand effect in nickel-catalyzed asymmetric three-component coupling of alkynes, imines, and organoboranes.

Entry	Ligand	Yield (%)	ee (%)
1	(R)-L1g	88	75
2	(R)-L1i	90	80
3	(R)-L1j	60	75
4	(R)-L1k	85	89
5	(R)-L1h	19	45
6	(R)-L1l	85	75
7	(R)-L1m	25	77
8	(R)-L1n	20	45
9	(R)-L1f	64	54
10	(R)-L1o	41	41
11	(R)-L1p	73	51

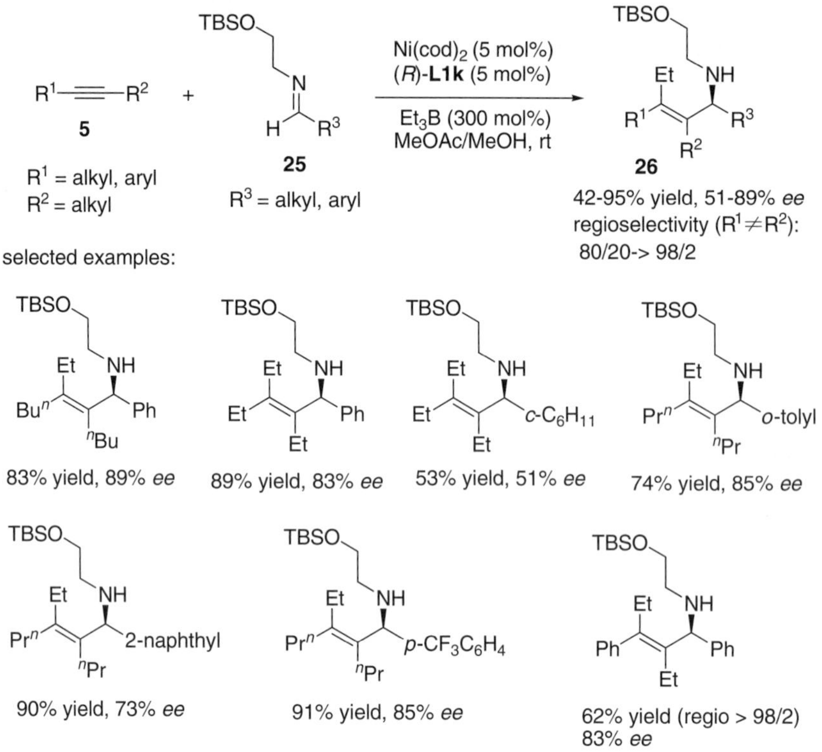

Scheme 3.6 Nickel-catalyzed asymmetric three-component coupling of alkynes, imines, and organoboranes.

group with TBAF followed by H_5IO_6 oxidative cleavage of the resulting amino alcohol. The primary allylic amines were obtained without loss of the optical purity and further recrystallized to optically pure as maleic acid salts (Scheme 3.6) [12].

3.3
Copper(I)-Catalyzed Asymmetric Allylic Alkylation Reactions

The copper-catalyzed allylic alkylation reaction is an important method for the construction of C—C bonds. Compared with the most extensively studied palladium catalyst, copper has the advantages of toleration of the use of hard "non-stabilized" nucleophiles and favoring the branched selectivity for unsymmetrical allylic substrates [13].

In 1999, Dübner and Knochel reported that a series of novel monosubstituted ferrocenyl amines were effective ligands in the copper(I)-catalyzed substitution of unsymmetrical allyl chloride with diorganozinc reagents. The synthesis is shown in Scheme 3.7. Starting from the readily prepared ferrocenyl ketones **27a–j**, followed by an enantioselective CBS reduction, the corresponding alcohols (*R*)-**29** were obtained

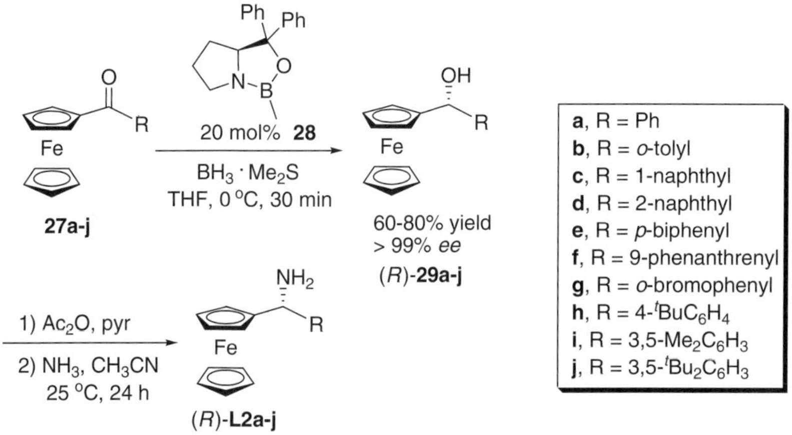
Ph
Ph
O
N
B
O
R
20 mol% 28
BH3 · Me2S
THF, 0 °C, 30 min
Fe
27a-j
OH
R
Fe
60-80% yield
> 99% ee
(R)-29a-j
NH2
R
Fe
1) Ac2O, pyr
2) NH3, CH3CN
25 °C, 24 h
(R)-L2a-j
a, R = Ph
b, R = o-tolyl
c, R = 1-naphthyl
d, R = 2-naphthyl
e, R = p-biphenyl
f, R = 9-phenanthrenyl
g, R = o-bromophenyl
h, R = 4-ᵗBuC6H4
i, R = 3,5-Me2C6H3
j, R = 3,5-ᵗBu2C6H3

Scheme 3.7 Ligands (*R*)-**L2a–j** and their general synthesis.

in excellent enantioselectivities (>99% *ee*). The ferrocenyl amines were obtained by acylation of the alcohols with acetic anhydride in pyridine, followed by treatment with aqueous ammonia in acetonitrile.

All the primary ferrocenyl amines were tested in the copper-catalyzed substitution of cinnamyl chloride with dineopentylzinc, and ferrocenyl amine (*R*)-**L2j** was found to be the optimal ligand, affording the alkylation product in excellent regioselectivity and enantioselectivity (branched/linear: 98/2, 92% *ee*) (Table 3.3). In addition, simultaneous addition of the diorganozinc and the allylic chloride over 3 h using

Table 3.3 Optimization of copper-catalyzed asymmetric allylic substitution of cinnamyl chloride with dineopentylzinc.

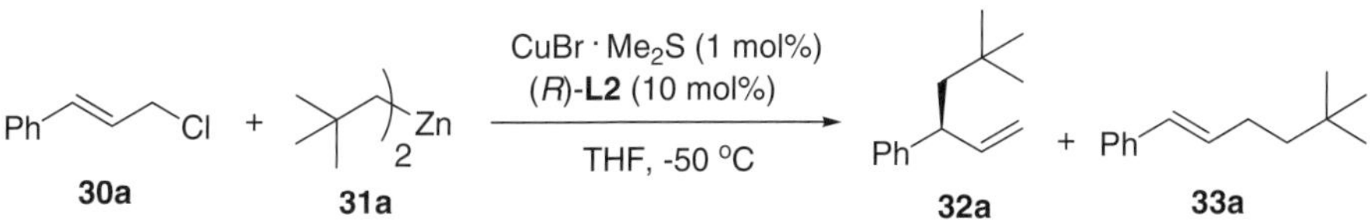
Ph
Cl
30a
+
Zn
2
31a
CuBr · Me2S (1 mol%)
(R)-L2 (10 mol%)
THF, -50 °C
Ph
32a
+
Ph
33a

Entry	Ligand	32a/33a	*ee* of 33a (%)
1	(*R*)-L2a	95/5	32
2	(*R*)-L2c	93/7	33
3	(*R*)-L2d	95/5	42
4	(*R*)-L2e	97/3	38
5	(*R*)-L2f	98/2	61
6	(*R*)-L2g	96/4	38
7	(*R*)-L2h	96/4	56
8	(*R*)-L2i	97/3	66
9	(*R*)-L2j	98/2	92

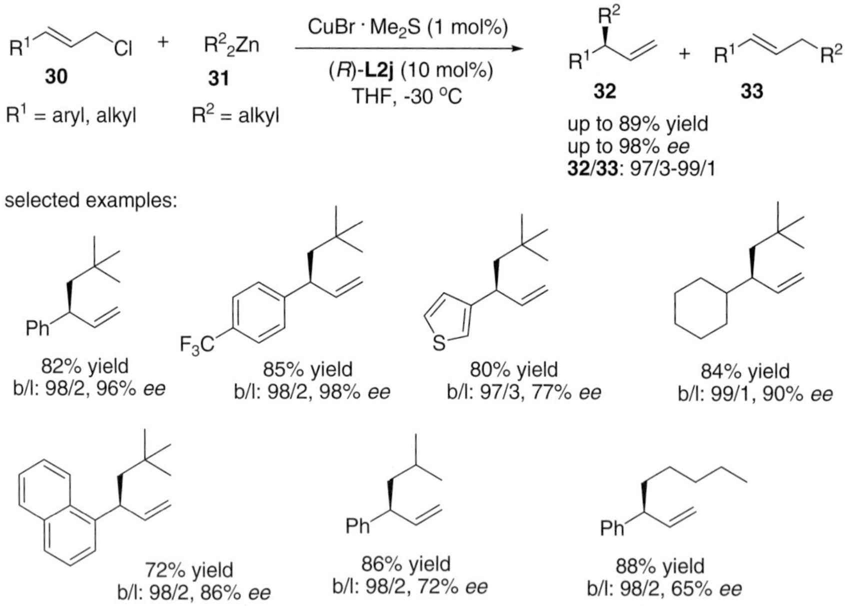

3 Monodentate Chiral Ferrocenyl Ligands

$$R^1\diagdown\diagdown Cl + R^2{}_2Zn \xrightarrow[\text{THF, -30 °C}]{\text{CuBr·Me}_2\text{S (1 mol\%)}\atop (R)\text{-L2j (10 mol\%)}} R^1\diagdown\underset{R^2}{\diagdown} + R^1\diagdown\diagdown R^2$$

30   31   32   33

R^1 = aryl, alkyl R^2 = alkyl   up to 89% yield
up to 98% ee
32/33: 97/3-99/1

selected examples:

Ph — 82% yield, b/l: 98/2, 96% ee

F_3C-aryl — 85% yield, b/l: 98/2, 98% ee

thiophene — 80% yield, b/l: 97/3, 77% ee

cyclohexyl — 84% yield, b/l: 99/1, 90% ee

naphthyl — 72% yield, b/l: 98/2, 86% ee

Ph — 86% yield, b/l: 98/2, 72% ee

Ph — 88% yield, b/l: 98/2, 65% ee

Scheme 3.8 Copper-catalyzed asymmetric allylic substitution.

a syringe pump further improved the enantioselectivity and avoided the using of very low reaction temperatures (−30 °C instead of −50 °C or −90 °C). In the presence of 1 mol% of CuBr Me$_2$S and 10 mol% of (R)-L2j, various diorganozincs and allylic chlorides were tested and the desired allylated products were obtained with excellent S_N2' regioselectivity and up to 98% ee for the branched products (Scheme 3.8). Both primary and secondary dialkylzincs are suitable substrates for the reaction [14].

3.4
Asymmetric Suzuki–Miyaura Reactions

Axially chiral biaryl compounds exist extensively as key structural units in biologically active compounds, as well as chiral auxiliaries and chiral ligands [15]. However, the direct catalytic asymmetric synthesis of the C–C biaryl bond from achiral substrates remains challenging since both the coupling partners are commonly two sterically hindered arenes. In this regard, the Suzuki–Miyaura reaction has emerged as an efficient method for the construction of the C–C biaryl bond due to the ready availability of the starting materials and mild reaction conditions [16].

In 2003, Johannsen and coworkers reported the synthesis of planar chiral monodentate ferrocenyl phosphines (Sp)-L3a–d, as shown in Scheme 3.9. Diastereoselective ortho-lithiation of (S)-34 followed by quenching with B(OMe)$_3$ and acid hydrolysis led to (S,Sp)-35 in excellent dr [17]. The Suzuki–Miyaura reaction between boronic acid (S,Sp)-35 and aryl iodide afforded the aryl ferrocenyl sulfoxides (S,Sp)-36a–d in moderate to excellent yields (41–73%). Sulfoxide cleavage of

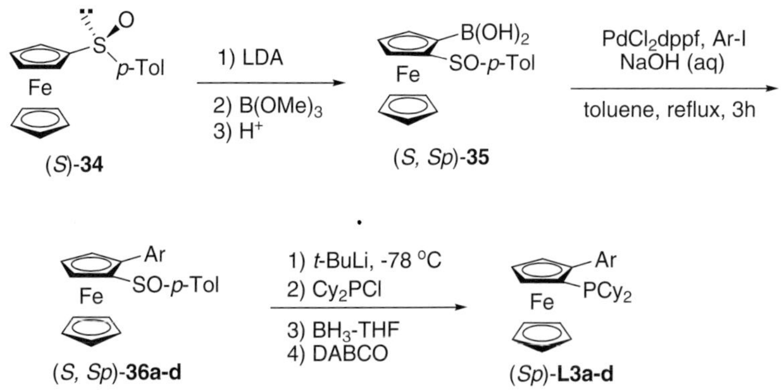

a, Ar = Ph (41%)
b, Ar = *o*-MeOC$_6$H$_4$ (69%)
c, Ar = 1-Naphthyl (60%)
d, Ar = 9-Phenanthrenyl (73%)

a, Ar = Ph (68%)
b, Ar = *o*-MeOC$_6$H$_4$ (58%)
c, Ar = 1-Naphthyl (63%)
d, Ar = 9-Phenanthrenyl (67%)

Scheme 3.9 Ligands (*Sp*)-**L3a–d** and their general synthesis.

(*S,Sp*)-**36** with *t*-BuLi followed by treatment of the chlorodicyclohexylphosphine gave arylferrocenylphosphines (aryl-MOPFs) (*S,Sp*)-**L3a–d** in 58–68% yields [18].

The performance of the aryl-MOPFs (*S,Sp*)-**L3b–d** was examined in the asymmetric coupling reaction between naphthyl bromide **37** and aryl boronic acid **38**. In the presence of 1 mol% of Pd$_2$dba$_3$, 5 mol% of (*Sp*)-**L3b–d**, and 2–3 equiv of K$_3$PO$_4$ as the base, the asymmetric Suzuki–Miyaura cross-coupling product **39** was obtained in modest yields (32–65%) and enantioselectivities (43–54% *ee*). (Scheme 3.10).

3.5
Addition of Organoaluminum to Aldehydes and Enones

Recently, a new class of ferrocenyl phosphonite ligands were synthesized and applied in the transition-metal catalyzed addition of organoaluminum to aldehydes and enones by Woodward and coworkers [19]. The ligands **L4** and **L5** were synthesized from (*S,Sp*)-**36** by diastereoselective lithiation and trapping with 1,1-binaphthyl- and

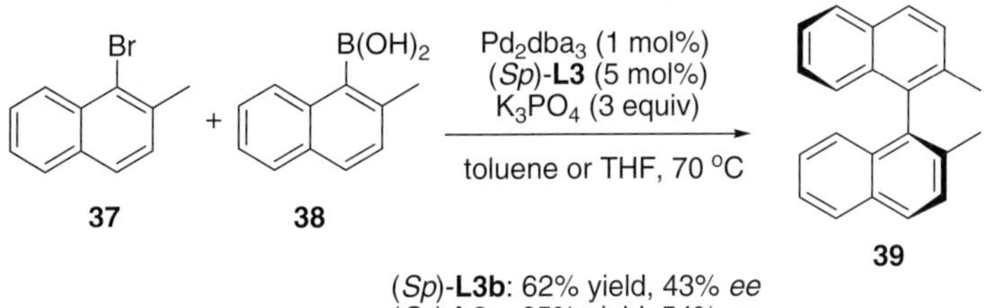

(*Sp*)-**L3b**: 62% yield, 43% *ee*
(*Sp*)-**L3c**: 65% yield, 54% *ee*
(*Sp*)-**L3d**: 32% yield, 45% *ee*

Scheme 3.10 Asymmetric Suzuki–Miyaura cross-coupling reaction.

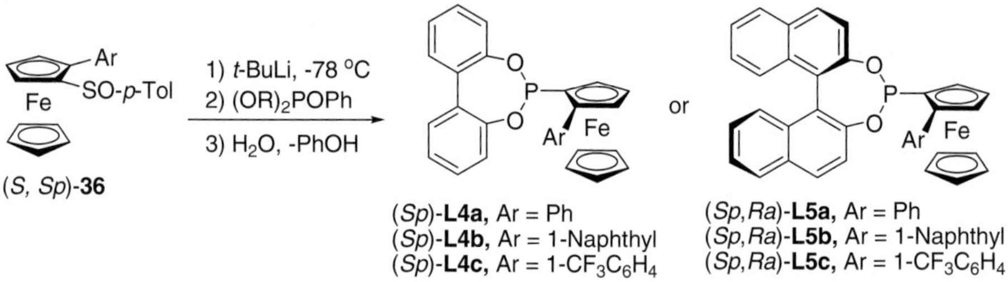

3 Monodentate Chiral Ferrocenyl Ligands

Scheme 3.11 Synthesis of ferrocenyl phosphonite ligands (Sp)-L4 and (Sp,Ra)-L5.

(S, Sp)-36

1) t-BuLi, −78 °C
2) (OR)$_2$POPh
3) H$_2$O, −PhOH

(Sp)-L4a, Ar = Ph
(Sp)-L4b, Ar = 1-Naphthyl
(Sp)-L4c, Ar = 1-CF$_3$C$_6$H$_4$

(Sp,Ra)-L5a, Ar = Ph
(Sp,Ra)-L5b, Ar = 1-Naphthyl
(Sp,Ra)-L5c, Ar = 1-CF$_3$C$_6$H$_4$

1,1-biphenyl-derived phosphites (Scheme 3.11). It should be noted that the use of chlorophosphite (RO)$_2$PCl species proved very inefficient, and only trace desired product was formed at best.

The efficacy of ligands L4 and L5 was tested in the nickel-catalyzed addition of AlMe$_3$ to benzaldehyde and copper-catalyzed additions of organoaluminum reagents to enones. In the presence of 1 mol% of Ni(acac)$_2$ and 2 mol% of the ferrocenyl phosphonite ligands, asymmetric addition of trimethylaluminum to benzaldehyde afforded 1-phenylethanol **41** in excellent yields with up to 77% ee (Table 3.4). In the presence of catalytic amounts of Cu(OTf)$_2$ (2 mol%) and the ferrocenyl phosphonite ligands (4 mol%), addition of triethylaluminum to cyclohexenone afforded the addition product **43** with up to 92% ee (Table 3.5).

3.6
Asymmetric Nucleophilic Catalysis

The design and synthesis of chiral variants of DMAP for asymmetric nucleophilic catalysis have gained increasing attention recently. Notably, Fu and coworkers have

Table 3.4 Nickel-catalyzed addition of trimethylaluminum to benzaldehyde.

AlMe$_3$ (2 equiv), Ni(acac)$_2$ (1 mol%), ligand (2 mol%), THF, −20 °C, 3h

40 → 41

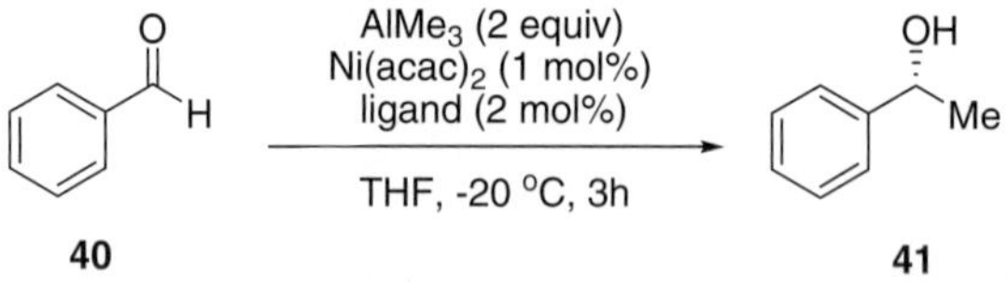

Entry	Ligand	Yield (%)	ee (%)
1	(Sp)-L4a	96	16
2	(Sp)-L4b	92	28
3	(Sp)-L4c	76	49
4	(Sp,Ra)-L5a	96	73
5	(Sp,Ra)-L5b	90	55
6	(Sp,Ra)-L5c	95	77

Addition of 2 equiv of AlMe$_3$ to benzaldehyde and Ni(acac)$_2$/ligand in THF at −20 °C followed by 3 h reaction at this temperature.

Table 3.5 Copper-catalyzed addition of triethylaluminum to cyclohexenone.

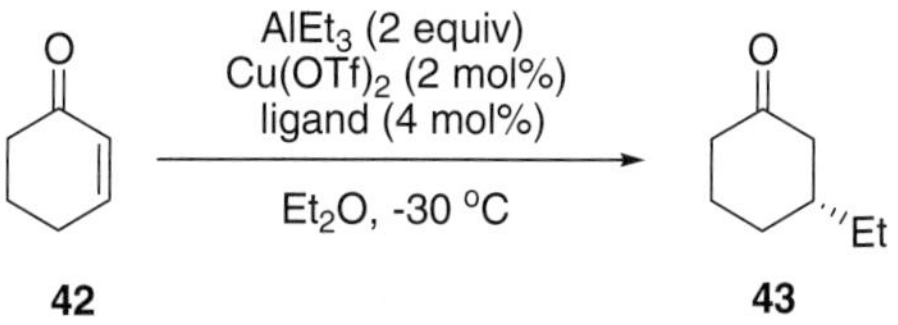

Entry	Ligand	Yield (%)	ee (%)
1	(Sp)-L4a	92	59
2	(Sp,Ra)-L5a	77	92
3	(Sp,Ra)-L5b	58	78
4	(Sp,Ra)-L5c	78	87

Addition of 1.4 equiv of AlEt$_3$ to 2-cyclohexenone and Cu(OTf)$_2$/ligand in ether at $-30\,°C$ followed by 20 min reaction at this temperature.

designed the ferrocene-based planar chiral DMAP by fusing the DMAP with one of the Cp rings in the same plane. This class of catalysts has been shown to perform very well for a variety of reactions (see Chapter 11). Different from Fu's design, Johannsen and coworkers recently reported a new type of ferrocene-based planar chiral DMAP analog by attaching a DMAP moiety to a planar chiral ferrocene backbone.

The synthesis is shown in Scheme 3.12. Treatment of (S,Sp)-36c with t-BuLi followed by electrophilic quenching with n-Bu$_3$SnCl afforded the tin intermediate, which was subjected to a Stille coupling to give (Sp)-44 in 57% overall yield.

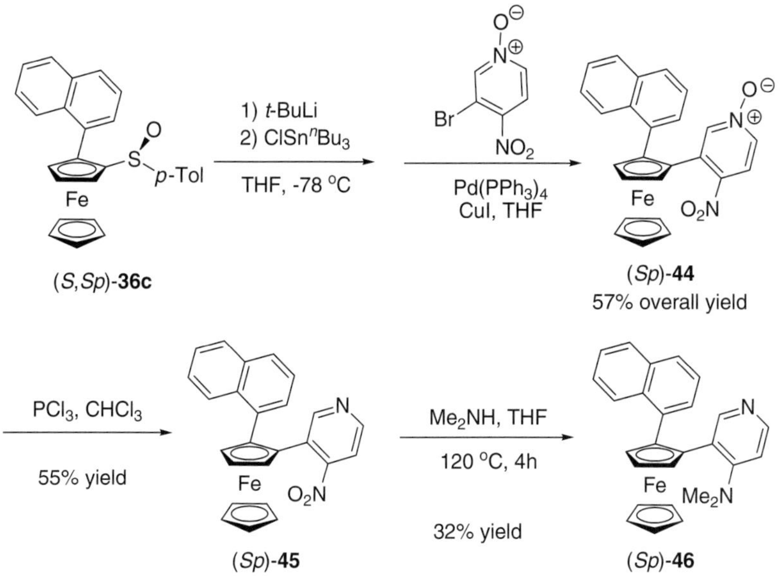

Scheme 3.12 Synthesis of ligand (Sp)-46.

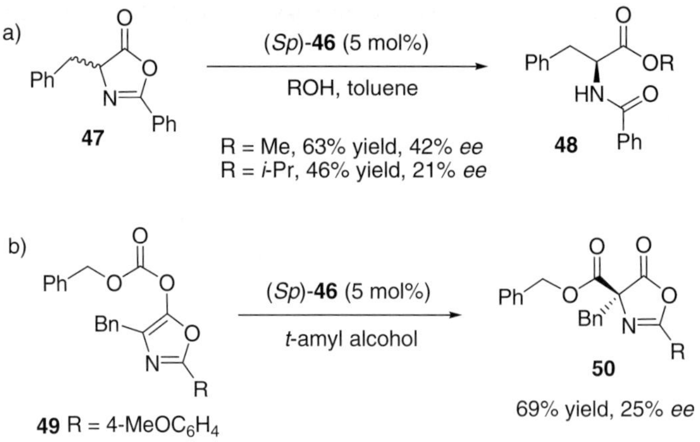

Scheme 3.13 (a) Kinetic resolution of azlactones and (b) rearrangement of O-acylated azlactone.

Deoxygenation of the oxo functionality in (Sp)-**44** by treatment with PCl₃ led to (Sp)-**45** in 55% yield. Then the planar chiral DMAP (Sp)-**46** was obtained by reacting the nitro compound (Sp)-**45** with dimethylamine in DMF [20].

With 5 mol% of (S)-**46** as catalyst, dynamic kinetic resolution of azlactones and rearrangement of azlactone were tested but unfortunately only modest enantioselectivities were obtained (Scheme 3.13).

In contrast to the commonly tedious synthesis of chiral DMAP, Richards and coworkers recently reported a relatively straightforward approach to ferrocene-based C_2-symmetric 4-pyrrolidinopyridine nucleophilic catalysts. As shown in Scheme 3.14, chiral DMAP analogs (R,R)-**53** and (R,R)-**54** were accessed from ferrocenyltributyltin

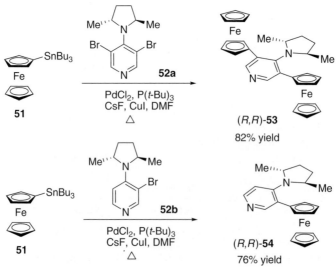

Scheme 3.14 Synthesis of (R,R)-**53-4**.

3.6 Asymmetric Nucleophilic Catalysis

Table 3.6 Kinetic resolution of aryl-alkyl secondary alcohols catalyzed by (R, R)-**53** and (R, R)-**54**.

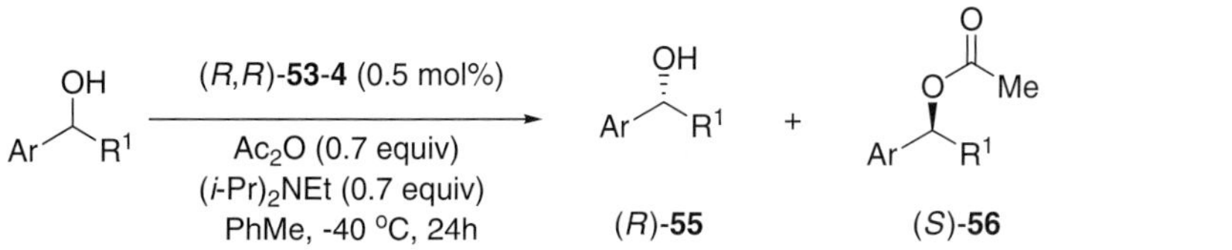

Entry	Catalyst	Ar, R	Conversion (%)	ee of 55	s
1	(R,R)-**53**	C_6H_5, Me	22	16%	4.2
2	(R,R)-**54**	C_6H_5, Me	95	43%	1.3
3[a]	(R,R)-**53**	C_6H_5, Me	23	20%	6.1
4	(R,R)-**53**	p-$NO_2C_6H_4$, Me	69	82%	5.0
5	(R,R)-**53**	p-$MeOC_6H_4$, Me	35	18%	2.4
6	(R,R)-**53**	1-naphthyl, Me	79	71%	2.7
7	(R,R)-**53**	C_6H_5, t-Bu	13	6%	2.5

[a] Pr_2O was used.

51 by a Stille coupling with (R,R)-pyrrolidinopyridine **52a** (82% yield) and **52b** (76% yield), respectively [21].

The catalytic activity of (R,R)-**53-54** in asymmetric nucleophilic catalysis was examined by application to the kinetic resolution of aryl-alkyl secondary alcohols. By utilizing the catalyst (R, R)-**53**, a selectivity factor up to 6.1 was obtained (s value) (Table 3.6).

In 2003 Kobayashi and coworkers reported that chiral alkyl aryl sulfoxides effectively mediate the enantioselective reaction of N-acylhydrazones with allyltrichlorosilanes and the corresponding homoallylic amines were obtained with high diastereo- and enantioselectivities [22]. Fernández, Khiar, and coworkers have recently reported the use of enantiomerically pure ferrocenyl sulfoxides as chiral Lewis bases to promote this reaction (Scheme 3.15). The allyl hydrazones were obtained in high yields within a short reaction time by using all the chiral ferrocenyl sulfoxides.

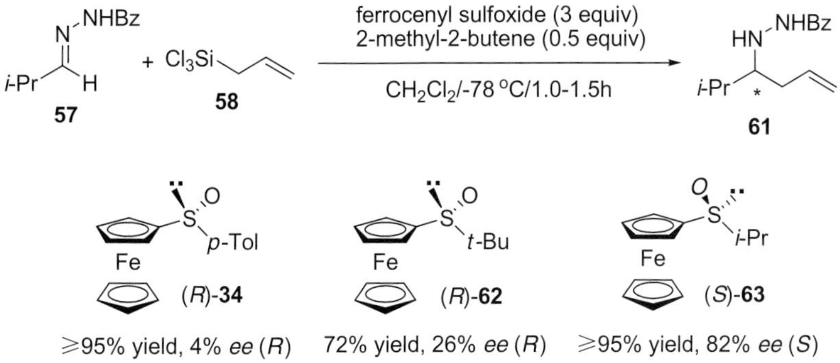

Scheme 3.15 Asymmetric allylation of hydrazone.

Isopropylsulfinylferrocene (S)-**63** gave the best result, and the product was obtained in a quantitative yield and 82% *ee*. Unfortunately, 3 equiv of the sulfoxides were required for maximum *ee* [23].

3.7
Conclusion and Perspectives

The synthesis and application of chiral monodentate ferrocenyl ligands in asymmetric catalysis have a relatively short history. Both the ligand types and suitable reactions are still limited. However, as seen from this chapter, chiral monodentate ferrocenyl ligands have been demonstrated to be efficient in several transition-metal catalyzed processes. For example, in the nickel-catalyzed reductive coupling reaction of alkynes with aldehydes, *P*-chiral mondentate phosphines have been demonstrated to be superior to all the other tested phosphine ligands. In addition, most of the chiral monodentate ferrocenyl ligands display synthetic facility. With more understanding of the enantioselective control with the chiral ferrocenyl monodentate ligands, it is reasonable to hypothesize that more efficient monodentate ferrocenyl ligands and suitable asymmetric reactions will be developed.

Acknowledgments

S.-L. You thanks the Chinese Academy of Sciences and the National Natural Science Foundation of China for financial support.

References

1 (a) Hoveyda, A.H., Evans, D.A., and Fu, G.C. (1993) *Chem. Rev.*, **93**, 1307–1370; (b) Johannsen, M. and Jøgensen, K.A. (1998) *Chem. Rev.*, **98**, 1689–1708.
2 Jugé, S., Stephan, M., Laffitte, J.A., and Genêt, J.P. (1990) *Tetrahedron Lett.*, **31**, 6357–6360.
3 Colby, E.A. and Jamison, T.F. (2003) *J. Org. Chem.*, **68**, 156–166.
4 For a review: Montgomery, J. (2004) *Angew. Chem.*, **116**, 3980–3998; (2004) *Angew. Chem., Int. Ed.*, **43**, 3890–3908.
5 (a) Oka, M., Iimura, S., Tenmyo, O., Sawada, Y., Sugawara, M., Ohkusa, N., Yamamoto, H., Kawano, K., Hu, S.-L., Fukagawa, Y., and Oki, T. (1993) *J. Antibiotics*, **46**, 367–373; (b) Jung, H.J., Lee, H.B., Kim, C.J., Rho, J.-R., Shin, J., and Kwon, H.J. (2003) *J. Antibiotics*, **56**, 492–496.
6 (a) Chan, J. and Jamison, T.F. (2003) *J. Am. Chem. Soc.*, **125**, 11514–11515; (b) Chan, J. and Jamison, T.F. (2004) *J. Am. Chem. Soc.*, **126**, 10682–10691.
7 Miller, K.M., Colby, E.A., Woodin, K.S., and Jamison, T.F. (2005) *Adv. Synth. Catal.*, **347**, 1533–1536.
8 Moslin, R.M. and Jamison, T.F. (2006) *Org. Lett.*, **8**, 455–458.
9 Miller, K.M. and Jamison, T.F. (2005) *Org. Lett.*, **7**, 3077–3080.
10 Patel, S.J. and Jamison, T.F. (2003) *Angew. Chem.*, **115**, 1402–1405; (2003) *Angew. Chem., Int. Ed.*, **42**, 1364–1367.

11 Miller, K.M., Huang, W.-S., and Jamison, T.F. (2003) *J. Am. Chem. Soc.*, **125**, 3442–3443.
12 Patel, S.J. and Jamison, T.F. (2004) *Angew. Chem.*, **116**, 4031–4034; (2004) *Angew. Chem., Int. Ed.*, **43**, 3941–3944.
13 For a recent review see: Falciola, C.A. and Alexakis, A. (2008) *Eur. J. Org. Chem.*, 3765–3780.
14 (a) Dübner, F. and Knochel, P. (1999) *Angew. Chem.*, **111**, 391–393; (1999) *Angew. Chem., Int. Ed.*, **38**, 379–381; (b) Dübner, F. and Knochel, P. (2000) *Tetrahedron Lett.*, **41**, 9233–9237.
15 For a recent review on the atroposelective synthesis of axially chiral biaryl compounds see: Bringmann, G., Mortimer, A.J.P., Keller, P.A., Gresser, M.J., Garner, J., and Breuning, M. (2005) *Angew. Chem.*, **117**, 5518–5563; (2005) *Angew. Chem., Int. Ed.*, **44**, 5384–5427.
16 For a review on asymmetric Suzuki coupling reactions to axially chiral biaryls see: Baudoin, O. (2005) *Eur. J. Org. Chem.*, 4223–4229.
17 Riant, O., Argouarch, G., Guillaneux, D., Samuel, O., and Kagan, H.B. (1998) *J. Org. Chem.*, **63**, 3511–3514.
18 (a) Jensen, J.F., Søtofte, I., Sørensen, H.O., and Johannsen, M. (2003) *J. Org. Chem.*, **68**, 1258–1265; (b) Jensen, J.F. and Johannsen, M. (2003) *Org. Lett.*, **5**, 3025–3028.
19 Albrow, V.E., Blake, A.J., Fryatt, R., Wilson, C., and Woodward, S. (2006) *Eur. J. Org. Chem.*, 2549–2557.
20 Seitzberg, J.G., Dissing, C., Søtofte, I., Norrby, P.-O., and Johannsen, M. (2005) *J. Org. Chem.*, **70**, 8332–8337.
21 Nguyen, H.V., Motevalli, M., and Richards, C.J. (2007) *Synlett*, 725–728.
22 Kobayashi, S., Ogawa, C., Konishi, H., and Sugiura, M. (2003) *J. Am. Chem. Soc.*, **125**, 6610–6611.
23 Fernández, I., Valdivia, V., Gori, B., Alcudia, F., Álvarez, E., and Khiar, N. (2005) *Org. Lett.*, **7**, 1307–1310.

4
Bidentate 1,2-Ferrocenyl Diphosphine Ligands
Hans-Ulrich Blaser and Matthias Lotz

4.1
Introduction

Ferrocene as a (at the time rather exotic) backbone for chiral ligands was introduced by Kumada and Hayashi [1], based on Ugi's pioneering work related to the synthesis of enantiopure ferrocenes (see Figure 4.1). PPFA as well as BPPFA and BPPFOH were not only the first ferrocene-based ligands but also the first ligands to exhibit both central and planar chirality and they proved to be very effective ligands for a variety of asymmetric transformations. From this starting point, several ligand families with a range of structural variations and types of chiral element have been developed in the last few years, very often starting from Ugi amine. In this chapter we will focus on the application of diphosphines developed over time where the two diphosphine moieties are attached to only one of the cyclopentadienyl (Cp) rings in a 1,2-arrangement. As a consequence these ligands can have a mix of chiral elements: planar chirality due to the ferrocene moiety, central chirality due to stereogenic carbon and/or phosphorus centers and, in one case, even axial chirality. Several recent reviews covering various aspects of ferrocene-based chiral ligands can serve to put the present account into a broader perspective [2]. Furthermore, we have recently summarized the synthetic procedures for most of the ligands described here [3].

This chapter is organized according to the five types of arrangement of the two diphosphine moieties at the Cp ring (see Figure 4.2).

Type A: Both PR_2 groups are directly attached to the Cp ring.

Type B: One PR_2 group is directly attached to the Cp ring, the second PR_2 group is attached to the α position of a side chain.

Type C: One PR_2 group is directly attached to the Cp ring, the second PR_2 group is attached to the β position of a side chain.

Chiral Ferrocenes in Asymmetric Catalysis. Edited by Li-Xin Dai and Xue-Long Hou
Copyright © 2010 WILEY-VCH Verlag GmbH & Co. KGaA, Weinheim
ISBN: 978-3-527-32280-0

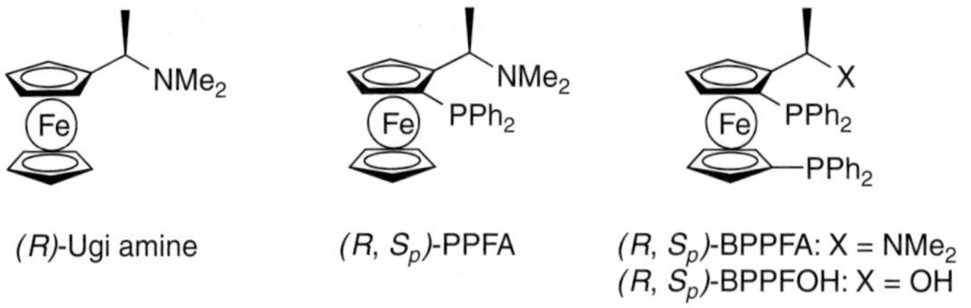

Figure 4.1 Structures of Ugi amine and the first ferrocene-based chiral phosphine ligands. Note that the first descriptor is for the central chirality, the second for the planar chirality of the ferrocene.

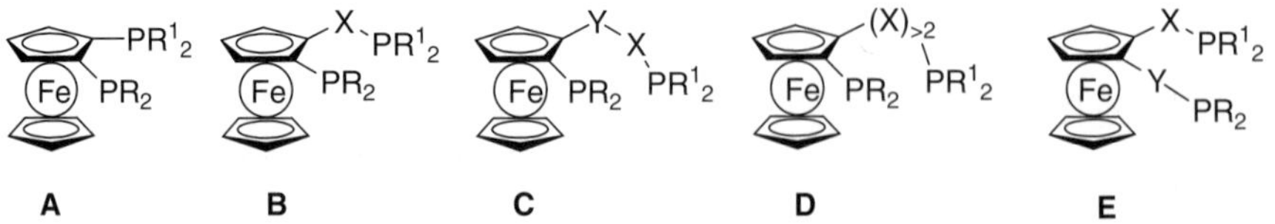

(R)-Ugi amine

(R, S_p)-PPFA

(R, S_p)-BPPFA: X = NMe₂
(R, S_p)-BPPFOH: X = OH

Figure 4.2 Classification of the 1,2-ferrocenyl diphosphines.

Type D: One PR_2 group is directly attached to the Cp ring, the second PR_2 group is attached to other positions of a side chain.

Type E: Both PR_2 groups are attached to the side chains.

A short note on the description of the catalytic results: Many of the ligands described in this chapter were tested on model test substrate under standard reaction conditions. Unless otherwise noted, this means that usually a substrate to catalyst ratio (s/c) of 100 was employed and that neither turnover number (TON) nor turnover frequency (TOF) were optimized. A few of the most frequently used reactions and standard conditions are depicted in Figure 4.3.

4.2
Type A Both PR₂ Groups Attached to the Cp Ring

Ligands of Type A (Figure 4.4) are the most simple chiral 1,2-ferrocene diphosphine. Obviously there is not much diversity possible and, indeed, only two derivatives of **L1** (R = Me, Cy) were prepared by Kagan [4], probably also because the catalytic results were rather modest. The two ligands were tested in two classical test reactions, namely the Rh-catalyzed hydrogenation of functionalized olefins and the Pd catalyzed allylic alkylation of allyl acetates with dimethyl malonate. While both ligands gave only modest *ees* for the allylic alkylation (best *ee* 47%, R = Me), enantioselectivities of 82% (R = Cy) and 95% (R = Me) were obtained for the Rh-catalyzed hydrogenation of itaconates and both ligands achieved 82% *ee* for the hydrogenation of a substituted enamide.

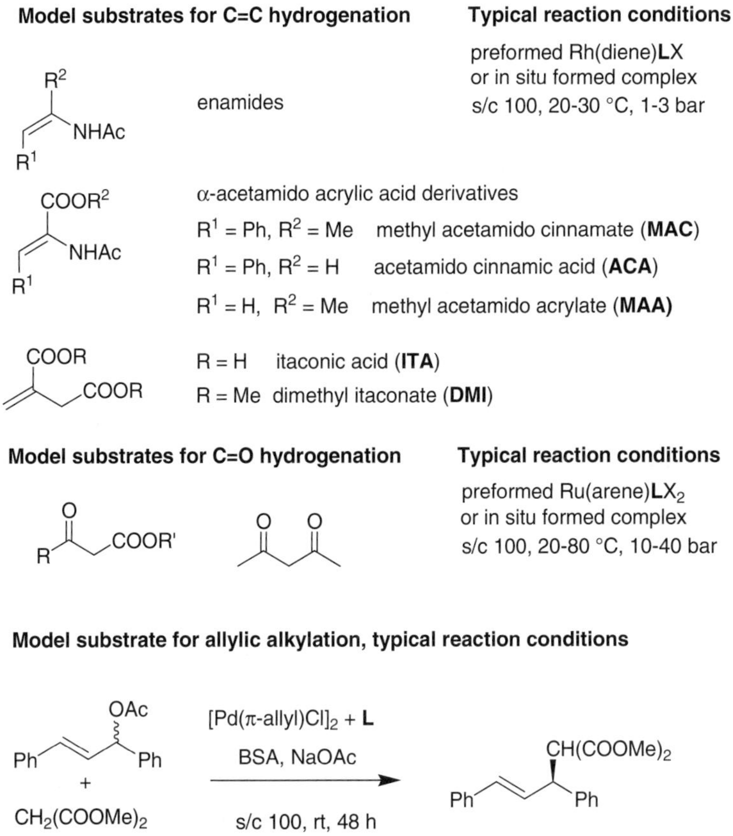

4.3 Type B One PR$_2$ Group Attached to the Cp Ring

Model substrates for C=C hydrogenation — **Typical reaction conditions**

enamides — preformed Rh(diene)LX or in situ formed complex, s/c 100, 20-30 °C, 1-3 bar

α-acetamido acrylic acid derivatives
- R^1 = Ph, R^2 = Me   methyl acetamido cinnamate (**MAC**)
- R^1 = Ph, R^2 = H   acetamido cinnamic acid (**ACA**)
- R^1 = H, R^2 = Me   methyl acetamido acrylate (**MAA**)

- R = H   itaconic acid (**ITA**)
- R = Me   dimethyl itaconate (**DMI**)

Model substrates for C=O hydrogenation — **Typical reaction conditions**

preformed Ru(arene)LX$_2$ or in situ formed complex, s/c 100, 20-80 °C, 10-40 bar

Model substrate for allylic alkylation, typical reaction conditions

[Pd(π-allyl)Cl]$_2$ + **L**, BSA, NaOAc, s/c 100, rt, 48 h

Figure 4.3 Model substrates and standard reaction conditions for ligand tests.

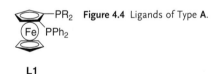

Figure 4.4 Ligands of Type **A**.

L1

4.3
Type B One PR$_2$ Group Attached to the Cp Ring, one PR$_2$ Group Attached to the α-Position of the Side Chain

The presence of a side chain allows a much wider structural variability (planar, central C and P chirality). Josiphos, the most important member of the Type B ligands (Figure 4.5), was developed only shortly after the publication of the PPFA ligands.

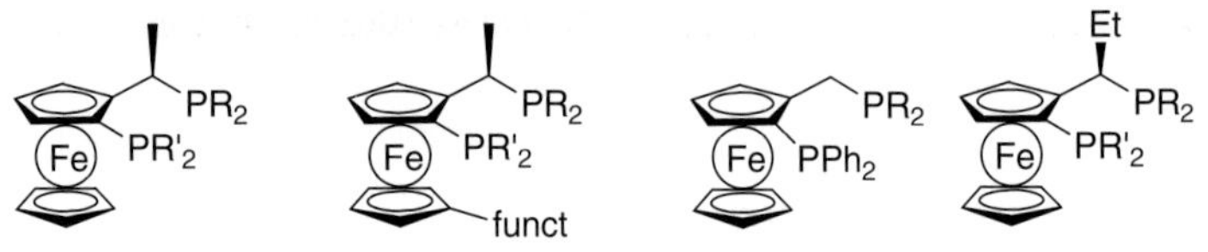

Josiphos | functionalized Josiphos L2-L6 | **L7** | **L8**

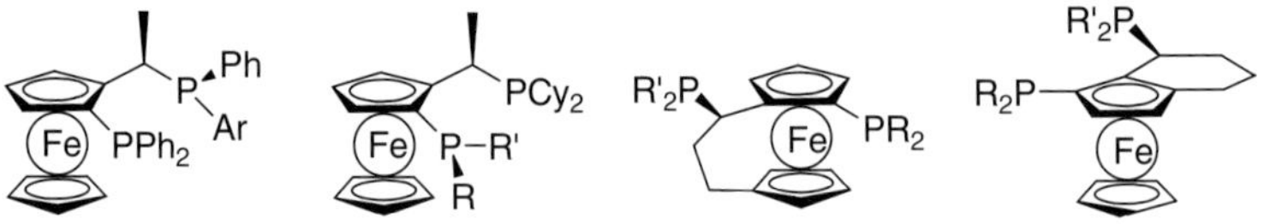

L9 | **L10** | **L11** | **12**

Figure 4.5 Ligands of type **B**.

4.3.1 Josiphos

The Josiphos ligands arguably constitute one of the most versatile and successful ligand families, second probably only to the Binap ligands. The two phosphine groups are introduced in consecutive steps starting from Ugi amine **1** via a PPFA derivative with very high yield [5, 6]. Due to this elegant synthesis, a variety of ligands are readily available with widely differing steric and electronic properties. Up to now, only the (R,S_p)-family (and its enantiomers) but not the (R,R_p) diastereomers have led to high enantioselectivities. At present about 150 different Josiphos ligands have been prepared and 40 derivatives are available in a ligand kit for screening and on a multi-kg scale for production (Table 4.1 [7]). The most successful ligands, their numbering and applications are depicted in Table 4.1. X-Ray analyses of both the free ligand **J001** and its Rh, Pd and Pt complexes were determined and confirmed both the structure and sense of the planar chirality [8]. In order to avoid interaction of the large PR'_2 group with the rest of the molecule, H and Me attached to the side chain are forced below the Cp plane while the PR'_2 group is above. This positions the two phosphine groups ideally for the formation of metal complexes, thereby dictating a similar coordination behavior to different metal centers while still allowing adaptation to different bond lengths, oxidation states and coordination geometries.

Catalytic applications of the Josiphos ligand family have been reviewed up to 2002 [7a]. For this reason, we provide only a summary of the most important processes operated with Josiphos ligands and give an update on new results. Up to now, Josiphos ligands are (or have been) applied in four production processes and about five or six pilot and bench scale processes involving Rh, Ir and Ru catalyzed hydrogenation reactions of C=C and C=N bonds [9] and selected applications are summarized in Figure 4.6. The most important process is undoubtedly the Ir/**J005**-catalyzed hydrogenation of a hindered *N*-aryl imine of methoxyacetone, the largest known enantioselective process operated for the enantioselective production of the herbicide (*S*)-

Table 4.1 Synthesis of Josiphos ligands, structure and naming of the most important derivatives.

(R)-1 → (R,S_p)-PPFA → (R,S_p)-Josiphos (R'$_2$PH, AcOH, 80 °C)

Name	R	R'	important applications
J001	Ph	Cy	process for jasmonate; hydrogenation of enamids, itaconantes; hydroboration; allylic alkylations; Michael additions; PMHS reduction of C=C; addition to meso anhydrides
J002	Ph	t-Bu	opening of oxabicycles; biotin and MK-0431 process
J003	Cy	Cy	hydrogenation of phoshinylimines
J004	Cy	Ph	Michael additions
J005	Ph	3,5-Xyl	metolachlor process; methoxycarboxylation
J009	Cy	t-Bu	hydrophosphination; cross coupling
J011	4-CF$_3$-Ph	t-Bu	hydrogenation of β-imino acid derivatives
J013	4-MeO-Xyl	t-Bu	dextromethorphane process

metolachlor [6, 10]. The hydrogenation of a tetrasubstituted C=C bond was the key step for two production processes developed for the synthesis of biotin by Lonza and of methyl dihydrojasmonate by Firmenich. Pilot processes were developed by Lonza for a building block of crixivan and for dextromethorphane (for more information see [11a]). Recently, Merck chemists reported the unprecedented hydrogenation of unprotected dehydro β-amino acid derivatives catalyzed by Rh/Josiphos with *ee*s up to 97%. It was shown that not only simple derivatives but also the complex intermediate for MK-0431 depicted in Figure 4.6 can be hydrogenated successfully. Regular production on a multi-ton/y scale with *ee*s up to 98% started in 2006 [11].

Feasibility studies were reported for the Rh-catalyzed asymmetric hydrogenation of a β,β-diarylsubstituted α,β-unsaturated acid, **J001** giving the highest enantioselectivity (93% *ee*) [12] and for the hydrogenation of N-sulfonylated-α-dehydroamino acids using Ru/**J002** (up to 99.1% *ee*), the first example of a Ru-catalyzed hydrogenation of a tetrasubstituted α-dehydroamino acid [13]. Merck's chemists also developed a mild and efficient route to chiral aryl alkyl amines via the Rh/Josiphos-catalyzed asymmetric hydrogenation of ene-trifluoroamides with 94–97% *ee*. The advantage of using trifluoroamides as substrates for this transformation is the mild deprotection (K$_2$CO$_3$, MeOH, H$_2$O) [14].

Rh and Ir complexes with chiral Josiphos ligands are highly selective, active and productive catalysts for various enantioselective reductions. Very high enantioselectivities have been described for the enantioselective hydrogenation of enamides, itaconic acid derivatives, acetoacetates as well as N-aryl imines (in the presence of acid and iodide) and phosphinylimines [7]. Recently, Baratta *et al.* [15] reported on the

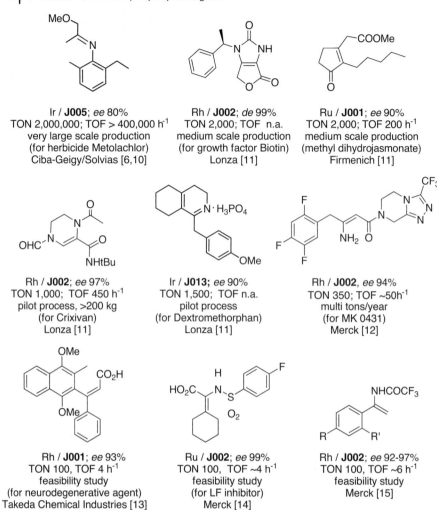

Figure 4.6 Important industrial applications of Josiphos ligands.

transfer hydrogenation of aryl methyl ketones using Ru and Os complexes of **J001** and R-substituted 1-pyridin-2-yl methanamine, RPyme, (Scheme 4.1). The reactions take place at 60 °C with exceptionally high TONs and TOFs leading to high conversion and excellent *ee* values.

Josiphos **J001** is the ligand of choice for the Cu-catalyzed reduction of activated C=C bonds with polymethylhydrosiloxane (PMHS) (Scheme 4.2) with very high enantioselectivities for nitro alkenes [16], αβ-unsaturated ketones [17a], esters [17b,c] and nitriles [18]. The same ligand has recently been reported to give high stereo-selectivity for the Ir-catalyzed hydrogenation of an α-amino acetophenone [19], while **J002** was preferred for the Rh-catalyzed hydrogenation of an α-amino-β-keto ester with moderate *ee* but high diastereoselectivity [20].

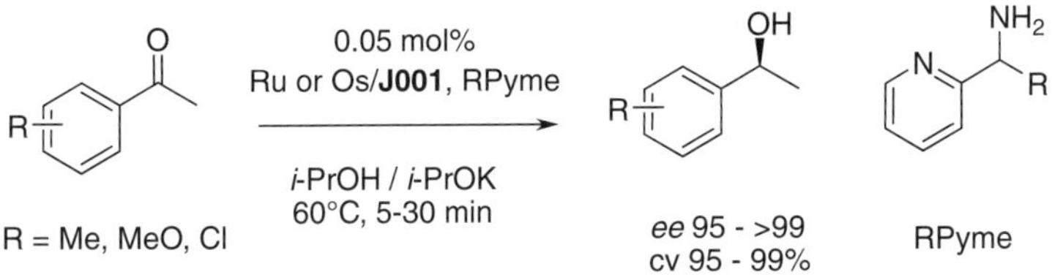

Scheme 4.1 Ru- and Os-catalyzed transfer hydrogenation of aryl ketones.

Josiphos complexes have been successfully applied to various asymmetric catalytic reactions such as allylic alkylation or hydroformylation [7]. Other recent examples are depicted in Schemes 4.3 and 4.4. Feringa's group reported high *ee*s for the Cu-catalyzed Michael addition of Grignard reagents to αβ-unsaturated esters [21a], thioesters [21b,c], and for selected cyclic enones [21d]. This reaction was extended to the 1,6-addition of Grignard reagents to linear dienoates, also with respectable to high enantioselectivities [21e]. Preferred ligands were **J001** and **J004**. The Rh/**J001**-catalyzed Michael addition of organotrifluoroborates to cyclic enones with enantioselectivities up to 99% was reported by Genêt and coworkers [22].

The same catalyst types were also very effective for the nucleophilic ring opening of various oxabicyclic substrates leading to interesting new dihydronaphthalene and cyclohexene derivatives (Scheme 4.4 [23a,b]). This reaction was scaled up to kg scale [23c]. The Pd-catalyzed opening of various cyclic anhydrides with Ph$_2$Zn was described by Bercot and Rovis to occur with very high *ee*s in the presence of **J001** (Scheme 4.4) [24]. The Hartwig group described the CpPd(allyl)/**J002**-catalyzed addition of acetylacetone to dienes with *ee*s up to 81% [25] and Lorman *et al.* reported up to 95% *ee* for an intramolecular Heck reaction using a Pd/**J001** catalyst (Scheme 4.4) [26]. Pd/Josiphos

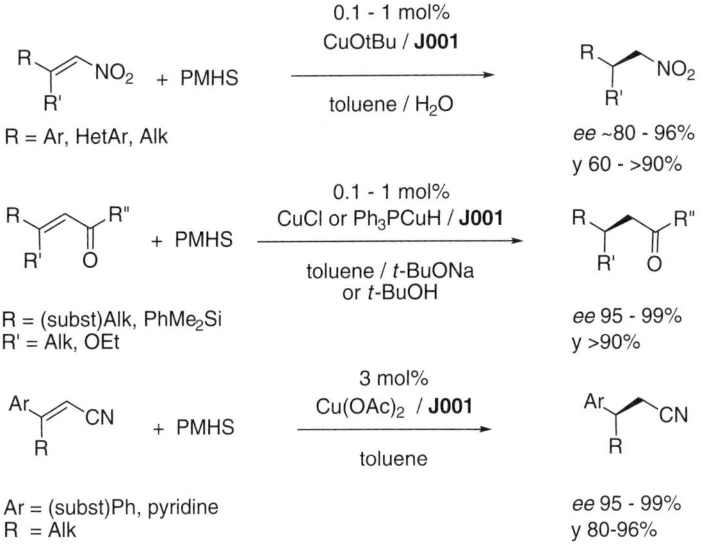

Scheme 4.2 Enantioselective reductions with Cu/Josiphos catalysts.

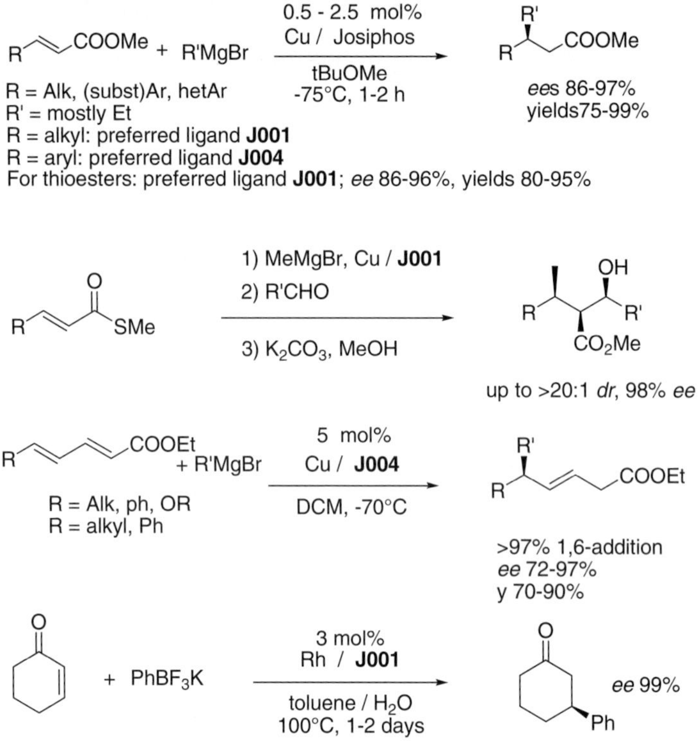

Scheme 4.3 Enantioselective Michael and 1,6-additions with Josiphos ligands.

catalysts are active for the Pd-catalyzed methoxycarbonylation of styrene (best ligand **J005**, low branched/linear ratio) [27] and for the hydrophosphonation of C=C bonds with *ee*s up to 88% for **J009** [28]. Finally, aryl and vinyl bromides can be cross-coupled with allyl trifluoroborate salts with high branched/linear ratios and enantioselectivities up to 90%; best catalyst Pd(OAc)$_2$/**J009** [29].

Josiphos ligands have not only been applied to enantioselective reactions but also showed very high activities for Pd-catalyzed coupling reactions. Hartwig's group reported that Pd/**J009** complexes were effective catalysts for the amination of aryl halides or sulfonates with ammonia [30a], the coupling of aryl halides or sulfonates with thiols [30b] and the Kumada coupling of aryl or vinyl tosylates with Grignard reagents [30c]. Merck chemists found Pd/Josiphos to be an efficient catalyst for carbonylation of arene sulfonates, giving over 90% isolated yields [31].

4.3.2
Immobilized Josiphos

The Josiphos backbone has been modified in order to covalently attach organic or inorganic polymeric supports, hydrophilic groups (to make the ligands water

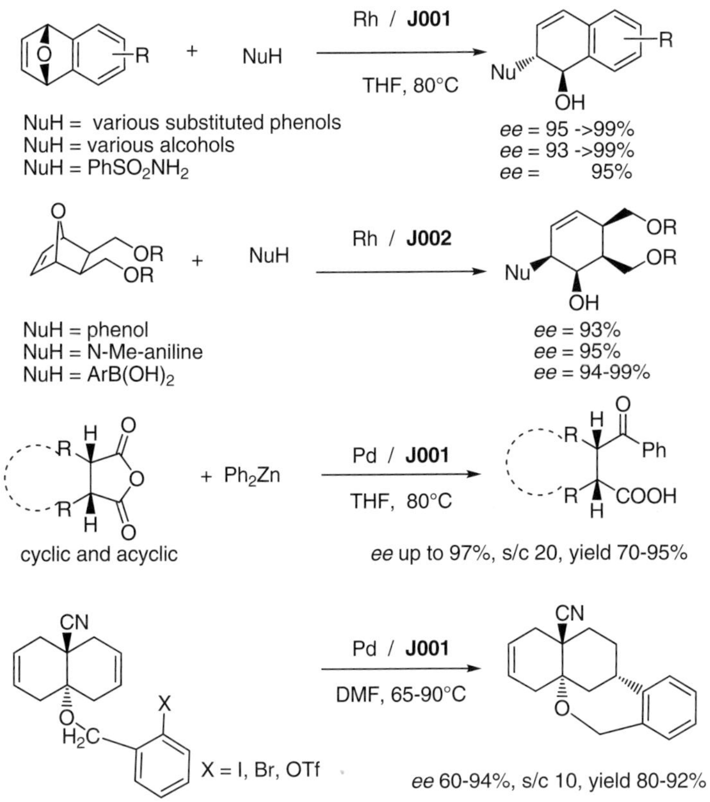

Scheme 4.4 Various transformations with Josiphos ligands.

Reactions shown:

- Oxabicyclic substrate + NuH, Rh / **J001**, THF, 80°C → ring-opened product with Nu and OH
  - NuH = various substituted phenols, ee = 95 ->99%
  - NuH = various alcohols, ee = 93 ->99%
  - NuH = PhSO$_2$NH$_2$, ee = 95%

- Diester oxabicyclic + NuH, Rh / **J002** → ring-opened diester product
  - NuH = phenol, ee = 93%
  - NuH = N-Me-aniline, ee = 95%
  - NuH = ArB(OH)$_2$, ee = 94-99%

- Cyclic and acyclic anhydride + Ph$_2$Zn, Pd / **J001**, THF, 80°C → keto-acid product
  - ee up to 97%, s/c 20, yield 70-95%

- CN-substituted bicyclic aryl halide (X = I, Br, OTf), Pd / **J001**, DMF, 65–90°C → cyclized product
  - ee 60-94%, s/c 10, yield 80-92%

---

soluble) or imidazolium tags (in order to immobilize the ligand in ionic liquids) (Figure 4.7 [32–34]).

The functionalized ligands were tested for various hydrogenation reactions. Ir/**L3** (bound to silica gel) as well as a water soluble complex Ir/**L6** gave TONs >100 000 and TOFs up to 20 000 h^{-1} for the Ir-catalyzed imine reduction while the polymer bound complex Ir/**L4** was much less active; in all cases ees were comparable to the homogeneous catalyst [33]. Dendrimer supported Rh/**L2** complexes hydrogenated DMI with ees up to 98.6% with similar activities to the mononuclear catalyst [34]. Similarly, ligands with an imidazolium tag **L5** had a comparable catalytic performance to the non-functionalized ligands for the Rh-catalyzed hydrogenation of MAA and DMI both in classical solvents and under two-phase conditions in ionic liquids but were easy to separate and to recycle [32]. Rh/Josiphos complexes were also ion-exchanged onto MCM 41 [35] and adsorbed onto a heteropolyacid modified alumina [36]. Both catalysts were active for the hydrogenation of DMI with good to very good ees and could easily be recycled or operated continuously in supercritical CO$_2$, respectively.

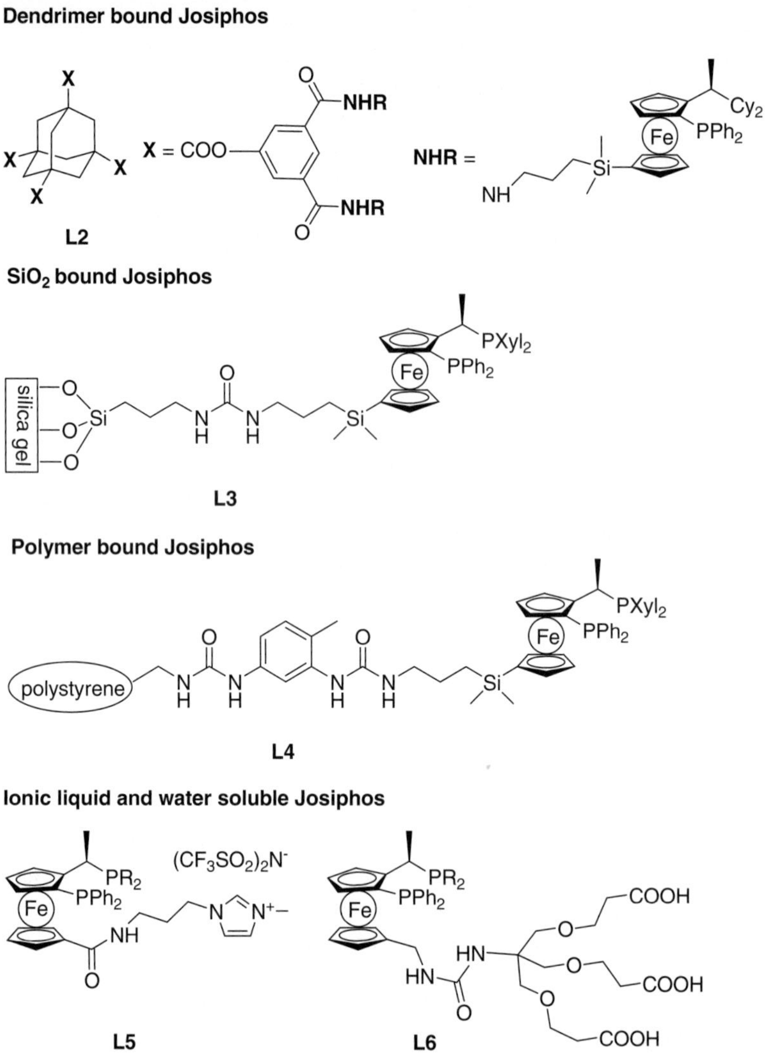

4 Bidentate 1,2-Ferrocenyl Diphosphine Ligands

Dendrimer bound Josiphos

L2

SiO₂ bound Josiphos

L3

Polymer bound Josiphos

L4

Ionic liquid and water soluble Josiphos

L5   L6

Figure 4.7 Structure of functionalized Josiphos ligands.

4.3.3
Josiphos Analogs

The homologous Josiphos analogs **L7** (only planar chirality) and **L8** were developed by Kagan [37] and Knochel [38], respectively. The catalytic results indicate that both homologs of Josiphos are much less effective. Various derivatives of **L7** were tested in the Rh catalyzed hydrogenation of functionalized olefins and the Pd catalyzed allylic alkylation of allyl acetates with dimethyl malonate. While only modest *ee*s were achieved for the allylic alkylation (best *ee* 82% for $R^1 = t$-Bu), enantioselectivities of 93–98% were obtained for the Rh-catalyzed hydrogenation of itaconates and a substituted enamide; best ees were usually observed with $R^1 = t$-Bu. Several derivatives of **L8** were tested for the asymmetric Rh-catalyzed hydroboration of styrene. The best results were achieved with R = Ph, R' = Cy giving 46% *ee* while **J001** provided 91.5% *ee*.

P chiral Josiphos analogs **L9** and **L10** were described by Togni *et al*. [8]. Both ligands were rather difficult to obtain in a pure form. Disappointingly, the P-chiral ligands **L9** only gave moderate results for the hydrogenation of various C=C bonds [8a,39] and for this reason, the investigation of **L10** was not pursued further [8b].

L11 [40] and **L12** [41], two constrained analogs of Josiphos, were prepared by Weissensteiner and colleagues in order to test the effect of restricting the conformational flexibility of the ligand on its catalytic performance. X-Ray structures of **L11** and **L12** show that while the phosphino groups are well positioned to form metal complexes, the ligands cannot adapt as easily as the Josiphos ligand to the requirements of a metal since the conformational space is restricted. In view of this structural information it was not unexpected that the enantioselectivity, and in many cases also the activity, of **L11** and **L12** metal complexes for the hydrogenation of various C=C, C=O and C=N-containing substrates [40] as well as for the allylic alkylation and amination reactions [42] were often significantly lower than Josiphos with the same phosphine moieties.

4.4
Type C One PR$_2$ Group Attached to the Cp Ring, one PR$_2$ Group Attached to the β-Position of the Side Chain

Several variants of this class of ligands have been prepared with both heteroaliphatic and aromatic moieties between the cp ring and the second PR$_2$ group (Figure 4.8). Up to now the most important representatives of this class are the BoPhoz ligands designed by Boaz [43], all other ligands have only been applied to one or two test reactions, albeit often with excellent enantioselectivities.

4.4.1
BoPhoz and Analogs

As Josiphos, BoPhoz are modular ligands with a PR$_2$ group on the Cp ring but an aminophosphine on the side chain. A new modular synthesis allows the incorporation of a wide range of nitrogen and phosphorus substituents R and R' [43c]. Even though a variety of ligands with different R and R' groups have been prepared, the preferred ligand up to now is the *N*-methyl derivative with R' = Ph (**L13**). Bophoz ligands are air stable but, depending on the solvent, the stability of the N–PR'$_2$ bond may be a critical issue. Selected BoPhoz ligands are available from Johnson Matthey [44]. A phosphonite analog **L13** was recently described by Chan *et al*. [45].

BoPhoz ligands are very effective for the Rh-catalyzed hydrogenation of a variety of activated C=C bonds such as enamides (*ee* 96–99%, s/c up to 10 000) and itaconates (*ee* 80–99%) [43a–c]. As observed for several ligands forming seven-membered

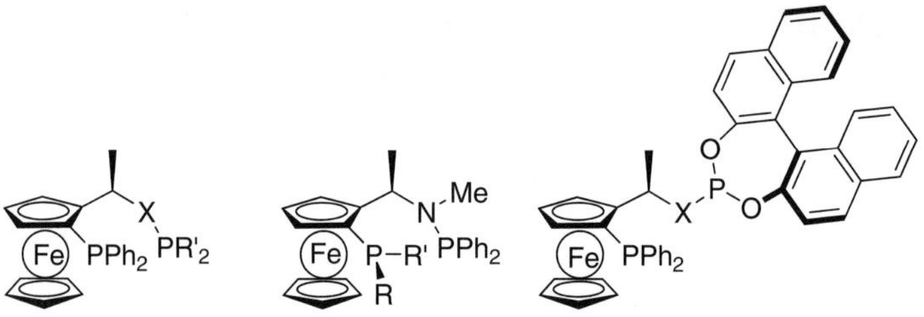

BoPhoz: X = NR
L13 X = O

L14

X = NH **L15a**
X = NMe **L15b**
X = O **L16**

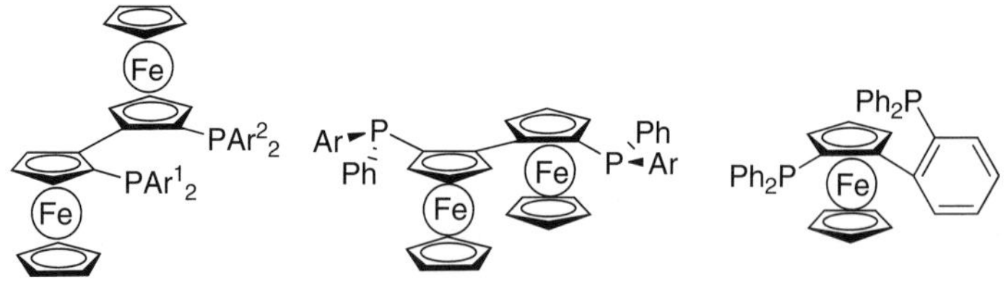

(S_p, S_p)-Bifep

(R_p, R_p, S_{phos}, S_{phos})-**L17**
a: Ar = 1-Np
b: Ar = 2-biphenyl

L18

Figure 4.8 Ligands of type C.

chelates, high activities can be reached (maximum TOFs for enamides up 68 000 h^{-1}) [43b]. While Boaz found that R′ = Ph gave the best results, Chan and coworkers [46] showed that Me or CF_3 substituents in the 3,5-position often had a beneficial effect on *ee* and ligand stability for the Rh/BoPhoz-catalyzed hydrogenation of a variety of enamides (*ee* up to 99.7%) and MAC (*ee* up to 99.5%). Similar results were reported for the hydrogenation of a variety of substituted acetamido acrylates (*ee* 95–99.5%) catalyzed by Rh/**L13** (Ar = Ph, 3,5-$(CF_3)_2$-Ph) [45]. Feasibility studies for the technical preparation of cyclopropylalanine [43d] and 2-naphthylalanine [43e] have been reported (see Scheme 4.5). BoPhoz is also suitable for the Rh-catalyzed hydrogenation of α-keto esters (*ee* 88–92%) and keto pantolactone (*ee* 97%). Recently, it was reported that various β-keto esters are hydrogenated with Ru/BoPhoz complexes with enantioselectivities of 94–95% (s/c 200, rt, 20 bar) [43f]. The addition of dialkyl zinc to an activated imine is catalyzed with high selectivities by a Cu/BoPhoz catalyst [47].

L14, a P-chiral analog of BoPhoz, shows significant matched/mismatched effects for the Rh-catalyzed hydrogenation of enamides [48]. On average, **L14** with R = 1-Np and R′ = Ph gives about 2–3% better *ees* and slightly higher activities than the analogous BoPhoz derivative.

L15 and **L16**, analogs with an additional chiral moiety, were independently developed by Zheng [49] and Chan [45]. The ligands were tested for the Rh-catalyzed hydrogenation of various acetamido olefins. Significant matched/mismatched effects were observed for **L15** (best combination R,S_p,R_a), less pronounced for **L16**. **L15a** is the ligand of choice for the hydrogenation of a variety of dehydro β-amino acid derivatives (*ee* 96 – >99%, s/c up to 5000); *E*- and *Z*- substrates give the opposite enantiomer with similar stereoselectivity [49a]. **L15b** is very effective for the hydrogenation of substituted aryl enamides (*ee* 98–99.6%, s/c up to 5000), for DMI and

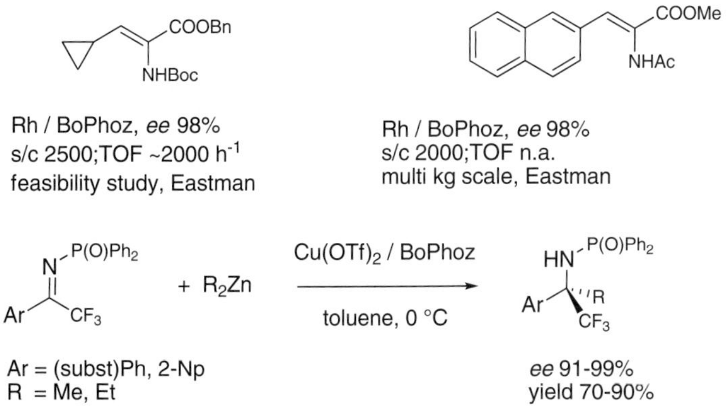

4.5 Type D, One PR₂ Group Attached to the Cp Ring

Scheme 4.5 Synthetic applications of BoPhoz.

MAC (*ee* 99–99.9%, s/c up to 10 000) [49b] as well as ACA (*ee* 99.9%). Finally, **L16** achieved *ee*s of 97–99.6% for the hydrogenation of various substituted MAC derivatives [45].

Bifep (an analog of binap where the naphthalene unit is replaced by a ferrocene) [50] and **L18** [51] exhibit only planar chirality while **L17** is a P-chiral Bifep [52]. The structure and the expected cis-coordination of Bifep was confirmed by the X-ray analysis of a Pd complex of the ligand. Several derivatives were tested in the Ru-catalyzed hydrogenation of ethyl acetyl acetate with good TONs but moderate *ee*s (82% *ee*, s/c 1000, 80 bar, 80 °C) [50]. Ligands **L17** were tested in the standard allylic alkylation and amination reactions of diphenyl allyl acetate with methyl malonate and benzyl amine, respectively. Best results were obtained for Pd/**L17a** which achieved 88% for the alkylation and 93% for the amination reaction [52]. For details of Bifep please see Chapter 10. **L18** proved to be very effective for the Pd-catalyzed reaction of the standard diphenylallyl acetate with dimethyl malonate (*ee* 98%) as well as with potassium amides KNHR (R = Bz, Ts, *ee* 95–97%) [51].

4.5
Type D, One PR₂ Group Attached to the Cp Ring, One PR₂ Group Attached to Other Positions of the Side Chain

The most important Type D ligands (Figure 4.9) with the largest structural variability (R, R′, X, Y can be varied) are the Taniaphos ligands, designed by Knochel [53].

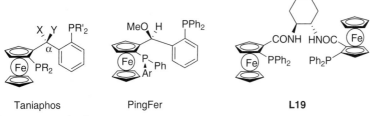

Figure 4.9 Ligands of type **D**.

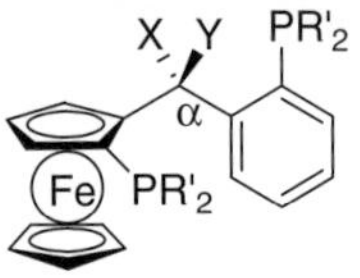

X	Y	
H	NR$_2$	first generation
H	MeO	second generation
H	Alkyl	third generation

	NR$_2$	R'
T001	NMe$_2$	Ph
T002	NMe$_2$	Cy
T003	NMe$_2$	4-MeO-Xyl
T021	NBu$_2$	Cy
T022	morph	Xyl

Figure 4.10 Structure and numbering of selected Taniaphos ligands [54e].

4.5.1
Taniaphos

Compared to Josiphos, the Taniaphos ligands have an additional phenyl ring inserted at the side chain of the Ugi amine. Besides the two phosphine moieties the substituent at the stereogenic center can also be varied and up to now, three generations of Taniaphos ligands with different substituents at the α-position have been prepared (Figure 4.10). Several first generation Taniaphos ligands are being marketed by Solvias in collaboration with Umicore [7b].

A variety of Taniaphos ligands have been shown to be very selective in a number of model hydrogenation reactions [53, 54]. Both the nature of the two phosphine moieties R, R', the substituent (X, Y) and the configuration of the stereogenic center at the α-position have strong effects on the catalytic performance, even the sense of induction. With very few exceptions, relatively electron-rich all-aryl substituted derivatives (R', R' = Ph, Xyl, MeO-Xyl) gave the best performance. All three generations of Taniaphos are highly active and stereoselective for the Rh-catalyzed hydrogenation of MAC (*ee* 94~99.5%) and DMI (*ee* 91–99.5%), and in the Ru-catalyzed hydrogenation of β-ketoesters (*ee* 94~98%), cyclic β-ketoesters (*ee* 85~94%, *de* ~99% of anti-product, s/c up to 25 000) and 1,3-diketones (*ee* 97~99.4%, *de* ~99% of anti-product). For the hydrogenation of enol acetates, *ee*s up to 98% but low activities were achieved. Enamides are hydrogenated with 92–97% *ee*s but low TOFs (Rh/**T021**) and a β-dehydro acetamido acid with 99.5% (Rh/**T003**).

Recently, a variety of highly enantioselective transformations have been described using Taniaphos complexes; selected reactions are depicted in Scheme 4.6. **T001** was found to be very selective for the Rh-catalyzed nucleophilic ring opening of an azabicycle [55] and for the Cu-catalyzed Michael addition of Grignard reagents to cyclohexenone (*ee* 94–96% [21c]). Cu/Taniaphos complexes were also very effective for the reductive addition of aldehydes [56a] and methyl ketones [56b] to methyl acrylate and to allene carboxyl ester [57] (see Scheme 4.6). Cu/**T021** catalyzed the addition of aryl methyl ketones to silylated ketene acetal with *ee*s up to 92% [58] and Cu/**T001** was effective for the cycloaddition of azomethine ylides to vinyl sulfones (Scheme 4.6 [59]) and for the reaction of various allylic bromides with Grignard reagents with 92–97% *ee* [60a, b].

PingFer is a P-chiral version of second generation Taniaphos described by Chen et al. [60c]. As expected significant matched/mismatched effects are observed for the Rh-catalyzed hydrogenation of enamides. On average, PingFer with Ar = 1-Np gives

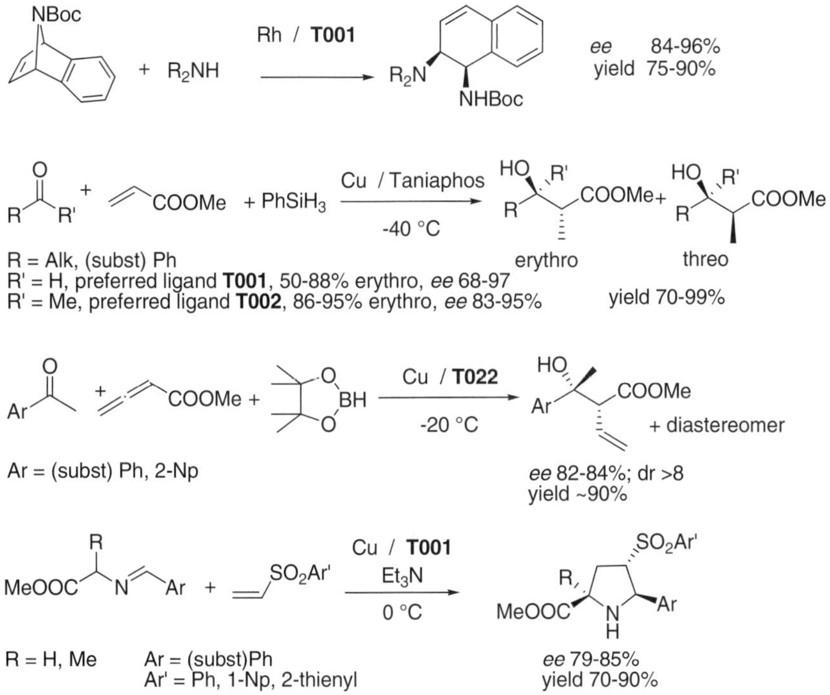

Scheme 4.6 Synthetic applications of Taniaphos ligands.

about 2–3% better *ees* than the corresponding Taniaphos. Best results were reported for MAC (99.6% *ee*) and brominated ACA (99.9% *ee*).

L19 was independently developed by the groups of Zhang [61] and Hou [62]. **L19** has been tested for a variety of Pd-catalyzed allylic alkylation reactions with moderate to very high enantioselectivities. Reactions of synthetic interest are shown in Scheme 4.7. Pd/**L19** catalyzed the allylation of methyl tetralone [62b] and the kinetic resolution of cyclohexenyl acetate [61a] with high selectivity. The Ag(I)/**L19** complex was shown to be effective in the 1,3-dipolar cycloaddition of azomethine ylides with dimethylmaleate, leading to complete endo-selectivity and excellent enantiocontrol [61b].

4.6
Type E, Both PR$_2$ Groups Attached to Side Chains

Up to now, only very few 1,2-diphosphine ligands of Type **E** (Figure 4.11) have been synthesized, probably because it was thought that the chelate size would be too large to give effective control of the stereoinduction. The success of the Walphos ligands, developed by Walter Weissensteiner [63] proved this concern to be wrong.

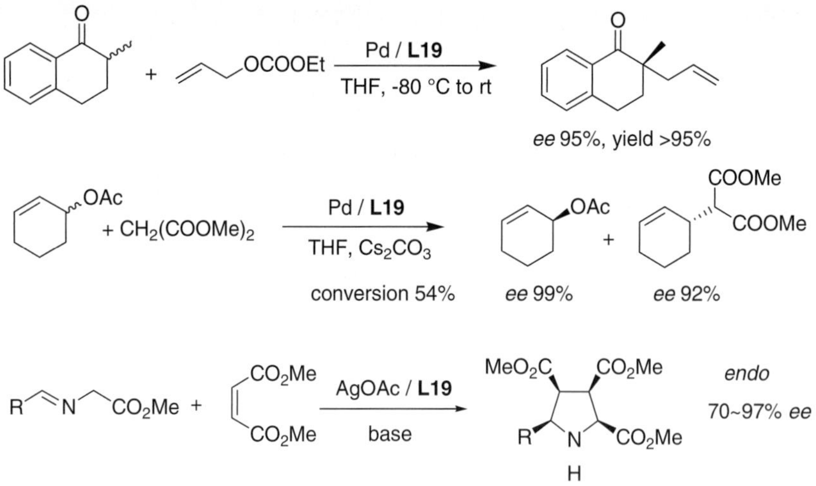

Scheme 4.7 Synthetic applications of ligand **L19**.

4.6.1
Walphos

Like Josiphos, Walphos ligands are modular but form 8-membered metallocycles due to the additional phenyl ring attached to the cyclopentadiene ring. There are noticeable electronic effects but the scope of this ligand family is still under investigation; several derivatives are available from Solvias on a technical scale (Figure 4.12 [7b]).

Walphos ligands show promise for various enantioselective hydrogenations. Rh/Walphos catalysts gave good results for dehydro amino and itaconic acid

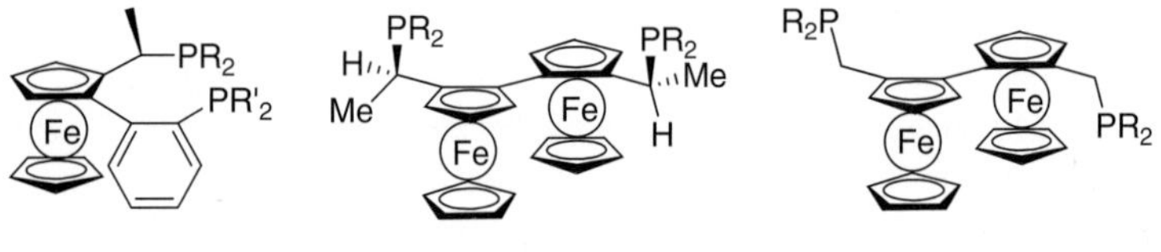

Walphos TRAP TRAP-H

Figure 4.11 Ligands of type **E**.

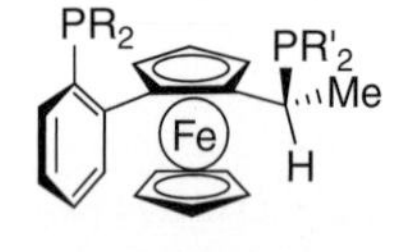

(*R*, R_p)-**Walphos**

	R	R'
W001	Ph	3,5-$(CF_3)_2$-Ph
W002	Ph	Ph
W003	Ph	Cy
W005	4-MeO-Xyl	3,5-$(CF_3)_2$-Ph
W006	Ph	Xyl
W008	Cy	3,5-$(CF_3)_2$-Ph
W009	Xyl	Xyl

Figure 4.12 Important Walphos derivatives.

derivatives (*ee* 92–95%, preferred ligands **W001**, **W002**, **W004**, **W006** and **W009** [39, 64]) and of vinyl boronates (see Scheme 4.8 [64, 65]). Ru/Walphos complexes were highly selective for the hydrogenation of β-keto esters (*ee* 91–95%, **W001**, **W002**, **W003** and **W005**) and acetyl acetone (*ee* >99.5%, s/c 1000, **W001** [39, 64]). Interestingly, it was claimed that a Ru$_4$–Walphos cluster is able to hydrogenate a number of αβ-unsaturated acids with enantioselectivities of 68–93% [66a]. The evidence for cluster catalysis is quite convincing but not unequivocal.

The first industrial application has just been realized in collaboration with Speedel/Novartis for the hydrogenation of SPP100-SyA, a sterically demanding αβ unsaturated acid intermediate of the renin inhibitor SPP100 [67](see Figure 4.13).

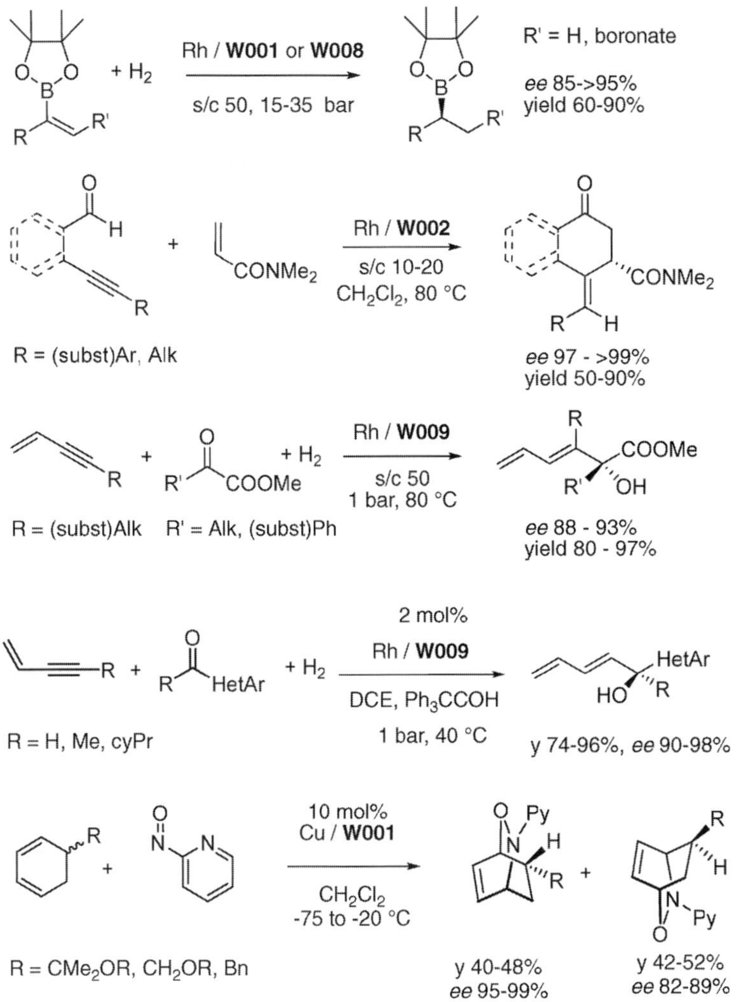

Scheme 4.8 Synthetic applications of Walphos ligands.

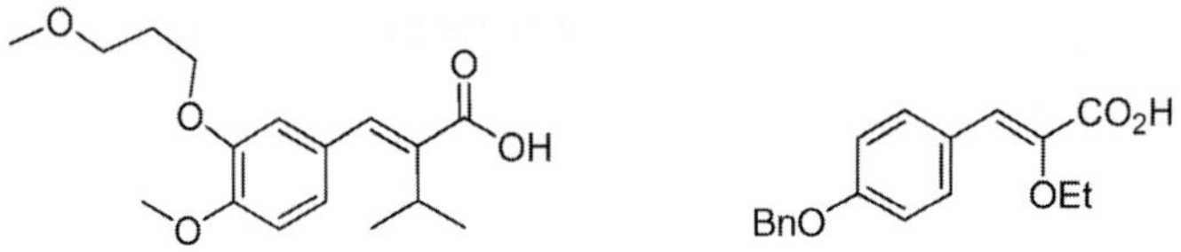

Rh / Walphos **W001**; ee 95%
TON 5000; TOF ~800 h⁻¹
medium scale production
(for renin inhibitor SPP 100)
Novartis / Solvias [68]

Rh / **W001**; ee 92%
TON ~300, TOF ~20 h⁻¹
feasibility study
(for PPAR agonist)
Lilly [69]

Figure 4.13 Industrial application of Walphos ligands.

The process has already been operated on a multi-ton scale. Lilly's chemists developed a process of a PPAR agonist using the Rh-catalyzed asymmetric hydrogenation of (Z)-cinnamic acid as the key step. A screen of over 250 catalysts and conditions revealed Rh/**W001** as the most effective ligand, giving the product in 92% ee [68].

Scheme 4.8 depicts a series of other reactions catalyzed by Walphos complexes. The Cu-catalyzed enantioselective reduction of αβ-unsaturated ketones with PMHS was carried out with 92–95% ee (s/c 100, −78 °C), preferred ligand **W001** [17a]. Several novel transformations were reported to be catalyzed by Rh/Walphos complexes with high enantioselectivities, namely the [4 + 2]-cycloaddition of 4-alkynals with an acryl amide by Tanaka and coworkers [69], the reductive coupling of enynes with α-keto esters [70a] and with heteroaromatic aldehydes and ketones [70b] by the Krische group and the very rare case of a divergent kinetic resolution of a Cu/Walphos-catalyzed nitroso Diels–Alder reaction to give potential intermediates for the synthesis of aza sugar analogs [71].

4.6.2
TRAP

The TRAP (trans-chelating phosphines) ligands developed by Ito and coworkers [72] form 9-membered metallocycles. The X-ray structures of several metal complexes have been determined and show that the major isomer has indeed trans-configuration. However, NMR experiments have shown that cis-isomers are also present.

Until now only a few different PR_2 fragments have been tested, but it is clear that the choice of R strongly affects the level of enantioselectivity and sometimes even the sense of induction. The Rh/TRAP-catalyzed hydrogenation of MAA, MAC and itaconates leads to enantioselectivities of 92–96% if carried out at very low pressures of 0.5–1 bar [73a–c]. The best ligands were Ph, Et, n-Pr and i-Bu derivatives. A number of difficult substrates, depicted in Figure 4.14, such as NAc-indole derivatives [73e,f] (the first examples of heteroaromatic substrates with high ees), protected β-hydroxy-α-amino and α,β-diamino acid derivatives [73a–c] and an indinavir intermediate [73d] are effectively hydrogenated by Rh/TRAP complexes. A number of N-Boc-indole [73h] and NTs-indole [73g], derivatives were hydrogenated with 87–96% ee

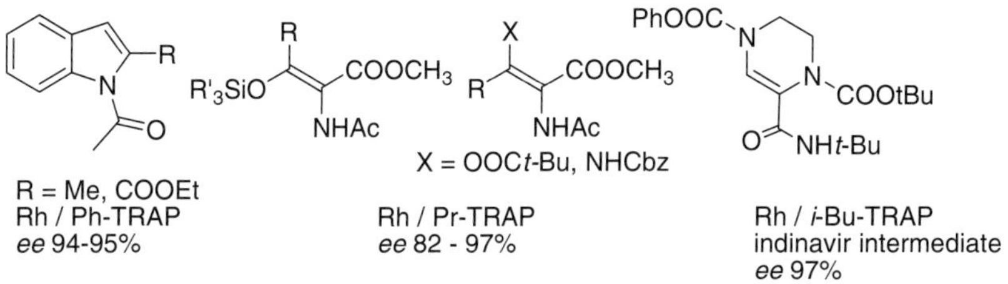

Figure 4.14 Various hydrogenations with Rh/TRAP complexes.

using Ru/Ph-TRAP catalysts. Rh/Alkyl-TRAP complexes were shown to be effective catalysts for the hydrosilylation of substituted acetophenones (ee 80–92%), acetylferrocene (ee 97%), α-disubstituted β-keto esters (ee 98%) and various diketones (ee 89–99%) [74]. Rh/Ph-TRAP complexes catalyzed various Michael additions with up to 93% ee [75] as well as the aldol condensation of 2-cyanopropionates with aldehydes in high yields and up to 93% ee [76]. TRAP was also effective for the Pd-catalyzed allylation of cyanoesters [77] and cycloisomerization of 1,6-enynes [78].

TRAP-H lacks the stereogenic centers in the side chains. Only one ligand of this family, EtTRAP-H, has been reported so far [79]. The ligand was tested in the Rh-catalyzed hydrogenation of various α-acetamido acrylates but ees were much lower than with the analogous TRAP ligands. The sense of induction was controlled by the planar chirality. Better success was obtained for the Rh-catalyzed hydrosilylation of a variety of aryl and alkyl methyl ketones with Ar_2SiH_2. Best enantioselectivities were reported for acetophenone (94% ee at −40 °C).

References

1 For an account see Hayashi, T. (1995) *Ferrocenes* (eds A. Togni and T. Hayashi) VCH, Weinheim, p. 105.

2 (a) Colacot, T.J. (2003) *Chem. Rev.*, **103**, 3101; (b) Dai, L.-X., Tu, T., You, S.-L., Deng, W.-P., and Hou, X.-L. (2003) *Acc. Chem. Res.*, **36**, 659; (c) Barbaro, P., Bianchini, C., Giambastiani, G., and Parisel, S.L. (2004) *Coord. Chem. Rev.*, **248**, 2131; (d) Atkinson, R.C.J., Gibson, V.C., and Long, N.J. (2004) *Chem. Soc. Rev.*, **33**, 313; (e) Arrayás, R.G., Adrio, J., and Carretero, J.C. (2006) *Angew. Chem. Int. Ed.*, **45**, 7674.

3 Chen, W. and Blaser, H.U. (2008) *Trivalent Phosphorus Compounds in Asymmetric Catalysis: Synthesis and Applications* (ed. A. Börner) Wiley-VCH Verlag GmbH, Weinheim, p. 345 and 359.

4 Argouarch, G., Samuel, O., Riant, O., Daran, J.-C., and Kagan, H.B. (2000) *Eur. J. Org. Chem.*, 2893.

5 Togni, A., Breutel, C., Schnyder, A., Spindler, F., Landert, H., and Tijani, A. (1994) *J. Am. Chem. Soc.*, **116**, 4062.

6 For a scalable synthesis of Ugi amine see Blaser, H.U., Buser, H.P., Coers, K., Hanreich, R., Jalett, H.P., Jelsch, E., Pugin, B., Schneider, H.D., Spindler, F., and Wegmann, A. (1999) *Chimia*, **53**, 275.

7 (a) Blaser, H.U., Brieden, W., Pugin, B., Spindler, F., Studer, M., and Togni, A. (2002) *Top. Catal.*, **19**, 3; For more information see (b) Thommen, M. and Blaser, H.U.(July/August (2002)) *PharmaChem*, 33 and http://www.solvias.com/english/

products-and-services/chemicals/ligands/chiral-ligand-kit.html.

8 (a) Togni, A., Breutel, C., Soares, M.C., Zanetti, N., Gerfin, T., Gramlich, V., Spindler, F., and Rihs, G. (1994) *Inorg. Chim. Acta*, **222**, 213; (b) Gambs, C., Consiglio, G., and Togni, A. (2001) *Helv. Chim. Acta*, **84**, 3105.

9 Blaser, H.U., Spindler, F., and Thommen, M. (2007) *Handbook of Homogeneous Hydrogenation* (eds J.G. de Vries and C.J. Elsevier), Wiley-VCH Verlag GmbH, Weinheim, p. 1279.

10 Blaser, H.U. (2002) *Adv. Synth. Catal.*, **344**, 17.

11 (a) Rouhi, M. (2004) *C&EN*, **82** (37), 28; (b) Hsiao, Y., Rivera, N.R., Rosner, T., Krska, S.W., Njolito, E., Wang, F., Sun, Y., Armstrong, J.D., Grabowski, E.J.J., Tillyer, R.D., Spindler, F., and Malan, C. (2004) *J. Am. Chem. Soc.*, **126**, 9918; (c) Kubryk, M. and Hansen, K.B. (2006) *Tetrahedron: Asymmetry*, **17**, 205; (d) Clausen, A.M., Dziadul, B., Cappuccio, K.L., Kaba, M., Starbuck, C., Hsiao, Y., and Dowling, T.M. (2006) *Org. Proc. Res. Devel.*, **10**, 723.

12 Ikemoto, T., Nagata, T., Yamano, M., Ito, T., Mizuno, Y., and Tomimatsu, K. (2004) *Tetrahedron Lett.*, **45**, 7757.

13 Shultz, C.S., Dreher, S.D., Ikemoto, N., Williams, J.M., Grabowski, E.J.J., Krska, S.W., Sun, Y., Dormer, P.G., and DiMichele, L. (2005) *Org. Lett.*, **7**, 3405.

14 Allwein, S.P., McWilliams, J.C., Secord, E.A., Mowrey, D.R., Nelson, T.D., and Kress, M.H. (2006) *Tetrahedron Lett.*, **47**, 6409.

15 Baratta, W., Ballico, M., Del Zotto, A., Siega, K., Magnolia, S., and Rigo, P. (2008) *Chem. Eur. J.*, **14**, 2557; Baratta, W., Cheluccci, G., Herdtweck, E., Magnolia, S., Siega, K., and Rigo, P. (2007) *Angew. Chem. Int. Ed.*, **46**, 7651.

16 (a) Czekelius, C. and Carreira, E.M. (2003) *Angew. Chem. Int. Ed.*, **42**, 4793; (b) Czekelius, C. and Carreira, E.M. (2004) *Org. Lett.*, **6**, 4575.

17 (a) Lipshutz, B.H. and Servesko, J.M. (2003) *Angew. Chem. Int. Ed.*, **42**, 4789; (b) Lipshutz, B.H., Servesko, J.M., and Taft, B.R. (2004) *J. Am. Chem. Soc.*, **126**, 8352; (c) Lipshutz, B.H., Tanaka, N., Taft, B.R., and Lee, C.-T. (2006) *Org. Lett.*, **8**, 1963; For a synthetic appliaction seeLipshutz, B.H., Lee, C.-T., and Serversko, J.M. (2007) *Org. Lett.*, **9**, 4713.

18 Lee, D., Kim, D., and Yun, J. (2006) *Angew. Chem. Int. Ed.*, **45**, 2785.

19 Andresen, B.M., Caron, S., Couturier, M., DeVries, K.M., Do, N.M., Dupont, K., Gosh, A., Girardin, M., Hawkins, J.M., Makowski, T.M., Riou, M., Sieser, J.E., Tucker, J.L., Vanderplas, B.C., and Watson, T.J.N. (2006) *Chimia*, **60**, 554.

20 Makino, K., Fijii, T., and Hamada, Y. (2006) *Tetrahedron: Asymmetry*, **17**, 481.

21 (a) Lopez, F., Harutyunyan, S.R., Meetsma, A., Minnaard, A.J., and Feringa, B.L. (2005) *Angew. Chem. Int. Ed.*, **44**, 2752; (b) Des Mazery, R., Pullez, M., Lopez, F., Harutyunyan, S.R., Minnaard, A.J., and Feringa, B.L. (2005) *J. Am. Chem. Soc.*, **127**, 9966; (c) Howell, G.P., Fletcher, S.P., Geurts, K., ter Horst, B., and Feringa, B.L. (2006) *J. Am. Chem. Soc.*, **128**, 14977; (d) Feringa, B.L., Badorrey, R., Pena, D., Harutyunyan, S.R., and Minnaard, A.J. (2004) *Proc. N. Y. Acad. Sci.*, **101**, 5834; (e) den Hartog, T., Harutyunyan, S.R., Font, D., Minnaard, A.J., and Feringa, B.L. (2008) *Angew. Chem. Int. Ed.*, **47**, 398.

22 Pucheault, M., Darses, S., and Genêt, J.-P. (2002) *Eur. J. Org. Chem.*, 3552.

23 (a) Lautens, M., Fagnou, K., and Hiebert, S. (2003) *Acc. Chem. Res.*, **36**, 48; (b) Lautens, M. and Fagnou, K. (2004) *Proc. N. Y. Acad. Sci.*, **101**, 5455; (c) Solvias AG unpublished results.

24 Bercot, E.A. and Rovis, T. (2004) *J. Am. Chem. Soc.*, **126**, 10248.

25 Leitner, A., Larsen, J., Steffen, C., and Hartwig, J.F. (2004) *J. Org. Chem.*, **69**, 7552.

26 Lormann, M.E.P., Nieger, M., and Bräse, S. (2006) *J. Organomet. Chem.*, **691**, 2159.

27 Goddard, C., Ruiz, A., and Claver, C. (2006) *Helv. Chim. Acta*, **89**, 161.

28 Xu, G. and Han, L.-B. (2006) *Org. Lett.*, **8**, 2099.
29 Yamamoto, Y., Takada, S., and Miyaura, N. (2006) *Chem. Lett.*, **35**, 1368.
30 (a) Shen, Q. and Hartwig, J.F. (2006) *J. Am. Chem. Soc.*, **128**, 10028; (b) Fernandez-Rodriguez, M.A., Shen, Q., and Hartwig, J.F. (2006) *J. Am. Chem. Soc.*, **128**, 2180; (c) Limmert, M.E., Roy, A.H., and Hartwig, J.F. (2005) *J. Org. Chem.*, **70**, 9364.
31 Cai, C., Rivera, N.R., Balsells, J., Sidler, R.R., McWilliam, J.C., Shultz, C.S., and Sun, Y. (2006) *Org. Lett.*, **8**, 5161.
32 Feng, X., Pugin, B., Küsters, E., Sedelmeier, G., and Blaser, H.U. (2007) *Adv. Synth. Catal.*, **349**, 1803.
33 Pugin, B., Landert, H., Spindler, F., and Blaser, H.-U. (2002) *Adv. Synth. Catal.*, **344**, 974.
34 Köllner, C., Pugin, B., and Togni, A. (1998) *J. Am. Chem. Soc.*, **120**, 10274.
35 Hems, W.P., McMorn, P., Riddel, S., Watson, S., Hancock, F.E., and Hutchings, G.J. (2005) *Org. Biomol. Chem.*, **3**, 1547.
36 Stephenson, P., Kondor, B., Licence, P., Scovell, K., Ross, S.K., and Poliakoff, M. (2006) *Adv. Synth. Catal.*, **348**, 1605.
37 Argouarch, G., Samuel, O., and Kagan, H.B. (2000) *Eur. J. Org. Chem.*, 2885.
38 Yasuike, S., Kofink, C.C., Kloetzing, R.J., Gommermann, N., Tappe, K., Gavryushin, A., and Knochel, P. (2005) *Tetrahedron: Asymmetry*, **16**, 3385.
39 Solvias AG unpublished screening results.
40 Sturm, T., Weissensteiner, W., Spindler, F., Mereiter, K., Lopez-Agenjo, A.M., Manzano, B.R., and Jalon, F.A. (2002) *Organometallics*, **21**, 1766.
41 (a) Cayuela, E.M., Xiao, L., Sturm, T., Manzano, B.R., Jalon, F.A., and Weissensteiner, W. (2000) *Tetrahedron: Asymmetry*, **11**, 861; (b) Mernyi, A., Kratky, C., Weissensteiner, W., and Widhalm, M. (1996) *J. Organomet. Chem.*, **508**, 209.
42 Sturm, T., Abad, B., Weissensteiner, W., Mereiter, K., Manzano, B.R., and Jalon, F.A. (2006) *J. Mol. Catal. A: General*, **255**, 209.
43 (a) Boaz, N.W., Debenham, D.D., Mackenzie, E.B., and Large, S.E. (2002) *Org. Lett.*, **4**, 2421; (b) Boaz, N.W., Mackenzie, E.B., Debenham, S.D., Large, S.E., and Ponasik, J.A. Jr (2005) *J. Org. Chem.*, **70**, 1872; (c) Boaz, N.W., Ponasik, J.A., and Large, S.E. (2005) *Tetrahedron: Asymmetry*, **16**, 2063; (d) Boaz, N.W., Debenham, S.D., Large, S.E., and Moore, M.K. (2003) *Tetrahedron: Asymmetry*, **14**, 3575; (e) Boaz, N.W., Large, S.E., Posniak, J.A., Moore, M.K., Barnett, T., and Nottingham, W.D. (2005) *Org. Proc. Res. Dev.*, **9**, 472; (f) Boaz, N.W., Ponasik, J.A., and Large, S.E. (2006) *Tetrahedron Lett.*, **47**, 4033.
44 http://www.jmcatalysts.com/pct/producttype.asp?producttypeid=265.
45 Jia, X., Li, X., Lam, W.S., Kok, S.H.L., Xu, L., Lu, G., Yeung, C., and Chan, A.S.C. (2004) *Tetrahedron: Asymmetry*, **15**, 2273.
46 Li, X., Jia, X., Xu, L., Kok, S.H.L., Yip, C.W., and Chan, A.S.C. (2005) *Adv. Synth. Catal.*, **347**, 1904.
47 Lauzon, C. and Charette, A.B. (2006) *Org. Lett.*, **8**, 2734.
48 (a) Chen, W., Mbafor, W., Roberts, S.M., and Whittall, J. (2006) *J. Am. Chem. Soc.*, **128**, 3922; (b) Chen, W., McCormack, P.J., Mohammed, K., Mbafor, W., Roberts, S.M., and Whittall, J. (2007) *Angew. Chem Int. Ed.*, **46**, 4141.
49 (a) Hu, X. and Zheng, Z. (2005) *Org. Lett.*, **7**, 419; (b) Hu, X. and Zheng, Z. (2004) *Org. Lett.*, **6**, 3585.
50 Xiao, L., Mereiter, K., Spindler, F., and Weissensteiner, W. (2001) *Tetrahedron: Asymmetry*, **12**, 1105.
51 Lotz, M., Kramer, G., and Knochel, P. (2002) *Chem. Comm.*, 2546.
52 (a) Nettekoven, U., Widhalm, M., Kamer, P.C.J., van Leeuwen, P.W.N.M., Mereiter, K., Lutz, M., and Spek, A.L. (2000) *Organometallics*, **19**, 2299; (b) Nettekoven, U., Widhalm, M., Kalchhauser, H., Kamer, P.C.J., van Leeuwen, P.W.N.M., Lutz, M., and Spek, A.L. (2001) *J. Org. Chem.*, **66**, 759.

53 (a) Ireland, T., Grossheimann, G., Wieser-Jeunesse, C., and Knochel, P. (1999) *Angew. Chem. Int. Ed.*, **38**, 3212; (b) Ireland, T., Tappe, K., Grossheimann, G., and Knochel, P. (2002) *Chem. Eur. J.*, **8**, 843; (c) Lotz, M., Polborn, K., and Knochel, P. (2002) *Angew. Chem. Int. Ed.*, **41**, 4708; (d) Tappe, K. and Knochel, P. (2004) *Tetrahedron: Asymmetry* **15**, 91. It has to be pointed out that the (R,S_p) stereochemistry for the first generation Taniaphos (as published in a, b, d) had to be corrected to (R,R_p), also see (e) Fukuzawa, S., Yamamoto, M., Hosaka, M., and Kikuchi, S., (2007) *Eur. J. Org. Chem.*, 5540.

54 Spindler, F., Malan, C., Lotz, M., Kesselgruber, M., Pittelkow, U., Rivas-Nass, A., Briel, O., and Blaser, H.U. (2004) *Tetrahedron: Asymmetry*, **15**, 2299.

55 Lautens, M., Fagnou, K., and Zunic, V. (2002) *Org. Lett.*, **4**, 3465.

56 (a) Chuzel, O., Deschamp, J., Chausteur, C., and Riant, O. (2006) *Org. Lett.*, **8**, 5943; (b) Deschamp, J., Chuzel, O., Hannedouche, J., and Riant, O. (2006) *Angew. Chem. Int. Ed.*, **45**, 1292.

57 Zhao, D., Oisaki, K., Kanai, M., and Shibasaki, M. (2006) *J. Am. Chem. Soc.*, **128**, 14440.

58 Oisaki, K., Zhao, D., Kanai, M., and Shibasaki, M. (2006) *J. Am. Chem. Soc.*, **128**, 7164.

59 Llamas, T., Gomez Arrayas, R., and Carretero, J.C. (2006) *Org. Lett.*, **8**, 1795.

60 (a) Lopez, F., van Zijl, A.W., Minnaard, A.J., and Feringa, B.L. (2006) *Chem. Comm.*, 409; (b) Geurts, K., Fletcher, S.P., and Feringa, B.L. (2006) *J. Am. Chem. Soc.*, **128**, 15572; (c) Chen, W., Roberts, S.M., Whittall, J., and Steiner, A. (2006) *Chem. Comm.*, 2916.

61 (a) Longmire, J.M., Wang, B., and Zhang, X. (2000) *Tetrahedron Lett.*, **41**, 5435; (b) Longmire, J.M., Wang, B., and Zhang, X. (2002) *J. Am. Chem. Soc.*, **124**, 13400.

62 (a) You, S.-L., Hou, X.-L., Dai, L.-X., Cao, B.-X., and Sun, J. (2000) *Chem. Comm.*, 1933; (b) You, S.-L., Hou, X.-L., and Zhu, X.-Z. (2001) *Org. Lett.*, **3**, 149; (c) You, S.-L., Zhu, X.-Z., Luo, Y.M., Hou, X.L., and Dai, L.-X. (2001) *J. Organomet. Chem.*, **637–639**, 762.

63 Sturm, T., Xiao, L., and Weissensteiner, W. (2001) *Chimia*, **55**, 688.

64 Morgan, J.B. and Morken, J.P. (2004) *J. Am. Chem. Soc.*, **126**, 15338.

65 Moran, W.J. and Morken, J.P. (2006) *Org. Lett.*, **6**, 2413.

66 Moberg, V., Haukka, M., Koshevoy, I.O., Ortiz, R., and Nordlander, E. (2007) *Organometallics*, **26**, 4090.

67 Sturm, T., Weissensteiner, W., and Spindler, F. (2003) *Adv. Synth. Catal.*, **345**, 160.

68 Houpis, I.N., Patterson, L.E., Alt, C.A., Rizzo, J.R., Zhang, T.Y., and Haurez, M. (2005) *Org. Lett.*, **7**, 1947.

69 Tanaka, K., Hagiwara, Y., and Noguchi, K. (2005) *Angew. Chem. Int. Ed.*, **44**, 7260.

70 (a) Kong, J.-R., Ngai, M.-Y., and Krische, M.J. (2006) *J. Am. Chem. Soc.*, **128**, 718; (b) Komanduri, V. and Krische, M.J. (2006) *J. Am. Chem. Soc.*, **128**, 16448.

71 Jana, C.K. and Studer, A. (2007) *Angew. Chem. Int. Ed.*, **46**, 6542–6544.

72 (a) Sawamura, M., Hamashima, H., and Ito, Y. (1991) *Tetrahedron: Asymmetry*, **2**, 593; (b) Sawamura, M., Hamashima, H., Sugawara, M., Kuwano, R., and Ito, Y. (1995) *Organometallics*, **14**, 4549; (c) Kuwano, R., Sawamura, M., Okuda, S., Asai, T., Ito, Y., Redon, M., and Krief, A. (1997) *Bull. Chem. Soc. Jpn.*, **70**, 2807.

73 (a) Kuwano, R., Okuda, S., and Ito, Y. (1998) *Tetrahedron: Asymmetry*, **9**, 2773; (b) Kuwano, R., Okuda, S., and Ito, Y. (1998) *J. Org. Chem.*, **63**, 3499; (c) Kuwano, R., Sawamura, M., and Ito, Y. (2000) *Bull. Chem. Soc. Jpn.*, **73**, 2571; (d) Kuwano, R. and Ito, Y. (1999) *J. Org. Chem.*, **64**, 1232; (e) Kuwano, R., Sato, K., Kurokawa, T., Karube, D., and Ito, Y. (2000) *J. Am. Chem. Soc.*, **122**, 7614; (f) Kuwano, R., Kashiwabara, M., Sato, K., Ito, T., Kaneda, K., and Ito, Y. (2006) *Tetrahedron: Asymmetry*, **17**, 521; (g) Kuwano, R., Kaneda, K., Ito, T., Sato, K., Kurokawa, T., and Ito, Y. (2004) *Org. Lett.*, **6**, 2213;

(h) Kuwano, R. and Kashiwabara, M. (2006) *Org. Lett.*, **8**, 2653.

74 (a) Sawamura, M., Kuwano, R., and Ito, Y. (1994) *Angew. Chem. Int. Ed.*, **33**, 111; (b) Kuwano, R., Sawamura, M., Shirai, J., Takahashi, M., and Ito, Y., (2000) *Bull. Chem. Soc. Jpn.*, **73**, 485.

75 (a) Sawamura, M., Hamashima, H., and Ito, Y. (1992) *J. Am. Chem. Soc.*, **114**, 8295; (b) Sawamura, M., Hamashima, H., and Ito, Y. (1994) *Tetrahedron*, **50**, 4439; (c) Sawamura, M., Hamashima, H., Shinoto, H., and Ito, Y. (1995) *Tetrahedron Lett.*, **36**, 6479; (d) Sawamura, M., Hamashima, H., and Ito, Y. (2000) *Bull. Chem. Soc. Jpn.*, **73**, 2559.

76 (a) Kuwano, R., Miyazaki, H., and Ito, Y. (1998) *Chem. Comm.*, 71; (b) Kuwano, R., Miyazaki, H., and Ito, Y. (2000) *J. Organomet. Chem.*, **603**, 18.

77 Sawamura, M., Sudoh, M., and Ito, Y. (1996) *J. Am. Chem. Soc.*, **118**, 3309.

78 Goeke, A., Sawamura, M., Kuwano, R., and Ito, Y. (1996) *Angew. Chem. Int. Ed*, **35**, 662.

79 Kuwano, R., Uemura, T., Saitoh, M., and Ito, Y. (2004) *Tetrahedron: Asymmetry*, **15**, 2263.

5
1,2-*P,N*-Bidentate Ferrocenyl Ligands
Yong Gui Zhou and Xue Long Hou

5.1
Introduction

1,2-Bidentate *P,N*-ferrocenes represent a large family of chiral ferrocene ligands. Since Hayashi and Kumada reported the synthesis of *N,N*-dimethyl-1-[2-(diphenylphosphino)ferrocenyl]ethylamine (PPFA), the first 1,2-*P,N*-ferrocene ligand, in 1974 [1], a variety of *P,N*-ferrocene ligands have been designed and synthesized (Figures 5.1 and 5.2). The diverse metal complexes with these ligands showed their unique catalytic activities as well as stereoselectivities in the carbon–carbon bond and carbon–heteroatom bond-forming reactions [2]. Most 1,2-*P,N*-ligands are synthesized by using an ortho-lithiation strategy with Ugi's amine and 2-ferrocenyl-1,3-oxazoline as starting materials. Some representative syntheses are given below (Schemes 5.1 and 5.2) [3]. The presence of TMEDA is critical for diastereoselectivity in the ortho-lithiation step of the oxazoline system [4].

This chapter will focus on the applications of 1,2-*P,N*-ferrocene ligands (Figure 5.1), while ligands **8–11** in Figure 5.2 will be discussed in Chapters 11 and 8, respectively.

5.2
Asymmetric Hydrogenation and Asymmetric Transfer Hydrogenation

5.2.1
Asymmetric Hydrogenation

Catalytic asymmetric hydrogenation is a fascinating area of research in which many highly innovative and imaginative systems have been developed [5]. Although much of the work has been curiosity driven, asymmetric hydrogenation is an important reaction, with applications ranging from small scale synthesis to industrial processes. Many *N,P*-type ligands have been successfully applied in asymmetric hydrogenation,

---

Chiral Ferrocenes in Asymmetric Catalysis. Edited by Li-Xin Dai and Xue-Long Hou
Copyright © 2010 WILEY-VCH Verlag GmbH & Co. KGaA, Weinheim
ISBN: 978-3-527-32280-0

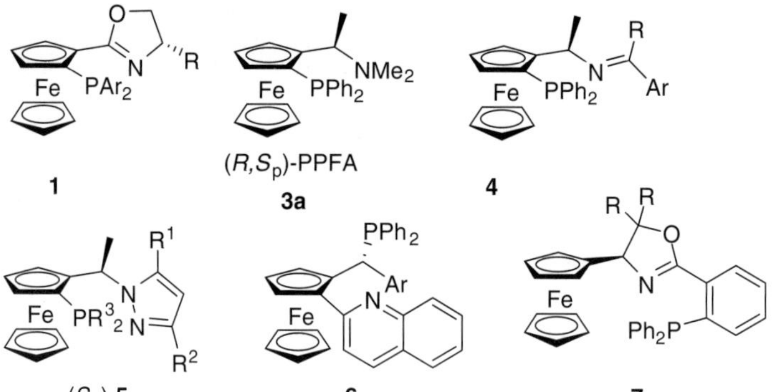

Figure 5.1 Different types of 1,2-P,N-ferrocene ligands.

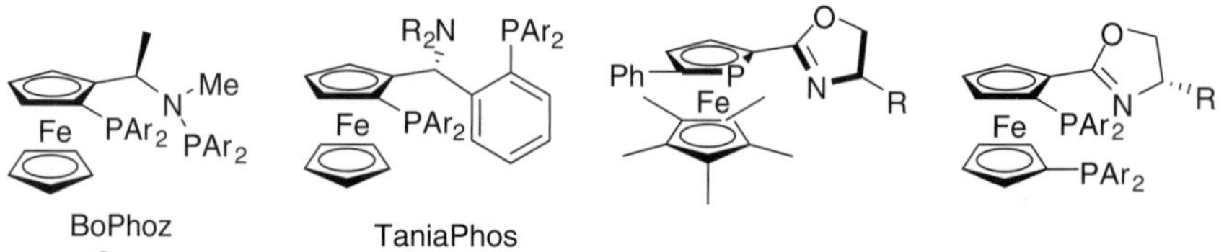

Figure 5.2 Other types of 1,2-P,N-containing ferrocene ligands.

thus showing that these ligands, based on the ferrocene framework, can also be considered as excellent structures for these reactions.

5.2.1.1 Asymmetric Hydrogenation of Alkenes

In 2001, Pfaltz first showed the usefulness of ferrocene-derived phosphinite-oxazoline ligands as N,P-ligands in the Ir-catalyzed asymmetric hydrogenation of unfunctionalized alkenes [6]. These ligands **14** were known as phosphinite-oxazoline ligands, which were readily prepared in three to four steps, starting from imidates or

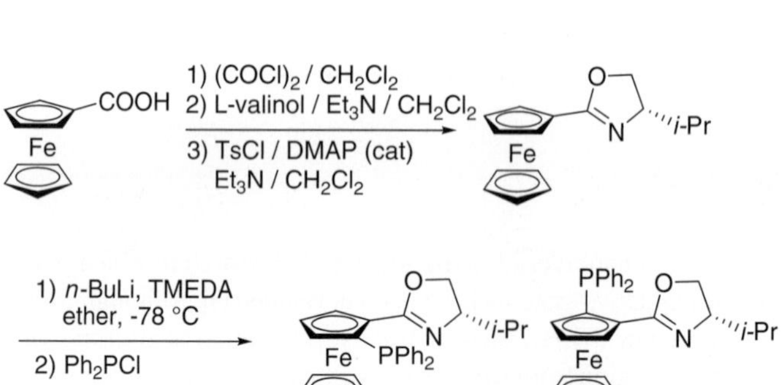

Scheme 5.1 Synthesis of 1,2-P,N-ferrocene ligands via ortho-lithiation.

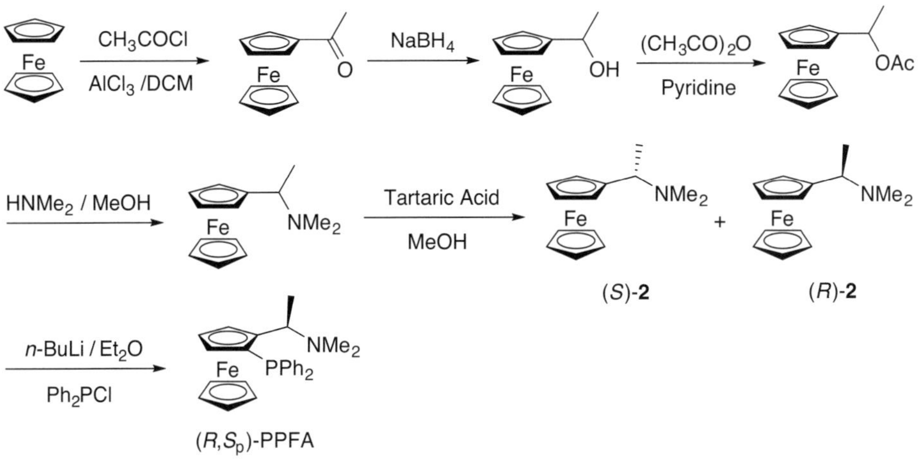

Scheme 5.2 Synthesis of Ugi's amine and chiral PPFA.

carboxylic acids and L-serine methyl ester hydrochloride (Scheme 5.3). Various unfunctionalized alkenes could be efficiently hydrogenated in the presence of catalytic amounts of iridium complexes with ferrocene phosphinite-oxazoline ligands (Scheme 5.4). High reactivities and good to excellent enantioselectivities were obtained.

Kim's group, in 2005, developed chiral (iminophosphoranyl)-ferrocene, **17a** and **17b**, as a new class of practical ligands for rhodium-catalyzed asymmetric hydrogenation of α,β-unsaturated carboxylic acids and α-dehydroamino acid derivatives. These ligands were easily prepared from N,N-dimethyl-1-ferrocenylethylamine with moderate yields [7]. The reaction was conducted at ambient temperature under H_2 pressure of 1 or 10 bar in the presence of 2 mol% of catalysts prepared *in situ* from [Rh(NBD)$_2$]BF$_4$ and 1.1 equiv of chiral ligand. High enantioselectivities (up to 94% *ee*) and catalytic activities were achieved (Scheme 5.5). These results are remarkable. The studies on the Rh/ligand complexes showed that ligand **17c** is a cis-*P,N*-chelation through PPh$_2$ and NMe$_2$.

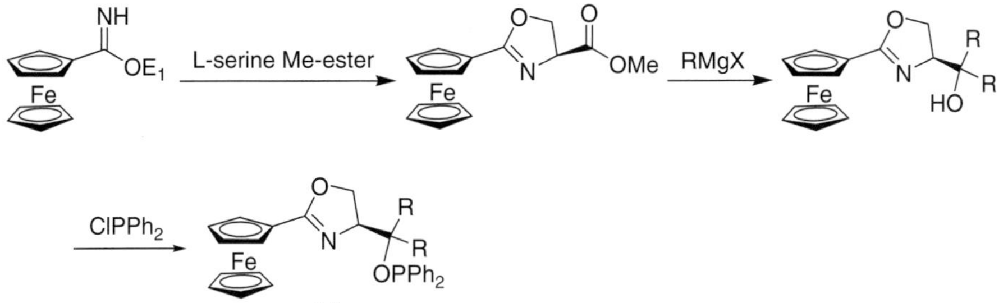

Scheme 5.3 Synthesis of phosphinite-oxazolinyl ferrocene **14**.

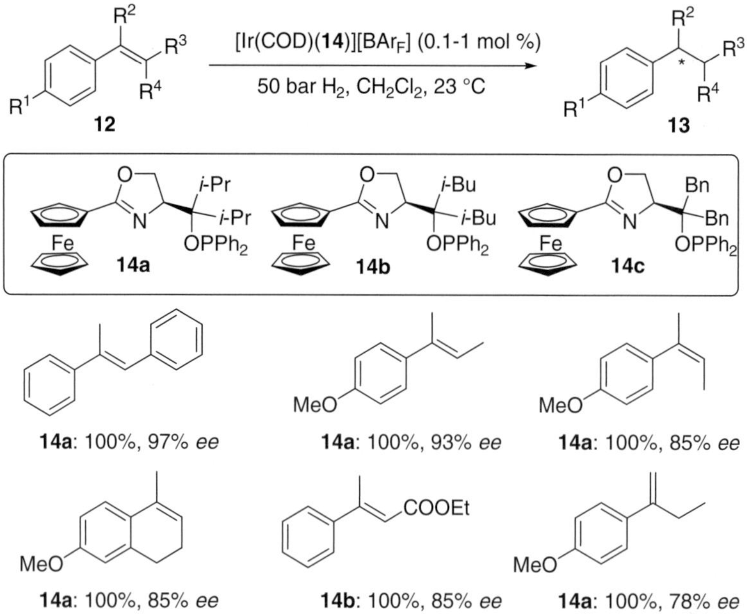

5 1,2-P,N-Bidentate Ferrocenyl Ligands

Scheme 5.4 Hydrogenation of unfunctionalized alkenes with phosphinite-oxazoline ligands.

Following these results, the same group reported, in 2006, the usefulness of the above chiral iminophosphoranylferrocene as *P,N*-ligands in the rhodium and iridium-catalyzed enantioselective hydrogenation of unfunctionalized olefins [8]. Both iridium and rhodium complexes catalyzed hydrogenation of unfunctionalized olefins with moderate to good enantioselectivities (Scheme 5.6).

In 2007, Li and coworkers also reported the asymmetric hydrogenation of unfunctionalized olefins with chiral iridium ferrocenyloxazolinylphosphine (Fc-Phox) complexes. The influence of the relative configuration of the chirality in the oxazoline ring and the planar chirality of the ferrocene ring was studied. The best result was obtained by using **1a** as chiral ligand (Table 5.1). Fc-Phox (S, S_p)- **1a** afforded the best results, while ligands with mis-matched (S, R_p) afforded lower *ee* values.

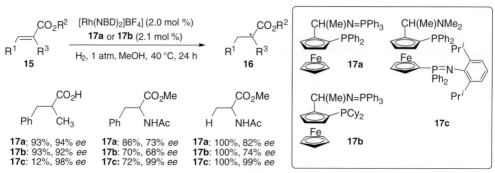

Scheme 5.5 Asymmetric hydrogenation of functionalized alkenes.

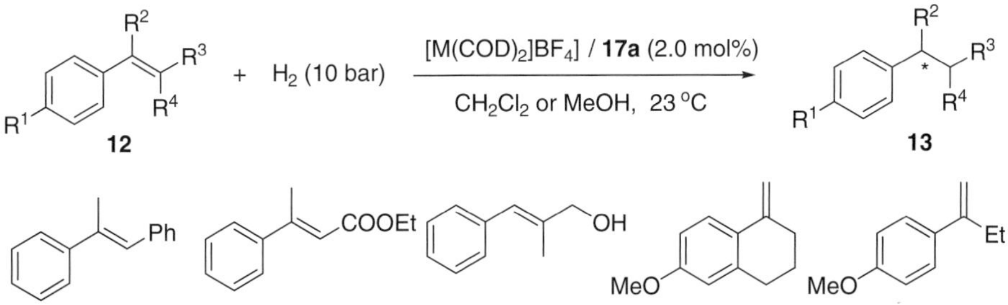

5.2 Asymmetric Hydrogenation and Asymmetric Transfer Hydrogenation

Scheme 5.6 Asymmetric hydrogenation of unfunctionalized alkenes.

Ir: 100%, 74% ee   Ir: 52%, 74% ee   Ir: 88%, 92% ee   Ir: 99%, 48% ee   Ir: 100%, 80% ee
Rh: 98%, 46% ee   Rh: 88%, 46% ee   Rh: 100%, 59% ee   Rh: 78%, 50% ee   Rh: 99%, 69% ee

Other unfunctionalized olefins could also be hydrogenated using complex **1a** as a catalyst to afford the products in high yields and moderate to high enantioselectivity [9].

5.2.1.2 Asymmetric Hydrogenation of Ketones

In 1997, Sammakia and coworkers showed that ruthenium complexes of Fc-phox, chiral (phosphinoferrocenyl) oxazolines, could be used in the transfer hydrogenation of aryl ketones [10]. Subsequently, Naud and coworkers reported, in 2006, the hydrogenation of aryl ketones using the same catalytic systems (Scheme 5.7) [11].

The catalyst showed extremely high reactivity and enantioselectivity for the hydrogenation of aryl ketones with *ee*s up to 99% and substrate to catalyst ratios of 10 000 to 50 000. This was the first system which could be applied to both transfer

Table 5.1 Asymmetric hydrogenation of alkenes with Fc-phox.

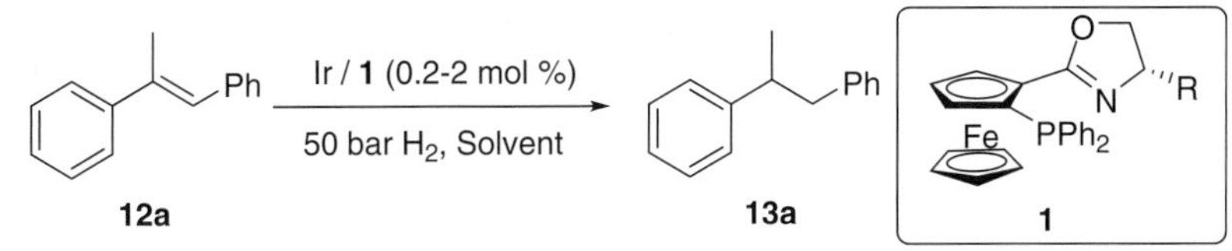

Entry	R	Solvent	Conversion (%)	ee (%)
1	**1a** (Me)	CH_2Cl_2	100	87
2	**1a** (Me)	$(CH_2Cl)_2$	61	85
3	**1a** (Me)	$CHCl_3$	30	ND
4	**1a** (Me)	Et_2O	0	—
5	**1a** (Me)	THF	0	—
6	**1a** (Me)	Toluene	5	ND
7	**1b** (*i*-Pr)	CH_2Cl_2	66	72
8	**1c** (*t*-Bu)	CH_2Cl_2	38	78
9	**1d** (Ph)	CH_2Cl_2	74	62
10	**1e** (Bn)	CH_2Cl_2	100	74

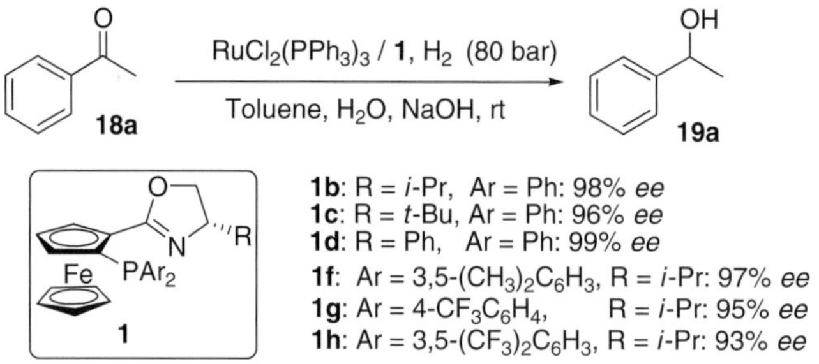
RuCl2(PPh3)3 / 1, H2 (80 bar)
Toluene, H2O, NaOH, rt
18a
19a
1
1b: R = i-Pr, Ar = Ph: 98% ee
1c: R = t-Bu, Ar = Ph: 96% ee
1d: R = Ph, Ar = Ph: 99% ee
1f: Ar = 3,5-(CH3)2C6H3, R = i-Pr: 97% ee
1g: Ar = 4-CF3C6H4, R = i-Pr: 95% ee
1h: Ar = 3,5-(CF3)2C6H3, R = i-Pr: 93% ee

Scheme 5.7 Asymmetric hydrogenation of ketones with Fc-phox.

hydrogenation and hydrogenation of aryl ketones with high *ee* and high efficiency. In 2007, Spindler and coworkers developed a pilot process for the asymmetric hydrogenation of 3,5-bistrifluoromethylacetophenone (BTMA) using a Ru/Fc-phox complex in toluene in the presence of aqueous NaOH. This catalytic system tolerates high substrate concentrations, and was carried out twice on a 140-kg scale at 20 bar of H_2 and at 25 °C and enantioselectivity was up to 95%. After crystallization, the product was obtained with an *ee* between 97.7 and 98.6% in 70% yield (Scheme 5.8) [12]. The product BTMP is an important intermediate for a number of pharmaceutically interesting targets like the NK-1 receptor antagonist. It seems to be the first industrial catalytic process with a ferrocenyl oxazoline ligand. A research group at Merck has also employed Fc-phox ligands in the enantioselective asymmetric hydrogenation of alpha-alkoxy-substituted aryl ketone (Scheme 5.9) [13].

In the same year, Chen's group prepared several ferrocene-based aminophosphine ligands and showed them to be effective in the Ru(II)-catalyzed asymmetric hydrogenation of ketones. The use of chiral ruthenium complexes in the presence of *t*-BuOK led to the desired product with full conversion. However, the enantioselectivity of the product was somewhat low. The enantioselectivity was mainly determined by the carbon-centered chirality of the ligands, but the planar chirality was also important, such that (R,S_p)- or (S,R_p)- were the matched chiralities (Scheme 5.10) [14].

In 2007, Palmer and Nettekoven reported the preparation of tricyclic imidazopyridines by asymmetric ketone hydrogenation of a key intermediate in the presence of ruthenium phosphino-oxazoline catalyst. The hydrogenation reaction proceeded well, with up to 90% *ee* with full conversion and high TOF value. However, the

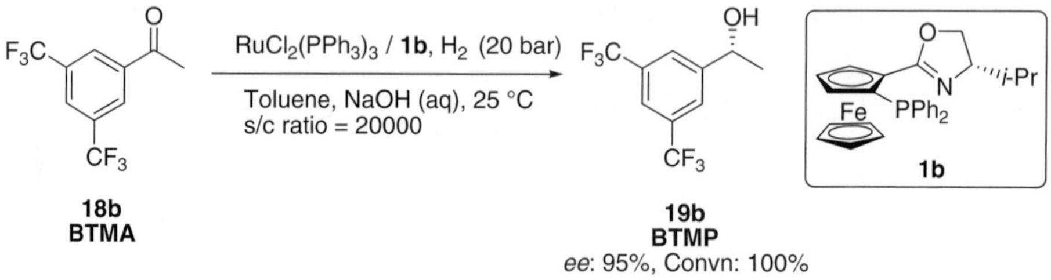
RuCl2(PPh3)3 / 1b, H2 (20 bar)
Toluene, NaOH (aq), 25 °C
s/c ratio = 20000
18b
BTMA
19b
BTMP
ee: 95%, Convn: 100%
1b

Scheme 5.8 Asymmetric hydrogenation of BTMA.

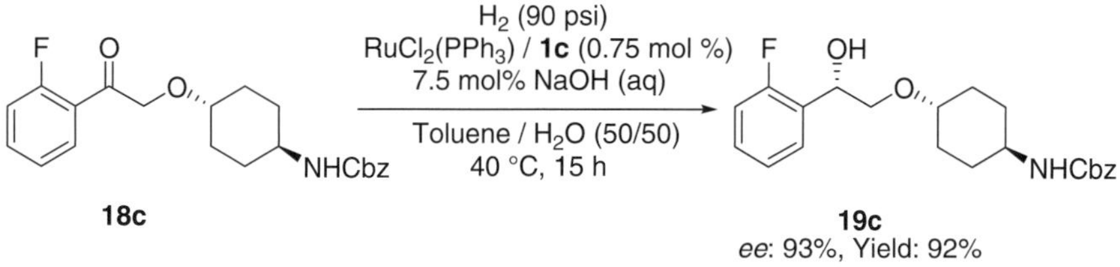

Scheme 5.9 Ru-catalyzed asymmetric hydrogenation of α-alkoxy-ketones.

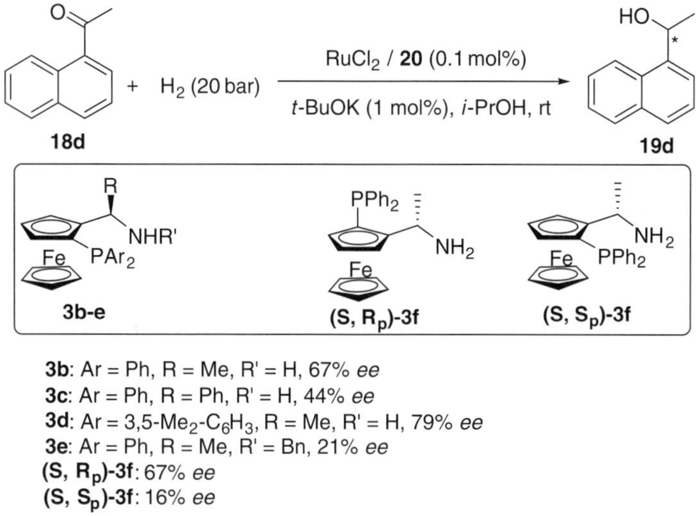

3b: Ar = Ph, R = Me, R' = H, 67% ee
3c: Ar = Ph, R = Ph, R' = H, 44% ee
3d: Ar = 3,5-Me$_2$-C$_6$H$_3$, R = Me, R' = H, 79% ee
3e: Ar = Ph, R = Me, R' = Bn, 21% ee
(S, R$_p$)-3f: 67% ee
(S, S$_p$)-3f: 16% ee

Scheme 5.10 Asymmetric hydrogenation of 1-acetonaphthone.

reaction pressure of hydrogen was somewhat high (80 bar). After deprotection, the final product was synthesized by a subsequent Mitsunobu cyclization reaction (Scheme 5.11) [15], which is a potential potassium-competitive acid blocker.

5.2.1.3 Asymmetric Hydrogenation of Heteroaromatic Compounds

Asymmetric hydrogenation of heteroaromatic compounds is a challenging and attractive research area as it provides convenient access to numerous saturated or partially saturated chiral heterocyclic compounds, whose synthesis by direct

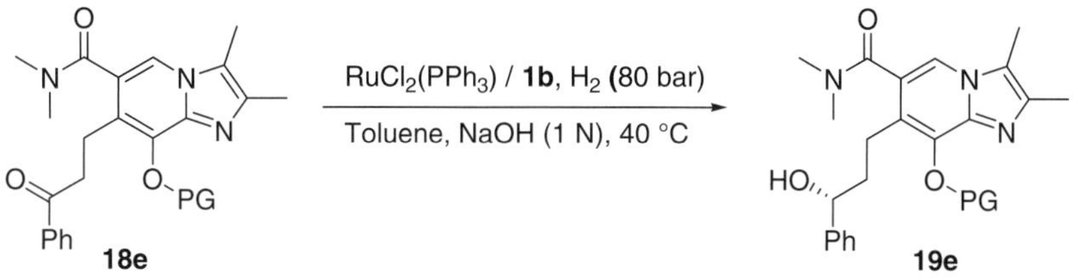

Scheme 5.11 Asymmetric hydrogenation of a prochiral aromatic ketone.

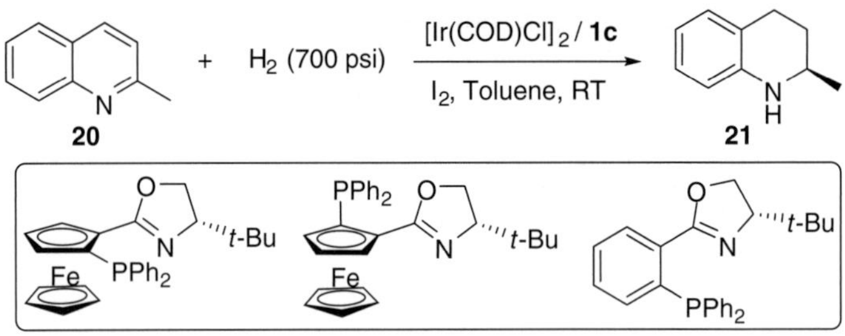

Scheme 5.12 Ir-catalyzed asymmetric hydrogenation of 2-methylquinoline.

cyclization is often difficult. In 2004, Zhou first reported the usefulness of Fc-phox as an N,P-ligand in the Ir-catalyzed asymmetric hydrogenation of quinolines. Tetrahydroquinolines were obtained as the sole products with up to 92% ee [16], a similar result to that obtained with their iridium/biphosphine catalytic system developed in 2003 [17]. The role of the planar and central chirality in the N,P ligands was also addressed, the central chirality on the oxazoline group playing the dominant role in the enantiocontrol of the process (Scheme 5.12), while **1c** and **1i** give the product with the same configuration. In this case, Fc-phox showed better enantioselectivity than phox.

After studying the effects of the solvent, hydrogen pressure and ligands, a variety of substituted quinoline derivatives were smoothly hydrogenated using the above catalytic system with full conversion and good to excellent enantioselectivities (Figure 5.3).

5.2.1.4 Asymmetric Hydrogenation of Imines

In 2007, Knochel's group reported new N,P-ferrocenyl ligands for the Ir-catalyzed asymmetric hydrogenation of imines (Table 5.2). The ligands were synthesized in 8 steps starting from (S)-ferrocenyl sulfoxide (Scheme 5.13) [18].

After synthesis of the ligands, the hydrogenation of some typical imine substrates was studied. Good to excellent *ee* were obtained with full conversion. Interestingly,

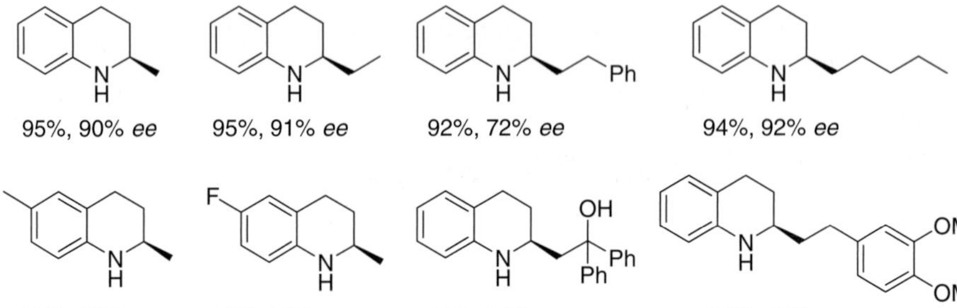

Figure 5.3 Ir-catalyzed asymmetric hydrogenation of quinolines.

Table 5.2 Ir-catalyzed asymmetric hydrogenation of imines.

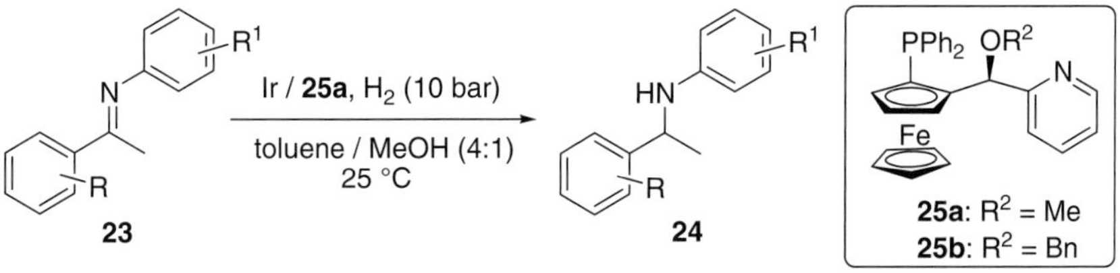

Entry	R/R¹	Time (h)	ee (%)
1	H/H	2	84
2	H/4-MeO	2	85
3	H/3,5-dimethyl	2	94
4	H/3,5-dimethyl-4-methoxy	2	94
5	4-Cl/3,5-dimethyl-4-methoxy	2-4	92

hydrogenation of imine ($R^1 = 3,5$-dimethyl) gave the corresponding secondary amine in 94% *ee* (Table 5.2, entry 3). The deprotection of the 3,5-dimethyl-4-methoxyphenyl moiety of the amines occurs smoothly with cerium ammonium nitrate (CAN; $Ce(NH_4)_2(NO_3)_6$) in a 6:1 $MeOH/H_2O$ mixture, providing the corresponding primary amines in good yields without loss of enantioselectivity (Scheme 5.14).

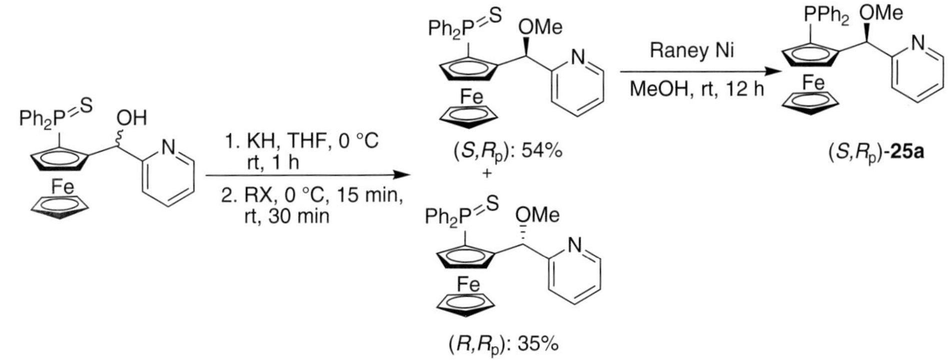

Scheme 5.13 Synthesis of pyridine-containing ferrocene ligands.

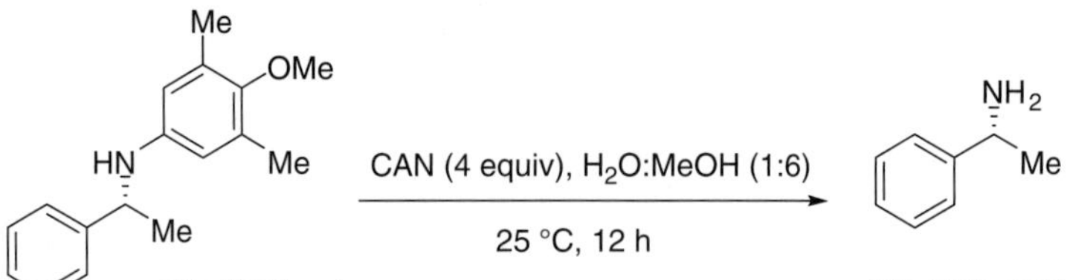

Scheme 5.14 The deprotection of hydrogenation product **24a**.

5.2.2
Asymmetric Transfer Hydrogenation

Asymmetric transfer hydrogenation as an attractive supplement to catalytic reduction with H_2 has been an intensely interesting area of asymmetric synthesis in recent years [19]. Ru, Rh, and Ir chiral catalysts are particularly useful and have been intensively studied. Ferrocene-based N,P-ligands have proved to be effective in the transfer hydrogenation [20].

In 1997, Sammakia reported the first hydrogen-transfer reactions of alkyl and aryl ketones using the catalyst prepared *in situ* from $RuCl_2(PPh_3)_3$ and oxazolinylferrocenyl phosphine at 80 °C [10]. NMR spectral studies of the reaction mixture of $RuCl_2(PPh_3)$ and ligand **1** indicated that the catalysts produced consisted of two diastereomers in 5 : 1 ratio. Acetophenone was reduced by the use of this catalyst with up to 94% *ee* (Table 5.3).

In the meanwhile, the group of Dai and Hou [21] reported that the planar chiral ferrocenes **1b** and their diastereoisomer **1j** were effective ligands for the ruthenium-catalyzed asymmetric transfer hydrogenation of ketones with *i*-PrOH as hydrogen source under reflux in the presence of sodium hydroxide. The results showed that the

Table 5.3 Ru-catalyzed asymmetric transfer hydrogenation of acetophenone using ligands **1**.

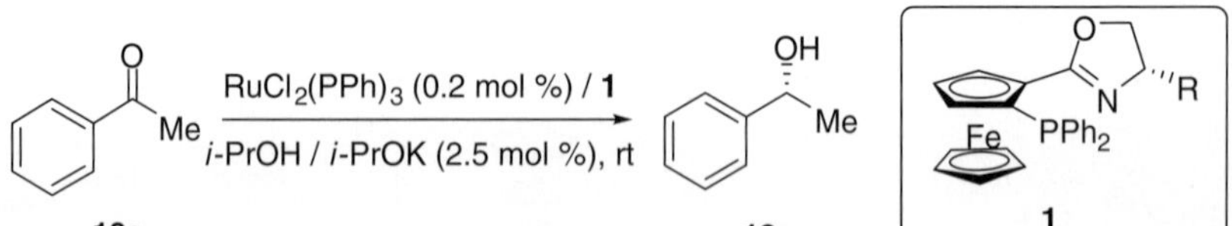

Entry	R	Time (h)	Conversion (%)	ee (%)
1	Me (**1a**)	8	92	92
2	*i*-Pr (**1b**)	3	93	92
3	*t*-Bu (**1c**)	6	51	94
4	Ph (**1d**)	6	93	94
5	Bn (**1e**)	6	93	90

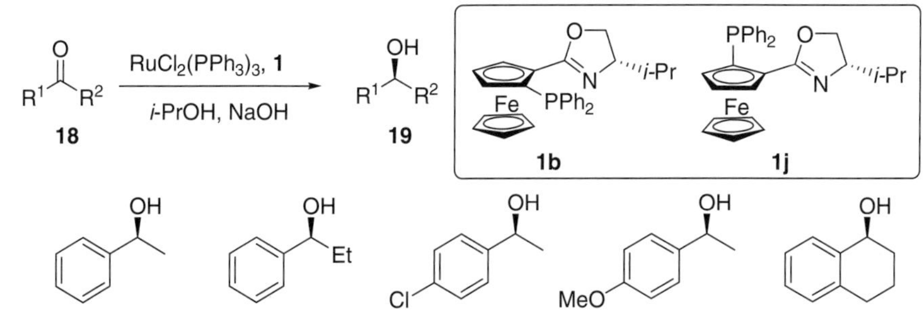

5.2 Asymmetric Hydrogenation and Asymmetric Transfer Hydrogenation | 107

Scheme 5.15 Ru-catalyzed asymmetric transfer hydrogenation of ketones.

absolute configuration of alcohol seemed to be governed by the central chirality in the oxazoline ring instead of the planar chirality. When acetophenone was used as substrate, 99% yield and 63% *ee* were obtained. However, when the diastereoisomer **1j** was used as ligand under essentially identical conditions, almost the same *ee* and the same absolute configuration were obtained. Moderate to good yield and *ee* were furnished with *p*-chloro, *p*-methoxyacetophenone and α-tetralone (Scheme 5.15).

In 1999, Uemura and coworkers succeeded in isolating the diastereomerically pure form of the RuCl$_2$(PPh$_3$) (oxazolinylferrocenylphosphine), solved the crystal structure, and studied the transfer hydrogenation of ketones with *i*-PrOH [22]. Treatment of acetophenone (1 mmol) in the presence of a catalytic amount of **1b** (0.5 mol %) and *i*-PrONa (2.0 mol %) in anhydrous *i*-PrOH (50 mL) at room temperature for 2 h afforded 1-phenylethanol in 94% GC yield with >99.6% *ee* (*R*). Almost the same result (>99.7% *ee*) was obtained in the case of **1d**, bearing a phenyl-substituted oxazoline (Scheme 5.16). On the contrary, the Ru-phox catalyzed reaction gave lower catalytic activity (43% conversion) and lower selectivity (55% *ee*).

At the same year, Uemura and Hidai reported results of the Ru(II)-catalyzed asymmetric transfer hydrogenation using several chiral oxazolinylferrocenylphosphines [23]. The reaction scarcely proceeded at lower temperature, such as 50 °C. The best enantioselectivity (83% *ee*) was obtained by use of ligand **1c**, while the use of ligands **1d**, **1e**, **1k** and **27a** resulted in lower enantioselectivity (66% *ee*, 50% *ee*, 54% *ee*, 47% *ee*) (Table 5.4).

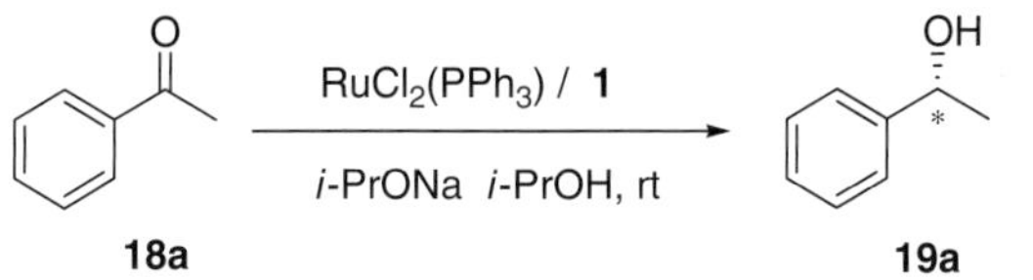

Scheme 5.16 Ru-catalyzed asymmetric transfer hydrogenation of acetophenone.

Table 5.4 Ligand screening of Ru-catalyzed asymmetric transfer hydrogenation of acetophenones.

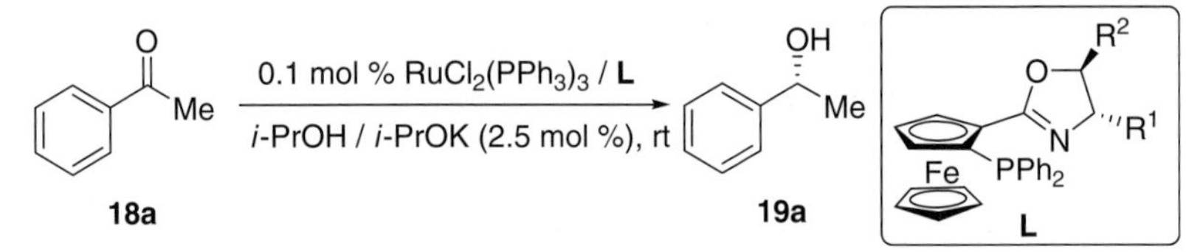

Entry	R¹	R²	Time (min)	Yield (%)	ee (%)
1	i-Pr (1b)	H	1	66	66
2	t-Bu (1c)	H	60	56	83
3	Bn (1e)	H	5	64	54
4	s-Bu (1k)	H	5	43	50
5	Ph (27a)	Ph	5	57	47

In 2003 Zheng and coworkers reported a series of P,N,O-ligands **4** for the ruthenium-catalyzed asymmetric transfer hydrogenation of simple ketones, up to 92% ee with full conversion was obtained (Scheme 5.17) [24].

In 2008, Lastra and coworkers [25] reported the diastereoselective synthesis of ruthenium complexes *mer–trans*-[RuCl$_2$[P(MeO)$_3$]$_2${K^2(P,N)-FcPN}] and *fac*-[RuCl$_2$(PTA)$_2${K^2(P,N)-FcPN}] (Table 5.5 entry 8 and 9). The catalytic activity of these complexes in the asymmetric transfer hydrogenation of ketones along with that of six-coordinate (Table 5.5, entries 1–3) and half-sandwich (Table 5.5, entries 4–7 Ru(II) complexes was also described [26, 27]. Table 5.5 shows the catalytic activity of the studied complexes in asymmetric transfer hydrogenation. The best result (94% ee) was achieved using complex *mer–trans*-[RuCl$_2$[P(MeO)$_3$]$_2${K^2(P,N)-FcPN}] (Table 5.5, entry 8). The catalytic transformations of a series of methyl-arylketones have also been studied. The results are collected in Figure 5.4. The best ee values for all the ketones, were obtained using complex *mer–trans*-[RuCl$_2$[P(MeO)$_3$]$_2${K^2(P,N)-FcPN}] (Table 5.5, entry 8) as catalyst, ee values being up to 92% for 2-acetylanisole and 94% for 3-acetylanisole.

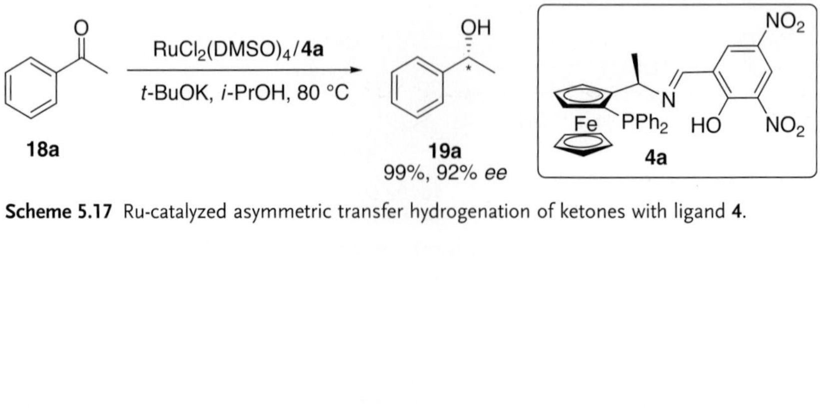

Scheme 5.17 Ru-catalyzed asymmetric transfer hydrogenation of ketones with ligand **4**.

Table 5.5 Catalytic transfer hydrogenation of acetophenone with octahedral Ru(II) complexes.

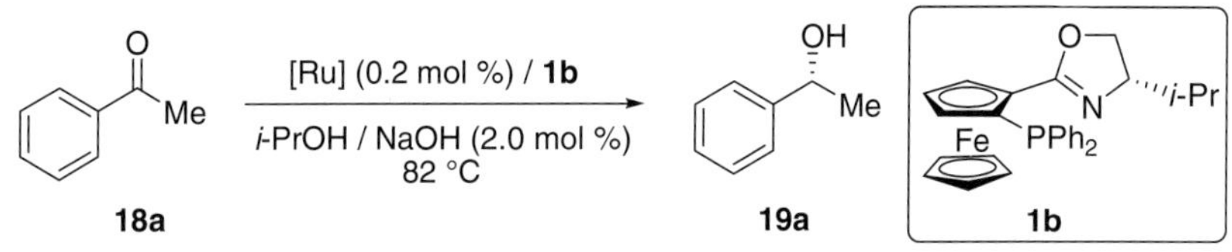

[Scheme: 18a (acetophenone) + [Ru] (0.2 mol %) / 1b, i-PrOH / NaOH (2.0 mol %), 82 °C → 19a (1-phenylethanol); 1b = ferrocenyl oxazoline-PPh$_2$ with i-Pr substituent]

Entry	Complex	Time (min)	Conversion (%)	ee (%)	TOF (h^{-1})
1	fac-[RuCl$_2$(PMe$_3$)$_2$]{K^2(P,N)-FcPN}]	3	96	72	9600
2	fac-[RuCl$_2$(PMe$_2$Ph)$_2$]{K^2(P,N)-FcPN}]	4	97	81	7275
3	fac-[RuCl$_2$(dppm)] {K^2(P,N)-FcPN}]	150	92	47	184
4	[Ru(η^5-C$_9$H$_7$)(PCl$_3$) {K^2(P,N)-FcPN}][PF$_6$]	120	92	76	230
5	[RuCl (η^5-C$_9$H$_7$) {K^2(P,N)-FcPN}]	90	92	31	307
6	[RuCl (η^6-cymene) {K^2(P,N)-FcPN}][PF$_6$]	50	93	47	560
7	[RuH(η^6-cymene) {K^2(P,N)-FcPN}][PF$_6$]	60	91	41	455
8	mer–trans-[RuCl$_2$[P(MeO)$_3$]$_2${K^2(P,N)-FcPN}	120	96	94	240
9	fac-[RuCl$_2$(PTA)$_2${K^2(P,N)-FcPN}	10	54	23	1633

5.2.3
Asymmetric Hydrosilylation Reaction

The development of asymmetric reactions catalyzed by transition metals is a field commanding great attention in current organic synthesis, and the design and preparation of optically active ligands which coordinate with transition metals are important key factors for obtaining excellent enantioselectivity [28]. Numerous efforts have been devoted to this field and great progress has been made. The development of new 1,2-N,P-bidentate ferrocenyl ligands has attracted much atten-

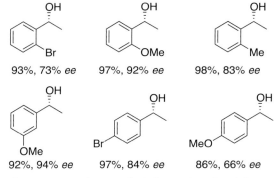

Figure 5.4 Ru-catalytic transfer hydrogenation of methyl-arylketones with mer-trans [RuCl$_2$[P(MeO)$_3$]$_2${K^2(P,N)-FcPN}].

tion, and their application in palladium-, rhodium-, iridium- and ruthenium-catalyzed asymmetric hydrosilylation of alkenes, ketones, imines and oximes has been extensively investigated. Enantioselective addition of hydrosilanes across double bonds, including C=C, C=O and C=N, is termed an asymmetric hydrosilylation (AHS) reaction. Herein a brief review of the development of asymmetric hydrosilylation with transition metals and 1,2-P,N-bidentate ferrocenyl ligands is given.

5.2.3.1 Palladium-Catalyzed Asymmetric Hydrosilylation of C=C Bond

The asymmetric functionalization of prochiral olefins is a challenging area in organic synthesis that continues to attract widespread interest [29]. Asymmetric hydrosilylation is a straightforward approach to the synthesis of optically active organosilicon compounds. In particular, optically active allylsilanes, available by asymmetric hydrosilylation of conjugated dienes, are useful synthetic intermediates for various carbon–carbon bond-forming reactions such as cross-coupling. In addition, asymmetric hydrosilylation provides a useful gateway to optically active alcohols via the Tamao oxidation [30], in which the carbon–silicon bond is converted to a carbon–oxygen bond with retention of configuration at the carbon center.

In 1980, Hayashi and coworkers reported the asymmetric hydrosilylation of styrene and norbornene with a palladium complex and (R,S_p)-PPFA **3** (Scheme 5.18) [31]. The reaction at 70 °C gave a >95% yield of benzylic silane, which was found to have an ee of 52% upon oxidation to α-methylbenzyl alcohol. When norbornene was subjected to the above conditions, similar results were also obtained.

In 1985, this chiral phosphine-palladium complex [PdCl$_2$·**3a**] was also applied to catalyze the asymmetric hydrosilylation of acyclic conjugated dienes 1-arylbutadienes (Scheme 5.19) [32]. The reaction proceeded with triethylamine in ethanol at 80 °C to afford allyltriethoxysilanes **34**. Methylation of **34** with MeLi gave optically active allyltrimethylsilanes **35** with ee of 64%.

Hayashi subsequently introduced polyfluoroalkyl groups onto the amine donor, and converted to the catalyst [PdCl$_2$·(**3**)] (Scheme 5.20) [33]. It was found that whilst catalyst [PdCl$_2$·(**3a**)] was inactive in the AHS of cyclic dienes, cyclopentadiene, catalysts [PdCl$_2$·(**3g**)] and [PdCl$_2$·(**3h**)] gave significant yields (73 and 41%) and

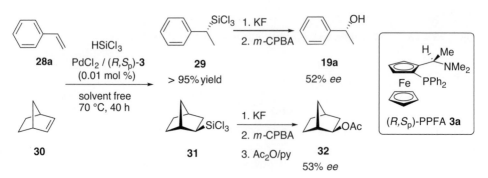

Scheme 5.18 Asymmetric hydrosilylation with (R,S_p)-PPFA.

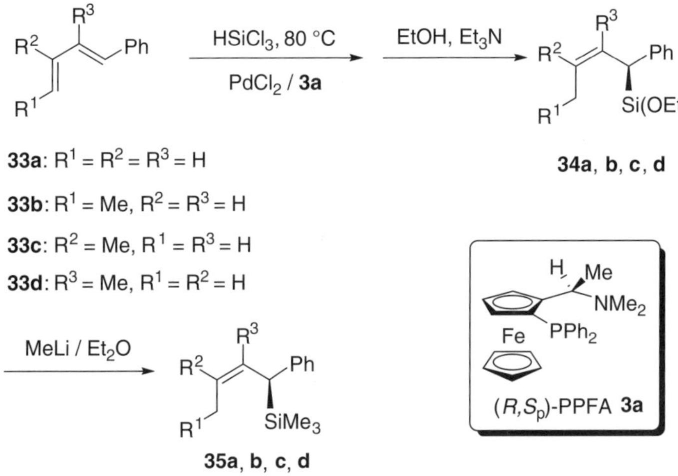

Scheme 5.19 Pd-catalyzed asymmetric hydrosilylation of 1-aryl-1,3-butadienes.

enantioselectivities (57 and 55%) operating at temperatures as low as 25 °C. Hence, the activity of the catalyst increased greatly, the authors attributed this to the improved solubility of the catalyst at ambient temperature.

Further modifications of the chiral side chain of PPFA-type ligands were reported by Togni and coworkers, several pyrazole-containing ligands were synthesized, and their activity in the AHS reaction of norbornene was examined (Table 5.6) [34]. Increasing the steric bulk of the substituent R^3, which might decrease the coordinative power of the nitrogen, allowed the reaction to be operated at lower temperatures with a concomitant increase in selectivity (entries 1–3). Reducing the electron-donating power of R^3 by replacing the 2,4-6-trimethoxyphenyl group with a 2,4,6-trimethylphenyl group, gave a palladium complex with high enantioselectivity (91% ee, entry 4) while lowering the coordinative ability of phosphorus by the introduction of electron-withdrawing 3,5-bis-(trifluoromethyl)phenyl groups gave a catalyst that delivered the highest enantioselectivity (99% ee, entry 5).

They also probed the relationship between the electronic nature of the substrate and stereoselectivity [34]. A series of para-substituted styrenes was subjected to asymmetric hydrosilylation using the palladium complex of ligand **5d** (Scheme 5.21). It was found that the electronic nature of the para-substituent plays a crucial role in the sense of asymmetric induction. Electron-rich 4-dimethylaminostyrene gave the

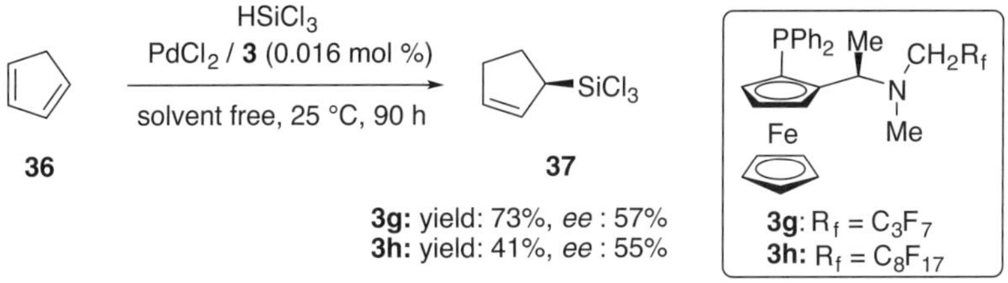

Scheme 5.20 Asymmetric hydrosilylation of cyclopentadiene.

Table 5.6 Scan of ligands **5** in asymmetric hydrosilylation of norbornene.

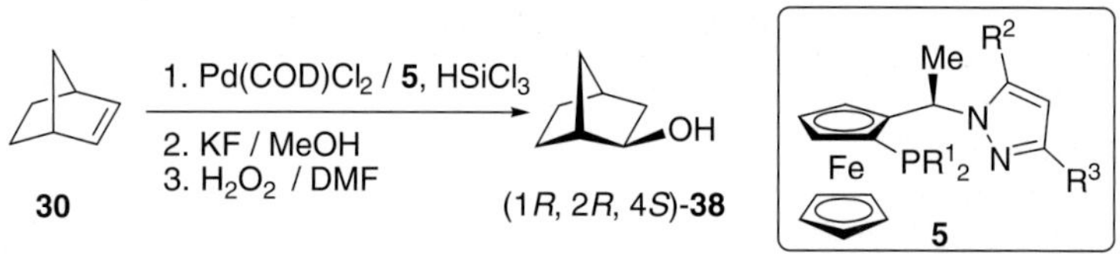

Entry	L	R^1	R^2	R^3	T [°C]	Yield [%]	ee [%]
1	5a	Ph	Me	Me	70	54	10
2	5b	Ph	Me	Ph	50	47	39
3	5c	Ph	H	2,4,6-$(OMe)_3C_6H_2$	0	30	82
4	5d	Ph	H	2,4,6-$(Me)_3C_6H_2$	0	56	91
5	5e	3,5-$(F_3C)_2C_6H_3$	H	2,4,6-$(Me)_3C_6H_2$	0	59	99

highest R-selectivity (64% ee), whereas 4-chlorostyrene afforded the S-product with 67% ee. The authors remarked that this was "a rare example of enantioselectivity reversal due to a remote substituent in the substrate". In any case, the result serves to underline the sensitivity of the reaction to steric and electronic influences of both ligand and substrate. Togni and coworkers subsequently extended their investigation to the catalytic cycle and mechanistic details for the AHS of styrene with a palladium complex of ligand **5d** [35].

5.2.3.2 Asymmetric Hydrosilylation of a C=O Bond

Hydrosilylation of ketones is useful for the reduction of ketones since asymmetric hydrosilylation followed by hydrolysis provides an effective route to optically active secondary alcohols.

In 1995 Hayashi and coworkers reported the asymmetric hydrosilylation of ketones with chiral ferrocenylphosphine-imine ligand **4** [36]. The ligands were conveniently prepared from ligand **3a** by a sequence of reactions as shown in Scheme 5.22. A slightly higher enantioselectivity was observed in the reactions with **4c, d** and **e**, which contain electron-withdrawing groups on the phenyl ring of the ligand (Table 5.7). The hydrosilylation was faster with these ligands (entries 3–5). The

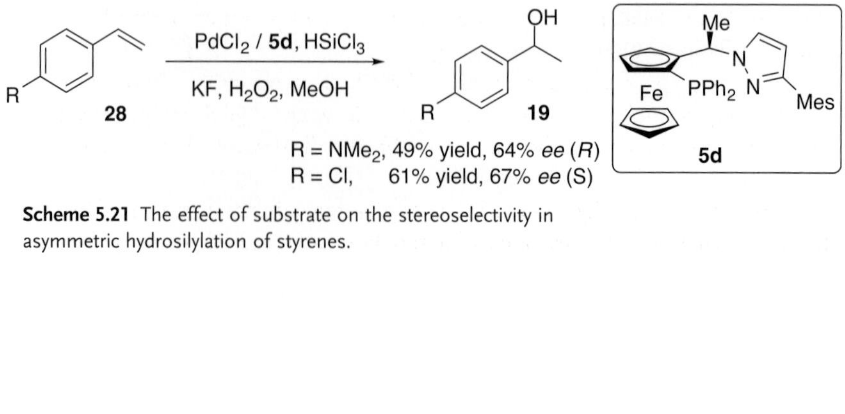

R = NMe_2, 49% yield, 64% ee (R)
R = Cl, 61% yield, 67% ee (S)

Scheme 5.21 The effect of substrate on the stereoselectivity in asymmetric hydrosilylation of styrenes.

5.2 Asymmetric Hydrogenation and Asymmetric Transfer Hydrogenation

Table 5.7 Scan of ligands in asymmetric hydrosilylation of ketones.

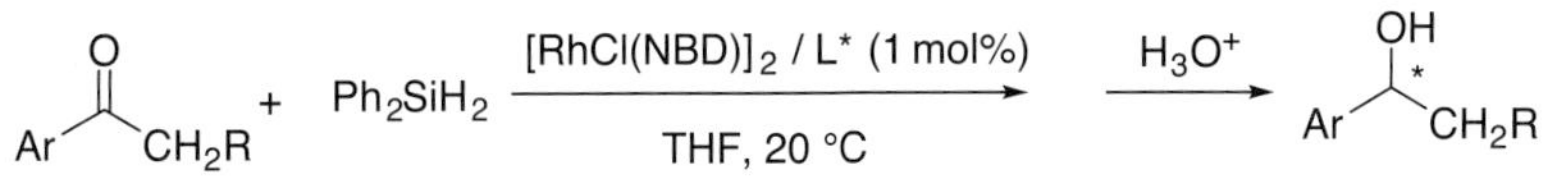

Ar–CO–CH$_2$R + Ph$_2$SiH$_2$ → [RhCl(NBD)]$_2$ / L* (1 mol%), THF, 20 °C → H$_3$O$^+$ → Ar–CH(OH)–CH$_2$R

18a: Ar = Ph, R = H
18f: Ar = Ph, R = Me
18g: Ar = p-ClC$_6$H$_4$, R = H

19

Ligand structure: Ar–N=CH(Me)–[Cp-Fe-Cp with Ph$_2$P]

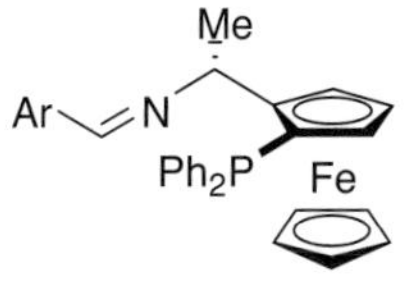

4b: Ar = Ph
4c: Ar = o-CF$_3$C$_6$H$_3$
4d: Ar = p-CF$_3$C$_6$H$_3$
4e: Ar = C$_6$F$_5$

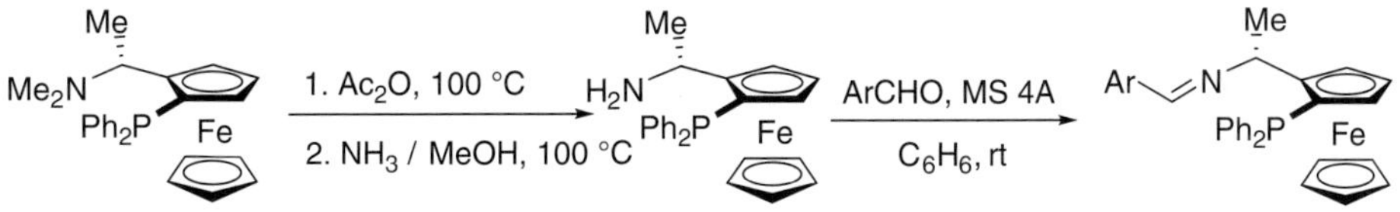

Entry	Ketone 18	Ligand	Reaction time	Yield [%]	ee [%]
1	18a	4b	<1 h	90	87 (S)
2	18f	4b	<1 h	97	86 (S)
3	18a	4c	<10 min	90	90 (S)
4	18a	4d	<10 min	94	89 (S)
5	18a	4e	<10 min	86	89 (S)
6	18g	4e	<1 h	90	81 (S)
7	18a	3f	<1 h	64	16 (R)
8	18a	3a	2 h	50	39 (R)

(R,R$_p$)-PPFA **3a** [Me$_2$N–CH(Me)–Cp-Fe-Cp–Ph$_2$P] → 1. Ac$_2$O, 100 °C; 2. NH$_3$ / MeOH, 100 °C → **3f** [H$_2$N–CH(Me)–Cp-Fe-Cp–Ph$_2$P] → ArCHO, MS 4A, C$_6$H$_6$, rt → **4** [Ar–N=CH(Me)–Cp-Fe-Cp–Ph$_2$P]

Scheme 5.22 Synthesis of ligand **4**.

importance of the imino group on the AHS is demonstrated by the reaction with **3a** and **3f**, which gave 1-phenylethanol with opposite configuration with much lower enantioselectivity (entries 7–8).

In the same year, Nishibayashi and coworkers reported the asymmetric hydrosilylation of simple ketones such as acetophenone in the presence of a rhodium and oxazolinylferrocene-phosphine hybrid ligand [37]. Based on the results of different ligands in the reaction, introduction of another phenyl group at the 5-position of the oxazoline ring of Fc-phox was essential for the achievement of high enantioselectivity. Thus, **27a** was revealed to be the most effective ligand for rhodium-catalyzed asymmetric hydrosilylation (Scheme 5.23).

The results of hydrosilylation of a variety of ketones in the presence of **27a** are shown in Scheme 5.24. Reactions of a variety of aryl and alkyl methyl ketones

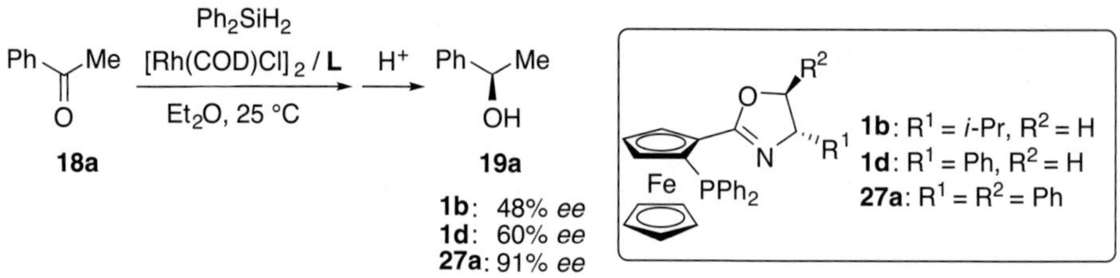

Scheme 5.23 Asymmetric hydrosilylation of acetophenone with ligand **27a**.

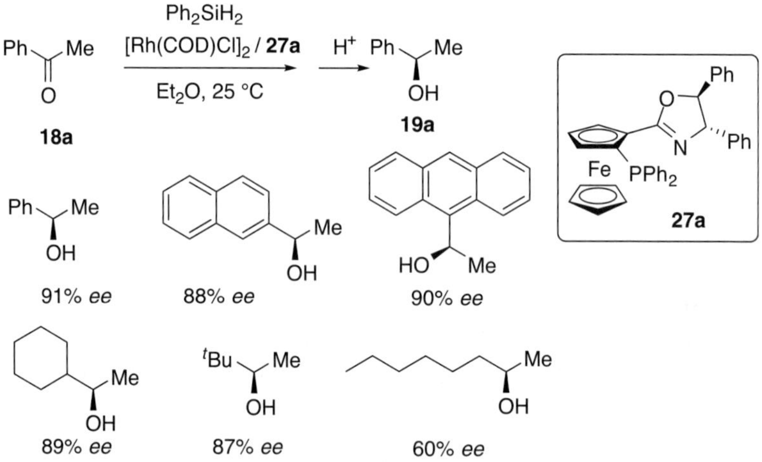

Scheme 5.24 Rh-catalyzed asymmetric hydrosilylation of ketones with ligand **27a**.

proceeded smoothly to give the corresponding alcohols in good yields with moderate to good enantioselectivity. Surprisingly, when [Ir(COD)Cl]$_2$ was used as a catalyst instead of [Rh(COD)Cl]$_2$, hydrosilylation of acetophenone with **27a** afforded 1-phenylethanol with even higher enantioselectivity (96% ee) in the (S)-configuration, while products in (R)-configuration were afforded by using the Rh-catalyst (Scheme 5.25).

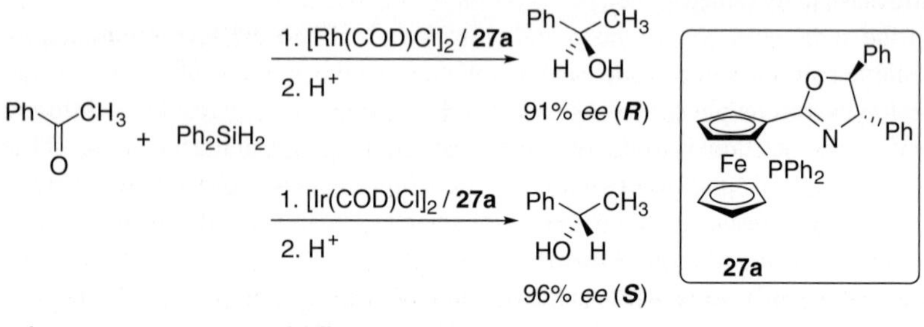

Scheme 5.25 Comparison of different metals in the asymmetric hydrosilylation of acetophenone.

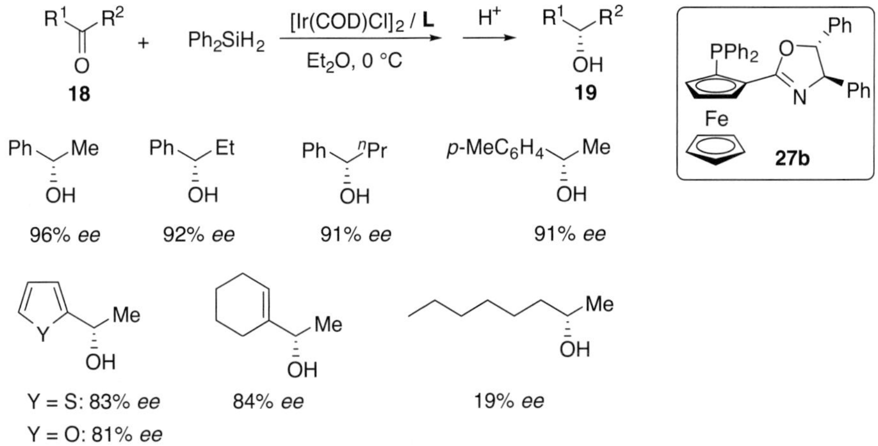

5.2 Asymmetric Hydrogenation and Asymmetric Transfer Hydrogenation | 115

Scheme 5.26 Ir-catalyzed asymmetric hydrosilylation of ketones.

Based on the results of the former investigation, in 1996 the same group reported the asymmetric hydrosilylation of a variety of ketones with [Ir(COD)Cl]$_2$ and **27b** [38]. As seen from the results shown in Scheme 5.26, the catalyst system was very effective for both alkyl aryl ketones and heterocyclic methyl ketones with high reactivity and good to excellent enantioselectivity. This is the first example of Ir-catalyzed, highly enantioselective hydrosilylation of ketones.

They then developed the first highly enantioselective ruthenium-catalyzed asymmetric hydrosilylation of both ketones and an imine [39]. Addition of metal triflates, such as copper triflate, was also essential for the achievement of a high enantioselectivity. Typical results are shown in Scheme 5.27. In the case of propiophenone, the best enantioselectivity of 97% ee was achieved with high yield. This catalytic system can also be applied to the asymmetric hydrosilylation of an imine **39** with moderate yield and high enantioselectivity. In contrast, the addition of AgOTf to the present system completely inhibited the hydrosilylation of the imine.

In 2002, Fu developed a heterocyclic P,N-ferrocene-type ligand **41** with only planar chirality, which is an excellent ligand in the Rh-catalyzed hydrosilylation of aryl alkyl ketones (95–99% ee) and dialkyl ketones (72–96% ee) (Scheme 5.28) [40]. In terms of both scope and stereoselectivity, this method compares favorably with the previously described catalysts for this process. The use of a bulky silylation reagent is also unique. This type of ligand will be discussed in detail in Chapter 11.

In 2003, Bolm and coworkers designed and synthesized a number of novel planar chiral phosphinocyrhetrenyloxazoline ligands and applied them in a series of catalytic reactions [41]. Comparison of those ligands with the 1,2-N,P-ferrocene ligand **1c** in the catalytic hydrosilylation of acetophenone, showed that ligand **1c** gave poor activity and moderate enantioselectivity (Scheme 5.29). However **1c** showed better performance in the Ru-catalyzed transfer hydrogenation and Pd-catalyzed allylic alkylation reaction than cyrhetrenyl-phox.

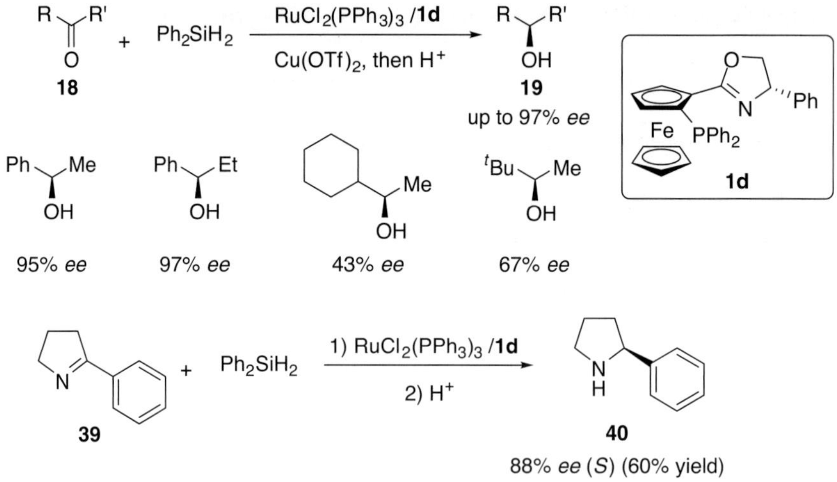

5 1,2-P,N-Bidentate Ferrocenyl Ligands

Scheme 5.27 Ru-catalyzed asymmetric hydrosilylation of ketones and imines.

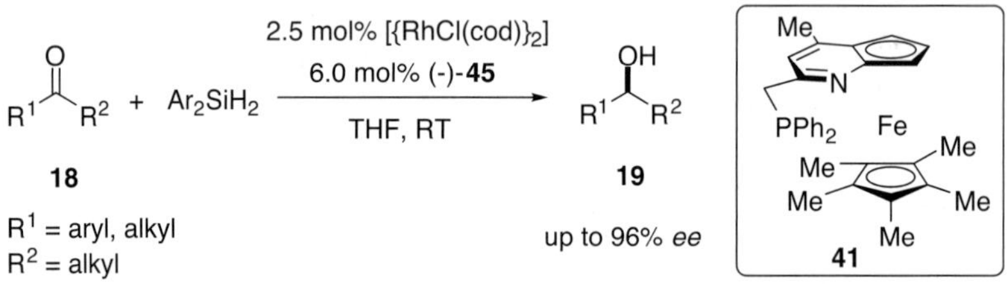

Scheme 5.28 Asymmetric hydrosilylation of ketones with **41**.

Zheng's group developed a series of new ferrocenylphosphine-ketimine ligands **4f–n** and investigated the difference in the catalytic activity between *P*-ketimines and *P*-aldimines in asymmetric catalysis [42]. In Rh-catalyzed asymmetric hydrosilylation of prochiral ketones, ketimine **4j** gives lower enantioselectivity, only 42% *ee* of alcohol was obtained, compared to an *ee*-value of 90% with the corresponding aldimine **4f**. They also indicated that (*R*)-central chirality and (S_p)-planar chirality in these ferrocenylphosphine-ketimine ligands was matched for the test reactions (Table 5.8).

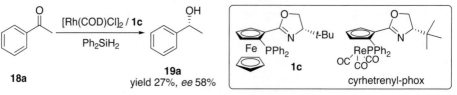

Scheme 5.29 Asymmetric hydrosilylation of acetophenone with **1c**.

Table 5.8 Comparison of ligands in asymmetric hydrosilylation of acetophenone.

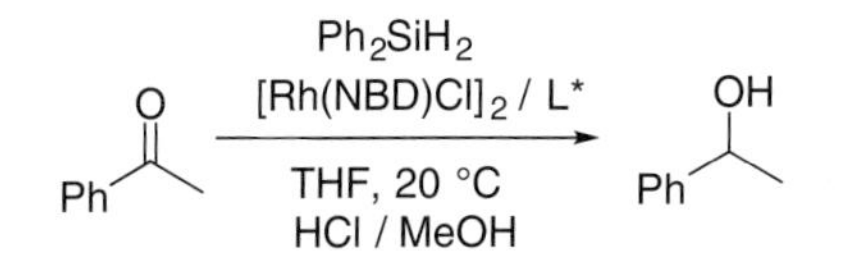
Ph2SiH2
[Rh(NBD)Cl]2 / L*
THF, 20 °C
HCl / MeOH
O
Ph
OH
Ph

Entry	Ligand	Yield [%]	*ee* [%]
1	**4f**	92	90 (*R*)
2	**4g**	90	80 (*R*)
3	**4h**	89	81 (*R*)
4	**4i**	90	90 (*R*)
5	**4j**	76	42 (*R*)
6	**4k**	71	38 (*R*)
7	**4l**	59	29 (*R*)
8	**4m**	74	40 (*R*)
9	**4n**	54	31 (*R*)
10	**4o**(S,S_p)	51	26 (*R*)

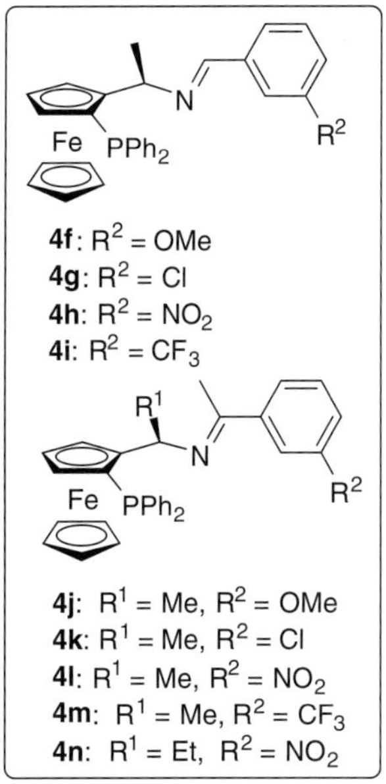
Fe
PPh2
N
R2
4f: R2 = OMe
4g: R2 = Cl
4h: R2 = NO2
4i: R2 = CF3
R1
Fe
PPh2
N
R2
4j: R1 = Me, R2 = OMe
4k: R1 = Me, R2 = Cl
4l: R1 = Me, R2 = NO2
4m: R1 = Me, R2 = CF3
4n: R1 = Et, R2 = NO2

5.2.3.3 **Asymmetric Hydrosilylation of the C=N Bond**

Asymmetric hydrosilylation of C=N, which gives the corresponding optically active amines after hydrolysis, is useful in organic synthesis. Hydrosilylation of oximes followed by hydrolysis provides the corresponding primary amines, while hydrolysis of hydrosilylated products of imines gives secondary amines.

In 1999 Uemura and Hidai reported the asymmetric hydrosilylation of imines by the iridium complex with ligand **1d** [43]. In sharp contrast to the hydrosilylation of ketones, the use of the iridium complex was essential for obtaining a high enantioselectivity. This is the first example of an Ir-catalyzed highly enantioselective hydrosilylation of imines (Scheme 5.30). This result is comparable with that of the Ru-catalyzed reaction (Scheme 5.27).

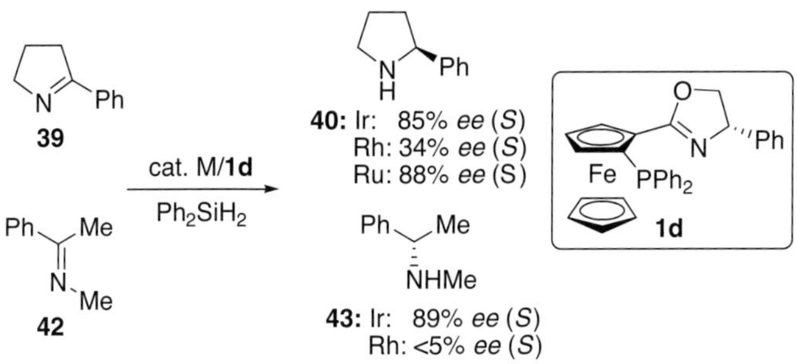
N
Ph
39
cat. M/1d
Ph2SiH2
N
H
Ph
40: Ir: 85% ee (S)
Rh: 34% ee (S)
Ru: 88% ee (S)
Ph
Me
N
Me
42
Ph
Me
NHMe
43: Ir: 89% ee (S)
Rh: <5% ee (S)
O
N
Ph
Fe
PPh2
1d

Scheme 5.30 Ir-catalyzed asymmetric hydrosilylation of imines with Fc phox and comparison of the metal used.

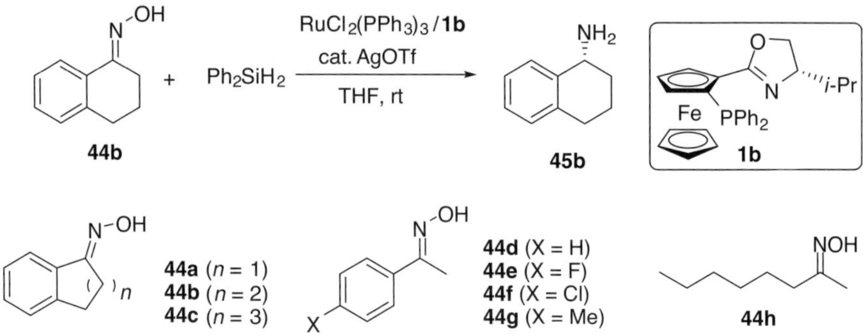

Scheme 5.31 Ru-catalyzed asymmetric hydrosilylation of oximes.

In 2001, Uemura and Hidai reported the asymmetric hydrosilylation of 1-tetralone oxime in the presence of a catalytic amount of the ruthenium complex with **1b**, affording the corresponding amine in good yield with high enantioselectivity (Scheme 5.31) [44]. Addition of a catalytic amount of silver triflate slightly increased the yield. Although the preparation of optically active primary amines by the hydrogenation of oximes has also been reported, the enantioselectivity is not satisfactory. This ruthenium-catalyzed system is useful as a direct and catalytic preparative method for primary amines without protection of the nitrogen atom. Several ketoximes have been investigated and moderate to excellent enantioselectivities were obtained. In the case of acetophenone oxime (**44d**), the best enantioselectivity of 89% *ee* was achieved. This may provide a versatile method for straightforward synthesis of chiral primary amines because of the ready accessibility of ketoxime by reaction of ketones with hydroxylamine.

5.2.4
Asymmetric Hydroboration

The asymmetric hydroboration of vinyl arene is a very valuable reaction for the enantioselective synthesis of a variety of chiral compounds, such as alcohols, amines and carboxylic acids (Scheme 5.32).

Actually, the hydroboration of alkenes proceeds even in the absence of catalysts, but great attention has been paid to the development of Rh-catalyzed asymmetric hydroboration using chiral ligands since the report that Nöth's group employed rhodium-phosphine complexes to catalyze this reaction [45]. Recently, great progress had been made using Rh complexes with chiral bidentate ferrocenyl ligands in asymmetric hydroboration.

P,N-bidentate ferrocenyl ligands are one of the largest classes of ligands, and have been effectively employed in asymmetric catalysis. However, successful examples of their application in asymmetric hydroboration are limited. In 1995, Togni and coworkers synthesized a variety of pyrazolyl ligands **5** in moderate yields starting from a ferrocenyl amine of type **3** and the corresponding pyrazoles in glacial acetic acid, as illustrated in Scheme 5.33.

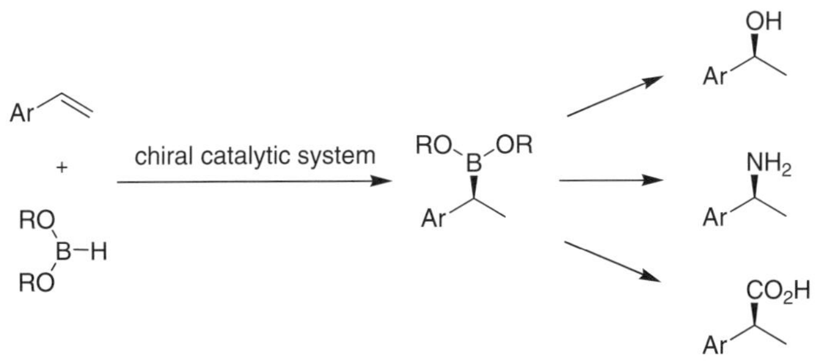

Scheme 5.32 Asymmetric hydroboration.

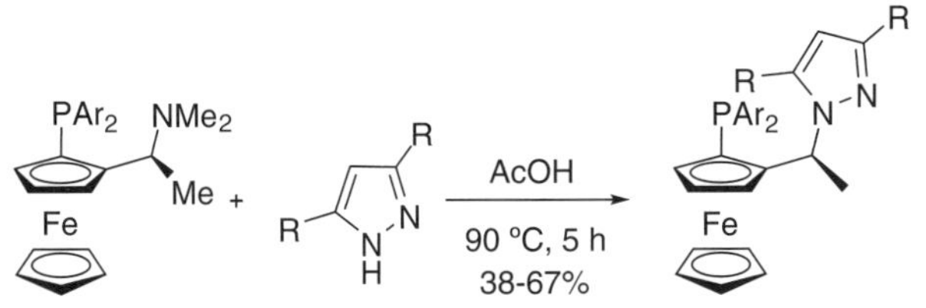

Scheme 5.33 The synthesis of ferrocene derived pyrazolyl-P,N-ligands.

The novel pyrazole-containing ferrocenyl phosphines have been successfully applied in the asymmetric Rh-catalyzed hydroboration of styrene using catecholborane, giving high enantioselectivity (Table 5.9). Compared to the application of Josiphos to the same transformation, the ligands (S,R_p)-**5** were highly effective at ambient temperature, albeit the regioselectivity was not good. Ligand (S,R_p)-**5a**, containing the 3,5-dimethylpyrazoly fragment, afforded an *ee* of 95%. Increasing the size of the pyrazole substituents from methyl to isopropyl gave a slightly lower, 92%, *ee* (Table 5.9, entry 3 vs. entry 1). On replacement of the methyl group by the strong σ-accepting trifluoromethyl fragment at positions 3 and 5 of the pyrazole, the regioselectivity dropped by only a few percent from 66:34 to 61:39 branched to linear, the enantioselectivity decreased dramatically from 95% to 33% (Table 5.9, entry 2). When the Ar group was 4-trifluoromethylphenyl, an unexpected 98% *ee* was obtained. When an electron-donating substituent OMe was introduced, lower enantioselectivity was observed (Table 5.9, entry 5). The combination of an electron-poor pyrazole and a relatively electron-rich phosphine, as in ligand (S,R_p)-**5k**, afforded almost a racemic product (Table 5.9, entry 7) [46]. In 1996, Togni developed the pentamethyl analog (S,R_p)-**5b**, which gave (R)-1-phenylethanol with 76% regioselectivity and with 94% *ee* [47]. Details of the preparation, characterization, and structure of this new class of ligands were reported in another paper by Togni's group [48].

Recently, Knochel and coworkers described a set of chiral ferrocenyl phosphine-amine ligands **6** (Figure 5.5) [49]. These ligands were tested in the Rh-catalyzed asymmetric hydroboration of styrene, and representative examples are shown in Table 5.10.

Table 5.9 Rh-catalyzed hydroboration of styrene with the pyrazolyl P,N-ferrocenyl ligands.

Entry	Ligand	R	Ar	Regioselectivity[a]	ee (%)[b]	Yield (%)
1	5a	Me	Ph	66:34	95.1	91
2	5f	CF_3	Ph	61:39	33.4	78
3	5g	^{i}Pr	Ph	36:64	91.6	55
4	5h	Me	4-CF_3-C_6H_4	60:40	98.0	68
5	5i	Me	4-MeO-C_6H_4	61:39	90.0	61
6	5j[c]	Me	3,5-(CF_3)-C_6H_3	not quoted	98.5	63
7	5k[c]	CF_3	4-MeO-C_6H_4	not quoted	5.0	28

[a]Ratio of 19a/46.
[b]ee of 19a.
[c]1 mol% catalyst formed in situ from [$Rh_2(CO)_2Cl_2$] and 2 equiv of ligand.

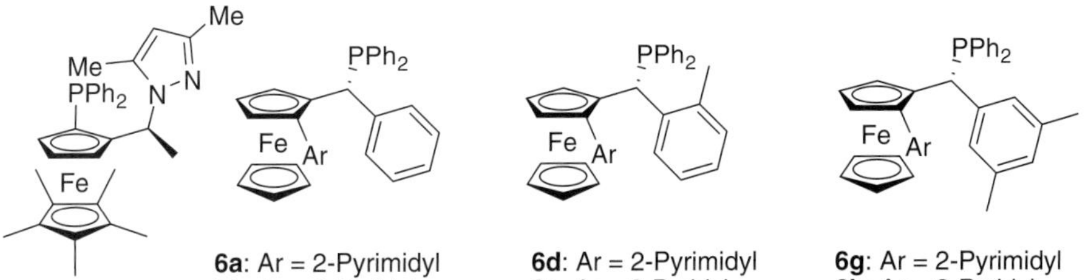

6a: Ar = 2-Pyrimidyl
6b: Ar = 2-Pyridyl
6c: Ar = 2-Quinoyl

6d: Ar = 2-Pyrimidyl
6e: Ar = 2-Pyridyl
6f: Ar = 2-Quinoyl

6g: Ar = 2-Pyrimidyl
6h: Ar = 2-Pyridyl
6i: Ar = 2-Quinoyl

Figure 5.5 P,N-Ferrocene ligands containing N-heteroaromatic rings.

The ligand **6a**, having the pyrimidyl moiety, showed good regioselectivity (97:3), while the enantioselectivity was only moderate (Table 5.10 entry 1). However, for the ligand **6b** containing the pyridyl group, moderate to good enantioselectivity and reasonable regioselectivity were obtained. Rhodium complexes of ligands **6g**, **6h** and **6i** with the sterically demanding xylyl group gave good enantioselectivity and modest regioselectivity. The optimal result was 92% *ee* with full conversion using the rhodium/ligand **6g** complex.

5.3
Formation of a C—C Bond

Carbon–carbon bond formations are among the most important reactions in organic synthesis. One of the most powerful approaches is the reaction using transition-metal complexes as catalyst. Especially, the catalysts with a chiral ligand show unique power in the carbon–carbon bond-forming reactions accompanied by the formation of a new chiral center. 1,2-P,N-Ferrocene ligands have found a role in many types of carbon–carbon bond-forming reactions, realizing high catalytic activity and high enantioselectivity.

5.3.1
Pd-Catalyzed Asymmetric Allylic Substitution Reaction

Transition-metal catalyzed allylic substitution reaction represents one of the most powerful and diverse tools in carbon–heteroatom and carbon–carbon bond-forming reactions in asymmetric catalysis. During the past decades, hundreds of ferrocene ligands have been applied in this reaction, many 1,2-P,N-ferrocenes showed high catalytic activity and excellent enantioselectivity.

The reaction of 1,3-diphenylpropenyl acetate with amines or a "soft" carbanion, such as malonate ester, is very often considered as a standard reaction to test the efficiency of the ligands, though the products are not very useful in organic synthesis. Many 1,2-P,N-ferrocenes have been used in this reaction (Figure 5.6) [50], some of which provided excellent enantioselectivities in this benchmark reaction (Table 5.11). Ahn [51] reported that 99% *ee* was obtained by using Fc-PHOX **1c** as ligand in

Table 5.10 Rh-catalyzed asymmetric hydroboration of styrene with ligand **6**.

Entry	Ligands	Conversion (%)	ee (%)	19a/46 (%)
1	6a	74	57	97:03
2	6b	52	55	87:13
3	6c	53	59	90:10
4	6d	75	61	87:13
5	6e	54	80	84:16
6	6f	22	32	57:43
7	6g	>99	92	64:36
8	6h	92	82	68:22
9	6i	>99	86	65:35

alkylation (entry 1), Jin [52] provided another example by using the imidazoline-containing ligand **56**, in which 99.6% ee was realized (entry 2). Phosphine-triazine ligand **55** with an additional heterocycle developed by Zheng showed high asymmetric induction in this reaction (entry 3) [53b]. Interestingly, the position of nitrogen in the pyridine subunit influences the reaction greatly. The 4-pyridinyl ligand **4r** gave the best result while no reaction took place when ligand **4p** was used (entry 5 vs. entry 4) [53a]. The electronic property of Ar in ligand **4** influences the results. Higher ees were obtained when the ligand **4h** with an electron-withdrawing group on the phenyl ring was used and the best results were given when the Ar of **4** was 3-NO_2 phenyl [54]. Ferrocene-substituted PHOX **7b** and homoannularly bridged ferrocene **58** also gave high ees in the reaction (entries 6 and 7) [55]. Introducing an additional chiral center at the substituent on the oxazoline ring of Fc-PHOX produced a new type of ligand **59**, the reaction using it as ligand provided the product in high ee as well (entry 8) [56]. Pyrazole-containing ferrocenyl phosphine ligands **5** developed by Togni demonstrated their excellent asymmetric induction in the amination reaction. Reaction provided the product in 99% ee using ligand **5l** with a very rigid, compact, and bulky 1-adamantyl group (entry 9) and in 96% ee using **5b** with phenyl as the substituent on the pyrazole ring (entry 10) [57]. Fc-PHOX ligands **1c** and **1b** were also used in the amination reaction, giving the product with 37% and 72% ee, respectively [58].

Polymer-supported ferrocene ligand **54** was prepared and used in the reaction. The catalytic activity and enantioselectivity were decreased when the catalyst was recovered and reused for a second time [59].

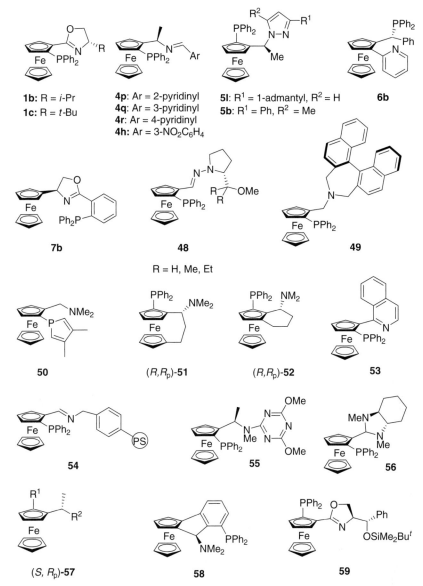

Figure 5.6 Representative examples of 1,2-P,N-ferrocene ligands in Pd-catalyzed asymmetric allylic substitution reactions.

Table 5.11 1,2-P,N-Ferrocene ligands showing better performance in Pd-catalyzed asymmetric allylic substitution reactions.

$$\text{Ph}\overset{\text{OAc}}{\diagup\!\!\!\diagdown}\text{Ph} + \text{Nu-H} \xrightarrow[\text{solvent}]{\substack{[Pd(C_3H_5)Cl]_2 \ (2 \text{ mol \%}) \\ L \ (4\text{-}6 \text{ mol \%}) \\ \text{base, additive}}} \text{Ph}\overset{\text{Nu}}{\overset{*}{\diagup\!\!\!\diagdown}}\text{Ph}$$

Entry	Ligand	NuH	Conditions	Yield (%)	ee (%)	Ref.
1	1c	$CH_2(CO_2Me)_2$	BSA, KOAc, CH_2Cl_2, rt	99	99	[51]
2	56	$CH_2(CO_2Me)_2$	BSA, LiOAc, benzene, rt	99	99.6	[52]
3	55	$CH_2(CO_2Me)_2$	BSA, KOAc, CH_2Cl_2, 10 °C	93	99	[53b]
4	4p	$CH_2(CO_2Me)_2$	BSA, KOAc, toluene, rt	No reaction	—	[53a]
5	4r	$CH_2(CO_2Me)_2$	BSA, KOAc, toluene, rt	95	98	[53a]
6	7b	$CH_2(CO_2Me)_2$	BSA, LiOAc, CH_2Cl_2, rt	63	96	[55a]
7	58	$CH_2(CO_2Me)_2$	BSA, KOAc, CH_2Cl_2, rt	92	96	[55c]
8	59	$CH_2(CO_2Me)_2$	NaH, CH_2Cl_2, 36 °C	82	97	[56]
9	51	$BnNH_2$	$Pd_2(dba)_3$, THF, 40 °C	90–95	99	[57]
10	5b	$BnNH_2$	$Pd_2(dba)_3$, THF, 40 °C	90–95	96	[57]

The reaction of 1,3-diphenylpropenyl acetate is also used to test some properties of ferrocene ligands. One of the important characteristics of 1,2-disubstituted ferrocenes is that they possess planar chirality when the two substituents are different. The role of planar chirality of ferrocene ligands with different patterns of substitution has been studied. Hou and Dai synthesized 1,2-P,N- and 1,2-S,P-ligands with and without planar chirality [60]. Based upon a detailed study of X-ray diffraction and/or solution NMR spectra of the palladium allylic complexes of these ligands, they found that the enantioselectivity of the reaction is mainly determined by the central chirality. However, to reach high asymmetric induction the matching of planar and central chiralities is important (Scheme 5.34).

Guiry synthesized ligand **60a** and compared the role of planar chirality with its counterpart **60b** (Scheme 5.35) [61]. The reaction provided the alkylated product with (R)-configuration using the benzene-based ligand **61** while that using the analogous ferrocene ligand gave the (S)-product though the enantioselectivity was lower. The results showed that the planar chirality in this system influences the stereochemistry outcome to a greater extent. The role of planar chirality was also demonstrated by another P,N-ferrocene ligand **62** with only planar chirality, synthesized by Anderson, which gave the alkylated product in 79% ee and 89% yield [62].

To take advantage of the above model reaction, Hou and Dai also established the sequence of the trans-effect is C=N > P > S for ligands **57** (Figure 5.6), having the same skeleton and chiralities but different disposition of the coordinated atoms, by analysis of the reaction results, molecular modeling at the PM3 level and study of X-ray diffraction and ^1H, ^{13}C NMR spectra of the complexes [63].

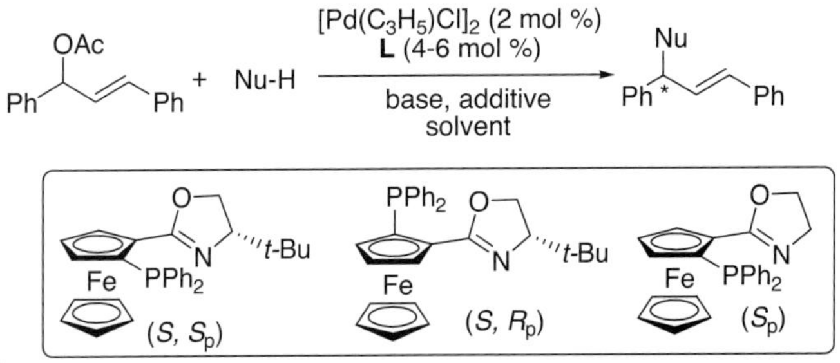

5.3 Formation of a C–C Bond | 125

Scheme 5.34 The role of planar chirality of Fc-phox in asymmetric allylic substitutions.

	Ligand (S, S_p)	Ligand (S, R_p)	Ligand (S_p)
Alkylation: (BSA, KOAc, CH_2Cl_2, rt)	92.3% ee (S)	94.6% ee (S)	76.6% ee (R)
Amination: (Bu_4NF, THF 50 °C)	91.3% ee (R)	97.2% ee (R)	73.7% ee (S)

Different from the reaction of 1,3-diphenylpropenyl acetate, those of cyclic alkenyl acetates have found applications in natural products synthesis [64]. Helmchen developed phosphinooxazoline pentamethylferrocenes **63** and showed their unique asymmetric induction ability in this Pd-catalyzed allylic alkylation reaction [65]. Zheng found that the reaction of cyclohexenyl acetate afforded the alkylated product in 83% ee when the phosphino imine ferrocene ligand **4h** was used [66]. However, products with lower ee were given by the ligand with the nitro group at the ortho- and para-positions [67]. Interestingly, many ligands, such as **7b** afforded products in 11% ee in this reaction while the ee is high (99.6%) when they are used in the reaction of 1,3-diphenylpropenyl acetate. However, the ee value of the product increased from 11% to 58% when the ligand **7a** with two Me group on oxazoline ring was used (Scheme 5.36) [68].

The regioselectivity of monosubstituted allyl substrates in Pd-catalyzed asymmetric allylic substitution reactions has been a challenge for a long time because the nucleophile usually attacks the terminal carbon of the Pd-allyl intermediate. Several approaches, including the use of another transition metal catalyst, have been

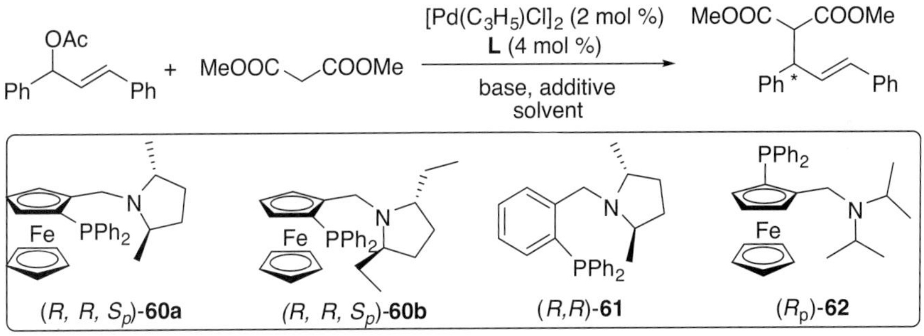

(R, R, S_p)-**60a**	(R, R, S_p)-**60b**	(R,R)-**61**	(R_p)-**62**
69% yield 36% ee (S)	71% yield 20% ee (S)	62% yield 55% ee (R)	89% yield 79% ee (R)

Scheme 5.35 The role of planar chirality in 1,2-P,N-ferrocene ligands, phosphinamine type.

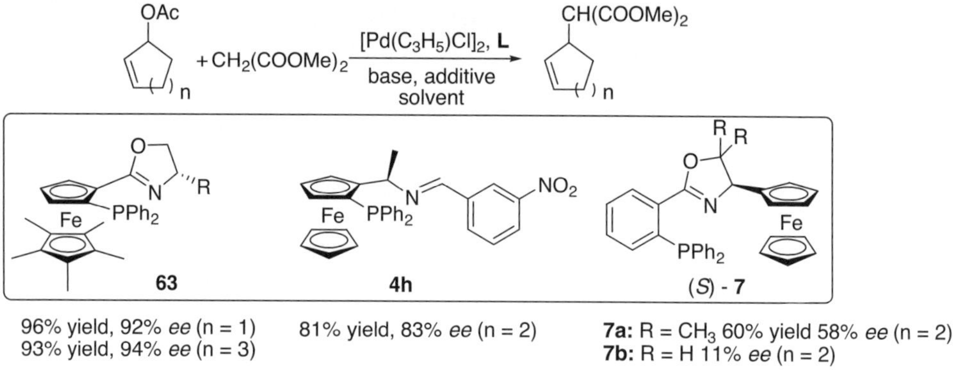

5 1,2-P,N-Bidentate Ferrocenyl Ligands

Scheme 5.36 Pd-catalyzed allylic alkylation reaction of cyclic alkenyl acetates.

96% yield, 92% ee (n = 1)
93% yield, 94% ee (n = 3)

81% yield, 83% ee (n = 2)

7a: R = CH_3 60% yield 58% ee (n = 2)
7b: R = H 11% ee (n = 2)

developed to solve the problem. The Ir-catalyst is one choice among others [69]. Gavrilov carried out the reaction of chlorophenyl-substituted allyl substrate **64** with malonate by using Ir-catalyst with the ferrocene derivative **67** as ligand to obtain the branched product in excellent regioselectivity and moderate enantioselectivity (Scheme 5.37) [70]. A highly regioselective and enantioselective product was obtained by using 1,1′-PN ligands (see Chapter 8).

5.3.2
Asymmetric Heck Reaction and Cross-Coupling Reactions

The asymmetric Heck reaction has found wide application in organic synthesis, both inter- and intramolecular reactions have well been documented. Among them, the reaction of dihydrofuran with aryl and alkenyl reagents has also been used as a model to test the efficacy of chiral ligands. Guiry achieved 98% *ee* with 61% yield for kinetic product **69** by using Fc-Phox **1c** with a *t*-Bu group on the oxazoline ring though the activity of the catalytic system was low (14 days was needed to complete the reaction).

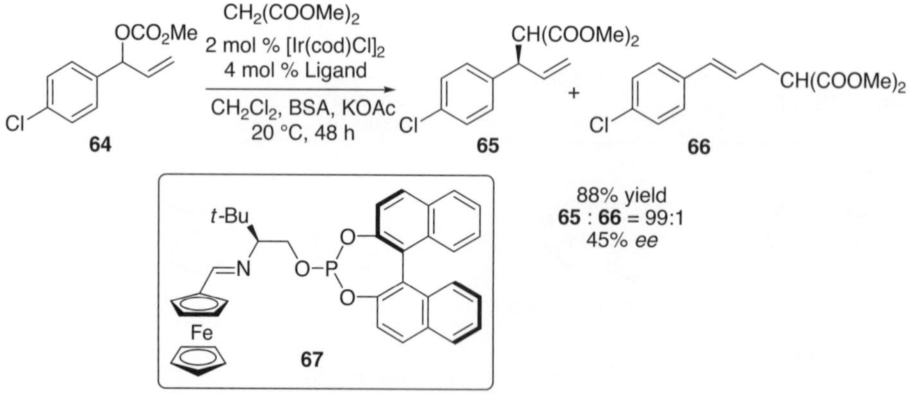

88% yield
65 : 66 = 99:1
45% ee

Scheme 5.37 Ir-catalyzed allylic alkylation reaction of chlorophenyl substituted allyl substrate.

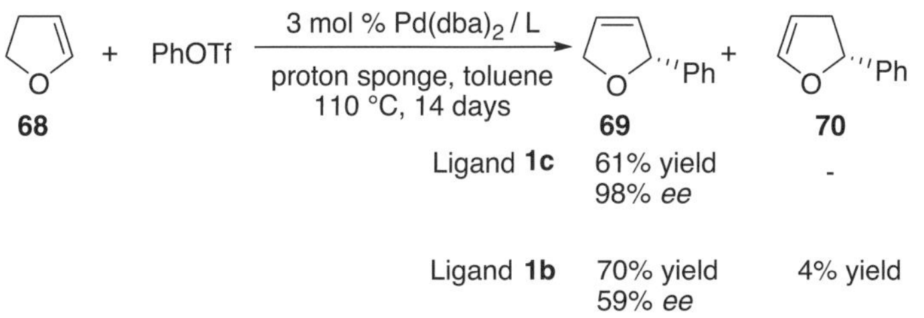

5.3 Formation of a C—C Bond | 127

Scheme 5.38 Intermolecular asymmetric Heck reaction of 2, 3-dihydrofuran.

	Ligand **1c**	61% yield 98% ee	—
	Ligand **1b**	70% yield 59% ee	4% yield

Reaction: **68** + PhOTf → (3 mol % Pd(dba)$_2$/L, proton sponge, toluene, 110 °C, 14 days) → **69** (···Ph) + **70** (···Ph)

Even lower catalytic activity, as well as selectivity, was obtained using ligand **1b** with *i*-Pr as the substituent on oxazoline (Scheme 5.38). Ligand **1c** was also used in the reaction of 2,2-dimethyl-2,3-dihydrofuran, affording 98% *ee* with 90% yield of product [71].

Compared to the intermolecular Heck reaction, the intramolecular version of the reaction is more diverse and has become an important protocol in natural products synthesis. Guiry showed the usefulness of ligand **1c** in the asymmetric intramolecular Heck reaction to synthesize spiro lactam **74** and *cis*-decalins **72**, though the yield in the latter case is lower, the enantioselectivity in both cases is high (Scheme 5.39) [72].

Cross-coupling reactions have attracted increasing attention in recent years. However, there are relatively few studies on the asymmetric version of the reaction. Hayashi used PPFA as the ligand in the coupling of the Grignard reagent **75** with vinyl bromide **76** to afford the corresponding coupling products **77** (Scheme 5.40) [73]. The product with (*S*)-configuration was provided when the ligands (*R*,*S*$_p$)-PPFA or (*S*$_p$)-PPFA was used while that with (*R*)-configuration was given if (*R*,*R*$_p$)-PPFA or (*S*,*R*$_p$)-PPFA was the ligand. These results clearly demonstrated that the

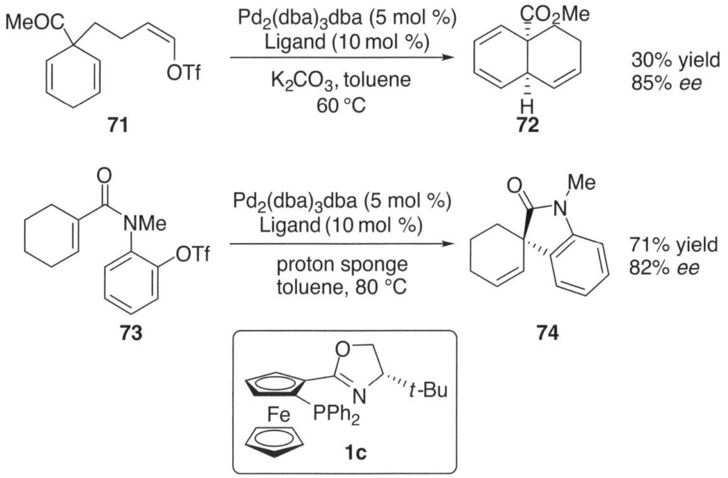

71 → (Pd$_2$(dba)$_3$dba (5 mol %), Ligand (10 mol %), K$_2$CO$_3$, toluene, 60 °C) → **72** (30% yield, 85% ee)

73 → (Pd$_2$(dba)$_3$dba (5 mol %), Ligand (10 mol %), proton sponge, toluene, 80 °C) → **74** (71% yield, 82% ee)

Ligand **1c**: ferrocene with oxazoline bearing *t*-Bu and PPh$_2$

Scheme 5.39 Pd-catalyzed intramolecular asymmetric Heck reaction.

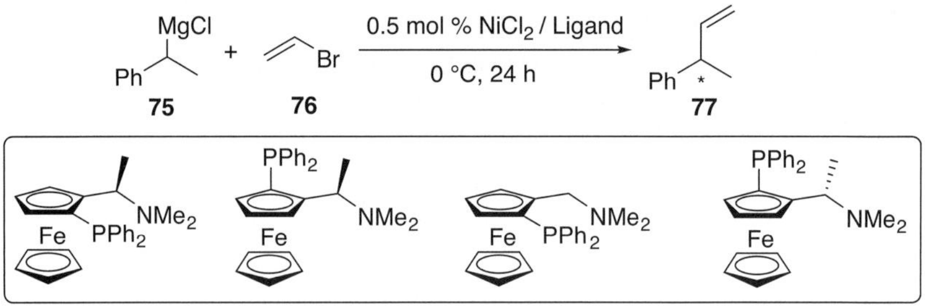

Scheme 5.40 Coupling of Grignard reagent with vinyl bromide.

Ph-CH(MgCl)-CH₃ (**75**) + CH₂=CH-Br (**76**) → [0.5 mol % NiCl₂ / Ligand, 0 °C, 24 h] → Ph-CH*(CH₃)-CH=CH₂ (**77**)

Ligands:
- (R, S_p)-PPFA — 68% ee (S)
- (R, R_p)-PPFA — 54% ee (R)
- (S_p)-PPFA — 65% ee (S)
- (S, R_p)-PPFA — 63% ee (R)

configuration of the product depends upon the planar chirality of the ligand. Another point worth mentioning is that the reaction gave the product in 65% ee when the ligand with only planar chirality was used, this value is almost the same as that using (R, S_p)-PPFA. Though the enantioselectivities were moderate, these results provided an excellent example of the importance of the planar chirality of the ligand in asymmetric induction of the reaction as well as in control of the configuration of the product. Richards tried the same reaction using styrene as reagent catalyzed by Pd-Fc-PHOX **1b** as catalyst to obtain the product in 74% yield and in 45% ee [74].

Uemura used the Fc-PHOX ligand **1b** in this Ni-catalyzed cross-coupling of allylic compounds with Grignard reagents. The corresponding products in 85–91% ee were provided by using Fc-PHOX ligand with planar and central chiralities which is better than that achieved using PHOX with only central chirality (Scheme 5.41) [75].

The binaphthylene structure represents an important class of compounds used widely in asymmetric catalysis as ligands or auxiliaries. One of the simplest and most straightforward procedures to gain these products is the cross-coupling reaction. Because of the difficulty of the inherent steric problem in the coupling of two substrates, only a few asymmetric versions of coupling reactions have been applied to synthesize this class of compounds. Cammidge provided a successful example using P,N-ferrocene. The 1,1′-binaphthalene derivative was provided in 85% ee and in 60% yield (Scheme 5.42) [76]. In this Suzuki coupling reaction only the P,N-ligand, (S, R_p)-PPFA showed high catalytic activity. This was explained by the presence of NMe₂ favoring the coordination to the boron atom because of its stronger donor property. Interestingly, the similar ligand, PPF-OMe, was very effective in Kumada coupling by

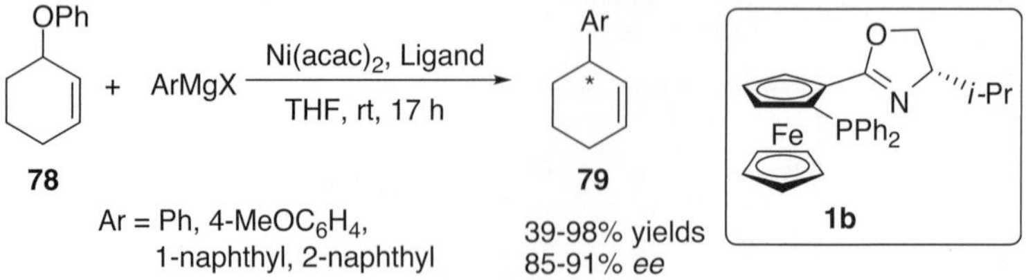

78 (cyclohexenyl OPh) + ArMgX → [Ni(acac)₂, Ligand, THF, rt, 17 h] → **79** (cyclohexenyl Ar*)

Ar = Ph, 4-MeOC₆H₄, 1-naphthyl, 2-naphthyl
39–98% yields
85–91% ee

Ligand **1b**: Fc-PHOX (Fe-cyclopentadienyl with PPh₂ and oxazoline-iPr)

Scheme 5.41 Ni-catalyzed cross-coupling of allylic compounds with Grignard reagents.

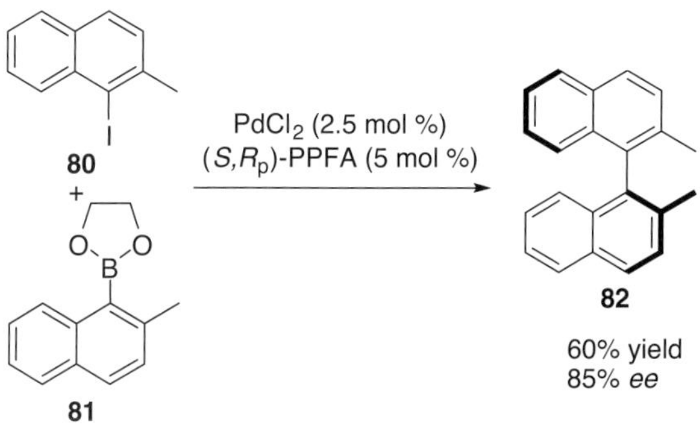

Scheme 5.42 Synthesis of 1,1′-binaphthalene derivative by Suzuki coupling.

using corresponding Grignard reagent and bromide to give the same product in 95% ee and in 60% yield while it was ineffective in the Suzuki–Miyaura coupling reaction [77].

The ligand PPFA was also used by Bringmann to prepare the antileishmanial alkaloids ancistroealanine A **85** via Suzuki coupling reaction of arylboronic acid **83** with chiral aryl iodide **84** in up to 50% de (Scheme 5.43). The use of $PdCl_2$ failed to give the product [78].

5.3.3
Addition of Organometallic Reagents to a C=X Bond

Sammakia studied the use of Fc-PHOX ligands in asymmetric catalysis and found that they were effective in the Cu-catalyzed conjugated addition of Grignard reagent to cyclic enone (Scheme 5.44). Among the Fc-PHOX with different substituents on the oxazoline ring, Fc-PHOX **1d** was the best, providing 3-butyl cyclic ketones in 65–92% ee, the ratio of 1,4-addition/1,2-addition being 100 : 1 [79].

Wang used PPFA as ligand in the conjugated addition of diethylzinc reagent to chalcones. The reaction afforded products in low to moderate ees when Ar^1 of the

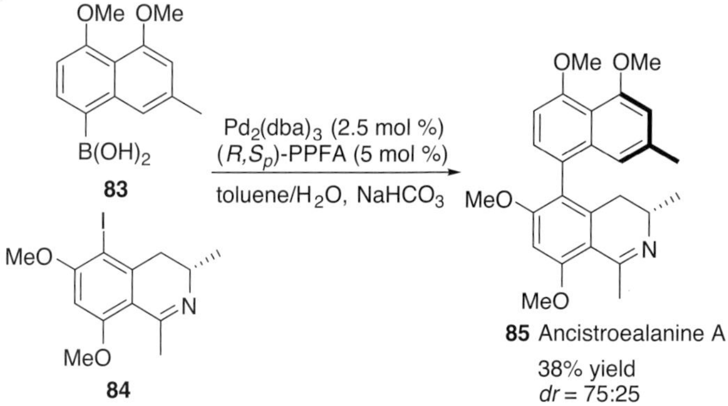

Scheme 5.43 Synthesis of ancistroealanine A **85** via Suzuki coupling.

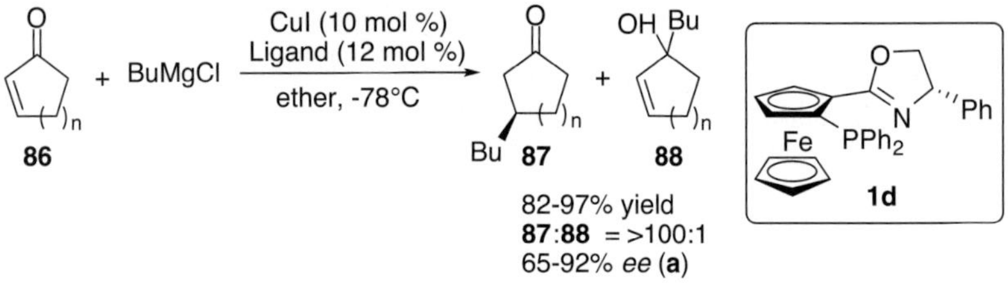

Scheme 5.44 Cu-catalyzed conjugated addition of Grignard reagent to cyclic enone.

substrate was phenyl or phenyl with a substituent at the meta- or para-position. A dramatic increase in enantioselectivity was found when halogen or MeO group was presented at the ortho-position on Ar^1 in substrates, providing addition products in high yields and in high ee (>90%). No change in ee was found if the ortho-substituent was Me. It was postulated that the non-bonding electron pair of the halogen or oxygen plays a role in the reaction (Table 5.12) [80].

The effectiveness of PPFA was also demonstrated in the addition reaction of diethylzinc to N-diphenylphosphinoyl aromatic aldimines **91**, providing corresponding products **92** in high enantioselectivities (Scheme 5.45). Much lower enantioselectivity was obtained when the N,P,P-ligand BPPFA was used in the reaction [81a]. The same group also studied the addition reaction of diethylzinc to N-Ts aldimines by using N-ferrocenoyl-2-[(diphenylphosphino)methyl]-pyrrolidine **96** that they developed as a ligand (Figure 5.7). When 4 mol% of $Cu(OTf)_2$ and 3 mol% of ligand was used as catalyst, products in 79–86% ee were provided [81b]. Interestingly the ee decreased sharply when the amide group of the ligand was reduced to amine and used as the ligand.

Table 5.12 Cu-catalyzed conjugated addition of Et_2Zn to chalcones, effect of ortho-substituent.

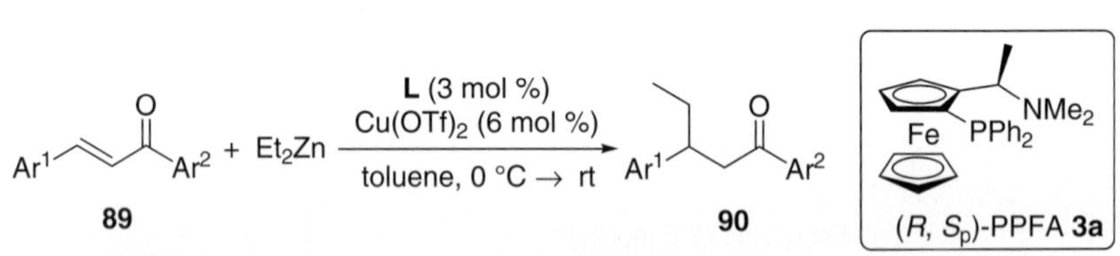

Entry	Ar^1	Ar^2	Yield (%)	ee (%)
1	Ph	p-MeOC$_6$H$_4$	51	27
2	p-ClC$_6$H$_4$	Ph	59	38
3	o-ClC$_6$H$_4$	Ph	95	92
4	o-MeOC$_6$H$_4$	ferrocenyl	88	92
5	o-BrC$_6$H$_4$	ferrocenyl	80	91
6	o-MeC$_6$H$_4$	ferrocenyl	50	20

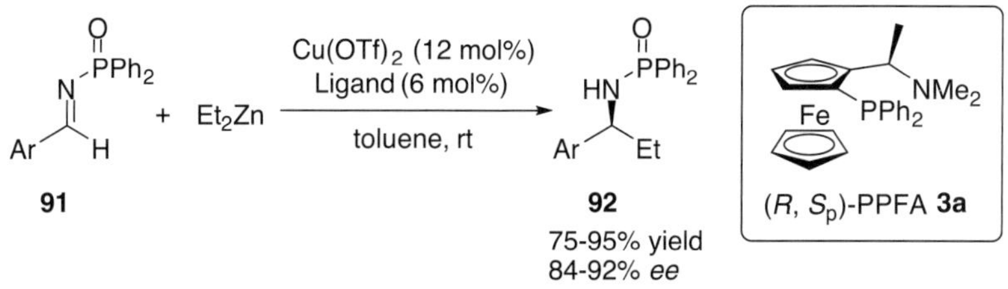

Scheme 5.45 Cu-catalyzed addition of Et$_2$Zn to aldimines.

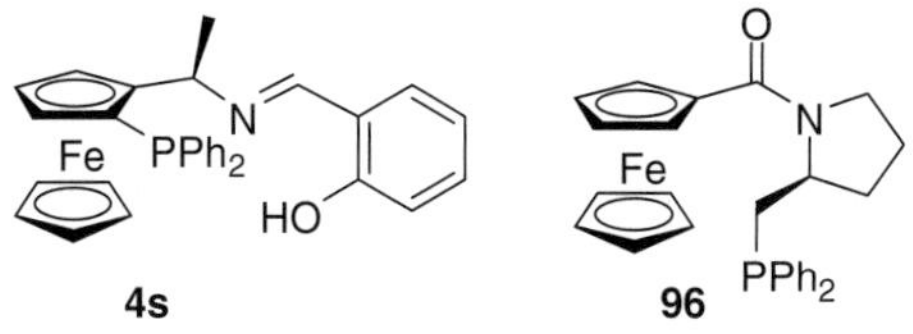

Figure 5.7 Structure of ligands **4s** and **96**.

Lautens realized high enantioselective Pd-catalyzed ring opening reaction of [2,2,1]- and [3,2,1]-oxabicyclic alkenes with organozinc using FcPHOX as ligand to afford products in high enantioselectivities (Scheme 5.46). The size of substituent on the oxazoline ring influenced the stereoselectivity, better results were given using a ligand with *i*-Pr as the substituent. The planar chirality has little effect on the stereochemistry of the reaction [82].

Kim synthesized ligand **4s** (Figure 5.7) with a hydroxy group on the phenyl ring and used it in the addition of diethylzinc reagent to benzaldehydes to obtain the product in 40% yield and 98% *ee*. Low to good enantioselectivity with low to good conversion were provided for other arylaldehydes [83].

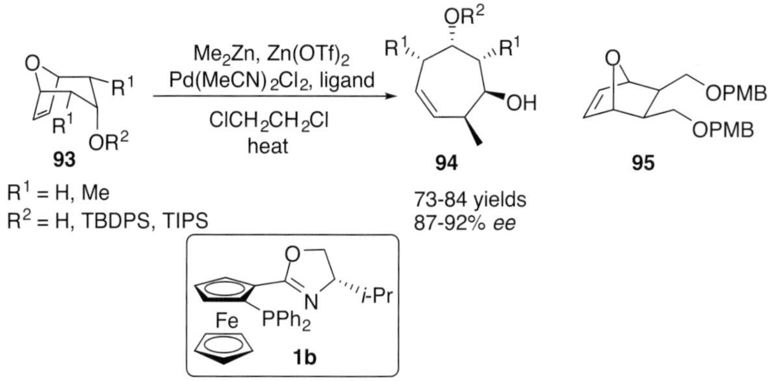

Scheme 5.46 Pd-catalyzed ring opening reaction of [2,2,1]- and [3,2,1]-oxabicyclic alkenes.

5.4
Cycloaddition Reactions

Cycloaddition reactions represent one of the fastest growing areas of research in homogeneous catalysis for the synthesis of fine chemicals and pharmaceutical materials. Although numerous reports are available today on various cycloadditions, ferrocene-derived P,N-based catalysts played a significant role in the asymmetric cycloaddition reactions. Descriptions of some of the most important work in this area follow.

5.4.1
[3 + 2] Cycloaddition Reactions

1,3-Dipolar cycloaddition is a classic reaction in organic chemistry, consisting of the reaction of a dipolarophile with a 1,3-dipolar compound that allows the production of various five-membered heterocycles. Metal-catalyzed asymmetric 1,3-dipolar cycloadditions have only recently become an important research field [84, 85]. The efficiency of chiral catalysts relies not only on the capability of the enantiopure catalyst to help discriminate between the two p-faces of the dipolarophile, but also on its ability to control both the exo/endo selectivity and the regiochemistry as well as the yield.

The most recent advance in the chemistry of azomethine ylides is the use of chiral catalysts in the stereoselective synthesis of pyrrolidine derivatives via metallo-azomethine ylides. The most effective Lewis acids in the cycloaddition reactions of metallo-azomethine ylides are unarguably Ag and Cu salts. Several chiral ligands have been used in the catalytic asymmetric cycloaddition of azomethine ylides. Zhou and coworkers achieved excellent levels of asymmetric induction using the AgOAc complex with a chiral ferrocenyloxazoline-derived P,N ligand **1k** (Scheme 5.47) [86].

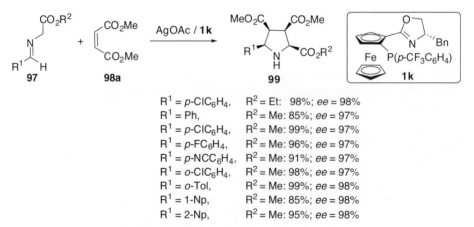

Scheme 5.47 Ag-catalyzed 1, 3-dipolar cycloadditions of azomethine ylides.

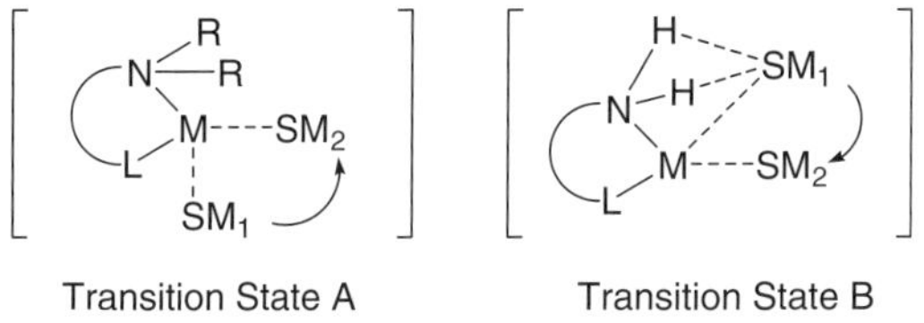

Figure 5.8 Hydrogen-bonding directed change of transition state.

It is noteworthy that, comparing the enantioselectivity of the reaction catalyzed by **1k** with the counterpart of the phenyloxazoline *i*-Pr-Phox, low *ee* (14%) was obtained. An extra base, such as a tertiary amine, was not necessary in this case, since the reactive metal-bound azomethine ylide dipole was formed by deprotonation with acetate that played the role of base. Indeed, it was postulated that AgOAc, bearing a weakly basic-charged acetate ligand, facilitated the deprotonation of iminoesters to generate the azomethine ylides.

In 2007, Zhou and coworkers reported a hydrogen-bonding-directed reversal of enantioselectivity in the AgOAc-catalyzed [3 + 2] cycloaddition of azomethine ylides employing similar chiral ferrocene-derived *P,N* ligands **3i** and **3j** [87]. On using ligand **3i** containing a free NH_2, the product was just the enantiomer of that obtained with ligand **3j** having no free NH_2 group but an NMe_2 group. The reversal of absolute configuration could be rationalized by the difference in the reaction transition states (Figure 5.8). Both substrates SM1 and SM2 coordinate with the central metal M, but in transition state B, SM2 can coordinate with the central metal M, while the other substrate SM1 has an interaction with NH_2 of the ligand through hydrogen bonding. Thus, different enantiofacial attack was afforded, leading to the reversal of enantioselectivity.

A variety of iminoester substrates derived from aldehydes with different steric and electronic properties have been examined (Table 5.13). All reactions went to completion in high isolated yields with high enantioselectivity. Reversal of the absolute configuration was realized in all the reactions of the various iminoesters and dimethyl maleate, regardless of the steric hindrance and electronic properties of the benzene ring of the iminoester. Job's method and ^{1}H-NMR titration experiments were employed to probe the hydrogen bonding between the complex AgOAc-**3i** and dimethyl maleate. A significant change in N–H chemical shift was observed, and an ∼1 : 1 complex was indicated. DFT studies proposed a reasonable mechanism for the reversal of the enantioselectivity. The strategy may provide some useful hints for ligand design.

In recent years, Cu–Lewis acids in combination with ferrocene-derived chiral ligands have been evaluated in the enantioselective cycloadditions of azomethine ylides. Zhang *et al.* demonstrated that the combination of copper(I) salts and a chiral phosphinooxazoline *P,N* ligand can efficiently catalyze [3 + 2] cycloaddition of azomethine ylieds (Scheme 5.48) [88]. The exo adduct was obtained as the major product in all cases with high enantioselectivity (up to 98% *ee*).

In 2006, Cu-catalyzed asymmetric cycloaddition of azomethine ylides to nitroalkenes was realized for the first time by Hou and coworkers using the Cu(I) complexes

Table 5.13 Ag-Catalyzed asymmetric cycloaddition of azomethine ylides.

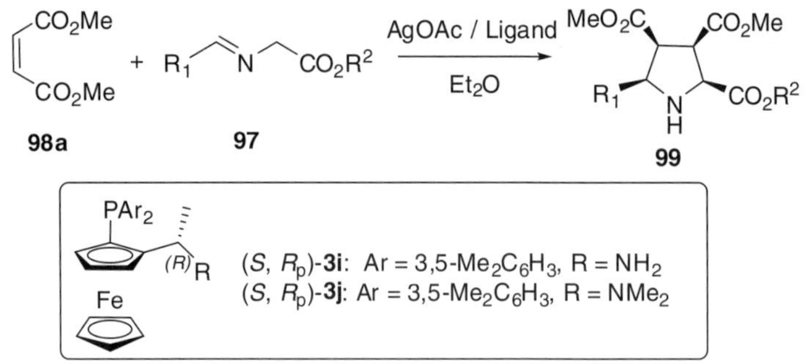

$98a + 97 \xrightarrow{\text{AgOAc / Ligand}}_{\text{Et}_2\text{O}} 99$

98a: maleate with CO_2Me/CO_2Me
97: $R_1\text{CH=N-CH}_2\text{CO}_2R^2$
99: pyrrolidine product with MeO_2C, CO_2Me, R_1, CO_2R^2, NH

Ligand: Ferrocene with PAr_2 and $(R)R$ substituents
(S, R$_p$)-3i: Ar = 3,5-Me$_2$C$_6$H$_3$, R = NH$_2$
(S, R$_p$)-3j: Ar = 3,5-Me$_2$C$_6$H$_3$, R = NMe$_2$

Entry	R^1/R^2	Ligand	Yield (%)	ee (%)
1	p-ClC$_6$H$_4$/Et	3j	95	−92
2	p-ClC$_6$H$_4$/Et	3i	90	92
3	Ph/Me	3i	95	90
4	Ph/Me	3j	96	−85
5	p-Anisyl/Me	3i	93	90
6	p-Anisyl/Me	3j	98	−87
7	p-ClC$_6$H$_4$/Me	3i	96	88
8	p-ClC$_6$H$_4$/Me	3j	91	−91
9	o-Toluyl/Me	3i	95	88
10	o-Toluyl/Me	3j	95	−85
11	2-Naphthyl/Me	3i	98	91
12	2-Naphthyl/Me	3j	91	−87

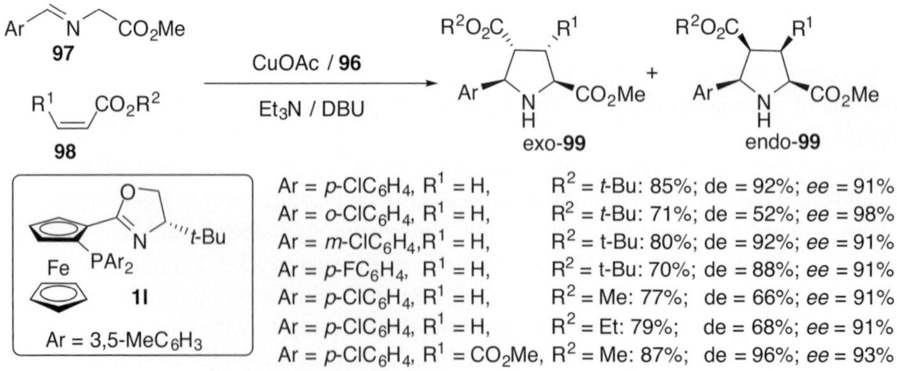

$97: \text{Ar-CH=N-CH}_2\text{CO}_2\text{Me}$
$98: R^1\text{CH=CH-CO}_2R^2$

$\xrightarrow[\text{Et}_3\text{N / DBU}]{\text{CuOAc / 96}}$ exo-99 + endo-99

Ligand 1l: Ferrocene with PAr$_2$ and oxazoline-t-Bu, Ar = 3,5-MeC$_6$H$_3$

Ar = p-ClC$_6$H$_4$, R^1 = H, R^2 = t-Bu: 85%; de = 92%; ee = 91%
Ar = o-ClC$_6$H$_4$, R^1 = H, R^2 = t-Bu: 71%; de = 52%; ee = 98%
Ar = m-ClC$_6$H$_4$, R^1 = H, R^2 = t-Bu: 80%; de = 92%; ee = 91%
Ar = p-FC$_6$H$_4$, R^1 = H, R^2 = t-Bu: 70%; de = 88%; ee = 91%
Ar = p-ClC$_6$H$_4$, R^1 = H, R^2 = Me: 77%; de = 66%; ee = 91%
Ar = p-ClC$_6$H$_4$, R^1 = H, R^2 = Et: 79%; de = 68%; ee = 91%
Ar = p-ClC$_6$H$_4$, R^1 = CO$_2$Me, R^2 = Me: 87%; de = 96%; ee = 93%

Scheme 5.48 Cu-catalyzed [3 + 2] cycloadditions of azomethine ylides.

with modified Fc-Phox ligand [89]. Subtle variation of the electronic properties of the aryl group on the phosphorus atom of the chiral P,N-ferrocene ligands led to a dramatic switch in exo/endo selectivity (entries 1–5) in the 1,3-dipolar cycloaddition of nitroalkenes to an iminoester derived from glycine (Table 5.14). **1b** was used as ligand, a high yield and enantioselectivity of exo adducts was achieved, while endo adducts were obtained with high enantioselectivity when **1h** was used as ligand,. A qualitative model has been proposed to rationalize the observed stereoselectivity. These results point to a new possibility for switching the stereochemistry of a reaction by varying the electronic properties of the ligands and give useful hints for ligand design.

Table 5.14 Cu FcPhox catalyzed [3 + 2] cycloadditions of nitroalkenes and iminoesters.

Entry	R^1	R^2	Ligand	exo/endo	Yield (%)	ee (%)
1	Ph	Ph	1b	Only exo	87	95
2	Ph	Ph	1g	73:27	67	95
3	Ph	Ph	1m	Only exo	65	98
4	Ph	Ph	1f	Only exo	49	98
5	Ph	Ph	1h	14:86	85	98
6	Ph	p-MeOC$_6$H$_4$	1b	Only exo	77	96
7	Ph	p-MeOC$_6$H$_4$	1h	30:70	79	95
8	Ph	m-ClC$_6$H$_4$	1b	Only exo	74	95
9	Ph	m-ClC$_6$H$_4$	1h	11:89	82	92
10	Ph	i-Pr	1b	Only exo	75	98
11	Ph	i-Pr	1h	06:94	88	97
12	p-MeOC$_6$H$_4$	Ph	1b	89:11	96	97
13	p-MeOC$_6$H$_4$	Ph	1h	18:82	79	96
14	m-ClC$_6$H$_4$	Ph	1b	88:12	97	92
15	m-ClC$_6$H$_4$	Ph	1h	17:83	98	84
16	p-BrC$_6$H$_4$	Ph	1b	86:14	77	83
17	p-BrC$_6$H$_4$	Ph	1h	18:82	71	88
18	2-Naphthyl	Ph	1b	92:8	92	92
19	2-Naphthyl	Ph	1h	19:81	98	97

5.4.2
Asymmetric Cyclopropanation Reactions

Catalytic asymmetric cyclopropanation of olefins with diazoacetates is an important method for the production of chiral cyclopropane compounds. The transition metal-catalyzed decomposition of diazoacetate derivatives in the presence of alkenes is one of the most efficient procedures.

In 2004, Zheng and coworkers reported the application of ferrocene-derived P,N,N phosphine-heteroaryl imine ligands in the Ru(II)-catalyzed asymmetric cyclopropanation of styrene with ethyl diazoacetate (Table 5.15) [90]. Both ligands **4p** and **4t** gave the major cyclopropanation products with (1S, 2R) configuration. Surprisingly, replacement of the aminomethyl group of **4p** with an aminopropyl (**4u**) or an aminobenzyl group (**4v**) led to dramatic decrease in reactivity and enantioselectivity. Interestingly, the product formed on ligand (**4v**) showed the reverse configuration (entry 4). The best enantioselectivity (up to 95% for the cis-isomer) was obtained with the ligand **4p** at high reaction temperature (entry 5).

5.4.3
Asymmetric Diels-Alder Reactions

Diels–Alder reactions are classical pattern reactions that play an important role in the construction of complicated molecules with stereochemical control. However, few ferrocene-based P,N-ligands have demonstrated high performance in this transformation. Carmona and coworkers have applied ferrocene-derived phosphino-oxazo-

Table 5.15 Ru-catalyzed asymmetric cyclopropanation.

4p: R = Me, R' = H
4t: R = Me, R' = Me
4u: R = Et, R' = H
4v: R = Ph, R' = H

Entry	Ligand	Yield (%)	Cis/Trans	ee of cis (%)	ee of trans (%)
1	4p	93	25:75	85 (1S,2R)	83 (1S,2S)
2	4t	90	20:80	64 (1S,2R)	50 (1S,2S)
3	4u	52	20:80	50 (1S,2R)	45 (1S,2S)
4	4v	32	20:80	3 (1R,2S)	10 (1R,2R)
5[a]	4p	99	19:81	95 (1S,2R)	90 (1S,2S)

[a]Run in 60 °C.

Table 5.16 Rh- And Ir-catalyzed asymmetric Diels–Alder reactions.

Entry	Precatalyst	T (°C)	Yield (%)	exo: endo	ee of exo (%)	ee of endo (%)
1	—	RT	0.5	—	—	—
2	Rh/1b	RT	91	86:14	4	14
3	Rh/1b	−20	75	91:9	18	36
4	Ir/1b	RT	94	87:13	8	10
5	Ir/1b	−20	92	91:9	16	38

line ligands in the Rh- and Ir-catalyzed asymmetric Diels–Alder reaction of cyclopentadiene. However, poor diastereoselectivity and enantioselectivity were obtained (Table 5.16) [91].

5.5
Miscellaneous Reactions

In addition to the reduction reactions, allylic amination reactions, and carbon–carbon bond-formation reactions, 1,2-*P,N*-ferrocenes have also played a role in some other reactions.

Although the hydroesterification of vinyl aromatics produces valuable intermediates for perfumes and pharmaceuticals, there are only limited *N,P*-ferrocenyl ligands employed in the hydroesterification of styrene. In 1997, Inoue *et al.* described a Pd-catalyzed asymmetric hydroesterification using a chiral ferrocene ligand containing aminophosphine **107** with good enantioselectivity (86% *ee*), but with low yield (17%) and poor regioselectivity (b/n = 44/56) [92]. Then, Kollár's group used Pt-complexes with a ferrocene-based chiral aminophosphine **52** with high yield (99%) and good regioselectivity (b/c = 99/1), but rather low enantioselectivity (0.5%) (Scheme 5.49) [93a].

Recently, Chan and coworkers reported the palladium-catalyzed methoxycarbonylation of styrene using *N,P*-ferrocenyl ligands (**107**, **52**, **11a**) in the presence of *p*-TsOH and $CuCl_2$ at 50 °C. In the case of ligand **107**, excellent yield (>99%) and good enantioselectivity (63% *ee*) were obtained, but the regioselectivity to the branched ester was poor. When ligand **11b** was employed in the reaction, although the enantiomeric excess was relatively high (up to 64% *ee*), low regioselectivity (b/n = 40/60) and poor yield (14%) were obtained. In addition, the use of $CuCl_2$ as a cocatalyst enhanced the regioselectivity (b/n) of this reaction [93b].

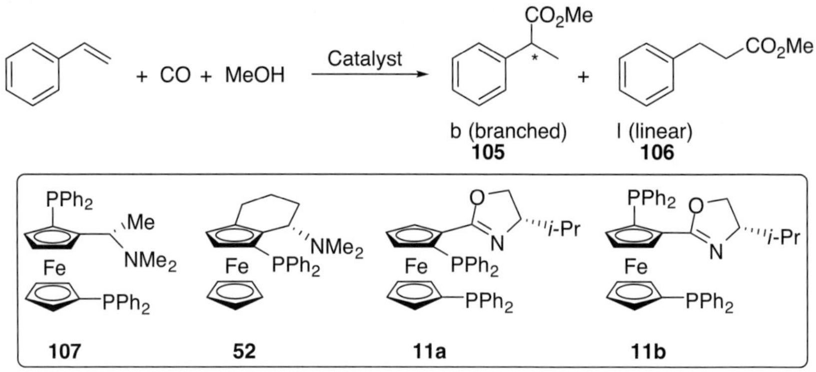

Scheme 5.49 Asymmetric hydroesterification of styrene.

Though asymmetric allylic oxidation of olefins catalyzed by chiral metal complexes is well established, there are few reports of chiral N,P-bidentate ferrocenyl ligands applied in this particular reaction. Very recently, Kim's group reported that cyclic olefins were oxidized by the copper complex with ligand **17a** in MeCN, excellent enantioselectivities (91–98%) were obtained (Scheme 5.50). Among them, the cyclopentene gave the highest enantioselectivity of 98% and the best isolated yield of 85% while the cyclooctene showed lower reactivity. 1,5-Cyclooctadiene gave perfect enantio-selectivity (99%) and complete conversion. In all cases, 1,2-disubstituted ligands showed better catalytic activity and stereoselectivity than their 1,1′-counterparts [94].

Uemura developed a Pd-catalyzed arylation of *tert*-cyclobutanols via C−C bond cleavage to provide γ-aryl ketones. When the modified PPFA **3k** was used, chiral products were given in excellent yields and in good enantioselectivity (Scheme 5.51) [95].

Using Fc-PHOX, Uemura developed a Ru-catalyzed oxidative kinetic resolution of secondary alcohols in the presence of sodium isopropoxide and acetone (Scheme 5.52). A variety of secondary alcohols are suitable in this reaction, providing

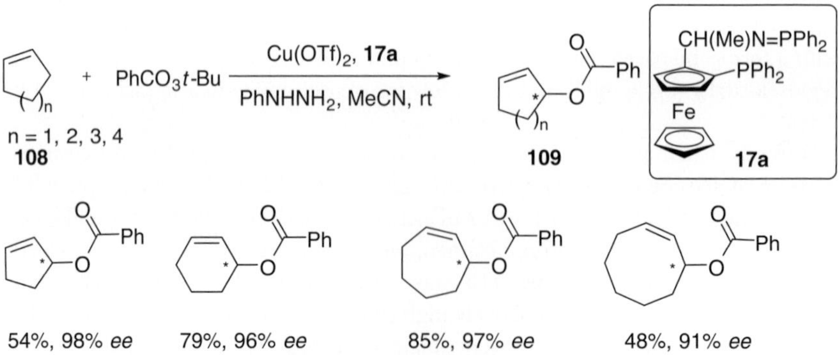

Scheme 5.50 Cu-catalyzed asymmetric allylic oxidation.

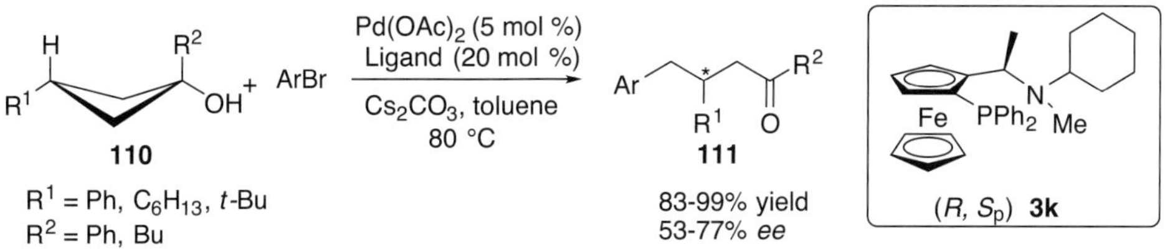

Scheme 5.51 Pd-catalyzed arylation of tert-cyclobutanols.

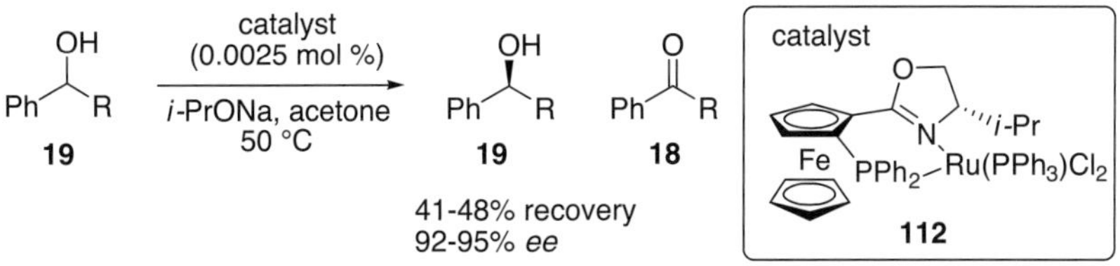

Scheme 5.52 Ru-catalyzed oxidative kinetic resolution of secondary alcohols.

optically alcohols in high *ee*. The catalytic activity of the Ru-catalyst is high, especially for the reaction of 1-indanol where the TOF exceeded $80\,000\,h^{-1}$ and TON value approached 40 000. The reaction of 1-indanol also proceeded on the 1 mol-scale (134 g of racemic 1-indanol was used). It is interesting that the opposite configuration of alcohols was obtained by transfer hydrogenation and oxidative kinetic resolution when the same ligand was used [96].

Fc-PHOX was used as a Lewis base to catalyze the aza-Baylis–Hillman reaction of N-sulfonated imines with activated olefins (Scheme 5.53). When (R, R_p)-Fc-PHOX **1n** was the catalyst, the chiral adduct **115** was afforded in 45% yield and 65% *ee* [97].

Schmalz developed a desymmetrization protocol to prepare planar chiral tricarbonylchrominium complexes via the Pd-catalyzed methoxycarbonylation reaction (Scheme 5.54). When (R,S_P)-PPF-pyrrolidine was used as ligand, the planar-chiral products were obtained in up to 95% *ee*. A strong dependence of enantioselectivity on the reaction time (conversion) was observed. It is believed that the initial enantioselectivity is enhanced by a subsequent kinetic resolution connected to the formation of the bis-methoxycarbonylated byproduct [98].

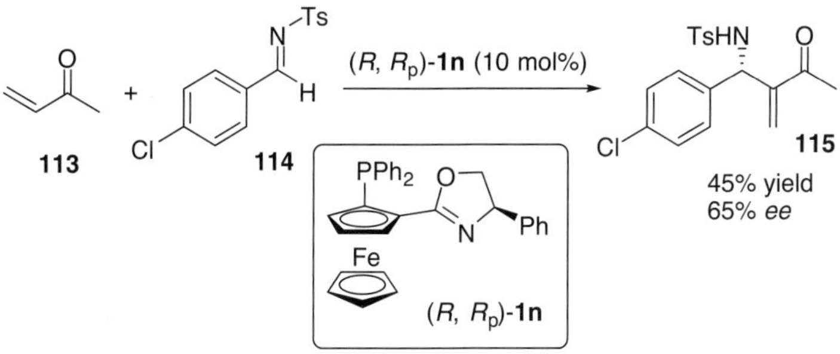

Scheme 5.53 Fc-Phox catalyzed aza-Baylis–Hillman reaction.

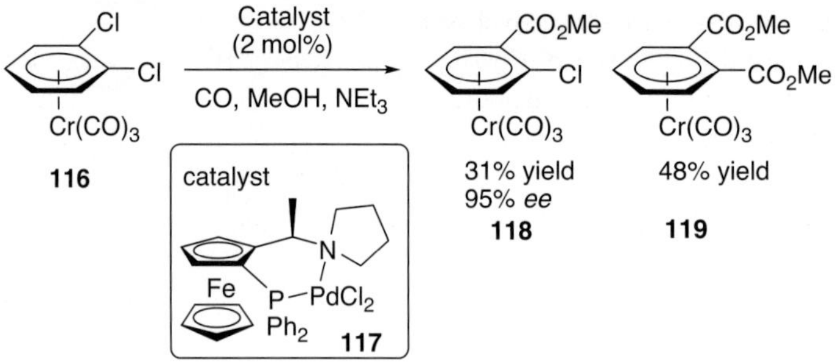

Scheme 5.54 Pd-Catalyzed desymmetrization reaction.

5.6
Conclusion and Perspectives

As shown in this chapter, over the past decades, many different types of chiral 1,2-P, N-ferrocene ligands have been developed and applied successfully in carbon–carbon and carbon–heteroatom bond-forming reactions with high enantioselectivity. The focus is not only on exploring the application of ligands in different types of reactions but also on the modification of ligands to improve the catalytic activity of the metal–ligand complexes as well as the stereoselectivities of the reactions. With more understanding of steric and electronic factors of ferrocene ligands in asymmetric catalysis, even more new types of chiral 1,2-P,N-ferrocene ligands will be designed and used in asymmetric catalysis.

5.7
Experimental: Selected Procedures

5.7.1
Ugi's Amine Synthesis: [99]

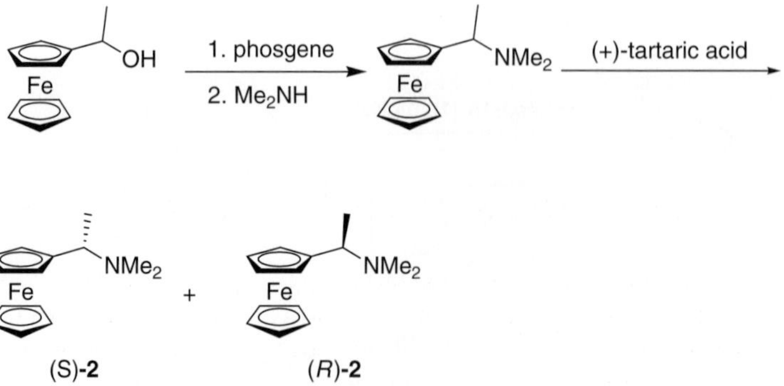

5.7.1.1 Synthesis of α-Ferrocenylethyldimethylamine (2)

At −20 °C a solution of 23.0 g of l-ferrocenylethano1 in 150 mL of toluene was added dropwise to a stirred solution of 12.5 g of phosgene in 100 ml of toluene. 30 min after the addition was complete the reaction mixture was allowed to warm to 20 °C and, without isolation of the chloride, added to a −20 °C solution of 22.5 g of dimethylamine in 200 mL of isopropyl alcohol. The temperature was allowed to rise to +20 °C. The reaction mixture was filtered and evaporated to dryness. The residue was taken up in benzene, extracted with 8.5% phosphoric acid, washed with benzene, neutralized with Na_2CO_3, extracted with benzene, dried, and evaporated; crude yield: 24.4 g (95.7%); yield after distillation (bp 110 °C/0.45 mm Hg), with some decomposition): 17.5 g (68%). The undistilled product was sufficiently pure for the resolution of the antipodes of 2.

5.7.1.2 Resolution of 2

Solutions of 51.4 g of racemic 2 and 30.0 g of (R)-(+)-tartaric acid, each in 100 mL of methanol, were mixed at 55 °C with stirring. Seeding crystals were added. The temperature was lowered at a rate of 2 °C h^{-1}. After 24 h 30.0 g (75% of one antipode) of 2-tartrate was collected and 19.0 g of partially optically active 2 was set free, $[\alpha]_D^{25}$ −11.0° (c 1.5, ethanol). Solutions of this (−)-2 and 11.1 g of (R)-(+)-tartaric acid, each in 50 ml of methanol, were mixed at 55 °C and seeded. After slow cooling 27.5 g of (−)-2-tartrate was obtained; this was converted into 17.0 g (66% overall yield) of (S)-2, bp 120–121 °C (0.7 mm); $[\alpha]_D^{25}$ −14.1 (c 1.6, ethanol). The mother liquor of the first crystallization was concentrated to 1/4 of its original volume. Diethyl ether was added until no further precipitate was formed. After standing at 0 °C overnight 48.6 g of 2-tartrate was collected, $[\alpha]_D^{25}$ +8.0° (c 1.5, ethanol) and recrystallized from 800 ml of acetone–water (10:1) to yield 34.5 g (85%); $[\alpha]_D^{25}$ +12.0° (c 1, ethanol). By a second recrystallization from 500 mL of aqueous acetone 28.0 g (69%) was obtained; $[\alpha]_D^{25}$ +14.0. By working up the mother liquors an overall yield of 80–90% of both antipodes could easily be obtained.

(R)-2 → (n-BuLi / Et$_2$O, Ph$_2$PCl) → 3a (R,S$_p$)-PPFA

5.7.1.3 Synthesis of (S)-N,N-Dimethyl-1-[(R)-2-(diphenylphosphino)ferrocenyl]ethylamine (3a)

22.4 mL of 1.6 M butyllithium in hexane was added to a solution of 7.6 g (29.5 mmol) of (R)-2 in 20 mL of dry ether at 25 °C over a period of 30 min. The mixture was stirred at room temperature for 1.5 h and then 9.7 g (44 mmol) of chlorodiphenylphosphine in 20 mL of ether was added with heating under gentle reflux over 50 min. After 3 h reflux, aqueous sodium hydrocarbonate was slowly added with cooling in an ice-bath. The resulting organic layer and benzene extracts from the aqueous layer were combined, washed with water, dried over anhydrous sodium sulfate, and concentrated *in vacuo* to afford a red oil. The oil was chromatographed on alumina to give 8.1 g (62%) of (R,S_p)-**3a** as a red solid.

5.7.2
(S, S_p)-2-[(S)-2-(diphenylphosphino)ferrocenyl]-4-(1-methylethyl)oxazoline 1b [100]

5.7.2.1 Synthesis of (2S)-N-(1-hydroxy-3-methylbutyl)ferrocenamide

To a stirred suspension of ferrocenecarboxylic acid (1.033 g, 4.49 mmol) in CH_2Cl_2 (15 mL) at room temperature under nitrogen, was added via syringe oxalyl chloride (0.79 mL, 9 mmol). Gas evolution was accompanied by the formation of a dark red homogeneous solution after 10 min. The reaction mixture was stirred for an additional 10 min, followed by removal of the solvent *in vacuo*. The resultant crude ferrocenyl chloride, isolated as a dark oil that crystallized on standing, was taken up in CH_2Cl_2 (10 mL) and added via syringe to a solution of (S)-(+)-valinol (0.554 g, 5.37 mmol) and NEt_3 (1.25 mL, 9 mmol) in CH_2Cl_2 (7 mL) at room temperature under nitrogen. After stirring for 3 h, the dark reaction mixture was washed with H_2O (2 × 20 mL), dried over $NaSO_4$, filtered and evaporated *in vacuo*. The crude product was purified by column chromatography (petroleum ether/ethyl acetate 1/1) to give the amide as a yellow crystalline solid (1.184 g, 84%). m.p. 109–110 °C; $[\alpha]^D = -8$ (c 1.34, EtOH) ^{1}H-NMR: ($CDCl_3$) δ 0.97 (3 H, d, $J=6.8$), 0.98 (3 H, d, $J=6.8$), 1.93 (1 H, octet, $J=6.8$), 2.71 (1 H, brt), 3.65–3.83 (3 H, m), 4.16 (5 H, s), 4.30 (2 H, brs), 4.60 (1 H, brs), 4.62 (1 H, brs), 5.80 (1 H, brd); ^{13}C-NMR: δ ($CDCl_3$) 19.04, 19.69, 28.99, 57.01, 63.87, 67.91, 68.42, 69.75, 70.46, 70.51, 75.95), 171.37.

5.7.2.2 Synthesis of Ferrocenyloxazoline

To a light orange solution of (2S)-ferrocenamide (0.817 g, 2.59 mol) and PPh_3 (2.49 g, 9.5 mmol) in acetonitrile (60 mL), was added NEt_3 (1.6 mL, 11.5 mmol) followed by CCl_4 (2.2 mL, 22.8 mmol) and the resulting solution stirred at room temperature under nitrogen overnight. After quenching with H_2O (80 mL) the mixture was extracted with petroleum ether (5 × 50 mL), the organic layer was combined, dried ($MgSO_4$), filtered and evaporated. The crude product, which was contaminated by a substantial quantity of triphenylphosphine oxide, was purified by column chromatography (SiO_2, 30% EtOAc/40–60 petroleum ether) to give the pure ferrocenyl oxazoline as a dark yellow crystalline solid (0.685 g, 89%). [*Note - Removing the solvent from the reaction mixture before work-up results in a substantial reduction in yield of the oxazoline.*] m.p. 71.5–72.5 °C; $[\alpha]^D = 129$ (c 1.5, EtOH); ^{1}H-NMR (CDCl$_3$): δ 0.87 (3 H, d, J = 6.8), 0.94 (3 H, d, J = 6.8), 1.78 (1 H, hextet, J = 6.6), 3.41 (1 H, q, J = 7.0), 3.89–3.95 (1 H, m), 4.00 (1 H, t, J = 7.7), 4.06 (5 H, s), 4.26 (2 H, brs), 4.66 (1 H, brs), 4.70 (1 H, brs); ^{13}C-NMR (CDCl$_3$): δ 17.79, 18.81, 32.27, 68.92, 68.95, 69.27, 69.51, 70.06, 70.09, 70.60, 72.27, 165.56.

5.7.2.3 Synthesis of (S, S$_p$)-2-[(S)-2-(diphenylphosphino)ferrocenyl]-4-(1-methylethyl) oxazoline 1b

A yellow–orange stirred solution of ferrocenyloxazoline (0.158 g, 0.53 mmol) and TMEDA (0.10 mL, 0.7 mmol) in Et$_2$O (6 mL) under nitrogen was cooled to −78 °C resulting in the formation of a yellow precipitate. To this was added dropwise *n*-BuLi (0.38 mL, 0.7 mmol), the reaction mixture darkening to red–brown. After stirring at −78 °C for 2 h, the Schlenk tube containing the orange non-homogeneous reaction mixture was transferred to an ice bath and stirring was maintained for a further 5 min. To the resultant homogeneous orange–red solution was added PPh$_2$Cl (0.12 mL, 0.7 mmol) and the reaction mixture was allowed to warm to room temperature. After 15 min the reaction was quenched with saturated NaHCO$_3$ (10 mL) and diluted with Et$_2$O (10 mL). The two layers were separated and the aqueous phase was extracted with Et$_2$O (10 mL). The organic layer was combined, dried (MgSO$_4$), filtered and evaporated to give an orange crystalline solid, which was purified by column chromatography (10% EtOAc/petroleum ether) to afford 0.163 g (64%) of a yellow–orange crystalline solid. Recrystallization from hexane gave pure (S, S$_p$)-Fcphox **1b** as a single diastereoisomer. m.p. 157–158 °C; $[\alpha]^D = +112$ (c 0.1, EtOH); ^{1}H-NMR (CDCl$_3$): δ 0.68 (3 H, d, J = 7 Hz), 0.82 (3 H, d, J = 7 Hz), 1.61–1.69 (1 H, m), 3.61 (1 H, brs), 3.67 (1 H, t, J = 8 Hz), 3.83–3.90 (1 H, m), 4.22 (5 H, s), 4.22–4.30 (1 H, m), 4.37 (1 H, brs), 4.99 (1 H, brs), 7.18–7.24 (5 H, m), 7.36–7.37 (3 H, m), 7.46–7.51 (2 H, m); ^{13}C-NMR (CDCl$_3$): δ 17.52, 18.61, 32.05, 69.57, 70.72, 72.02, 72.14, 73.81, 73.85, 75.32, 78.55, 127.81, 127.92, 127.99, 128.10, 128.18, 128.89, 132.40, 134.86, 138.21, 139.54; ^{31}P-NMR (CDCl$_3$): δ −16.92.

Acknowledgment

Financial support by the Major Basic Research Development Program (2006CB806106), National Natural Science Foundation of China, Chinese Academy of Sciences, Croucher Foundation of Hong Kong, and Science and Technology Commission of Shanghai Municipality is gratefully acknowledged. We thank Dr. Wen Qiong Wu, Ms. Di Chen and Mr. Do-Liang Mo for their assistance in the preparation of the text.

References

1 Hayashi, T., Yamamoto, K., and Kumada, M. (1974) *Tetrahedron Lett.*, **15**, 4405–4408.
2 For some recent reviews on the 1,2-*P,N*-ferrocene ligands and their applications in asymmetric catalysis: (a) Gómez Arrayás, R., Adrio, J., and Carretero, J.C. (2006) *Angew. Chem. Int. Ed.*, **45**, 7674–7715; (b) Atkinson, R.C.J., Gibson, V.C., and Long, N.J. (2004) *Chem. Soc. Rev.*, **33**, 313–328; (c) Colacot, T.J. (2003) *Chem. Rev.*, **103**, 3101–3118; (d) Richards, C.J. and Locke, A.J. (1998) *Tetrahedron: Asymmetry*, **9**, 2377–2407.
3 (a) Nishibayashi, Y. and Uemura, S. (1995) *Synlett*, 79–81; (b) Richards, C.J., Damalidis, T., Hibbs, D.E., and Hursthouse, M.B. (1995) *Synlett*, 74–76; (c) Sammakia, T., Latham, H.A., and Schaad, D.R. (1995) *J. Org. Chem.*, **60**, 10–11.
4 Sammakia, T. and Latham, H.A. (1995) *J. Org. Chem.*, **60**, 6002–6003.
5 (a) Tang, W.-J. and Zhang, X.-M. (2003) *Chem. Rev.*, **103**, 3029–3070; (b) Cui, X.-H. and Burgess, K. (2005) *Chem. Rev.*, **105**, 3272–3296; (c) Zhou, Y.-G. (2007) *Acc. Chem. Res.*, **40**, 1357–1366.
6 Blankenstein, J. and Pfaltz, A. (2001) *Angew. Chem. Int. Ed.*, **40**, 4445–4447.
7 Co, T.T., Shim, S.C., Cho, C.S., and Kim, T.-J. (2005) *Organometallics*, **24**, 4824–4831.
8 Co, T.T. and Kim, T.-J. (2006) *Chem. Comm.*, 3537–3539.
9 Li, X.-S., Li, Q., Wu, X.-H., Gao, Y.-G., Xu, D.-C., and Kong, L.-C. (2007) *Tetrahedron: Asymmetry*, **18**, 629–634.
10 Sammakia, T. and Stangeland, E.L. (1997) *J. Org. Chem.*, **62**, 6104–6105.
11 Naud, F., Malan, C., Spindler, F., Rüggeberg, C., Schmidt, A.T., and Blaser, H.-U. (2006) *Adv. Synth. Catal.*, **348**, 47–50.
12 Naud, F., Spindler, F., Rüggeberg, C.J., Schmidt, A.T., and Blaser, H.-U. (2007) *Org. Process Res. Dev.*, **11**, 519–523.
13 Tellers, D.M., Bio, M., Song, Z.J., McWilliams, J.C., and Sun, Y.-K. (2006) *Tetrahedron: Asymmetry*, **17**, 550–553.
14 Chen, W., Mbafor, W., Roberts, S.M., and Whittall, J. (2006) *Tetrahedron: Asymmetry*, **17**, 1161–1164.
15 Palmer, A.M. and Nettkoven, U. (2007) *Tetrahedron: Asymmetry*, **18**, 2381–2385.
16 Lu, S.-M., Han, X.-W., and Zhou, Y.-G. (2004) *Adv. Synth. Catal.*, **346**, 909–912.
17 Wang, W.-B., Lu, S.-M., Yang, P.-Y., Han, X.-W., and Zhou, Y.-G. (2003) *J. Am. Chem. Soc.*, **125**, 10536–10537.
18 Cheemala, M.N. and Knochel, P. (2007) *Org. Lett.*, **9**, 3089–3092.
19 Zassinovich, G., Mestroni, G., and Gladiali, S. (1992) *Chem. Rev.*, **92**, 1051–1069.
20 (a) Noyori, R. and Hashiguchi, S. (1997) *Acc. Chem. Res.*, **30**, 97–102; (b) Clapham, S.E., Hadzovic, A., and Morris, R.H. (2004) *Coord. Chem. Rev.*, **248**, 2201–2237.
21 Du, X., Dai, L., and Hou, X. (1998) *Chinese J. Chem.*, **16**, 90–93.
22 Nishibayashi, Y., Takei, I., Uemura, S., and Hidai, M. (1999) *Organometallics*, **18**, 2291–2293.
23 Arikawa, Y., Ueoka, M., Matoba, K., Nishibayashi, Y., Hidai, M., and Uemura, S. (1999) *J. Organomet. Chem.*, **572**, 163–168.
24 Dai, H., Hu, X., Chen, H., Bai, C., and Zheng, Z. (2003) *Tetrahedron: Asymmetry*, **14**, 1467–1472.
25 Madrigal, C., Garcia-Fernández, A., Gimeno, J., and Lastra, E. (2008) *J. Organomet. Chem.*, **693**, 2535–2540.
26 Gimeno, J., Lastra, E., Madrigal, C., Graiff, C., and Tiripicchio, A. (2001) *J. Organomet. Chem.*, **637–639**, 463–468.
27 Garcia-Fernández, A., Gimeno, J., Lastra, E., Madrigal, C.A., Graiff, C., and Tiripicchio, A. (2007) *Eur. J. Inorg. Chem.*, 732–741.
28 Miyake, Y., Nishibayashi, Y., and Uemura, S. (2008) *Synlett*, 1747–1758.
29 Gibson, S.E. and Rudd, M. (2007) *Adv. Synth. Catal.*, **349**, 781–795.

30 Tamao, K., Kakui, T., and Kumada, M. (1978) *J. Am. Chem. Soc.*, **100**, 2268–2269.
31 Hayashi, T., Tamao, K., Katsuro, Y., Nakae, I., and Kumada, M. (1980) *Tetrahedron Lett.*, **21**, 1871–1874.
32 Hayashi, T. and Kabeta, K. (1985) *Tetrahedron Lett.*, **26**, 3023–3026.
33 Hayashi, T., Matsumoto, Y., Morikawa, I., and Ito, Y. (1990) *Tetrahedron: Asymmetry*, **1**, 151–154.
34 Pioda, G. and Togni, A. (1998) *Tetrahedron: Asymmetry*, **9**, 3903–3910.
35 (a) Togni, A., Bieler, N., Burckhardt, U., Köllner, C., Pioda, G., Schneider, R., and Schnyder, A. (1999) *Pure Appl. Chem.*, **71**, 1531–1537; (b) Woo, T.K., Pioda, G., Rothlisberger, U., and Togni, A. (2000) *Organometallics*, **19**, 2144–2152; (c) Magistrato, A., Woo, T.K., Togni, A., and Rothlisberger, U. (2004) *Organometallics*, **23**, 3218–3227; (d) Magistrato, A., Togni, A., and Rothlisberger, U. (2006) *Organometallics*, **25**, 1151–1157.
36 Hayashi, T., Hayashi, C., and Uozumi, Y. (1995) *Tetrahedron: Asymmetry*, **6**, 2503–2506.
37 Nishibayashi, Y., Segawa, K., Ohe, K., and Uemura, S. (1995) *Organometallics*, **14**, 5486–5487.
38 Nishibayashi, Y., Segawa, K., Takada, H., Ohe, K., and Uemura, S. (1996) *Chem. Comm.*, 847–848.
39 Nishibayashi, Y., Takei, I., Uemura, S., and Hidai, M. (1998) *Organometallics*, **17**, 3420–3422.
40 Tao, B. and Fu, G.C. (2002) *Angew. Chem. Int. Ed.*, **41**, 3892–3894.
41 Bolm, C., Xiao, L., and Kesselgruber, M. (2003) *Org. Biomol. Chem.*, **1**, 145–152.
42 Hu, X., Chen, H., Dai, H., and Zheng, Z. (2003) *Tetrahedron: Asymmetry*, **14**, 3415–3421.
43 Takei, I., Nishibayashi, Y., Arikawa, Y., Uemura, S., and Hidai, M. (1999) *Organometallics*, **18**, 2271–2274.
44 Takei, I., Nishibayashi, Y., Ishii, Y., Mizobe, Y., Uemura, S., and Hidai, M. (2001) *Chem. Comm.*, 2360–2361.
45 (a) Männig, D. and Nöth, H. (1982) *Angew. Chem. Int. Ed.*, **24**, 878–879; (b) Zhang, J.-F., Lou, B.-L., Guo, G.-Z., and Dai, L.-X. (1991) *J. Org. Chem.*, **56**, 1670–1672.
46 (a) Schnyder, A., Hintermann, L., and Togni, A. (1995) *Angew. Chem. Int. Engl. Ed.*, **34**, 931–933; (b) Schnyder, A., Togni, A., and Wiesli, U. (1997) *Organometallics*, **16**, 255–260.
47 Abbenhuis, H.C.L., Burckhardt, U., Gramlich, V., Martelletti, A., Spencer, J., Steiner, I., and Togni, A. (1996) *Organometallics*, **15**, 1614–1621.
48 Burckhardt, U., Hintermann, L., Schnyder, A., and Togni, A. (1995) *Organometallics*, **14**, 5415–5425.
49 Kloetzing, R.J., Lotz, M., and Knochel, P. (2003) *Tetrahedron: Asymmetry*, **14**, 255–264.
50 Ligand 48: Mino, T., Ogawa, T., and Yamashita, M., (2000) *Hetereocycles*, **55**, 453–456. Ligand 49: Widhalm, M., Mereiter, K., and Bourghida, M., (1998) *Tetrahedron: Asymmetry*, **9**, 2983–2986. Ligand 50: Mourgues, S., Serra, D., Lamy, F., Vincendeau, S., Daran, J.-C., Manoury, E., and Gouygou, M., (2003) *Eur. J. Inorg. Chem.*, 2820–2826. Ligand 51 and 52: Sturm, T., Abad, B., Weissensteiner, W., Mereiter, K., Manzano, B.R., and Jalón, F.A., (2006) *J. Mol. Catal. A: Chem*, **255**, 209–219. Ligand 53: Kloetzing, R.J. and Knochel, P., (2006) *Tetrahedron: Asymmetry*, **17**, 116–123.
51 Ahn, K.H., Cho, C.-W., Park, J., and Lee, S. (1997) *Tetrahedron: Asymmetry*, **8**, 1179–1185.
52 Jin, M.-J., Takale, V.B., Sarkar, M.S., and Kim, Y.-M. (2006) *Chem. Comm.*, 663–664.
53 (a) Hu, X., Dai, H., Bai, C., Chen, H., and Zheng, Z. (2004) *Tetrahedron: Asymmetry*, **15**, 1065–1068; (b) Hu, X.-P., Chen, H.-L., and Zheng, Z. (2005) *Adv. Synth. Catal.*, **347**, 541–548.
54 Hu, X., Dai, H., Hu, X., Chen, H., Wang, J., Bai, C., and Zheng, Z. (2002) *Tetrahedron: Asymmetry*, **13**, 1687–1693.

55 (a) Moreno, R.M., Bueno, A., and Moyano, A. (2002) *J. Organomet. Chem.*, **660**, 62–70; (b) Patti, A., Lotz, M., and Knochel, P. (2001) *Tetrahedron: Asymmetry.*, **12**, 3375–3380; (c) Fukuzawa, S.-i., Yamamoto, M., Hosaka, M., and Kikuchi, S. (2007) *Eur. J. Org. Chem.*, 5540–5545.

56 Manoury, E., Fossey, J.S., Aït-Haddou, H., Daran, J.-C., and Balavoine, G.G.A. (2000) *Organometallics*, **19**, 3736–3739.

57 Togni, A., Burckhardt, U., Gramlich, V., Pregosin, P.S., and Salzmann, R. (1996) *J. Am. Chem. Soc.*, **118**, 1031–1037.

58 Malone, Y.M. and Guiry, P.J. (2000) *J. Organomet. Chem.*, **603**, 110–115.

59 Park, H.-J., Han, J.W., Seo, H., Jang, H.-Y., Chung, Y.K., and Suh, J. (2001) *J. Mol. Catal. A: Chem.*, **174**, 151–157.

60 You, S.L., Hou, X.L., Dai, L.X., Yu, Y.H., and Xia, W. (2002) *J. Org. Chem.*, **67**, 4684–4695.

61 Farrell, A., Goddard, R., and Guiry, P.J. (2002) *J. Org. Chem.*, **67**, 4209–4217.

62 Anderson, J.C. and Osborne, J. (2005) *Tetrahedron: Asymmetry*, **16**, 931–934.

63 Tu, T., Zhou, Y.G., Hou, X.L., Dai, L.X., Dong, X.C., Yu, Y.H., and Sun, J. (2003) *Organometallics*, **22**, 1255–1265.

64 For review: Trost, B.M. and Crawley, M.L. (2003) *Chem. Rev.*, **103**, 2921–2944.

65 Geisler, F.M. and Helmchen, G. (2006) *J. Org. Chem.*, **71**, 2486–2492.

66 Hu, X., Chen, H., Dai, H., and Zheng, Z. (2003) *Tetrahedron: Asymmetry*, **14**, 3415–3421.

67 Hu, X., Bai, C., Dai, H., Chen, H., and Zheng, Z. (2004) *J. Mol. Catal. A*, **218**, 107–112.

68 Bueno, A., Moreno, R.M., and Moyano, A. (2005) *Tetrahedron: Asymmetry*, **16**, 1763–1778.

69 For review: Helmchen, G., Dahnz, A., Dübon, P., Schelwies, M., and Weihofen, R. (2007) *Chem. Commun.*, 675–691.

70 Gavrilov, K.N., Maksimova, M.G., Zheglov, S.V., Bondarev, O.G., Benetsky, E.B., Lyubimov, S.E., Petrovskii, P.V., Kabro, A.A., Hey-Hawkins, E., Moiseev, S.K., Kalinin, V.N., and Davankov, V.A. (2007) *Eur. J. Org. Chem.*, 4940–4947.

71 Kilroy, T.G., Hennessy, A.J., Connolly, D.J., Malone, Y.M., Farrell, A., and Guiry, P.J. (2003) *J. Mol. Catal. A*, **196**, 65–81.

72 (a) Kiely, D. and Guiry, P.J. (2003) *Tetrahedron Lett.*, **44**, 7377–7380; (b) Kiely, D. and Guiry, P.J. (2003) *J. Organomet. Chem.*, **687**, 545–561.

73 Hayashi, T., Konishi, M., Fukushima, M., Mise, T., Kagotani, M., Tajika, M., and Kumada, M. (1982) *J. Am. Chem. Soc.*, **104**, 180–186.

74 Richards, C.J., Hibbs, D.E., and Hursthouse, M.B. (1995) *Tetrahedron Lett.*, **36**, 3745–3748.

75 (a) Chung, K.-G., Miyake, Y., and Uemura, S. (2000) *J. Chem. Soc., Perkin Trans. 1*, 15–18; (b) Chung, K.-G., Miyake, Y., and Uemura, S. (2000) *J. Chem. Soc., Perkin Trans. 1*, 2725–2729.

76 (a) Cammidge, A.N. and Crépy, K.V.L. (2000) *Chem. Commun.*, 1723–1724; (b) Cammidge, A.N. and Crépy, K.V.L. (2004) *Tetrahedron*, **60**, 4377–4386.

77 Hayashi, T., Hayashizaki, K., Kiyoi, T., and Ito, Y. (1988) *J. Am. Chem. Soc.*, **110**, 8153–8156.

78 Bringmann, G., Hamm, A., and Schraut, M. (2003) *Org. Lett.*, **5**, 2805–2808.

79 Stangeland, E.L. and Sammakia, T. (1997) *Tetrahedron*, **53**, 16503–16510.

80 Liu, L.T., Wang, M.C., Zhao, W.X., Zhou, Y.L., and Wang, X.D. (2006) *Tetrahedron: Asymmetry*, **17**, 136–141.

81 (a) Wang, M.C., Liu, L.T., Hua, Y.Z., Zhang, J.S., Shi, Y.Y., and Wang, D.K. (2005) *Tetrahedron: Asymmetry*, **16**, 2531–2534; (b) Wang, M.C., Xu, C.L., Zou, Y.X., Liu, H.M., and Wang, D.K. (2005) *Tetrahedron Lett.*, **46**, 5413–5416.

82 Lautens, M., Hiebert, S., and Renaud, J.-L. (2000) *Org. Lett.*, **2**, 1971–1973.

83 Kim, T.-J., Lee, H.-Y., Ryu, E.-S., Park, D.-K., Cho, C.S., Shim, S.C., and Jeong, J.H. (2002) *J. Organomet. Chem.*, **649**, 258–267.

84 Huisgen, R. (1963) *Angew. Chem., Int. Ed.*, **2**, 565–598.
85 See recent review: Pellissier, H. (2007) *Tetrahedron*, **63**, 3235–3285.
86 Zeng, W. and Zhou, Y.G. (2005) *Org. Lett.*, **7**, 5055–5058.
87 Zeng, W., Chen, G.Y., Zhou, Y.G., and Li, Y.X. (2007) *J. Am. Chem. Soc.*, **129**, 750–751.
88 Gao, W., Zhang, X., and Raghunath, M. (2005) *Org. Lett.*, **7**, 4241–4244.
89 Yan, X.X., Peng, Q., Zhang, Y., Zhang, K., Hong, W., Hou, X.L., and Wu, Y.D. (2006) *Angew. Chem., Int. Ed.*, **45**, 1979–1983.
90 Dai, H.C., Hu, X.P., Chen, H.L., Bai, C.M., and Zheng, Z. (2004) *J. Mol. Cat: A. Chem.*, **211**, 17–21.
91 Carmona, D., Medrano, R., Dobrinovich, I.T., Lahoz, F.J., Ferrer, J., and Oro, L.A. (2006) *J. Organomet. Chem.*, **691**, 5560–5566.
92 Oi, S., Nomura, M., Aiko, T., and Inoue, Y. (1997) *J. Mol. Catal. A:*, **115**, 289–295.
93 (a) Jedlicka, B., Weissensteiner, W., Kégl, T., and Kollár, L. (1998) *J. Organomet. Chem.*, **563**, 37–41; (b) Wang, L.L., Kwok, W.H., Chan, A.S.C., Tu, T., Hou, X.L., and Dai, L.X. (2003) *Tetrahedron: Asymmetry*, **14**, 2291–2295.
94 Hoang, V.D.M., Reddy, P.A.N., and Kim, T.-J. (2008) *Organometallics*, **27**, 1026–1027.
95 Nishimura, T., Matsumura, S., Maeda, Y., and Uemura, S. (2002) *Chem. Comm.*, 50–51.
96 Nishibayashi, Y., Yamauchi, A., Onodera, G., and Uemura, S. (2003) *J. Org. Chem.*, **68**, 5875–5880.
97 Shi, M., Chen, L.-H., and Li, C.-Q. (2005) *J. Am. Chem. Soc.*, **127**, 3790–3800.
98 (a) Gotov, B. and Schmalz, H.-G. (2001) *Org. Lett.*, **3**, 1753–1756; (b) Böttcher, A. and Schmalz, H.-G. (2003) *Synlett*, 1595–1598.
99 Marquarding, D., Klusacek, H., Gokel, G., Hoffmann, P., and Ugi, I. (1970) *J. Am. Chem. Soc.*, **92**, 5389–5393.
100 Richards, C.J. and Mulvaney, A.W. (1996) *Tetrahedron: Asymmetry*, **7**, 1419–1430.

6
N,O-Bidentate Ferrocenyl Ligands
Anne Nijs, Olga García Mancheño, and Carsten Bolm

6.1
Introduction

This chapter focuses on applications of N,O-bidentate ferrocenyl ligands and gives an overview of the scope and limitations of their use in asymmetric catalysis. In contrast to the wide range of phosphino-substituted ferrocenes, relatively few N,O-bidentate derivatives have been studied. Most of them were prepared starting from chiral amino- or oxazolinyl-containing ferrocenes. They can be classified according to their substitution pattern at the ferrocene framework (Figure 6.1).

The reported mono-substituted ferrocenes serving as N,O-bidentate ligands are alcohols with 2-amino or 2-oxazolinyl substituents. A unique class of planar-chiral aza-heterocyclic ferrocenes with a pendant amino hydroxy chain has also been described. Most of the chiral ferrocenyl ligands have a 1,1'- or a 1,2-substitution pattern. Compounds of the latter class containing both central and planar chirality are the best studied compounds of this type in asymmetric catalysis. Other derivatives have either additional stereogenic elements (such as axial chirality) or are exclusively planar-chiral.

N,O-Bidentate ferrocenes have predominantly been examined in catalyzed enantioselective nucleophilic addition reactions of organozinc reagents to aldehydes, and this will be discussed in detail in the following sections. Although less studied, enantioselective aryl transfer reactions involving organoboron reagents have also been covered.

6.2
Addition of Organozinc Reagents to Aldehydes

The enantioselective 1,2-addition of organozinc reagents, mainly diethylzinc, to aldehydes providing optically active secondary alcohols has been extensively studied [1]. A wide variety of chiral catalysts such as amino alcohols, diamines and diols

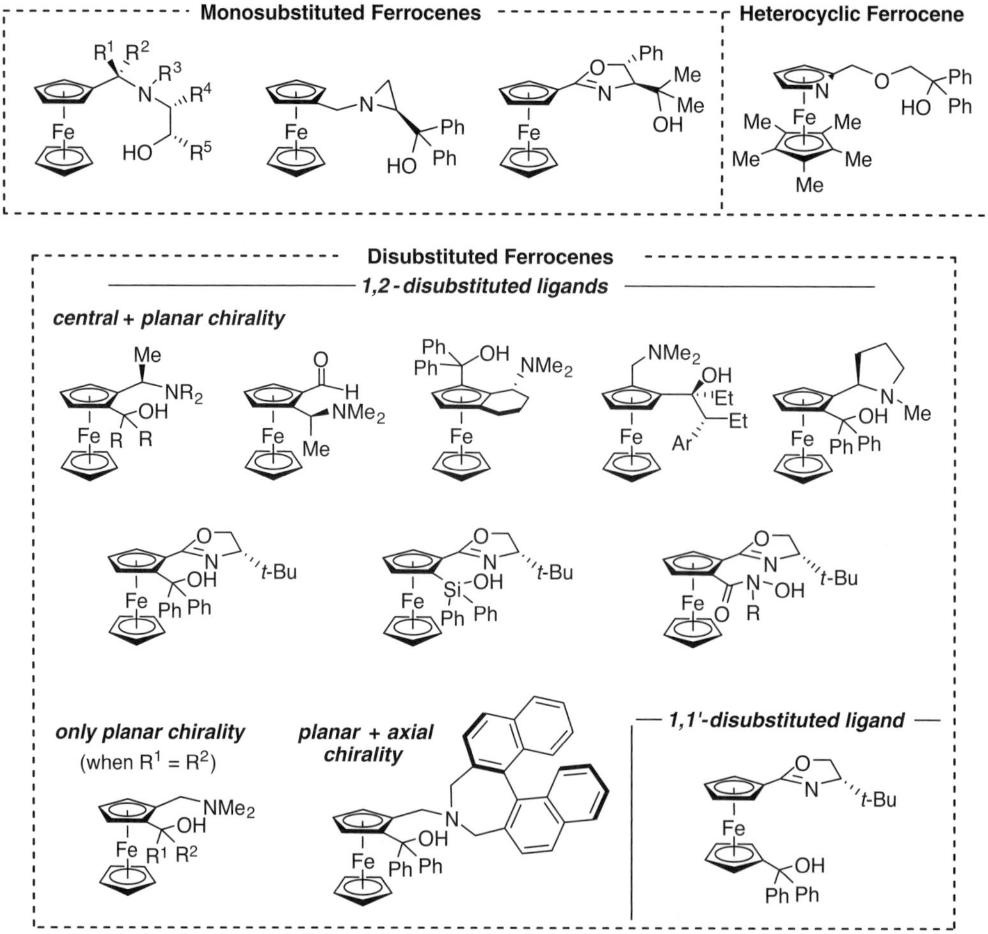

6 N,O-Bidentate Ferrocenyl Ligands

Figure 6.1 N,O-Bidentate ferrocenyl ligands employed in asymmetric catalysis.

have been shown to efficiently promote this C–C bond-forming reaction. In particular, chiral N,O-bidentate ferrocenes have emerged as a highly effective class of compounds for this and related organozinc additions.

6.2.1
Addition of Dialkylzinc

6.2.1.1 With Ferrocenyl Amino Alcohols

In 1989 the first catalytic applications of chiral ferrocenes in asymmetric diethylzinc additions to aldehydes were described by Butsugan and coworkers (Scheme 6.1) [2]. Starting from (R)-N,N–dimethyl-1-ferrocenylethylamine (Ugi's amine), 1-iodo-2-substituted ferrocenes containing various amino hydroxy side chains with stereogenic centers were prepared in high yields. Iodoferrocene **1** gave a poor enantioselectivity

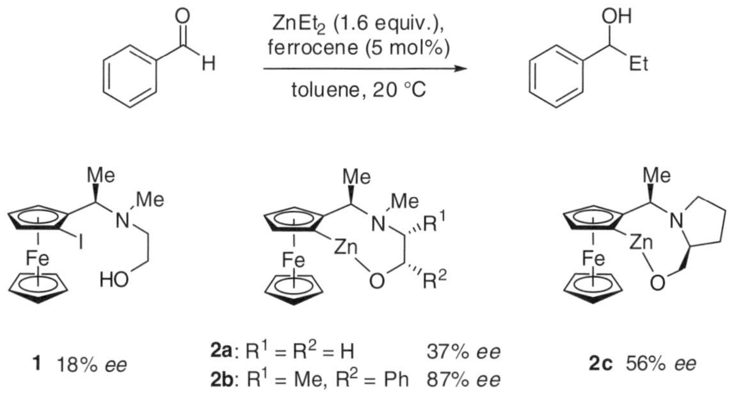

Scheme 6.1 Early ferrocenyl catalysts in the reaction between benzaldehyde and diethylzinc.

(18% ee) in the reaction between diethylzinc and benzaldehyde. Zinc reagents **2**, prepared from the corresponding iodoferrocenes by halogen–metal exchange with *n*-butyllithium followed by treatment with zinc chloride, proved to be more effective catalysts. Among various derivatives, ephedrine-derived **2b** was superior to the amino ethanol- and prolinol-derived catalysts **2a** and **2c**, respectively. Finally, the best result was obtained with a catalyst loading of 5 mol% of **2b**, generating the desired secondary alcohol in high yield (81%) and with an enantioselectivity of 87% *ee*.

Interestingly, diastereomeric ligands derived from the *S*-enantiomer of Ugi's amine, keeping the stereochemistry at the amino hydroxy moiety unchanged, gave a product with the same absolute configuration. From this result the authors deduced that the enantioselectivity depended predominantly on the structural details of the side chain and that the absolute configuration at the ferrocenylethyl group played only a minor role.

Shortly thereafter, the same group introduced *N*-(1-ferrocenylalkyl)-*N*-alkylnorephedrines **3** as catalysts for ethylations of benzaldeyde, and the effect of the substituents R^1, R^2, and R^3 on the stereoselectivity was investigated (Table 6.1) [3].

The secondary amine **3a** proved to be less catalytically active and enantioselective than the tertiary amines **3b–3e**, which furnished (*S*)-1-phenylpropanol in high yields and enantioselectivities of up to 99% *ee*. The absolute configuration at the α-carbon of the 1-ferrocenylethylamino moiety had a negligible effect on both enantioselectivity and chemical yield (entries 2 and 3). Ferrocenes lacking a substituent at that position also performed well (entries 4 and 5). Notably, the authors were able to carry out ethyl transfers onto an aliphatic aldehyde, heptanal, obtaining the corresponding alcohols with good enantioselectivities (71–80% *ee*).

To ease product isolation and work-up, chiral vinylferrocene-derived polymers with varying stoichiometries of *N*-ferrocenylmethylephedrine units were applied in the diethylzinc addition to benzaldehyde [4]. All polymers catalyzed the reaction and provided products in good yields (67–85%). However, only modest enantioselectivities were achieved (51–72% *ee*). The authors speculated that this was a result of their lower solubility compared to the parent compound **3**.

Table 6.1 Ethyl transfer onto benzaldeyde catalyzed by norephedrine-derived ferrocenes.

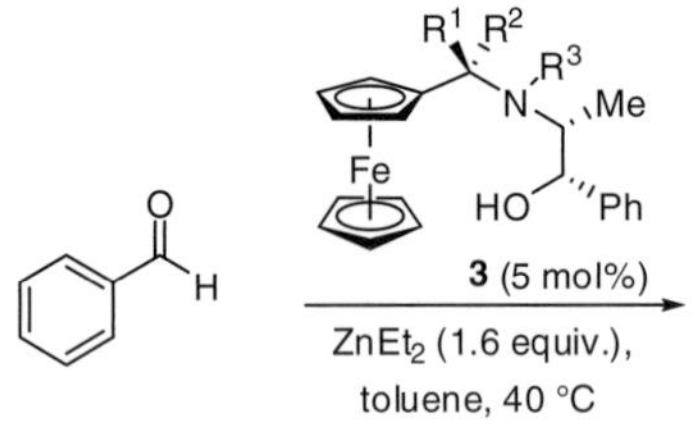

Entry	Ferrocene	R^1	R^2	R^3	Yield (%)	ee (%)
1[a]	3a	Me	H	H	58	47
2	3b	Me	H	Me	92	95
3	3c	H	Me	Me	92	94
4	3d	H	H	Me	91	94
5[a]	3e	H	H	n-Bu	85	99

[a]Reaction at room temperature.

In 1991 Butsugan and coworkers reported asymmetric alkylations of aldehydes catalyzed by chiral 1,2-disubstituted ferrocenyl amino alcohols **4** [5]. Over 20 catalysts with different substituents at the hydroxymethyl and amino position were synthesized and some of them were evaluated in the catalyzed addition of diethylzinc to benzaldehyde. Generally, high yields and excellent enantioselectivities (≥91% ee) were achieved. Among the ferrocenes incorporating a dimethylamino group, derivative **4a** with a diphenylhydroxymethyl moiety gave the best result (99%, 97% ee, Scheme 6.2). Subsequently, the enantioselectivity was further improved by incorporating this structural motif into a structure bearing a piperidinyl group **4b** (99%, 99% ee).

Various other aldehydes could also be efficiently ethylated in the presence of catalytic amounts of ferrocenyl amino alcohol **4b** (Scheme 6.3). Conversions of aryl-, heteroaryl-, α,β-unsaturated- and α-branched aliphatic aldehydes gave good to excellent enantioselectivities (87–100% ee). Use of other substrates without α-branching resulted in diminished stereoselectivities (62–63% ee).

Moreover, the addition of di-n-butylzinc to benzaldehyde (92%, 99% ee) and isobutyraldehyde (66%, >98% ee) in the presence of **4b** proceeded in a highly stereoselective manner.

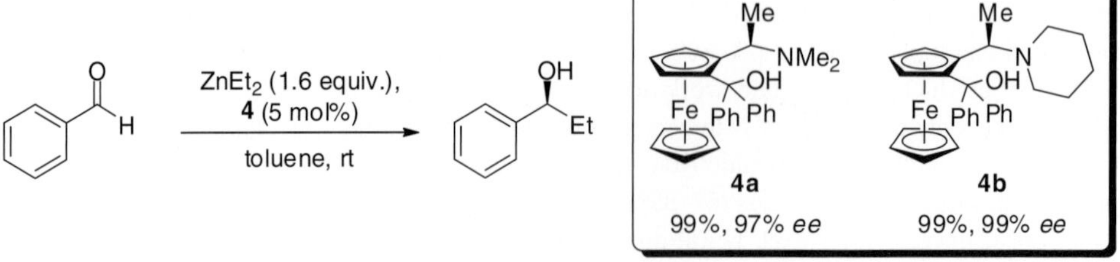

Scheme 6.2 Ferrocenyl amino alcohols **4** in the diethylzinc addition to benzaldehyde.

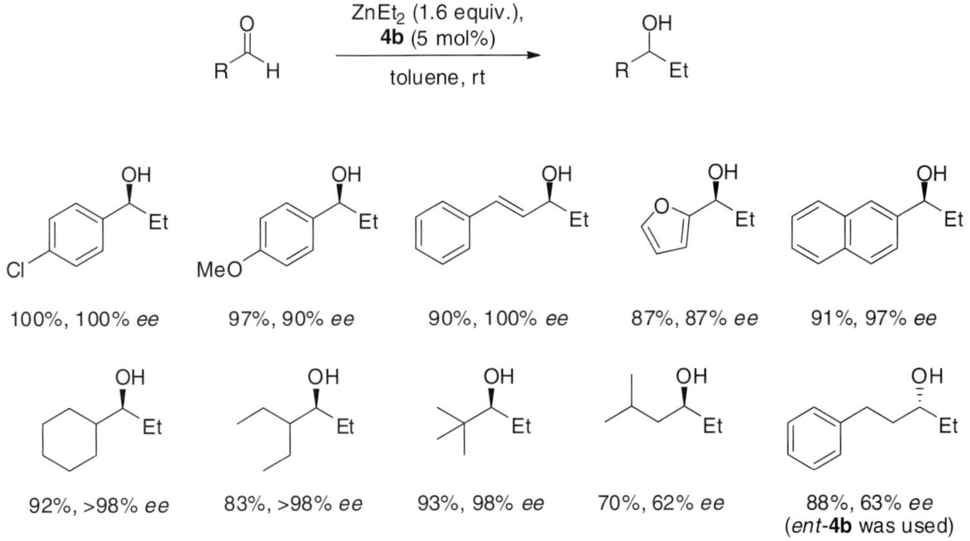

6.2 Addition of Organozinc Reagents to Aldehydes | 153

Scheme 6.3 Asymmetric alkylation of aldehydes with ferrocenyl amino alcohol **4b**.

Syntheses of 3-alkylphthalides based on enantioselective additions of dialkylzinc reagents to *o*-phthaldehyde using the previously reported ferrocenyl amino alcohols **4** were reported by Butsugan and coworkers in 1992 [6]. The best results involving diethyl- and di-*n*-butylzinc were obtained with the bis(*p*-chlorophenyl) derivative **4c**, which afforded the corresponding lactols with high enantioselectivity as a mixture of diastereomers (Scheme 6.4). The authors suggested that the *p*-chloro substituents in **4c** increased the Lewis acidity of the resulting zinc complex. Subsequently, the lactols were oxidized with silver oxide, giving the desired 3-alkylphthalides in high yields (80–81%) without racemization.

Furthermore, Butsugan and coworkers showed that the 1,2-disubstituted ferrocenyl amino alcohol **4b** catalyzed the diastereoselective addition of diethylzinc to racemic α-thio- and α-seleno-aldehydes [7] (Scheme 6.5). The aldehydes with *R*-configuration were ethylated at a faster rate, forming (3*S*,4*R*)-thio- and selenoalcohols with high diastereoselectivity (98%). Ethyl transfers onto the *S*-enantiomeric

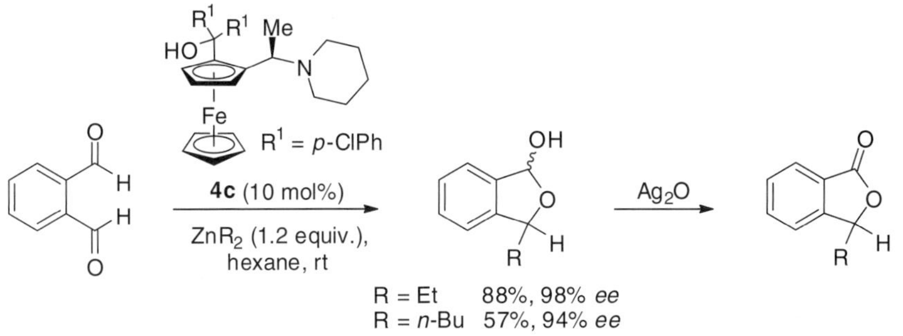

Scheme 6.4 Synthesis of 3-alkylphhalides in the presence of ligand **4c**.

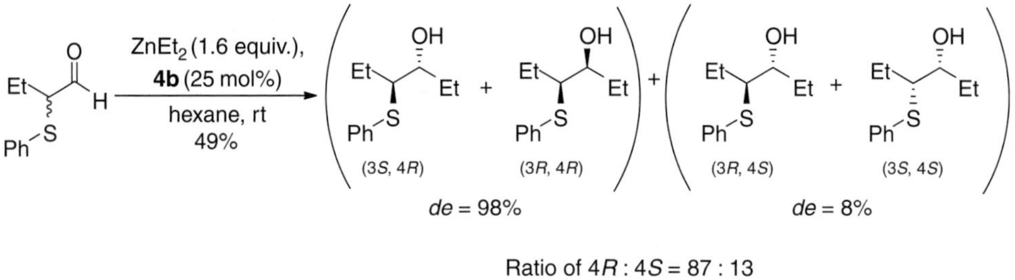

Scheme 6.5 Asymmetric alkylation of a racemic α-thio-aldehyde.

aldehydes were slower and proceeded with low diastereoselectivity (8%). Although this process represented a kinetic resolution, the unreacted aldehyde was racemic. The authors suggested that the base-sensitive aldehyde racemized during work-up.

In 1993 Schlögl and coworkers reported the asymmetric ethyl transfer onto aldehydes in the presence of 3–5 mol% of ferrocenyl amino alcohol **5** (Scheme 6.6) [8]. Both aromatic as well as aliphatic (linear and branched) aldehydes reacted with diethylzinc, affording the corresponding secondary alcohols in good to high yields with enantiomeric excesses of up to 97% ee. Moreover, it was possible to recover and reuse the ferrocene without substantial loss of stereoselectivity. Based on the observation that ferrocenyl amino alcohols **4** and **5** with identical central chirality and opposite planar chirality gave the same enantiomer as the major product, the authors suggested that the former element of chirality dominated the stereochemistry-determining step.

An enantioselective alkylation of aldehydes employing 1-hydroxymethyl-2-dimethyl aminomethylferrocene **6** possessing only planar chirality was described one year later by Nicolosi and coworkers (Scheme 6.7) [9]. Ferrocenyl amino alcohol **6** (10 mol%) catalyzed the diethylzinc addition onto benzaldehyde, generating the R-enantiomer of the alcohol in high yield and good enantioselectivity (94%, 82% ee). Electronically modified para-substituted aromatic aldehydes and α-branched substrates were ethylated in good yields, and the enantiomeric excesses were comparable

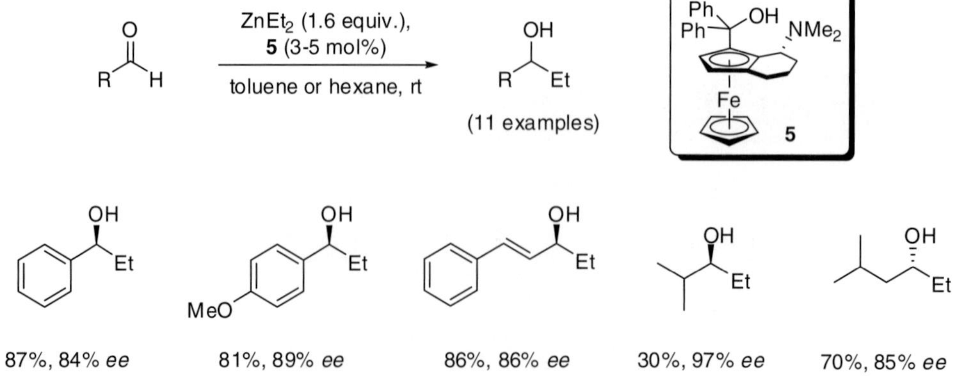

Scheme 6.6 Ferrocenyl amino alcohol **5** for the asymmetric alkylation of aldehydes.

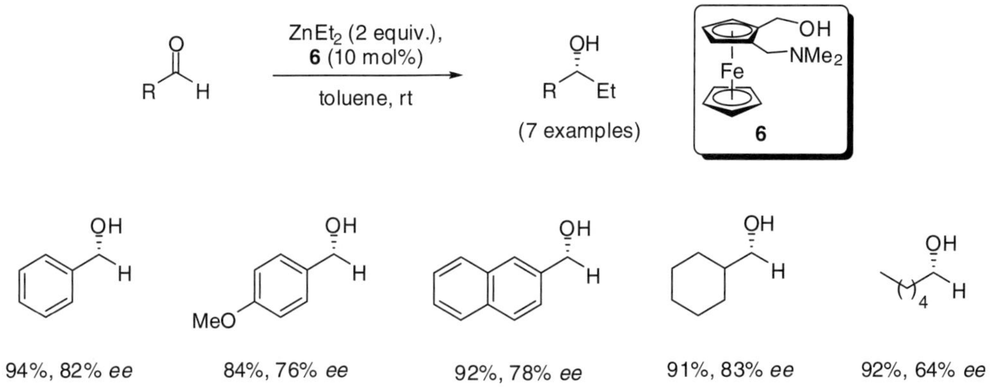

6.2 Addition of Organozinc Reagents to Aldehydes

Scheme 6.7 Asymmetric addition of diethylzinc to aldehydes in the presence of planar-chiral ferrocene **6**.

Results: 94%, 82% ee (phenyl); 84%, 76% ee (4-MeO-phenyl); 92%, 78% ee (naphthyl); 91%, 83% ee (cyclohexyl); 92%, 64% ee (aliphatic).

(76–83% ee). Use of linear aliphatic aldehydes led to slightly lower enantioselectivities (64–70% ee).

In 1998 Fukuzawa and Kato showed that 2-(1-dimethylaminoethyl)-ferrocenecarbaldehyde (**7**) was able to promote the addition of diethylzinc onto aldehydes (Scheme 6.8) [10]. Alkylation of aromatic, linear or branched aliphatic aldehydes in the presence of 5 mol% of **7** afforded the corresponding alcohols in good to high yields with up to 93% ee. The authors examined the origin of the catalytic activity of **7** and showed that the actual catalyst was a ferrocenyl zinc alkoxide formed *in situ* by an initial highly diastereoselective alkylation of **7**.

In 1999 Malézieux and coworkers reported the use of ferrocenyl amino alcohol (R_p,3R,4S)-**8** for the addition of diethylzinc to benzaldehyde, furnishing the *R*-enantiomer of the product alcohol in good yield and enantiomeric excess (78%, 98% ee) (Scheme 6.9) [11]. All possible diastereomers of **8** were prepared and, among the three elements of chirality present, the planar chirality was found to dominate the absolute configuration of the product.

Soon after, Brocard and coworkers described ferrocenyl amino alcohols **9** for the asymmetric alkylation of benzaldehyde with diethylzinc [12]. Diastereomeric mixtures of 1-(2-*N*,*N*-dimethyl amino methyl)ferrocenyl alcohols **9a–c** were separated using a highly stereoselective manganese dioxide oxidation in which one diastereomer was oxidized to its corresponding ketone. Following chromatographic separation, the ketone was useful as it could be converted into ferrocenes **9d–e** by the addition of alkyllithium reagents. All six ferrocenes proved to be effective and a

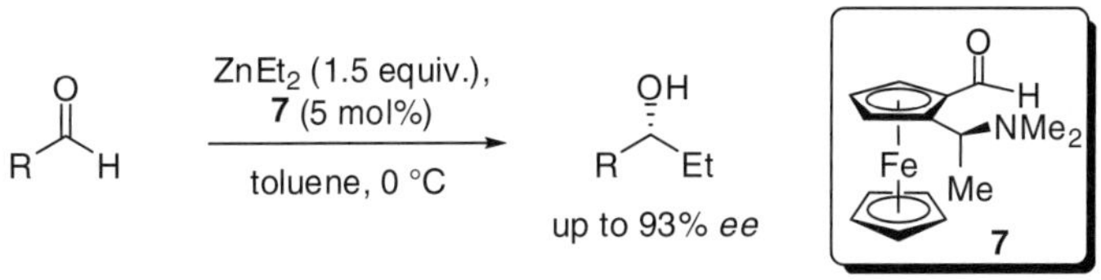

Scheme 6.8 Ferrocenecarbaldehyde **7** in the asymmetric ethylation of aldehydes.

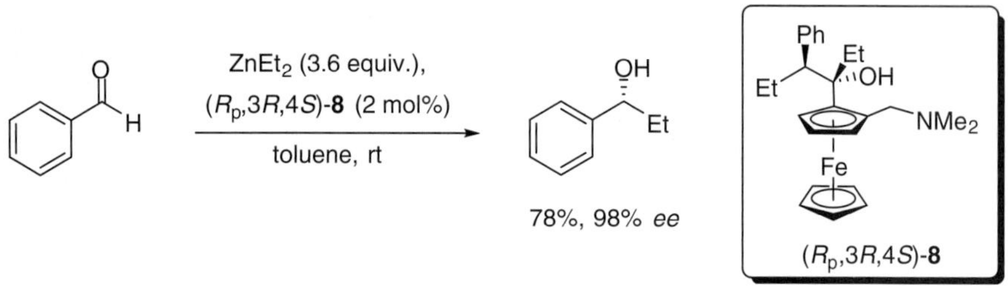

6 N,O-Bidentate Ferrocenyl Ligands

Scheme 6.9 Enantioselective ethylation of benzaldehyde with ferrocenyl amino alcohol 8.

catalyst loading of 10 mol% was sufficient to obtain (S)-1-phenylpropanol in excellent yields with enantiomeric excesses ranging from 77–88% ee (Scheme 6.10).

In 2000 the use of new ferrocenyl amino alcohols containing a 2,2′-bridged binaphthyl fragment in ethyl transfer onto aldehydes was described by Widhalm and coworkers (Scheme 6.11) [13]. By combining the planar chirality of the 1,2-disubstituted ferrocene unit with the axial chirality of the binaphthyl fragment, the authors aimed for possible synergistic effects. The test reaction employing benzaldehyde as substrate with 5 mol% of (S_a,S_p)-10 as catalyst furnished the secondary alcohol in high yield and with excellent enantioselectivity (79%, 97% ee). Diastereomer (S_a,R_p)-10 showed almost no asymmetric induction (79%, 2% ee) revealing a mismatch of the stereogenic elements. With (S_a,S_p)-10 the catalyst loading could be reduced to 0.3 mol% without substantial loss in enantioselectivity (95% ee). In the reaction of various aromatic aldehydes, high enantiomeric excesses ranging from 92–96% ee were observed. With the exception of cyclohexylcarbaldehyde (92% ee) α,β-unsaturated and aliphatic substrates gave only moderate enantioselectivities (47–63% ee).

Wang and coworkers introduced new ferrocenyl aziridino alcohols as chiral catalysts for the asymmetric addition of diethylzinc onto aldehydes [14]. They were conveniently prepared from the readily available amino acid L-serine and ferrocenecarbaldehyde. The authors examined the effects of the ligand structure on the enantioselectivity of the reaction with benzaldehyde as substrate. Although the yields of the resulting alcohols were high in all cases (71–98%), the enantiomeric excesses strongly depended on the substituents in the aziridino moiety (1–93% ee). The best level of asymmetric induction was achieved with ferrocenyl aziridino alcohol 11 (3 mol%), containing a diphenylhydroxymethyl group (Scheme 6.12). Various

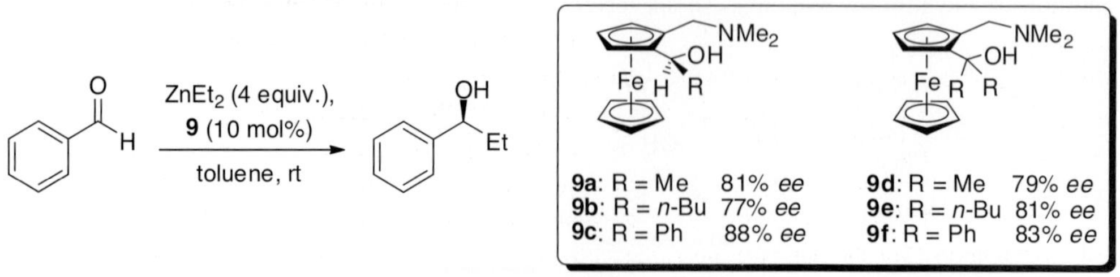

Scheme 6.10 Ferrocenyl amino alcohols 9 for the asymmetric ethylation of benzaldehyde.

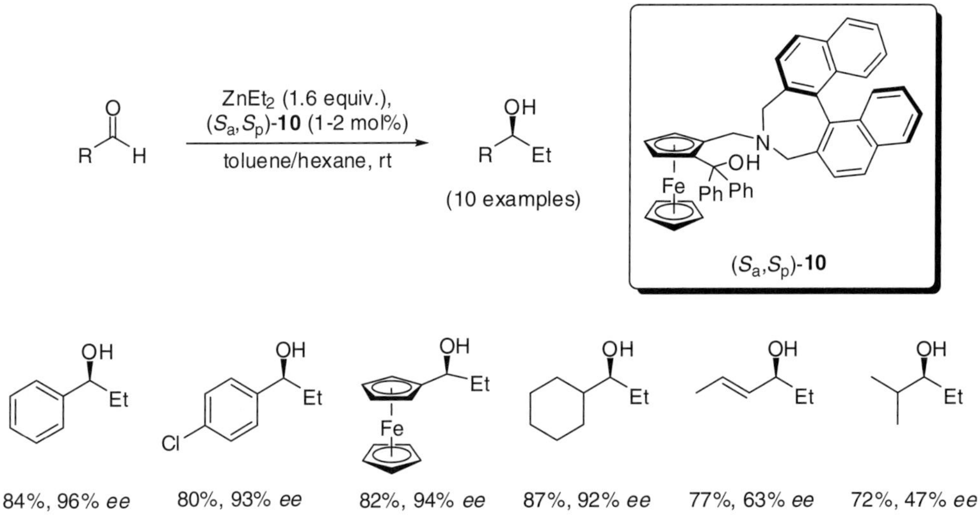

6.2 Addition of Organozinc Reagents to Aldehydes | 157

Scheme 6.11 Application of ferrocenyl amino alcohols **10** containing planar and axial chirality.

aromatic aldehydes were alkylated in high yields with enantioselectivities up to 99% ee, while aliphatic substrates gave lower enantioselectivities (79–80% ee).

In 2006 Guiry and coworkers described the synthesis of a series of planar-chiral pyrrolidinyl-containing ferrocenes [15]. In the ethylation of benzaldehyde the best result was obtained with **12**, incorporating a diphenylhydroxymethyl moiety and an N-methyl substituent (Scheme 6.13). (R,S_p)-**12** gave a higher asymmetric induction (92% ee) than its diastereomer (R,R_p)-**12** (84% ee). The planar chirality was the dominant stereochemistry-controlling element. Thus, the R_p-configured derivative furnished (R)-alcohols, while the catalyst of opposite planar chirality furnished the (S)-alcohol. Interestingly, removal of the N-methyl group resulted in a catalyst that gave the product with opposite configuration to that of the planar chirality of the

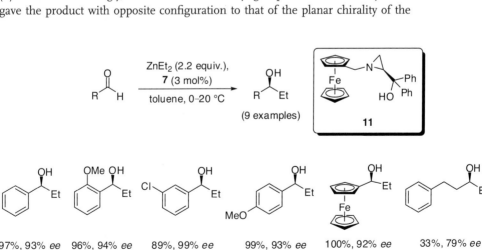

Scheme 6.12 Ferrocenyl aziridino alcohol **11** for the asymmetric diethylzinc addition to aldehydes.

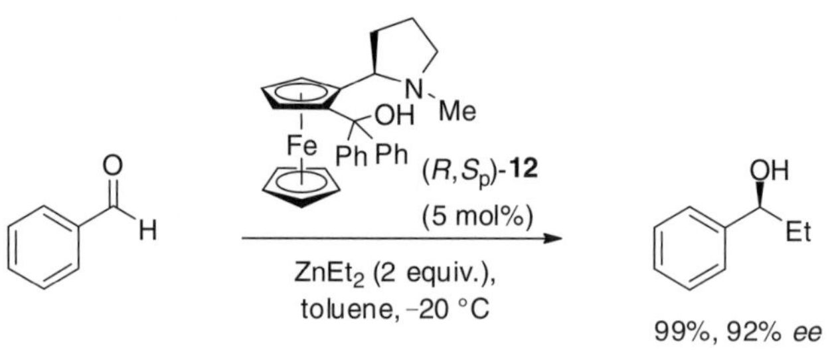

Scheme 6.13 Use of amino alcohol **12** in the ethylation of benzaldehyde.

catalyst. Conversions of substituted aromatic aldehydes showed lower stereoselectivities than those of benzaldehyde (35–71% *ee*). Aliphatic aldehydes proved to be unreactive.

6.2.1.2 With Ferrocenyl Oxazolinyl Alcohols

In 1997 Bolm and coworkers described the use of ferrocenyl oxazolinyl alcohols as catalysts in the diethylzinc addition to various aldehydes (Scheme 6.14) [16]. A 5 mol % catalyst loading of **13** was sufficient to obtain the resulting secondary alcohols with high enantioselectivities (up to 95% *ee*) in excellent yields (89–94%). Use of aromatic aldehydes furnished the best results (86–95% *ee*). Moreover, α,β-unsaturated and aliphatic substrates were efficiently alkylated (78–87% *ee*).

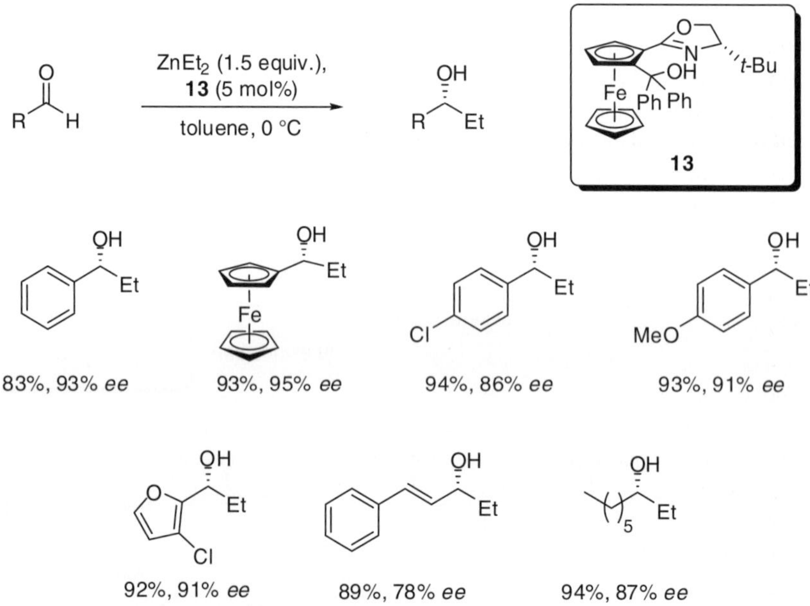

Scheme 6.14 Asymmetric additions of diethylzinc to aldehydes catalyzed by ferrocenyl oxazolinyl alcohol **13**.

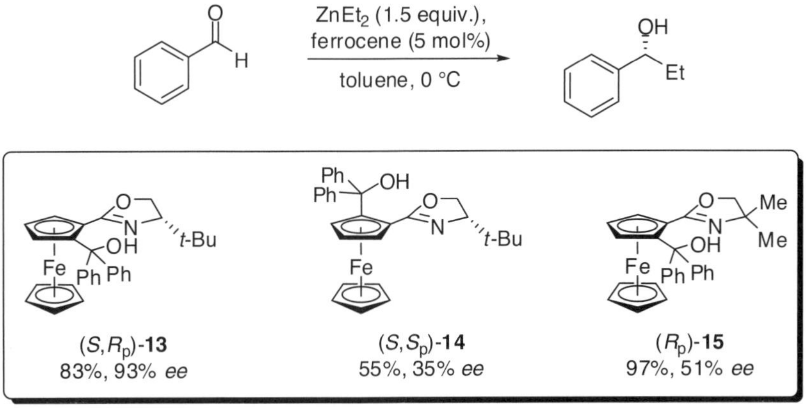

Scheme 6.15 Relevance of both central and planar chirality for asymmetric induction.

The authors extensively investigated the impact of planar chirality on the stereochemical outcome of the alkyl transfer reaction [17]. For example, they studied catalyzed diethylzinc additions to benzaldehyde employing ferrocenyl oxazolinyl alcohols (S,R_p)-**13** and (S,S_p)-**14** with identical central chirality but opposite planar chirality (Scheme 6.15). Whereas use of (S,R_p)-**13** gave a good result (83%, 93% *ee*), employment of (S,S_p)-**14** required a longer reaction time, leading to both a lower yield and reduced enantioselectivity (55%, 35% *ee*). Applying (R_p)-**15** lacking the stereogenic center resulted in moderate stereoselectivity and the reaction time was long (97% yield, 52% *ee*). In all cases the *R*-enantiomer of 1-phenylpropanol was obtained in excess. Thus, the planar chirality alone was not sufficient to lead to a product with high *ee*. Both elements of chirality were necessary and they had to be adjusted in such a manner that an internal group cooperativity resulted.

Bolm and co-workers also examined the catalytic properties of stereochemically inhomogeneous mixtures of chiral ferrocenes [18]. Amplification studies using nonequal mixtures of diastereomeric ferrocene (S,R_p)-**13** and (S,S_p)-**14** in the dimethylzinc addition to benzaldehyde revealed a nonlinear relationship between the enantiomeric excess of the product and the de of the ferrocene. Consequently, a high enantioselectivity (94% *ee*) was obtained using a 1:1 mixture of diasetero-isomers. The authors suggested that this nonlinear effect was not based on the formation of catalytically inactive heterochiral aggregates, but resulted from significantly different reaction rates of the diastereomers (S,R_p)-**13** and (S,S_p)-**14**.

1,1′-Disubstituted ferrocenes were introduced for the alkylation of aldehydes by Hou and coworkers in 1999 (Scheme 6.16) [19]. In the diethylzinc addition to benzaldehyde, ferrocene **16** furnished the resulting alcohol with an *ee* of 91% in almost quantitative yield. Various aldehydes were efficiently alkylated, and enantioselectivities up to 91% *ee* in combination with high yields were achieved. The presence of electron-donating substituents on the aromatic ring resulted in slightly reduced stereoselectivity (85–87% *ee*), and aliphatic aldehydes gave moderate levels of asymmetric induction (64–72% *ee*).

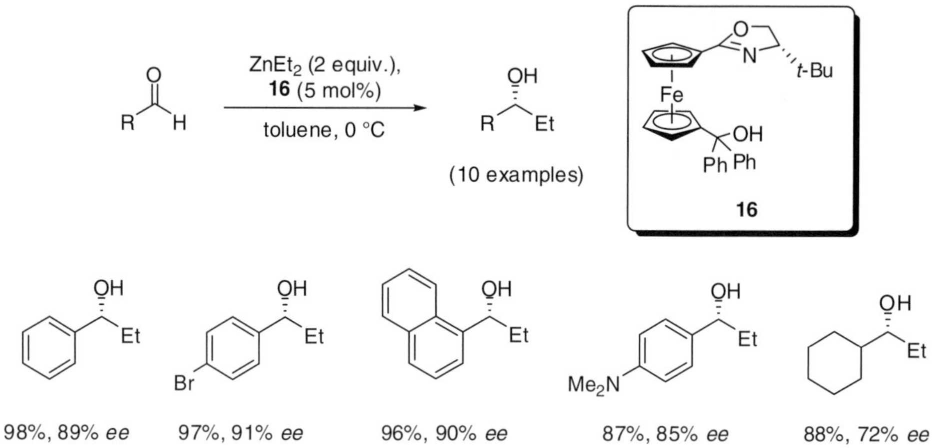

Scheme 6.16 A 1,1′-disubstituted ferrocene for the enantioselective alkylation of aldehydes.

In 2003 Bonini and coworkers reported the synthesis and application of various ferrocenyl oxazolinyl alcohols, which were prepared by iodide-mediated ring expansion of N-ferrocenoyl aziridinyl-2-carboxylic esters and subsequent modification of the ester function by reaction with methyl magnesium chloride [20]. Ferrocene **17** was tested in the asymmetric alkylation of benzaldehyde with diethylzinc, and a good yield and moderate enantioselectivity was achieved (46% *ee*, Scheme 6.17).

6.2.1.3 With Azaferrocenes

In 1997 Fu and coworkers introduced planar-chiral azaferrocenes for the enantioselective addition of organozinc reagents to aldehydes (Scheme 6.18) [21]. The reaction of benzaldehyde and diethylzinc in the presence of **18** furnished the resulting alcohol with only modest enantioselectivity (51% *ee*). However, azaferrocene **19**, generated from **18** by O-alkylation with 1,1-diphenyloxirane, proved to be more effective. With a catalyst loading of 3 mol% (S)-1-phenyl-1-propanol was isolated in 90% yield having 88% *ee*. Regardless of the electronic character of the aromatic ring, various para-substituted benzaldehydes were alkylated and high enantioselectivities were obtained (up to 94% *ee*). A significantly lower stereoselectivity was observed with the aliphatic aldehyde octanal (86%, 63% *ee*). Moreover, the addition of dimethylzinc to benzaldehyde was successfully achieved, affording (S)-1-phenylethanol with 83% *ee* in 82% yield.

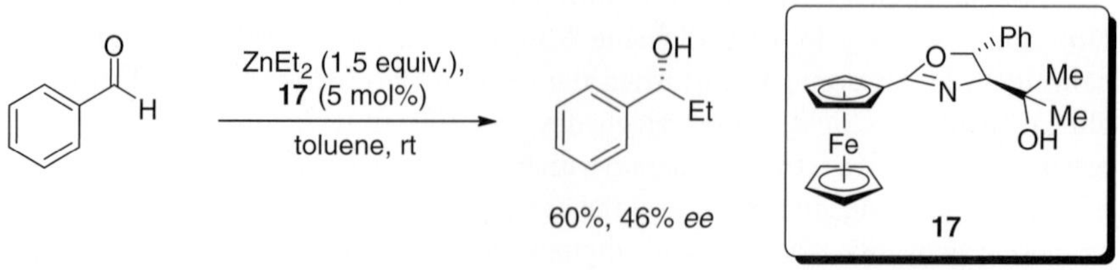

Scheme 6.17 A ferrocenyl oxazolinyl alcohol as catalyst for the diethylzinc addition to benzaldehyde.

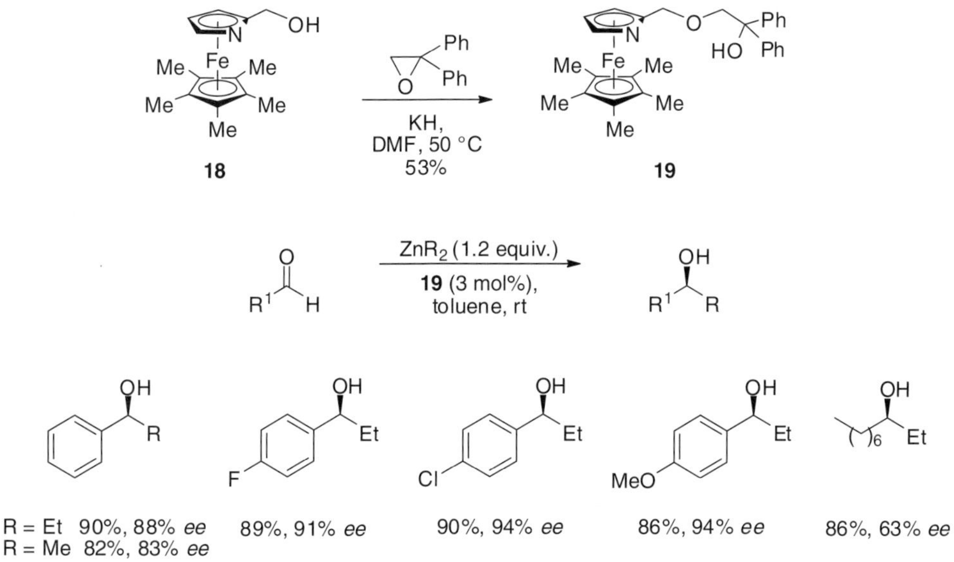

R = Et 90%, 88% ee   89%, 91% ee   90%, 94% ee   86%, 94% ee   86%, 63% ee
R = Me 82%, 83% ee

Scheme 6.18 Azaferrocenes in catalytic, asymmetric alkylations of aldehydes.

6.2.2
Addition of Arylzinc

The first catalytic, enantioselective addition of diphenylzinc to aldehydes was reported in 1997 by Fu and coworkers [21, 22]. Diphenylzinc, p-chlorobenzaldehyde and a catalytic amount of planar-chiral azaferrocene **19** led to the desired diarylmethanol with 57% ee in almost quantitative yield (Scheme 6.19).

In 1999 Bolm and coworkers used the planar-chiral ferrocenyl oxazolinyl alcohols, which they had previously applied in dialkylzinc additions [16–18], for the asymmetric aryl transfer onto aldehydes [23]. Ferrocenes **13** and **20** (10 mol%) proved to be equally efficient in the reaction of diphenylzinc and p-chlorobenzaldehyde furnishing the resulting alcohol with 88% ee in nearly quantitative yield. A wide range of aldehydes was tested with 5 mol% of **13** as catalyst. In all cases high yields (89–99%) were obtained, and the enantiomeric excesses depended on the substitution pattern of the aromatic aldehyde (Scheme 6.20). p-Chlorobenzaldehyde gave higher enantioselectivities (82% ee). The presence of *ortho*-substituents resulted in reduced

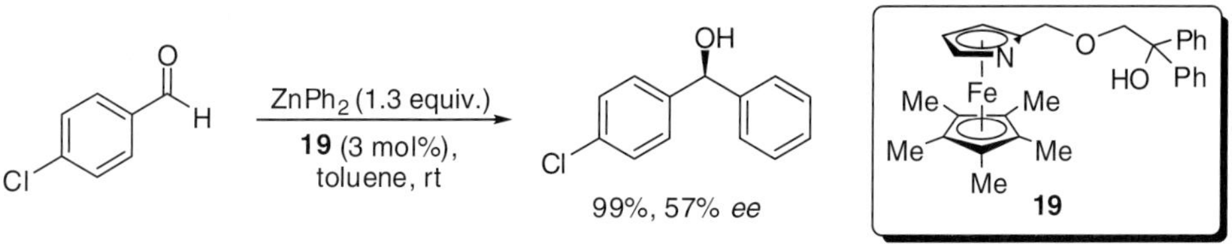

Scheme 6.19 Asymmetric phenyl transfer onto an aldehyde catalyzed by azaferrocene **19**.

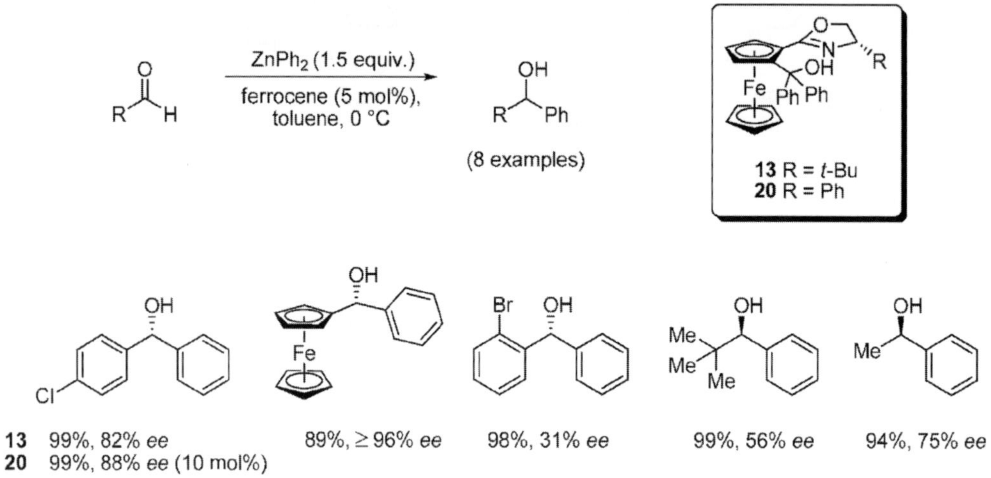
ZnPh2 (1.5 equiv.)
ferrocene (5 mol%),
toluene, 0 °C
(8 examples)
13 R = t-Bu
20 R = Ph
13 99%, 82% ee
20 99%, 88% ee (10 mol%)
89%, ≥96% ee
98%, 31% ee
99%, 56% ee
94%, 75% ee

Scheme 6.20 Asymmetric phenyl transfer onto aldehydes using $ZnPh_2$.

levels of asymmetric induction, as exemplified by *o*-bromobenzaldehyde (31% *ee*) and 1-naphthaldehyde (28% *ee*). Aliphatic aldehydes led to moderate results (50–75% *ee*). The best substrate was ferrocenylcarbaldehyde, which furnished the corresponding alcohol in high yield and ≥96% *ee*.

In 2000 the use of a modified phenylzinc reagent prepared *in situ* by mixing diphenylzinc and diethylzinc in a 1 : 2 ratio was reported by Bolm and coworkers (Scheme 6.21) [24]. With this protocol the amount of expensive, air- and moisture-sensitive diphenylzinc could be reduced to sub-stoichiometric quantities (0.65 equiv.). Here, the ethyl groups acted as the non-transferable groups on zinc, allowing

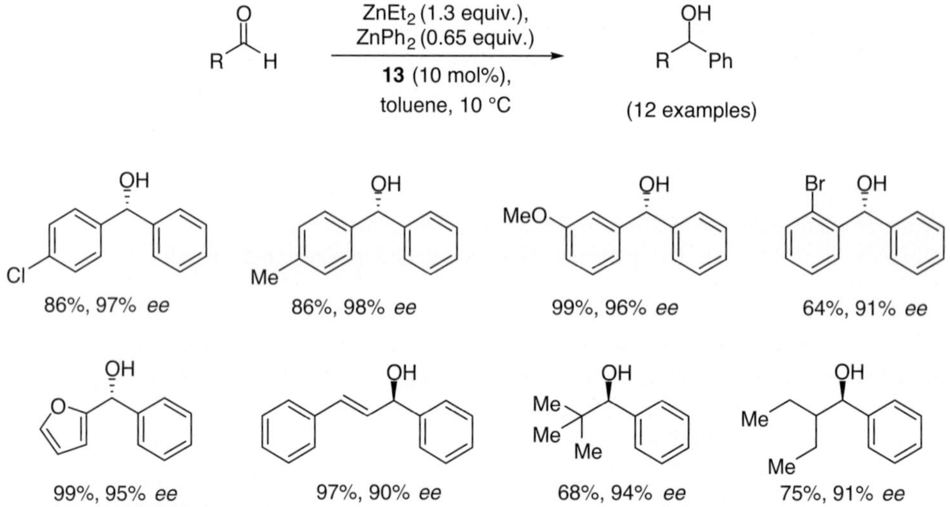
ZnEt2 (1.3 equiv.),
ZnPh2 (0.65 equiv.)
13 (10 mol%),
toluene, 10 °C
(12 examples)
86%, 97% ee
86%, 98% ee
99%, 96% ee
64%, 91% ee
99%, 95% ee
97%, 90% ee
68%, 94% ee
75%, 91% ee

Scheme 6.21 Phenyl transfer reactions by means of a mixed organozinc reagent.

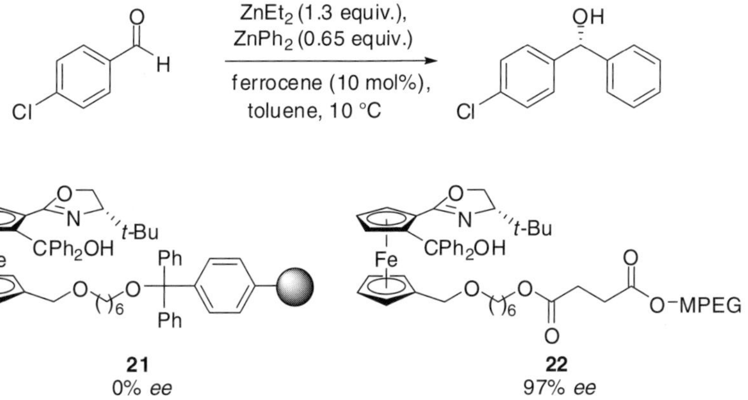

Scheme 6.22 Polymer-supported ferrocenes in organozinc additions.

both phenyl groups of phenylzinc to participate in the transfer reaction. The competitive background reaction became less pronounced, leading to a significant increase in the enantiomeric excesses of the products. The reaction of *p*-chlorobenzaldehyde afforded the desired diarylmethanol with 97% *ee*, which was significantly higher than the *ee* obtained under the original conditions (88% *ee*). Furthermore the substrate scope was no longer limited to *para*-substituted aromatic aldehydes. Using this improved procedure a wide range of aldehydes including *meta*- and *ortho*-substituted aromatic aldehydes as well as certain aliphatic aldehydes were phenylated with high stereoselectivities (up to 98% *ee*).

In order to facilitate the work-up and to recover and reuse the catalyst, polymer-supported ferrocenyl oxazolinyl alcohols were prepared and tested in the phenyl transfer reaction onto aldehydes (Scheme 6.22) [25]. Both soluble and insoluble polymeric supports (based on polyethylene glycol monomethyl ether [MeO–PEG–OH (MPEG), MW = 5000] and trityl chloride resins) were used for the immobilization via a linker at the 1′-carbon of the ferrocene. While the insoluble resin-bound ferrocene **21** gave racemic product in the addition of the mixed phenylzinc reagent to *p*-chlorobenzaldehyde, the soluble MeO–PEG-supported ferrocene **22** furnished the arylation product with high stereoselectivity (97% *ee*) in nearly quantitative yield. This result was identical to that obtained with the low molecular weight ferrocenyl oxazolinyl alcohol **13**. The recovery of ferrocene **22** proved to be facile, and the catalyst could be reused up to five times without significant loss in enantioselectivity.

The authors also examined the effect of using small amounts of dimethoxy polyethylene glycol (DiMPEG) as additive on the aryl transfer reaction to aldehydes [26]. In the reaction of *p*-methylbenzaldehyde with either pure diphenylzinc or the mixed ethyl–phenyl zinc reagent the addition of DiMPEG (10 mol%) led to reduced yields, but equal or better enantioselectivities. The authors suggested that both the presence of diethylzinc and DiMEPG reduced the quantity of the most reactive aryl transfer reagent diphenylzinc. Thereby the competitive uncatalyzed nonsymmetric (background) reaction was less pronounced, resulting in a higher stereoselectivity. In the reactions with *p*-chlorobenzaldehyde and benzaldehyde as substrates the amount of catalyst could be reduced by a factor of 10 (1 mol%) when 25 mol% of DiMPEG was added, and the product alcohols were obtained with >90% *ee*. Furthermore the presence of large quantities of $ZnBr_2$ was tolerated in the aryl transfer reaction with DiMPEG or MonoDMPEG as additives, and the authors hypothesized that this was due to their ability to deactivate (achiral) Lewis acidic species.

Density functional calculations were used to study the reactivity of mixed ethyl–phenyl zinc reagents and helped to rationalize the observed stereoselectivity of the aryl transfer reaction [27].

In 2005 Bolm and coworkers reported the syntheses of planar-chiral ferrocene-based organosilanols **23** and their catalytic use in the asymmetric phenyl transfer reaction onto substituted benzaldehydes (Scheme 6.23) [28]. These catalysts were easily prepared in four steps, starting from achiral ferrocene carboxylic acid. To examine their catalytic activity, a mixed ethyl–phenyl zinc reagent was used employing *p*-chlorobenzaldehyde as model substrate. The best results were obtained with ferrocenes **23a** and **23b** (10 mol%), affording the desired diarylmethanol with high enantiomeric excesses (89% and 91%, respectively) in good yields (82–84%). Other substituted aromatic aldehydes were also efficiently arylated with silanol **23a** as catalyst leading to diarylmethanols in good yields and high enantioselectivities (83–87% *ee*).

Encouraged by the successful application of ferrocene **12** in the diethylzinc addition to aldehydes (see Section 6.2.1), its catalytic activity was examined in aryl transfer reactions onto aldehydes (Scheme 6.24) [15]. Both diphenylzinc and a mixed ethyl–phenyl zinc reagent were used as aryl sources in the reaction with anisaldehyde in the presence of (R,S_p)-**12** or (R,R_p)-**12**. In the simple diphenylzinc addition with (R,R_p)-**12** as catalyst almost no asymmetric induction was observed (27% yield, 3%

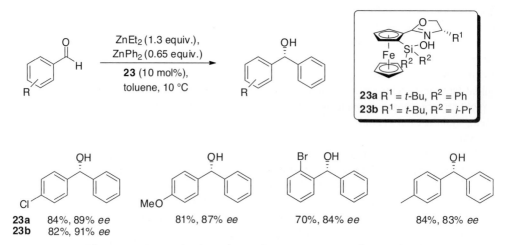

Scheme 6.23 Organosilanols used as catalysts in asymmetric arylations of aldehydes.

6.3 Addition to Aldehydes with Boron Reagents

ee). When using the modified zinc reagent low enantioselectivity was achieved (with (R,S_p)-**12**: 79% yield, 33% *ee*; with (R,R_p)-**12**: 49% yield, 30% *ee*). Notably, under those conditions a mixture of ethylation and arylation product was obtained.

6.2.3
Addition of Phenylacetylene

In 2004, the catalytic application of ferrocenyl oxazolinyl alcohols was expanded to the asymmetric alkynyl transfer reactions onto aldehydes. Hou and coworkers reported the use of 1,2- and 1,1'-disubstituted ferrocenes in the diethylzinc-mediated enantioselective phenylacetylene addition to aldehydes (Scheme 6.25) [29]. The substituents on the ligand framework had a great impact on the outcome. In the reaction between benzaldehyde and phenylacetylene, 1,2-substituted ferrocene **13** showed higher stereoselectivity than the corresponding 1,1'-derivative **16** (63% *ee* vs. 23% *ee*). Among the 1,2-ferrocenes tested, **24** gave the best result (90%, 87% *ee*). Furthermore, **24** (10–20 mol%) effectively catalyzed the alkynylation of various other aromatic aldehydes affording the targeted secondary propargylic alcohols with up to 93% *ee* in high yield. Aliphatic aldehydes reacted with comparable enantioselectivities. α,β-Unsaturated substrates provided propargylic alcohols with significantly lower enantiomeric excesses (54–59% *ee*).

6.3
Addition to Aldehydes with Boron Reagents

Most of the catalyzed asymmetric transfer reactions onto aldehydes imply the use of organozinc reagents. However, functionalized compounds of this type are often tedious to prepare or remain inaccessible. To overcome this limitation, the use of boron-based reagents was introduced.

6.3.1
Aryl Transfers with Boronic Esters

Arylboronic esters can easily be prepared from the corresponding aryl halides or aryl boronic acids, permitting a broader functionalization of the aryl group to be transferred. Consequently, Bolm and Rudolph studied the use of ferrocene **13** in

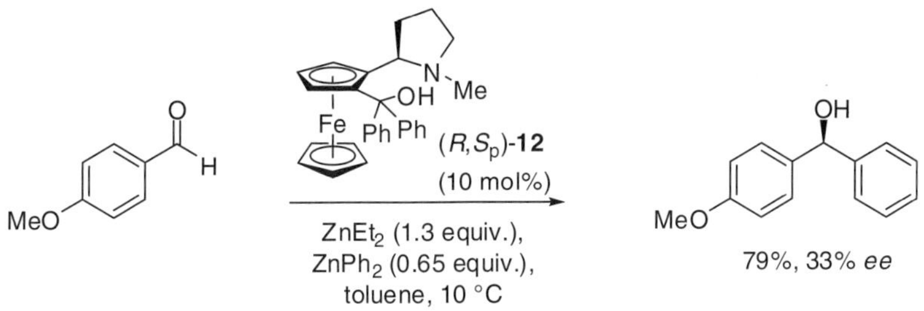

Scheme 6.24 Amino alcohol **12** as catalyst in the phenyl addition to aldehydes.

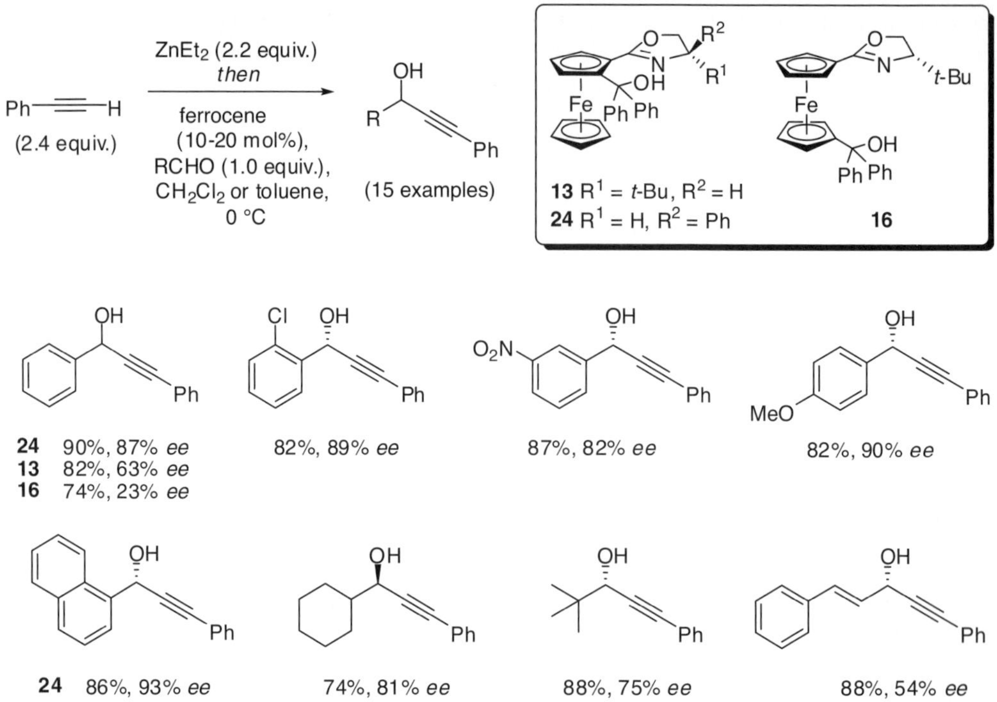

Scheme 6.25 Enantioselective alkynylations of aldehydes.

the phenyl transfer onto *p*-chlorobenzaldehyde using commercially available phenylboronic acid 1,3-propanediol ester as aryl source (Scheme 6.26) [30].

In combination with diethylzinc the corresponding diarylmethanol was obtained with excellent enantiomeric excess (97% *ee*) in high yield. The drawback of this protocol was the difficult purification of the desired alcohol due to the presence of large amounts of boron-containing by-products.

6.3.2
Aryl Transfer with Triphenylborane [Ph₃B]

The successful use of phenyl boronic ester encouraged Bolm and coworkers to investigate the aryl transfer capability of alternative boron reagents. Finally, employing triphenylborane in combination with an excess of diethylzinc and ferrocene **13** proved applicable (Scheme 6.27) [31].

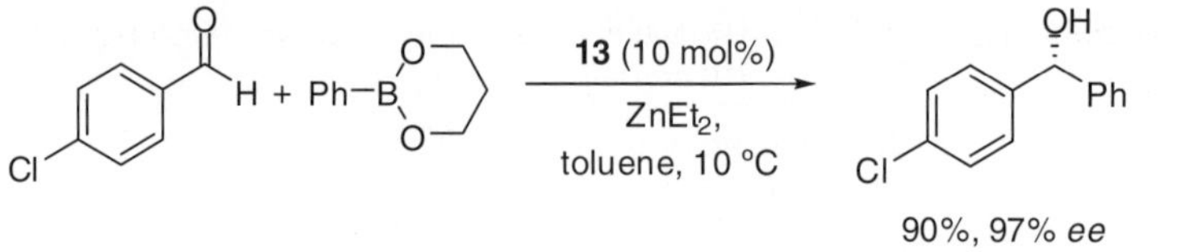

Scheme 6.26 Phenyl addition onto *p*-chlorobenzaldehyde using a phenyl boronic ester.

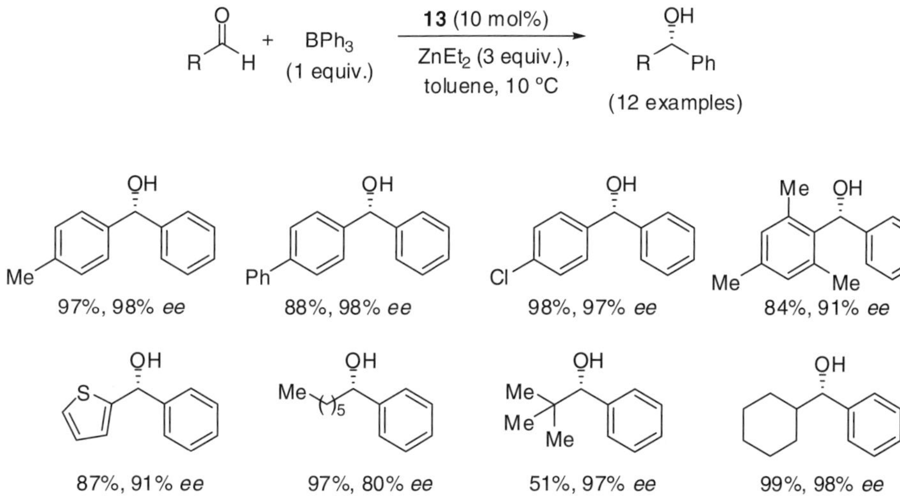

6.3 Addition to Aldehydes with Boron Reagents

$$R-CHO + BPh_3 \xrightarrow[\text{toluene, 10 °C}]{\text{13 (10 mol\%)}, \text{ZnEt}_2 \text{ (3 equiv.)}} R\text{-CH(OH)-Ph}$$

(1 equiv.) (12 examples)

- 4-Me-C6H4-CH(OH)-Ph: 97%, 98% ee
- 4-Ph-C6H4-CH(OH)-Ph: 88%, 98% ee
- 4-Cl-C6H4-CH(OH)-Ph: 98%, 97% ee
- 2,6-Me2-C6H3-CH(OH)-Ph: 84%, 91% ee
- 2-thienyl-CH(OH)-Ph: 87%, 91% ee
- Me(CH2)5-CH(OH)-Ph: 97%, 80% ee
- Me2CHCH2-CH(OH)-Ph: 51%, 97% ee
- Cy-CH(OH)-Ph: 99%, 98% ee

Scheme 6.27 Asymmetric phenyl transfer to aldehydes with triphenylborane.

This method allowed the synthesis of a variety of diarylmethanols with high enantioselectivities (up to 98% ee) in good yields. Moreover, under these conditions a phenyl transfer onto a heteroaromatic aldehyde, 2-thiophenecarbaldehyde, was achieved (91% ee). Additionally, the enantiomeric excesses obtained in the more demanding additions onto aliphatic aldehydes were remarkable (80–97% ee).

6.3.3
Boronic Acids as Aryl Source

The use of air-stable arylboronic acids as aryl sources has been reported by Bolm and Rudolph, and it broadened the substrate scope drastically [30]. The procedure involved a prior reaction of the arylboronic acid with an excess of diethylzinc in toluene at 60 °C for 12 h, followed by the addition of ferrocene **13** and the corresponding aldehyde at 10 °C. Since a beneficial effect on the enantioselectivity upon addition of polyethers had been observed [26], DiMPEG was used as additive. Then, in the presence of 10 mol% of this polyether, higher ee values were achieved for a variety of substrates. Overall this protocol led to diarylmethanols with enantiomeric excesses up to 98% in moderate to good yields (Scheme 6.28).

Subsequently, a scale-up of this procedure was demonstrated [32]. It worked efficiently with multigram quantities of starting materials, providing products with similar enantioselectivities and yields. Moreover, it was possible to recover and reuse the catalyst without negative effects on its activity. Lastly, these authors demonstrated the flexibility of this method by using a wide range of educts with various substitution patterns and a "reverse reagent combination", which allowed them to access both enantiomers of the product with a single enantiomer of the catalyst [33].

Ferrocenyl silanol **23a** has also been employed as catalyst in the aryl transfer reaction onto p-chlorobenzaldehyde starting from phenyl boronic acid as aryl source

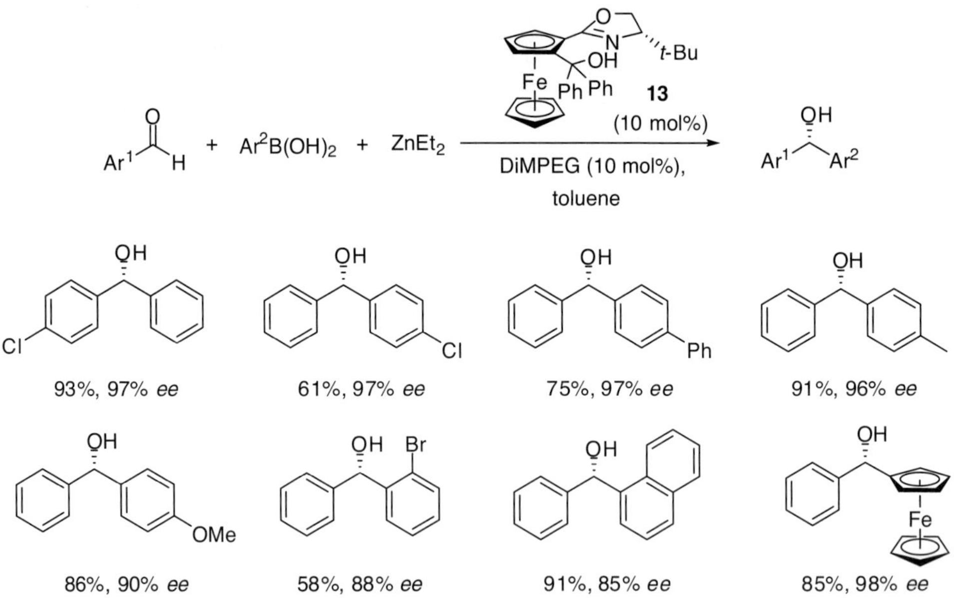

Scheme 6.28 Aryl transfer to aldehydes with boronic acids as aryl sources.

(Scheme 6.29) [28]. Although good reactivity and enantioselectivity were observed, both yield and ee were significantly lower than that obtained previously with its analogous ferrocenyl oxazolinyl alcohol **13** (67%, 83% ee vs. 93%, 97% ee).

6.3.4
Alkenylboronic Acids in Alkenyl Transfer Reactions

The applicability of alkenylboronic acids for the preparation of enantioenriched allyl alcohols has also been investigated [34]. This transformation was found to be highly dependent on the quality of the boronic acid. The best results were achieved when the boron reagents were pretreated with hot water to minimize the presence of boroxines. Moreover, the addition of isopropanol (1 equiv) to the reaction medium improved the enantiomeric excesses of the resulting allyl alcohols (up to 75% ee, Scheme 6.30).

6.4
Other Transformations: Asymmetric Epoxidation

As mentioned in the previous sections, N,O-bidentate ferrocenes have been used nearly exclusively as catalysts in asymmetric additions of organometallic reagents to aldehydes. An exception is the application of ferrocenyl oxazolinyl hydroxamic acids **25** in vanadium-catalyzed asymmetric epoxidations of allylic alcohols (Scheme 6.31) [35].

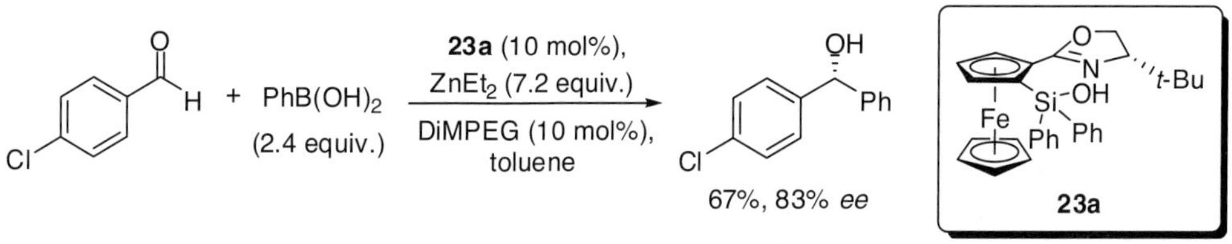

Scheme 6.29 Ferrocenyl silanol **23a** in aryl transfers with boronic acids.

Although good yields (72–90%) have been achieved, only modest enantioselectivities were reached (up to 21% ee). Worth mentioning is the fact that products with opposite absolute configurations have been obtained with the same ferrocene by changing the oxidant from cumene hydroperoxide (CHP) to *tert*-butyl hydroperoxide (TBHP).

6.5
Conclusion and Perspectives

As shown in this chapter, chiral N,O-bidentate ferrocenes can be highly effective catalysts in several asymmetric transformations. The major effort in this field has been focused on their use in enantioselective nucleophilic addition reactions to aldehydes. There, excellent enantioselectivities and high yields have been achieved. One interesting additional feature is the presence of the diphenylhydroxymethyl moiety in almost all active catalyst structures, where it proved to be crucial for achieving high asymmetric induction.

Apparently, the use of N,O-bidentate ferrocenes in asymmetric transformations is still relatively unexplored. Although excellent results have already been obtained in 1,2-additions to aldehydes, their successful application to other transformations remains a challenge. This leads to an open research area for the discovery and

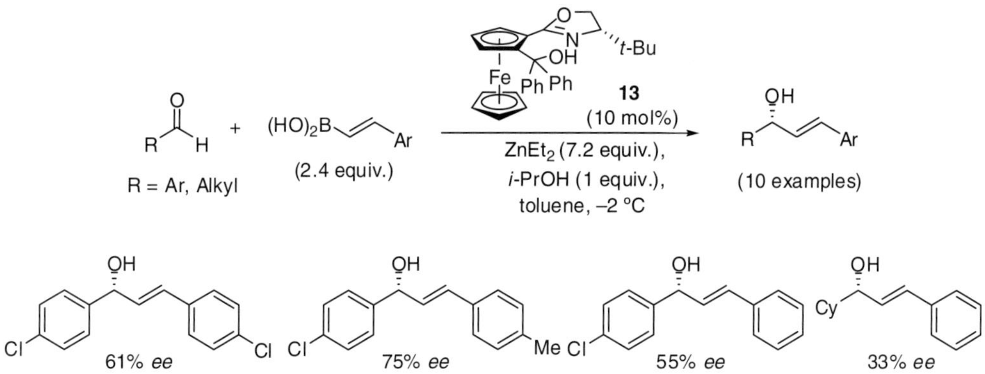

Scheme 6.30 Synthesis of enantioenriched allyl alcohols with alkenylboronic acids.

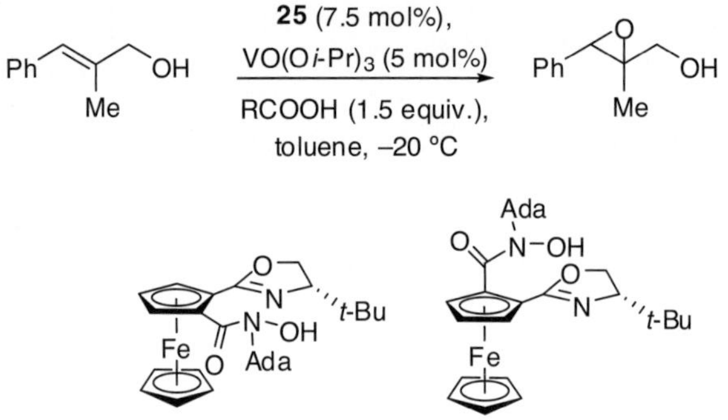

Scheme 6.31 Vanadium-catalyzed asymmetric epoxidations in the presence of **25**.

	25	**ent-25**
CHP	90%, 9% ee (R,R)	85%, 20% ee (S,S)
TBHP	72%, 21% ee (S,S)	89%, 21% ee (R,R)

development of new enantioselective processes with this catalyst type. Significant advances are therefore expected in this field in the very near future.

6.6
Experimental: Selected Procedures

6.6.1
Addition of Diethylzinc to Benzaldehyde with Ferrocene 13 [17]

A well-dried Schlenk flask under argon was charged with ferrocene **13** (5 mol%). The flask was evacuated twice and flushed with argon. Freshly distilled toluene was then added and the solution was cooled to 0 °C followed by addition of neat diethylzinc (1.5 equiv). The resulting solution was stirred for 20 min and benzaldehyde (1.0 equiv) was then added. The flask was sealed, and the progress of the reaction was monitored by TLC. For the work-up 2 M HCl was carefully added and the resulting mixture was extracted four times with dichloromethane. The combined organic phases were washed with brine, dried over $MgSO_4$ and the solvent was removed on a rotary evaporator. The crude product was purified by column chromatography (silica gel, hexanes/MTBE = 10/1) followed by distillation in a Kugelrohr apparatus, which yielded 1-phenyl-1-propanol as a colorless oil.

6.6.2
Phenylacetylene Addition to Aldehydes with Ferrocene 24 [29]

To a solution of phenylacetylene (123 mg, 1.2 mmol) in CH_2Cl_2 (2 mL) was added Et_2Zn (1.1 M in hexane, 1.1 mL, 1.2 mmol) at room temperature. The resulting mixture was stirred for 2 h after which ferrocene **24** (26 mg, 0.05 mmol) was added, and the reaction mixture stirred for an additional 30 min. The reaction system was cooled to 0 °C at which point the aldehyde (0.5 mmol) was added under an argon atmosphere. After complete consumption of the substrate (monitored by TLC), the reaction was quenched with saturated aqueous NH$_4$Cl. The mixture was then extracted with diethyl ether (10 mL × 3). The organic layer was washed with brine (10 mL × 2), dried over Na$_2$SO$_4$, and evaporated under reduced pressure to give an oily residue. Purification of the residue by column chromatography gave the corresponding optically active alcohol.

6.6.3
Phenyl Transfer (ZnEt$_2$ + ZnPh$_2$) to Aldehydes with Ferrocene 13 [24]

In a glove box a well-dried Schlenk flask was charged with diphenylzinc (36 mg, 0.16 mmol). The flask was sealed and removed from the glovebox. Freshly distilled toluene (3 mL) was added followed by diethylzinc (33 µL, 0.33 mmol). The mixture was stirred for 30 min at room temperature, ferrocene **13** (12 mg, 0.025 mmol) was added, and the resulting solution was cooled to 10 °C. Stirring was continued for an additional 10 min at this temperature, and the aldehyde (0.25 mmol) was then added directly in one portion. The Schlenk flask was sealed, and the reaction mixture was stirred at 10 °C overnight. Quenching with water followed by extracting with dichloromethane, drying of the combined organic phase over MgSO$_4$, and evaporating the solvent under reduced pressure gave the crude product. Column chromatography (silica gel; eluents: hexanes/diethyl ether) afforded the pure secondary alcohol.

6.6.4
Phenyl Transfer to Aldehydes with Triphenylborane on a Gram Scale [31]

A well-dried Schlenk flask was charged with triphenylborane (1.614 g, 6.67 mmol). Freshly distilled toluene (60 mL) and ZnEt$_2$ (1 M in heptane, 20 mL, 20 mmol) were added, and the mixture was stirred for 45 min at room temperature. A solution of ferrocene **13** (329 mg, 0.67 mmol) in toluene (65 mL) was transferred via a syringe into the first flask, the resulting mixture was stirred for 30 min at room temperature, then cooled to 10 °C and stirred for an additional 10 min at this temperature. Then, a solution of the aldehyde (6.67 mmol, 1 equiv) in toluene (65 mL) at 10 °C was transferred via syringe into the other reaction flask. After 12 h reaction time at 10 °C the mixture was quenched with H$_2$O and AcOH (20% in H$_2$O, 100 mL), and the mixture extracted with dichloromethane. The organic layer was washed with H$_2$O, dried (MgSO$_4$), filtered, and the solvent was removed under reduced pressure. The product was purified by column chromatography to afford pure optically active alcohol.

6.6.5
Aryl Transfer to Aldehydes with Boronic Acids on a Multigram Scale [32]

A well-dried Schlenk flask was charged with arylboronic acid (24 mmol) and dimethylpolyethyleneglycol ($M_w = 2000$ g mol^{-1}, 2.0 g, 1 mmol). After sealing with a septum, freshly distilled toluene (100 mL) was added, followed by ZnEt$_2$ (7.4 mL, 72 mmol). The mixture was heated to 60 °C, stirred for 12 h at this temperature and then cooled to room temperature. In another Schlenk flask, ferrocene **13** (500 mg, 1 mmol) was dissolved in toluene (35 mL) and transferred via syringe to the first solution. The mixture was stirred for 30 min at room temperature and then cooled to 10 °C. Stirring was continued for an additional 10 min at this temperature. A third Schlenk flask was charged with aldehyde (10 mmol) and toluene (35 mL) was added. After cooling to 10 °C, this solution was transferred via syringe into the other flask. The reaction mixture was stirred for 12 h at 10 °C, then the reaction was carefully quenched with H$_2$O (10 mL) and the mixture stirred for an additional 10 min. Subsequently, it was filtered through a pad of Celite and washed vigorously with CH$_2$Cl$_2$ (400 mL). The organic layer was washed with AcOH (20% in H$_2$O, 250 mL) and H$_2$O, dried (MgSO$_4$), filtered and the solvent removed under reduced pressure. The residue was purified by column chromatography to afford pure optically active alcohol.

Appendix: Synthesis of Bolm's Ferrocene 13 (S,R$_p$)-2-(α-Diphenylhydroxymethyl)ferrocenyl-5-*tert*-butyloxazoline [(S,R$_p$)-13] [17]

A solution of (S)-ferrocenyl-5-*tert*-butyloxazoline [17] (643 mg, 2.07 mmol) in freshly distilled THF (40 mL) was cooled to −78 °C and then treated dropwise with *sec*-BuLi (1.9 mL, 2.48 mmol, 1.3 M in cyclohexane). After stirring for 2 h at this temperature, benzophenone (527 mg, 2.89 mmol) was added at −60 °C. The resulting solution was allowed to warm to room temperature over night. The reaction mixture was quenched with distilled water and extracted with methyl *tert*-butyl ether. The combined organic phases were washed subsequently with brine and water, dried over MgSO$_4$, and evaporated under reduced pressure. The product was purified by column chromatography (hexanes/ethyl acetate 70 : 30) followed by recrystallization from *n*-hexane to give (S,R$_p$)-**13** (887 mg, 87%) as orange crystals.

Abbreviations

Ada	adamantyl
CHP	cumene hydroperoxide
Cy	cyclohexyl
DiMEPG	dimethoxy poly(ethylene glycol)
TBHP	*tert*-butyl hydroperoxide

Acknowledgments

The authors are grateful to the Fonds der Chemischen Industrie for financial support. O. M. G. thanks *Syngenta* for postdoctoral support.

References

1. (a) Noyori, R. and Kitamura, M. (1991) *Angew. Chem. Int. Ed. Engl.*, **30**, 49–69; (b) Soai, K. and Niwa, S. (1992) *Chem. Rev.*, **92**, 833–856; (c) Pu, L. and Yu, H.-B. (2001) *Chem. Rev.*, **101**, 757–824.
2. (a) Watanabe, M., Araki, S., Butsugan, Y., and Uemura, M. (1989) *Chem. Express*, **4**, 825–828; (b) see also: Watanabe, M., Araki, S., and Butsugan, Y. (1995) *Ferrocenes* (eds A. Togni and T. Hayashi), VCH Verlagsgesellschaft, Weinheim, pp. 143–169.
3. Watanabe, M., Araki, S., Butsugan, Y., and Uemura, M. (1990) *Chem. Express*, **5**, 661–664.
4. Watanabe, M., Araki, S., Butsugan, Y., and Uemura, M. (1990) *Chem. Express*, **5**, 761–764.
5. Watanabe, M., Araki, S., Butsugan, Y., and Uemura, M. (1991) *J. Org. Chem.*, **56**, 2218–2224.
6. Watanabe, M., Hashimoto, N., Araki, S., and Butsugan, Y. (1992) *J. Org. Chem.*, **57**, 742–744.
7. Watanabe, M., Komota, M., Nishimura, M., Araki, S., and Butsugan, Y. (1993) *J. Chem. Soc., Perkin Trans. 1*, 2193–2196.
8. Wally, H., Widhalm, M., Weissensteiner, W., and Schlögl, K. (1993) *Tetrahedron: Asymmetry*, **4**, 285–288.
9. Nicolosi, G., Patti, A., Morrone, R., and Piattelli, M. (1994) *Tetrahedron: Asymmetry*, **5**, 1639–1642.
10. Fukuzawa, S. and Kato, H. (1998) *Synlett*, 727–728.
11. Malézieux, B., Andrés, R., Gruselle, M., Rager, M.-N., and Thorimbert, S. (1999) *Tetrahedron: Asymmetry*, **10**, 3253–3257.
12. Delacroix, O., Picart-Goetgheluck, S., Maciejewski, L., and Brocard, J. (1999) *Tetrahedron: Asymmetry*, **10**, 4417–4425.
13. Arroyo, N., Haslinger, U., Mereiter, K., and Widhalm, M. (2000) *Tetrahedron: Asymmetry*, **11**, 4207–4219.
14. (a) Wang, M.-C., Wang, D.-K., Zhu, Y., Liu, L.-T., and Guo, Y.-F. (2004) *Tetrahedron: Asymmetry*, **15**, 1289–1294; (b) Wang, M.-C., Liu, L.-T., Zang, J.-S., Shi, Y.-Y., and Wang, D.-K. (2004) *Tetrahedron: Asymmetry*, **15**, 3853–3859.
15. Ahem, T., Müller-Bunz, H., and Guiry, P.J. (2006) *J. Org. Chem.*, **71**, 7596–7602.
16. Bolm, C., Muñiz Fernández, K., Seger, A., and Raabe, G. (1997) *Synlett*, 1051–1052.
17. Bolm, C., Muñiz-Fernández, K., Seger, A., Raabe, G., and Günther, K. (1998) *J. Org. Chem.*, **63**, 7860–7867.
18. (a) Bolm, C., Muñiz, K., and Hildebrand, J.P. (1999) *Org. Lett.*, **1**, 491–494; (b) See also: Muñiz, K., and Bolm, C. (2000) *Chem. Eur. J.*, **6**, 2309–2316.
19. Deng, W.-P., Hou, X.-L., and Dai, L.X. (1999) *Tetrahedron: Asymmetry*, **10**, 4689–4693.
20. Bonini, B.F., Fochi, M., Comes-Franchini, M., Ricci, A., Thijs, L., and Zwanenburg, B. (2003) *Tetrahedron: Asymmetry*, **14**, 3321–3327.
21. Dosa, P.I., Craig Ruble, J., and Fu, G.C. (1997) *J. Org. Chem.*, **62**, 444–445.
22. (a) For general overviews on aryl transfer reactions, see: Bolm, C., Hildebrand, J.P., Muñiz, K., and Hermanns, N. (2001) *Angew. Chem.*, **113**, 3382–3407; (2001) *Angew. Chem. Int. Ed. Engl.*, **40**, 3284–3308; (b) Schmidt, F., Stemmler, R.T., Rudolph, J., and Bolm, C. (2006) *Chem. Soc. Rev.*, **35**, 454–470.
23. Bolm, C. and Muñiz, K. (1999) *Chem. Commun.*, 1295–1296.
24. (a) Bolm, C., Hermanns, N., Hildebrand, J.P., and Muñiz, K. (2000) *Angew. Chem.*, **112**, 3607–3609; (2000) *Angew. Chem. Int. Ed.*, **39**, 3465–3467.
25. Bolm, C., Hermanns, N., Claßen, A., and Muñiz, K. (2002) *Bioorg. Med. Chem. Lett.*, **12**, 1795–1798.
26. (a) Rudolph, J., Hermanns, N., and Bolm, C. (2004) *J. Org. Chem.*, **69**, 3997–4000; (b) See also: Rudolph, J., Lormann, M., Bolm, C., and Dahmen, S. (2005) *Adv. Synth. Catal.*, **347**, 1361–1368.

27 (a) Rudolph, J., Rasmussen, T., Bolm, C., and Norrby, P.-O. (2003) *Angew. Chem.*, **115**, 3110–3113; (2003) *Angew. Chem. Int. Ed.*, **42**, 3002–3005; (b) Rudolph, J., Bolm, C., and Norrby, P.-O. (2005) *J. Am. Chem. Soc.*, **127**, 1548–1552.

28 Özçubukçu, S., Schmidt, F., and Bolm, C. (2005) *Org. Lett.*, **7**, 1407–1409.

29 Li, M., Zhu, X.-Z., Yuan, K., Cao, B.-X., and Hou, X.-L. (2004) *Tetrahedron: Asymmetry*, **15**, 219–222.

30 Bolm, C. and Rudolph, J. (2002) *J. Am. Chem. Soc.*, **124**, 14850–14851.

31 Rudolph, J., Schmidt, F., and Bolm, C. (2004) *Adv. Synth. Catal.*, **346**, 867–872.

32 Rudolph, J., Schmidt, F., and Bolm, C. (2005) *Synthesis*, 840–842.

33 Schmidt, F., Rudolph, J., and Bolm, C. (2007) *Adv. Synth. Catal.*, **349**, 703–708.

34 Schmidt, F., Rudolph, J., and Bolm, C. (2006) *Synthesis*, 3625–3630.

35 Bolm, C. and Kühn, T. (2001) *Isr. J. Chem.*, **41**, 263–269.

7
Symmetrical 1,1′-Bidentate Ferrocenyl Ligands
Wanbin Zhang and Delong Liu

7.1
Introduction

Ferrocene-based chiral ligands designed for asymmetric synthesis have attracted tremendous scientific interest over the past decades [1]. Some have also been successfully applied in industrial processes [2]. Apart from the properties of rigidity and stability, the reason why these ligands are a versatile class of compounds is that structural modification can be readily made by introduction of a desired functional group on the cyclopentadienyl rings according to the demand of the reaction type. Among the vast variety of ferrocene-based ligands, the design and preparation of C_2-symmetrical chiral ferrocenes are of great importance in the development of transition metal-catalyzed enantioselective reactions [1].

In the majority of scenarios for absolute stereochemical control, the presence of a C_2-symmetrical factor within the chiral complexes can serve a very important function by dramatically reducing the number of possible competing, diastereomeric transition states [3]. This was first suggested by Kagan who successfully designed and synthesized the C_2-symmetrical DIOP with central chirality in the 1970s (Figure 7.1) [4]. DIOP led to superior enantioselectivity compared to monodentate phosphines in asymmetric hydrogenation and thus the C_2-symmetrical factor was deemed to be an important structural feature for developing new efficient chiral ligands.

Subsequently, Knowles made his significant discovery of a C_2-symmetrical chelating diphosphine ligand DIPAMP with P-chirality [5], which was quickly employed in the industrial production of L-DOPA (Figure 7.1) [6].

The importance of the C_2-symmetrical factor has been further confirmed. In the 1980s, Noyori and Takaya reported an atropisomeric C_2-symmetrical diphosphine ligand BINAP (Figure 7.1) [7]. BINAP appears to be both the most used and the most useful ligand for asymmetric catalysis [8]. One of the successful applications of BINAP is the Takasago industrial production of a carbapenem key intermediate **4-AA**

Chiral Ferrocenes in Asymmetric Catalysis. Edited by Li-Xin Dai and Xue-Long Hou
Copyright © 2010 WILEY-VCH Verlag GmbH & Co. KGaA, Weinheim
ISBN: 978-3-527-32280-0

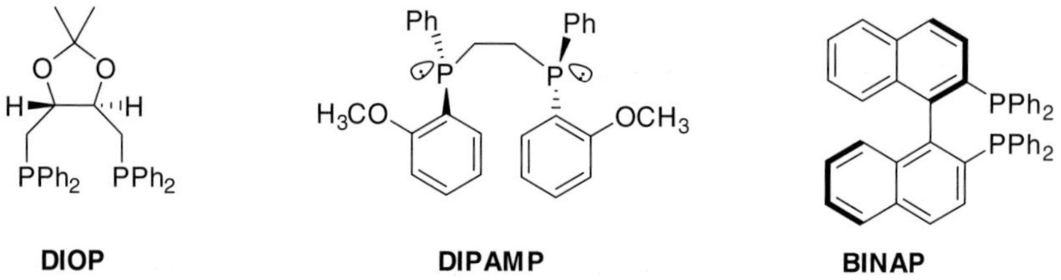

7 Symmetrical 1,1′-Bidentate Ferrocenyl Ligands

Figure 7.1 Typical examples of chiral C_2-symmetrical ligands.

by Ru-catalyzed asymmetric hydrogenation of (±)-methyl-2-(benzamidomethyl)-3-oxobutanoate.

Compared to the various chiral ligands, including those mentioned above, the C_2-symmetrical chiral ligands with ferrocene as a backbone did not receive much attention until the last decade. However, a large number of prominent C_2-symmetrical chiral ligands with a ferrocene backbone, for example, Ferro-TANE, FERRIPHOS, Ikeda's ligands, have been developed in recent years and afforded excellent enantioselective induction in asymmetric catalysis (Figure 7.2).

This chapter attempts to provide the reader with a concise summary of the synthetic routes and applications of most of the symmetrical 1,1′-bidentate ligands with ferrocene as a backbone. These chiral ferrocenes are sorted according to the number of substituted groups, for example, 1,1′-disubstituted and 1,1′,2,2′-tetrasubstituted ferrocenes. In addition, we mention, as the analogs of the corresponding chiral ferrocenes, the chiral ruthenocene ligands, although these have received much less attention than the ferrocene ones. Several successful examples of symmetrical 1,1′-bidentate ruthenocene ligands have been developed independently by Zhang's and Bolm's groups. The difference in structure between ferrocenyl and ruthenocenyl ligands is only the distance between the two Cp rings which might provide different, even opposite, catalytic behavior in asymmetric catalysis. For the purpose of further understanding the properties of chiral ferrocenyl ligands, a discussion of C_2-symmetrical chiral ruthenocenyl ligands is added at the end of this chapter.

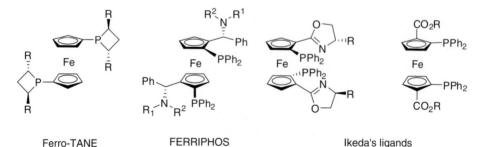

Figure 7.2 Examples of ferrocene-based chiral C_2-symmetrical ligands.

7.2
Symmetrical 1,1′-Disubstituted Ferrocenyl Ligands

7.2.1
P-Centered Chiral Diphosphine Ligands

Among the plenitude of chiral phosphines developed for application in asymmetric catalysis, examples possessing optically active phosphorus donors are rare [9]. Metal complexes of bidentate P-chiral compounds are considered to be excellent optical inducers in asymmetric catalysis because the chiral P-atoms are in the closest proximity to the catalytic center. Dramatic results from Knowles and coworkers with the C_2-symmetric P-chiral ligand DIPAMP for Rh-catalyzed enantioselective hydrogenations highlighted this field in asymmetric catalysis [5a]. Whereas, it took nearly two decades for scientists to develop efficient methods to prepare C_2-symmetric P-chiral diphosphine ligands, largely due to the synthetic difficulties in the construction of the P-chiral centers [10].

Jugé and Genêt's group have successfully worked on constructing P-chiral ferrocene compounds for a long time. Two synthetic approaches have been developed to asymmetrically synthesize the P-chiral 1,1′-bis(methylphenylphosphino)ferrocenes **L-1** either by P–C bond formation starting from ferrocene **1** (Scheme 7.1, pathway A), or by homocoupling of the cyclopentadienyl methyl phenyl phosphine **2** with FeCl$_2$ (Scheme 7.1, pathway B) [11]. The second approach afforded a better yield with simple operation whereas poor stereoselectivity was obtained with path A because this approach led to a diastereomeric mixture of the diphosphine **L-1** (55 : 45). This could be attributed to the effect of steric hindrance of the ferrocene mono- and/or dianion during the formation of two P–C bonds.

According to the method reported by Jugé and coworkers, van Leeuwen developed a series of symmetrical 1,1′-disubstituted P-chiral bidentate phosphine ferrocenyl ligands **L-2a~2f** (Scheme 7.2) [12]. Treatment of **3** with aryllithium reagents at −78 °C afforded the phosphine amide boranes **4** in 85–94% yields. Compounds **4** were subjected to acidic methanolysis, the methyl phosphinite borane complexes **5** being isolated, after column chromatography, in 66–94% yields. **L-2** were synthesized

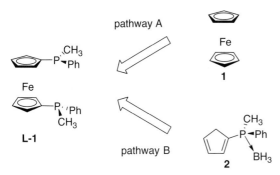

Scheme 7.1 Two synthetic approaches to **L-1**.

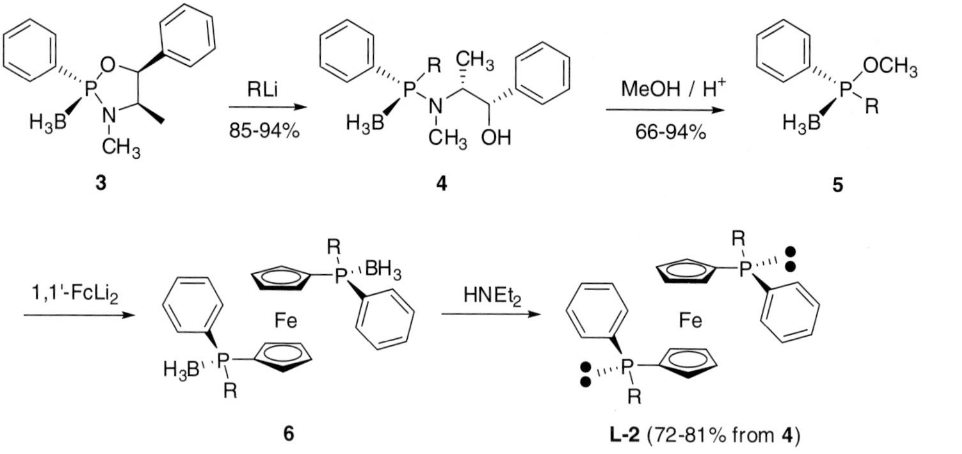

a: R=2-anisyl; **b**: R=1-naphthyl; **c**: R=2-naphthyl; **d**: R=9-phenanthryl; **e**: R=2-biphenyl; **f**: R=ferrocenyl.
Scheme 7.2 Synthetic route to **L-2**.

by addition of a suspension of 1,1′-dilithioferrocene (1,1′-FcLi$_2$) to **5** in one pot (72–81% yields from **5**, >98% ee) followed by decomplexation of the borane complex.

In the Pd-catalyzed asymmetric allylic substitution of 1,3-diphenylprop-2-en-1-yl acetate (**7a**), **L-2b**, **2d** ∼ **2f** with bulky 1-naphthyl, 9-phenanthryl, 2-biphenyl, or ferrocenyl moieties induced moderate to good enantioselectivities. However, **L-2a** and **2c** showed disappointing performance which emphasized the importance of the bulky aryl units in the chiral ligands. Furthermore, benzylamine and phthalimide potassium salt were used as nucleophiles in the reactions of **7a**, and excellent enantioselectivities were obtained using **L-2b**, **2d** ∼ **2f** as chiral ligands (Scheme 7.3). However, considerably lower enantioselection was obtained in the reactions of **7b** and **7c** with dimethyl malonate.

These ligands were also successfully applied in the Rh-catalyzed asymmetric hydrogenation of α-acylamino cinnamic acid and its derivatives **8** (Table 7.1) [13]. Excellent enantioselectivities (up to 98.7% ee) and high catalytic activity were obtained in this reaction using **L-2a, b, d**. However, disappointing performance was observed with ligands **L-2c** and **e**, which is assumed to originate from different steric interaction with the substrate during the hydrogenation reactions.

Van Leeuwen and coworkers also applied (*R,R*)-BPNF (**L-2b**) and its derivatives (**L-2f** and **L-2g**) in the asymmetric hydroformylation of styrene [14]. The reaction provided moderate enantioselectivity (Scheme 7.4). It was shown that regio- and enantioselective features in hydroformylation are governed by a subtle interplay of steric and electronic features linked to the substrate as well as the ligand properties. For example, in the hydroformylation of 1-octene, electronic perturbations of the ligand influenced only the reactivity, whereas linear/branched ratios were left unaffected.

Mezzetti improved van Leeuwen's synthetic method to prepare (*S,S*)-(**L-2a**) and (*S,S*)-(**L-2b**) [15]. The major difference in the synthetic process was that the deboranation

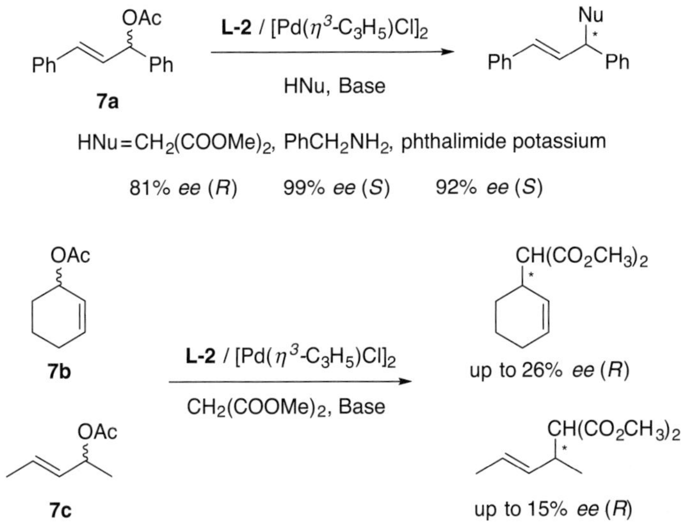

7.2 Symmetrical 1,1'-Disubstituted Ferrocenyl Ligands

Scheme 7.3 Application of **L-2** in asymmetric allylic substitution.

For substrate **7a**: Ph-CH=CH-CH(OAc)-Ph reacts with HNu/Base and **L-2** / [Pd(η^3-C$_3$H$_5$)Cl]$_2$ to give Ph-CH=CH-CH(Nu)-Ph.

HNu = CH$_2$(COOMe)$_2$, PhCH$_2$NH$_2$, phthalimide potassium

81% ee (R)   99% ee (S)   92% ee (S)

For substrate **7b** (cyclohexenyl acetate): reaction with CH$_2$(COOMe)$_2$, Base, **L-2** / [Pd(η^3-C$_3$H$_5$)Cl]$_2$ gives cyclohexenyl-CH(CO$_2$CH$_3$)$_2$, up to 26% ee (R).

For substrate **7c**: gives up to 15% ee (R).

Table 7.1 Rh-catalyzed asymmetric hydrogenations of **8**.

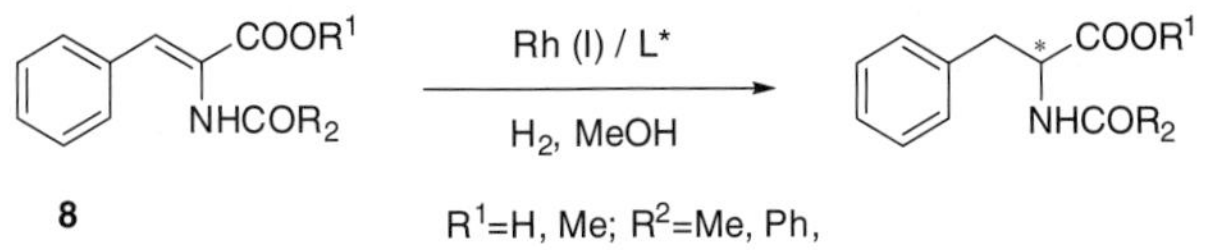

Substrate **8**: PhCH=C(COOR^1)(NHCOR^2) → PhCH$_2$-C*H(COOR^1)(NHCOR^2) with Rh(I)/L*, H$_2$, MeOH

R^1=H, Me; R^2=Me, Ph,

Entry	R^1	R^2	Ligands	ee (%)
1	H	Me	L-2b	98.2
2	H	Ph	L-2d	98.5
3	Me	Me	L-2b	97.3
4	Me	Me	L-2d	98.7
5	H	Ph	L-2c	21.0
6	H	Ph	L-2e	Trace product

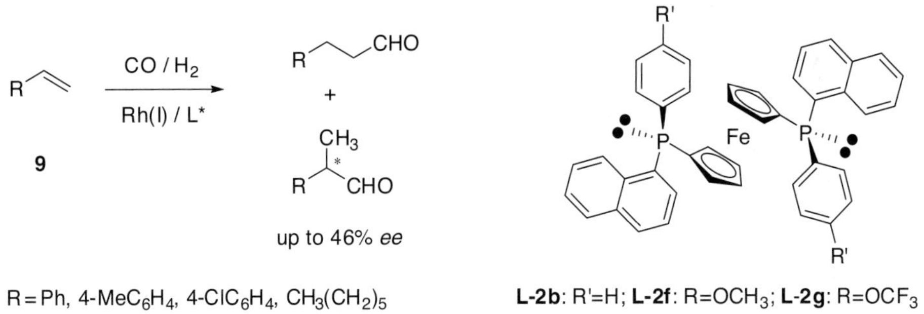

Substrate **9**: R-CH=CH$_2$ + CO/H$_2$, Rh(I)/L* → R-CH$_2$-CH$_2$-CHO + R-C*H(CH$_3$)-CHO

up to 46% ee

R = Ph, 4-MeC$_6$H$_4$, 4-ClC$_6$H$_4$, CH$_3$(CH$_2$)$_5$

L-2b: R'=H; **L-2f**: R=OCH$_3$; **L-2g**: R=OCF$_3$

Scheme 7.4 Use of **L-2b, 2f** and **2g** in asymmetric hydroformylation of styrene.

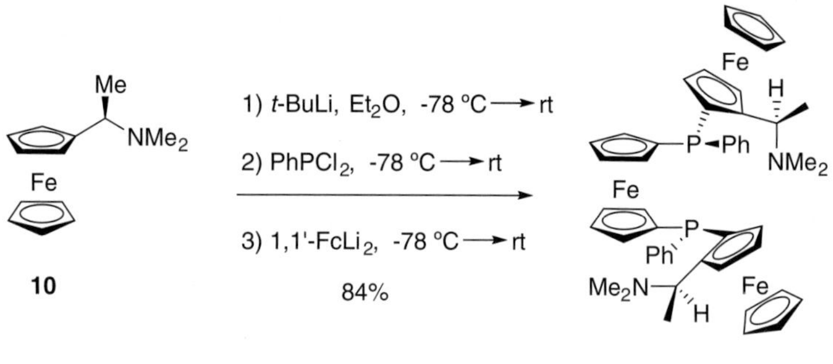

Scheme 7.5 Synthesis of TriFer.

was carried out in the presence of morpholine instead of HNEt$_2$. X-ray study of the complexes of these ligands with Pt and Rh showed that they had large bite angles (≈100°). However, the size of the bite angle does not improve the catalytic performance in the Rh-catalyzed asymmetrical hydrogenation of dehydroamino acids. Details are summarized in Table 7.2.

Very recently, Chen and McCormack reported the highly stereoselective synthesis of a new ferrocene-based C_2-symmetric diphosphine ligand **L-3** (TriFer), the first class of C_2-symmetric diphosphines that combines C-centered, P-centered, and planar chiralities (Scheme 7.5) [16d]. **L-3** can be prepared from the readily available Ugi's amine via three steps in one pot in good yield (84%) with excellent stereoselectivity (95 : 5).

A key intermediate in the synthesis of renin inhibitor Aliskiren was obtained by asymmetric hydrogenation of (*E*)-2-[3-(3-methoxypropoxy)-4-methoxybenzylidene]-3-methylbutanoic acid **11** which was difficult in general to hydrogenate with high enantioselectivity [16a–c]. **L-3** was successfully applied in the Rh-catalyzed asymmetric hydrogenation of **11** with over 99% conversion and unprecedented enantioselectivities of up to 99.6% *ee*. Satisfactory results were also obtained in the Rh-catalyzed asymmetric hydrogenation of 3-aryl-2-ethoxyacrylic acids **12** (up to 98.0% *ee*, Scheme 7.6) [16d].

7.2.2
C-Centered Chiral Diphosphine Ligands

The successful application of DIOP suggested that the P-centered chiral factor in diphosphine ligands was not necessary in order to achieve high enantioselectivity, and ligands with backbone chirality could also provide excellent *ee* values in asymmetric catalysis [4]. After this, the design of new chiral ligands turned to the synthesis of those with backbone chirality, mainly due to their ease of preparation and structural modification as well as excellent catalytic performance.

Watanabe reported a versatile synthetic method to prepare highly optically active 1,1′-bis(1-hydroxyalkyl)ferrocenes **14** utilizing enantio- and diastereo-selective addition of dialkylzincs to 1,1′-ferrocenyldicarboxaldehydes **13** catalyzed by 1,2-disubsti-

7.2 Symmetrical 1,1'-Disubstituted Ferrocenyl Ligands

Table 7.2 Rh-catalyzed hydrogenation of various substrates using **L-2a**, **2b** as chiral ligands.

Substrate	Product	ee (%)	Substrate	Product	ee (%)
Ph-CH=C(COOR)(NHCOCH₃)	Ph-CH₂-CH*(COOR)(NHCOCH₃)	96 (R=Me) 88 (R=H)	2,6-Me₂-C₆H₃-N(COCH₃)-C(=CH₂)-COOR	2,6-Me₂-C₆H₃-N(COCH₃)-CH*(CH₃)-COOR	47
Ph-CH=C(COOR)(NCOCH₃)	Ph-CH₂-CH*(COOR)(NCOCH₃)	97 (R=Me) 93 (R=Ph)	Ph-C(OAc)=CH₂	Ph-CH*(OAc)-CH₃	28
Ph-CH=C(COOH)(CH₃)	Ph-CH₂-CH*(COOH)(CH₃)	37	Ph-CO-COOCH₃	Ph-CH*(OH)-COOCH₃	26
Ph-CH=C(NHCOCH₃)(CH₃)	Ph-CH₂-CH*(NHCOCH₃)(CH₃)	47	CH₃-CO-COOCH₃	CH₃-CH*(OH)-COOCH₃	34
cyclohexylidene-C(COOR)(NHCOCH₃)	cyclohexyl-CH*(COOR)(NHCOCH₃)	28	3,3-dimethyl-2-oxo-γ-butyrolactone	3,3-dimethyl-2-hydroxy-γ-butyrolactone	29

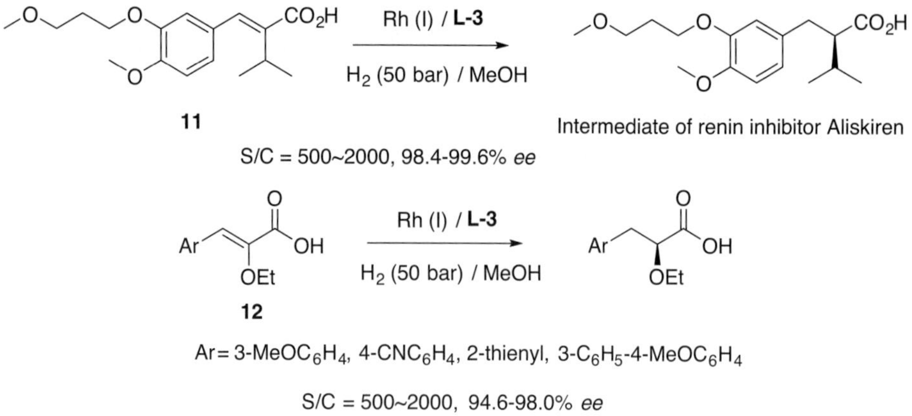

Scheme 7.6 Use of **L-3** in Rh-catalyzed enantioselective hydrogenation.

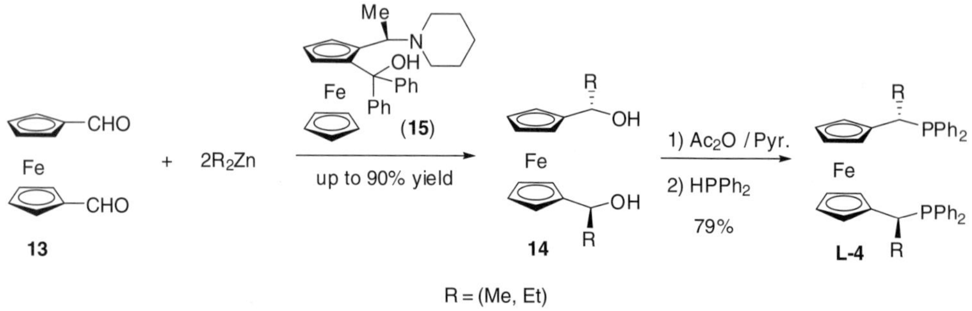

Scheme 7.7 Synthesis of **L-4**.

tuted ferrocenylaminoalcohol **15** (Scheme 7.7) [17]. The diastereomer ratio and the enantioselectivity of the diols were determined, by ^{1}H NMR analysis, to be 95 : 5 and >99% ee, respectively. These enantiopure diols **14** with an active hydroxy group could be further converted into many new ligands. One example is the preparation of C_2-symmetric ferrocenyl ligands with central chirality (**L-4**) with 79% yield by addition of an excess of diphenylphosphine to the diacetates derived from **14** with retention of the configuration.

New chiral C_2-symmetrical diphosphines **L-5** have been designed by Knochel's group using a convergent synthesis (Scheme 7.8) [18]. The reaction of the (S,S)-dizinc compound **16** with Cl_2PNEt_2 provided the bicyclic aminophosphine-borane complex **17** (50% yield) which furnished the chlorophosphine-borane complex **18** (95% yield) by reaction with HCl in ether (0 °C, 12 h). Dilithiated ferrocene was reacted with **18**, leading to the diphosphinoferrocene-borane complex **L-5a** in 30% isolated yield (Scheme 7.8, I). The related ferrocenyldiphosphine **L-5b** could be prepared by a similar approach from **20**, which was obtained via the reaction of N,N′-dimethylcyclohexyldiamine **19** with PCl_3. The overall yield from **19** was 44% (Scheme 7.8, II).

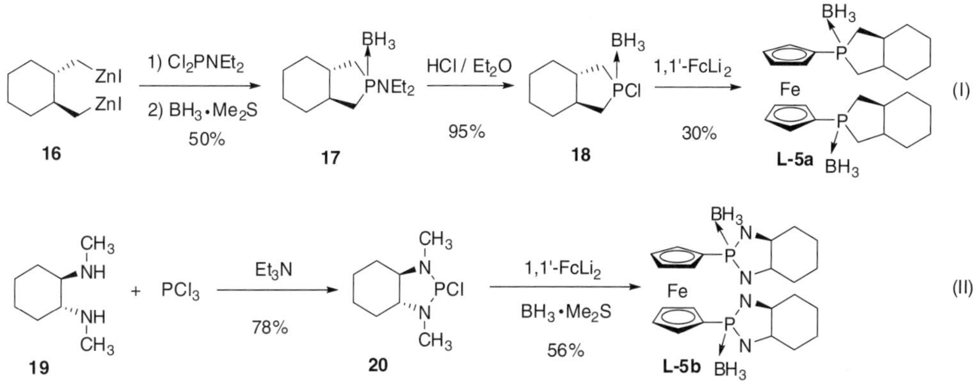

7.2 Symmetrical 1,1'-Disubstituted Ferrocenyl Ligands | 183

Scheme 7.8 Synthesis of **L-5**.

According to the methodology of synthesizing chiral 2,4-disubstituted phosphetane rings [19], Burk's and Marinetti's groups independently prepared enantiopure 2,4-disubstituted phosphetanes, 1,1'-diphosphetanylferrocene ligands (FerroTANE, **L-6**). As shown in Scheme 7.9, the reaction between diphosphanylferrocene **21** and the cyclic sulfates **22** provided facile access to the desired FerroTANE as a yellow–orange crystalline solid in moderate to good overall yield [20].

FerroTANE was surveyed for metal-catalyzed asymmetric reactions of a wide range of substrates. **L-6b** with an ethyl group was found to be the most efficient in most of these reactions. The details are summarized in Table 7.3.

After a new series of efficient chiral bisphospholane ligands BPE and DuPhos were developed with excellent results in the asymmetric hydrogenation of functionalized olefins and ketones [25], Burk introduced new chiral bisphospholane ligands **L-7** (DiPFc) possessing a ferrocenyl backbone (Figure 7.3), using the same method as for **L-6** [26]. **L-7** afforded 31–86% ee in the asymmetric Rh-catalyzed hydrogenation of several model substrates. It is remarkable that **L-7c** has been applied in the preparation of kilogram quantities of a substituted imidazole compound which is a potent inhibitor of thrombin fibrinolysis (Scheme 7.10). An integrated approach of the asymmetric hydrogenation and diastereomeric recrystallization was adopted. The quinidine salt of **23** was hydrogenated with high TON (s/c = 5000), followed by recrystallization of the diastereomeric salt to afford the final product in 76% yield and

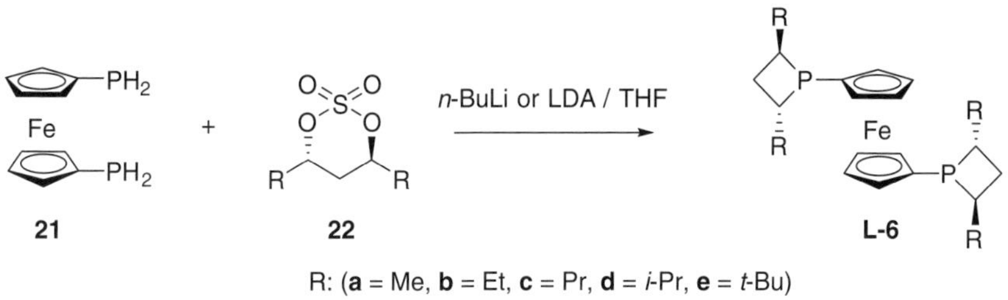

R: (**a** = Me, **b** = Et, **c** = Pr, **d** = *i*-Pr, **e** = *t*-Bu)

Scheme 7.9 Synthesis of FerroTANE.

Table 7.3 Asymmetric hydrogenations and P–C cross-coupling reactions with FerroTANE as chiral ligands.

Entry	Substrate	Product	Ligand	ee (%)	Ref.
1	MeO$_2$C–C(=CH$_2$)–CH$_2$CO$_2$Me	MeO$_2$C–*CH(CH$_3$)–CH$_2$CO$_2$Me	L-6b	98 (R)	[20b]
2	MeO$_2$C–C(=CH$_2$)–NHAc	MeO$_2$C–*CH(CH$_3$)–NHAc	L-6b	94 (R)	[20a]
3	CH$_3$–C(=O)–CH=C(OMe)–	CH$_3$–*CH(OH)–CH$_2$–C(=O)–OMe	L-6b	57 (S)	[20a]
4	Ph–CH=C(CO$_2$Me)(NHAc)	Ph–CH$_2$–*CH(CO$_2$Me)(NHAc)	L-6a	90 (S)	[20a]
5	HO$_2$C–C(Ar)=CH–C(=O)–N(morpholine)	HO$_2$C–*CH(Ar)–CH$_2$–C(=O)–N(morpholine)	L-6b	99 (R)	[20b]

7.2 Symmetrical 1,1′-Disubstituted Ferrocenyl Ligands

6	R—⟨⟩—I + Ar₁Ar₂PH	R—⟨⟩—P*(Ar₁)(Ar₂)	L-6b	93 (R)	[21]
7	(o-I-C₆H₄)C(O)NiPr + ArRPSiiPr₃	(o-P(S)ArR-C₆H₄)C(O)NiPr (R')	(R,R)-L-6b	97	[22]
8	R³C(O)NH–C(=CHR²)–C(O)OR¹	R³C(O)NH–CH(R²)–CH₂–C(O)OR¹	L-6b	>99 (S)	[23]
9	colchicine-type enamide (MeO, MeO, OMe, NHAc)	colchicine-type saturated (MeO, MeO, OMe, NHAc)	(S,S)-L-6d	92 (S)	[24]

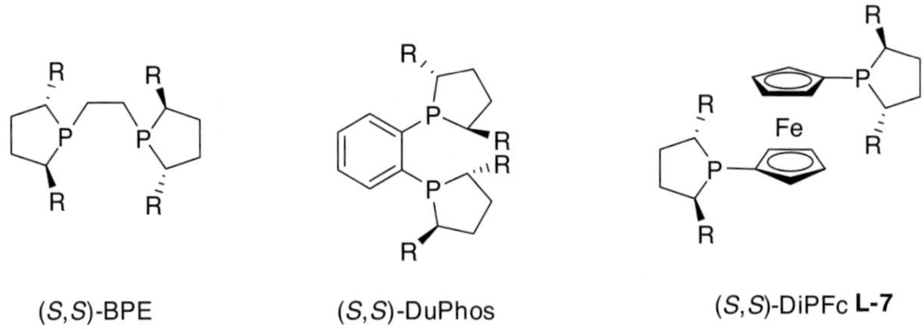

186 | *7 Symmetrical 1,1′-Bidentate Ferrocenyl Ligands*

(S,S)-BPE   (S,S)-DuPhos   (S,S)-DiPFc **L-7**

R: (**a**=Me, **b**=Et, **c**=*i*-Pr)

Figure 7.3 Structure of **L-7** and its analogues.

94% ee [27]. However, only 27% overall yield (97% ee) was obtained for this chiral inhibitor via a traditional resolution procedure and the work up is very difficult to handle.

Zhang reported the synthesis of a new C_2-symmetrical chiral ferrocenyl polysubstituted phospholane ligand **L-8** (Scheme 7.11), which was obtained with a similar procedure as that for FerroTANE [28]. Thus, the key intermediate 1,4-diol cyclic sulfate **27** was prepared from D-mannitol **24** in three steps, with **25** and **26** as intermediates, in an overall yield of 20%. The 1,1′-bis(phosphino)ferrocene **L-8** was then prepared, via nucleophilic attack on **27** by disubstituted diphosphanylferrocene **21** in the presence of *n*-BuLi, in 48% yield. Compared to their analogs phosphetane **L-6** and phospholane **L-7**, the bulky ketal substituent on both of the phospholane rings resulted in a significant improvement in the enantioselectivity of the Rh-catalyzed hydrogenation of α-acylamino cinnamic acid and its derivatives **8** as well as itaconic acid derivatives.

Reetz reported the synthesis of ferrocenyl diphosphonite ligand **L-9**, existing as an orange–brown crystalline compound, using ferrocene **1** and (R)-(+)-BINOL as cheap building blocks, via the intermediate **28**, in three steps (Scheme 7.12) [29]. The complex (R,R)-(**L-9**)Rh(cod)BF$_4$ exhibited excellent enantioselectivity and high cat-

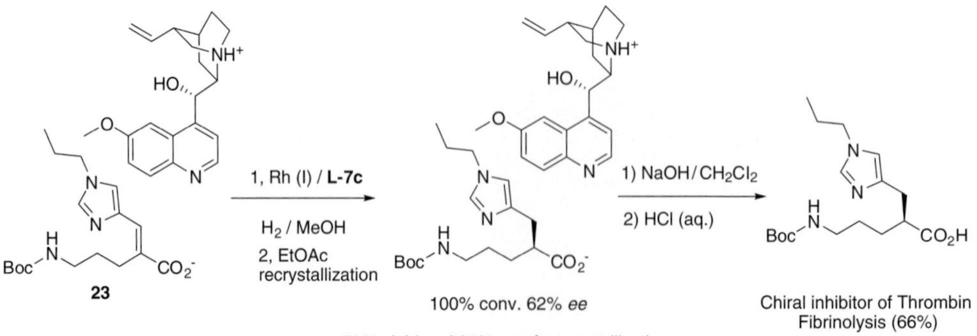

Scheme 7.10 Application of **L-7** in the synthesis of the inhibitor of thrombin fibrinolysis.

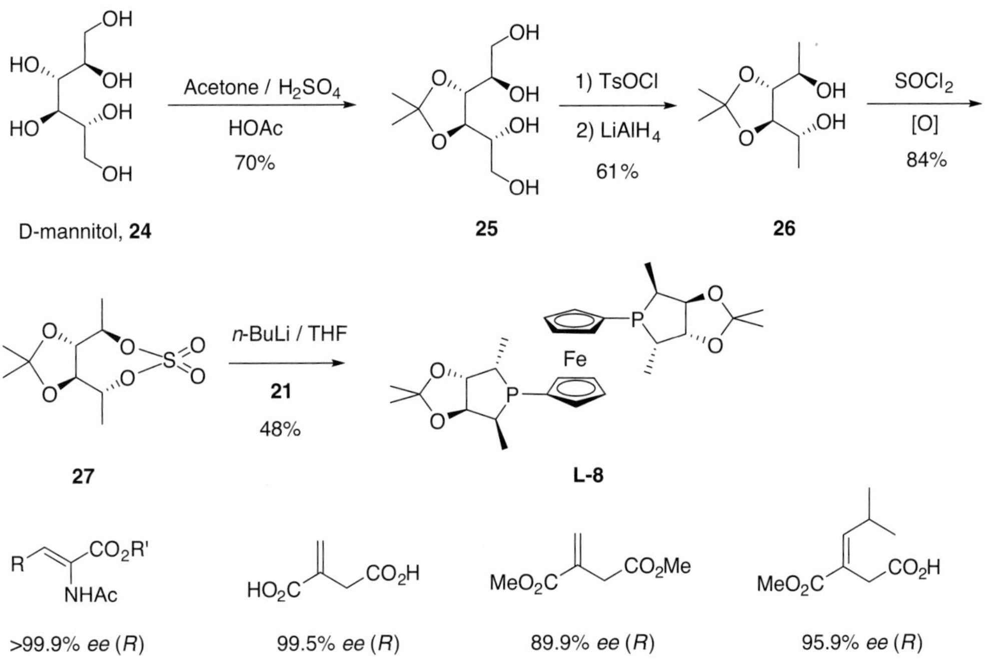

Scheme 7.11 Synthetic route to **L-8** and the substrates in asymmetric hydrogenation.

alytic activity in the asymmetric hydrogenation of dimethyl itaconate **29** and 2-acetamido methyl acrylate **30** (Scheme 7.13).

Among the various asymmetric hydrogenations, the hydrogenation of imines was still a major problem with the lack of highly enantioselective and reactive catalysts

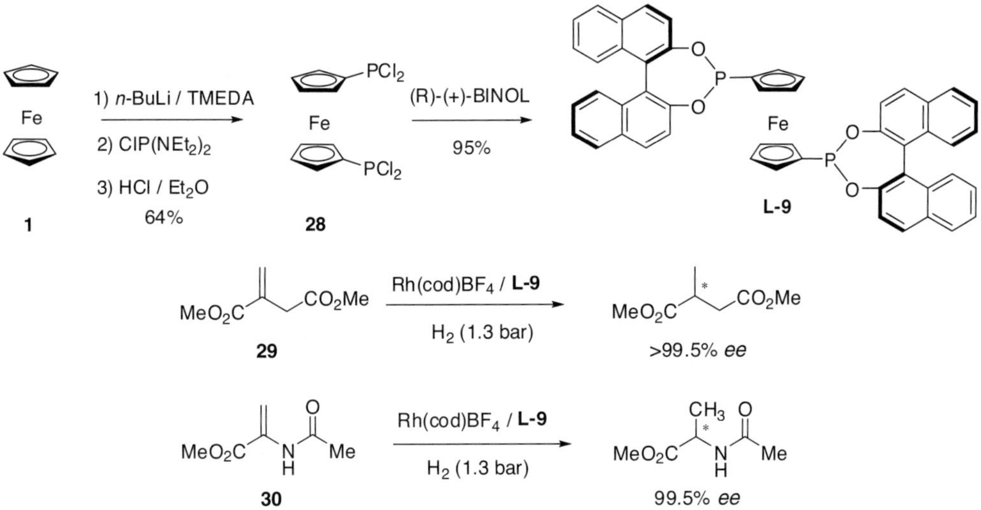

Scheme 7.12 Synthesis and application of **L-9**.

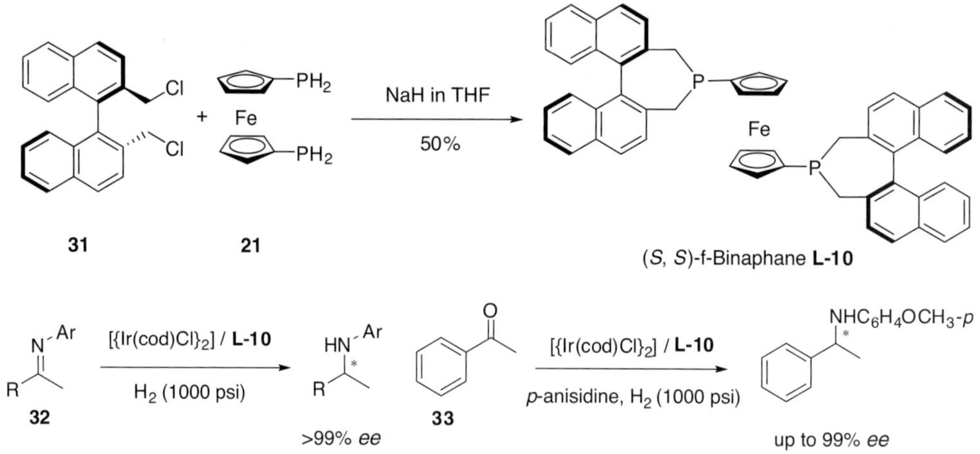

Scheme 7.13 Synthesis and application of **L-10**, f-Binaphane.

with broad substrate scopes. Zhang designed novel easily accessible C_2-symmetrical air-stable chiral 1,1′-bisphosphanoferrocene (**L-10**; abbreviated as f-Binaphane) with a ferrocene backbone (Scheme 7.13) [30]. f-Binaphane was tested to be an excellent ligand in the Ir-catalyzed asymmetric hydrogenation of acyclic imines **32**, an efficient method of synthesizing chiral secondary amines from N-arylimines [30a]. A new way to produce chiral amines was also achieved by f-Binaphane Ir-catalyzed asymmetric reductive amination of ketones **33** with primary amine [30b]. This method provided a simple and efficient way of preparing chiral secondary amines from aryl ketones.

Very recently, Zhou's group expanded the Pd-catalyzed hydrogenation of activated imines **34** using (S,S)-f-Binaphane as chiral ligand, and up to 99% ee was obtained (Scheme 7.14) [31]. The present method provides a new efficient route to the synthesis of enantiopure cyclic sulfamidates.

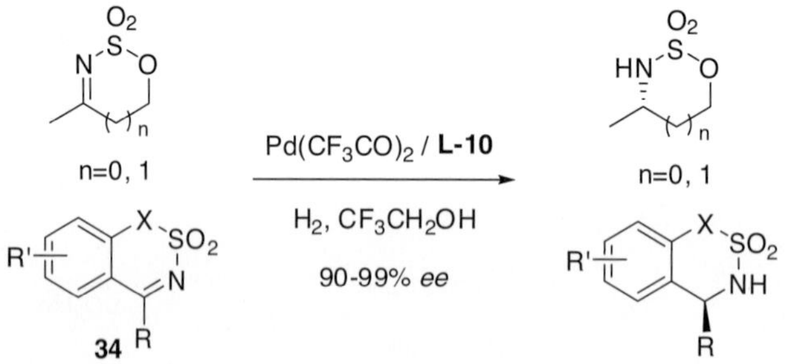

Scheme 7.14 Application of **L-10** in Pd-catalyzed hydrogenation.

7.2.3
Chiral Nitrogen-Containing Ligands

C_2-symmetrical chiral diphosphines are widely used ligands in asymmetric catalytic synthesis. However, attention has also been paid in recent years to the preparation of

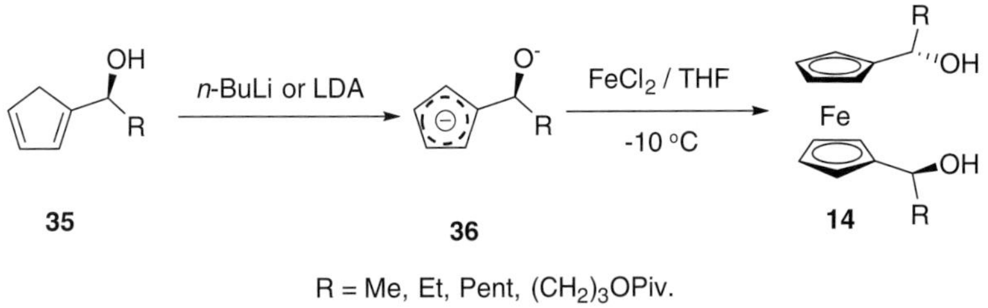

Scheme 7.15 Synthetic route to chiral diols **14**.

R = Me, Et, Pent, (CH$_2$)$_3$OPiv.

nitrogen-containing ligands. Because of their ready accessibility, low cost and great stability, nitrogen-containing ligands have become one of the most successful, versatile and commonly used classes of ligands for asymmetric catalysis. Many chiral diaminoferrocenyl ligands have been developed and applied in a wide range of metal-catalyzed asymmetric transformations.

One simple approach to chiral diamines was achieved from chiral diols **14** by transformation of functionalized groups. Apart from Watanabe's method [17], Knochel developed a new efficient enantioselective preparation of chiral diols **14** by reaction of the lithium dianion **36** obtained from chiral alcohols **35** with FeCl$_2$ in 43–74% yields (Scheme 7.15) [32].

The chiral diols **14** could also be synthesized via the CBS-catalyzed reduction of ferrocenyldiketones, which are further converted into the new chiral C$_2$-symmetrical diaminoferrocene derivatives **L-11** (Scheme 7.16) [33]. Four steps from ferrocene were required: acylation of ferrocene provided ferrocenyldiketones (71–92% yields) which were reduced enantioselectively with CBS-catalyst **37** to provide 1,1′-ferrocenyl diols **14** in 90–96% yields (>99% ee). After acetylation of these diols **14**, the diaminoferrocenes **L-11** were obtained by reaction with an excess of primary amines with retention of configuration in 56–93% yields. These ligands were applied in the transfer hydrogenation of ketones in i-PrOH or HCOOH/Et$_3$N in the presence of

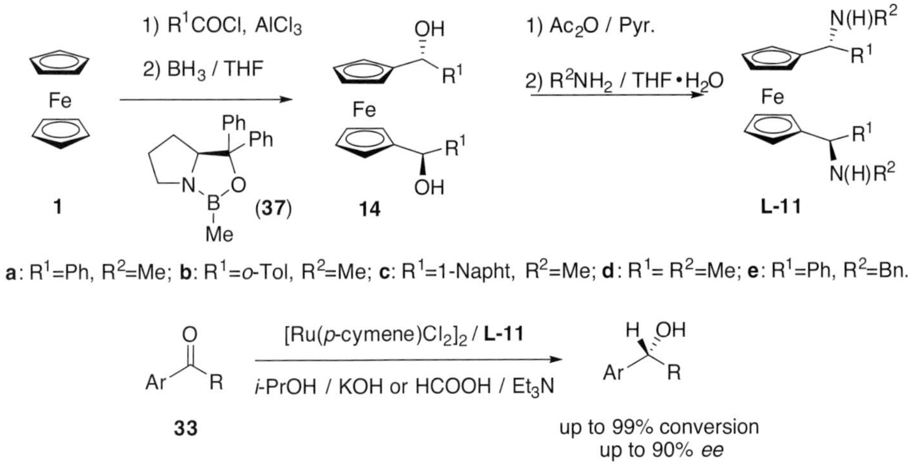

a: R^1=Ph, R^2=Me; b: R^1=o-Tol, R^2=Me; c: R^1=1-Napht, R^2=Me; d: R^1= R^2=Me; e: R^1=Ph, R^2=Bn.

up to 99% conversion
up to 90% ee

Scheme 7.16 Synthesis and application of **L-11**.

KOH (5 mol%) with high conversion (up to 99%) and good enantioselectivity (Scheme 7.16). A wide range of ketones was examined with 2-propanol as hydrogen source and those with a bulky group (e.g., 1-naphthyl) afforded higher *ee* values. Ligands with secondary diamines were able to form very active catalysts for these reactions compared to those with primary amines, while those with tertiary diamine did not give satisfactory results (Scheme 7.16).

Kim and Jeong have prepared a new ferrocenyl-1,1'-disulfonamide **L-12** using the same method developed by Knochel [34]. First, the chiral diols **14** were obtained from **38** in the presence of CBS-catalyst **37** with 93% yield. Then the formation of (*R,R*)-**L-12** was accomplished in high yield (86%) from the reaction of *p*-toluenesulfonyl chloride with (*R,R*)-ferrocenyl-1,1'-diamine which is produced by the acetylation of **14** followed by amination. **L-12** was used as ligand in the Ti-catalyzed enantioselective addition of diethylzinc to aromatic aldehydes **39** with excellent catalytic behavior (Scheme 7.17). Any substitution at the aromatic ring of benzaldehyde caused the *ee* values to drop significantly, regardless of the steric or electronic nature of the substituent. Ligands bearing an electron-donating group exhibited slightly higher enantioselectivities than those with electron-withdrawing groups. Furthermore, the enantiomeric excess of the products is heavily dependent on various reaction parameters including the solvent, the reaction temperature, and the nature of the ferrocene moieties.

Pélinski described the synthesis of a series of new chiral 1,1'-disubstituted ferrocenyl amino alcohols which were obtained in three steps, as depicted in Scheme 7.18 [35]. First, 1,1'- ferrocenyldicarboxaldehyde **13** was reacted with commercially available chiral aminoalcohols in the presence of neutral Al$_2$O$_3$ in THF, giving the corresponding imines. Then, the crude reaction mixture was reduced with sodium borohydride in methanol to provide the amino alcohols **40**

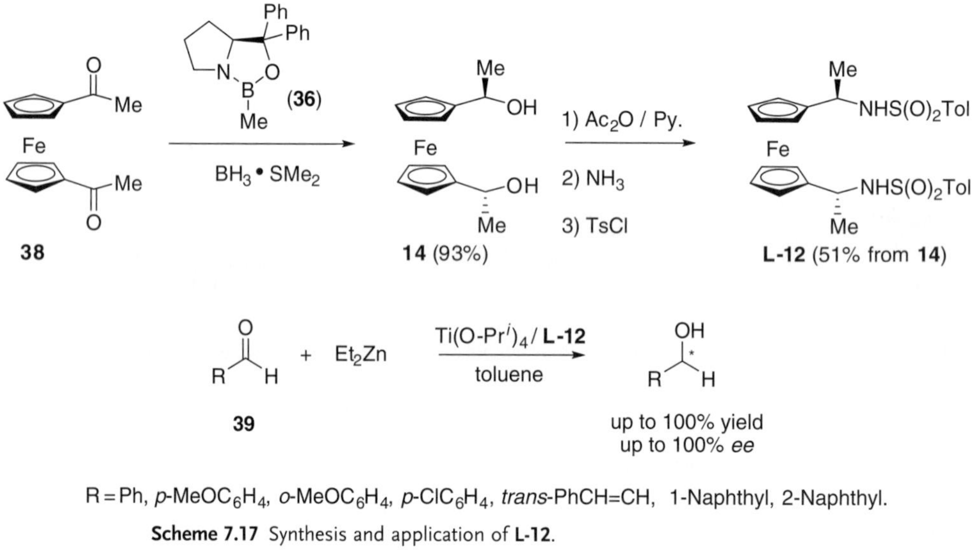

R = Ph, *p*-MeOC$_6$H$_4$, *o*-MeOC$_6$H$_4$, *p*-ClC$_6$H$_4$, *trans*-PhCH=CH, 1-Naphthyl, 2-Naphthyl.

Scheme 7.17 Synthesis and application of **L-12**.

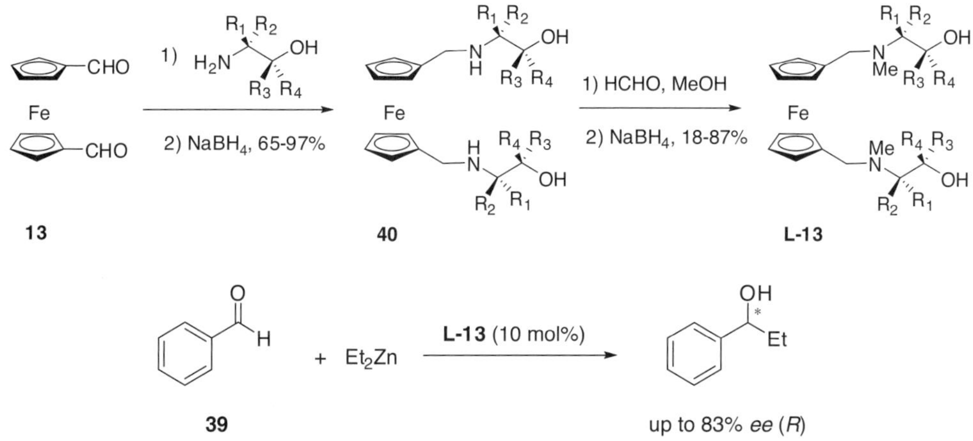

7.2 Symmetrical 1,1'-Disubstituted Ferrocenyl Ligands | 191

Scheme 7.18 Synthesis and application of **L-13**.

in 65–97% overall yields. Enantiopure ferrocenes **L-13** were finally obtained by N-methylation of the amino function of **40**. The enantioselective addition of diethylzinc to **39** was selected to test the asymmetric catalytic behavior of these catalysts. Up to 83% *ee* was obtained with complete conversion. It was shown that the steric hindrance near the hydroxy moiety was crucial for further enhancement of the enantioselectivity.

The chiral C_2-symmetrical 1,1'-disubstituted ferrocenyl aziridino alcohols **L-14** were prepared in three steps by Wang and coworkers (Scheme 7.19) [36]. First, 1,1'-ferrocenyldicarboxaldehyde **13** was reacted with methyl threonine ester hydrochloride in methanol in the presence of triethylamine, followed directly by reduction with sodium borohydride in methanol to provide compound **41** with a 69% overall

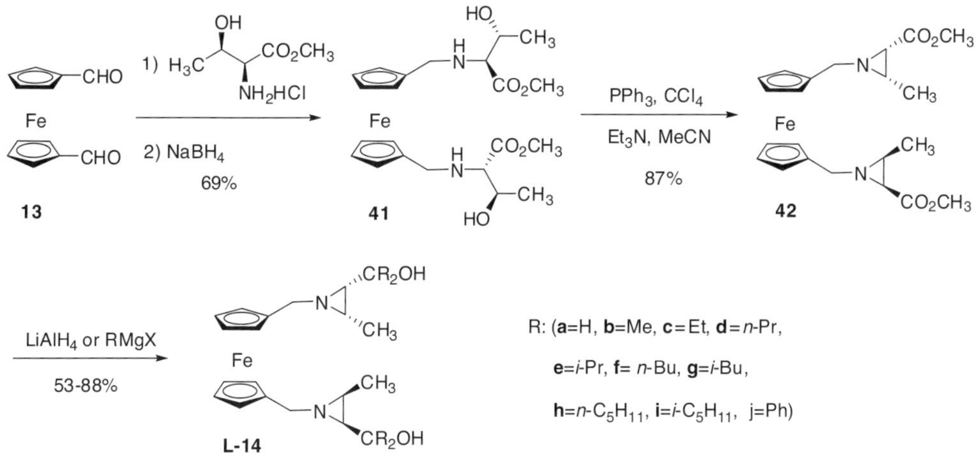

Scheme 7.19 Synthetic route to **L-14**.

yield. **41** was cyclized to aziridino ester **42** in 87% yield by reaction with triphenyl phosphine in the presence of tetrachloromethane in acetonitrile. Reduction with lithium aluminum hydride led to the aziridino alcohol **L-14a** in 66% chemical yield. Alternatively, treatment of **42** with an excess of Grignard reagent afforded the ligands **L14b-j** in 53–88% yields. These ligands were applied to promote enantioselective addition of diethylzinc to aldehydes with up to 99.5% ee. Compared to the mono-substituted ferrocenyl aziridino alcohol ligands, the introduction of a C_2-symmetrical factor led to a dramatic improvement in the enantioselectivity in the addition of diethylzinc to aromatic aldehydes (69 versus 92.6% ee).

Zhang and coworkers reported the preparation of the novel C_2-symmetrical bisoxazoline ligands **L-15** with a ferrocene backbone bearing a hydroxy group in the substituent of the oxazoline ring (Scheme 7.20) [37]. 1,1′-Ferrocenedicarboxyl dichloride **43** was reacted with L-serine methyl ester hydrochloride in the presence of triethylamine to give the amide compound **44** with 92% yield. Then, the hydroxy group of **44** was activated by Burgess reagent to afford the bisoxazoline compound **45** in 70%. Finally, reduction of **45** with LiAlH$_4$ at room temperature for 6 h afforded the desired ligand **L-15a** bearing a hydroxy group in a yield of 61%. Alternatively, treating **45** with Grignard reagent in THF at 0 °C afforded other analogous ligands **L-15b–d** in yields of 65–75%.

The asymmetric alkylation of arylaldehyde with diethylzinc was selected to examine the catalytic behavior. It was observed that lower temperatures gave higher enantioselectivity in the optimal solvent toluene (94% ee with **L-15b** at −25 °C). The bulkiness of ligand has some effect on the enantioselectivity. The electronic- or/and stereo-effects of the aromatic ring substituents also affected the asymmetric catalysis. Compared to the corresponding C_1-symmetrical monooxazoline ligand, **L-15** showed much better catalytic activity and enantioselectivity (41 versus 93% ee). A possible binuclear zinc catalytic mechanism was proposed by the authors (Scheme 7.21). The

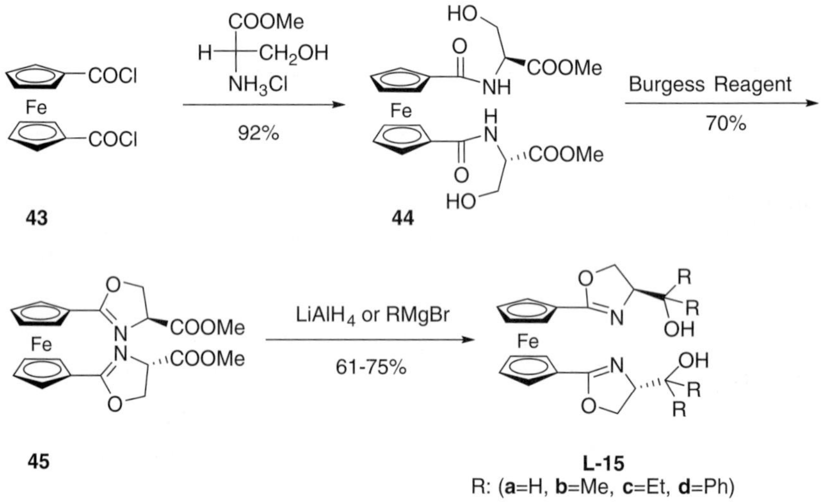

Scheme 7.20 Synthetic route to **L-15**.

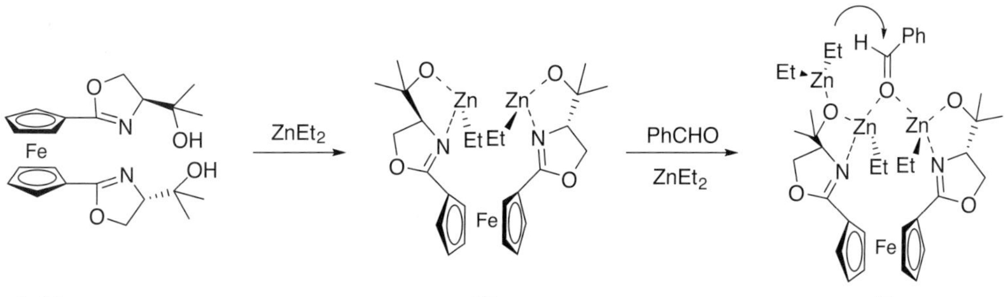

7.3 Symmetrical 1,1′,2,2′-Tetrasubstituted Ferrocenyl Ligands

Scheme 7.21 Possible mechanism of binuclear zinc catalytic system.

formed binuclear zinc intermediate **46** might coordinate with arylaldehyde and additional diethylzinc to form complex **47**. Then, the coordinated diethylzinc would attack the carbonyl group stereoselectively from the back of benzaldehyde, for less steric hindrance, affording the corresponding chiral alcohol.

7.3
Symmetrical 1,1′,2,2′-Tetrasubstituted Ferrocenyl Ligands

Planar chiral ferrocenes have found increasing importance for metal-catalyzed asymmetric organic synthesis in recent years because these stabilized ligands possess the advantages of adequate rigidity, steric bulkiness and are hardly racemized [38]. Among the vast variety of planar chiral ferrocene-based ligands, the design and preparation of C_2-symmetrically planar chiral ferrocenes are of great importance in the development of transition metal-catalyzed enantioselective reactions.

7.3.1
Tetrasubstituted Ferrocenyl Diphosphine Ligands with Multi-Chiralities

Hayashi and coworkers reported the first example of C_2-symmetrically chiral 1,1′,2,2′-tetrasubstituted ferrocene ligand **L-16** (Scheme 7.22) [39]. For its preparation, the starting diamine had to be used as a racemic mixture because it could hardly be resolved. So the tetrasubstituted ferrocene was obtained as a mixture of racemic

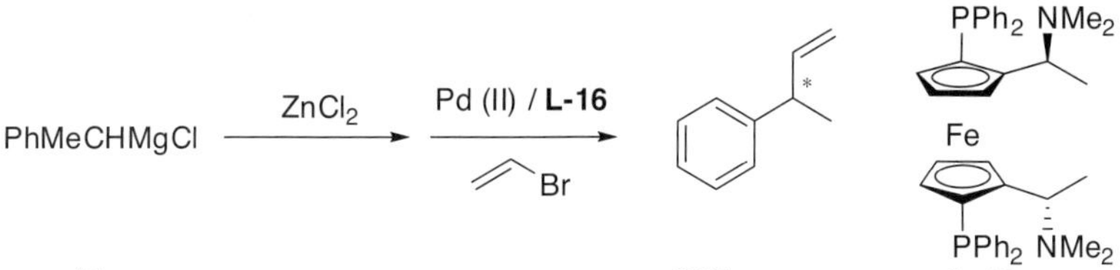

Scheme 7.22 Application of **L-16**.

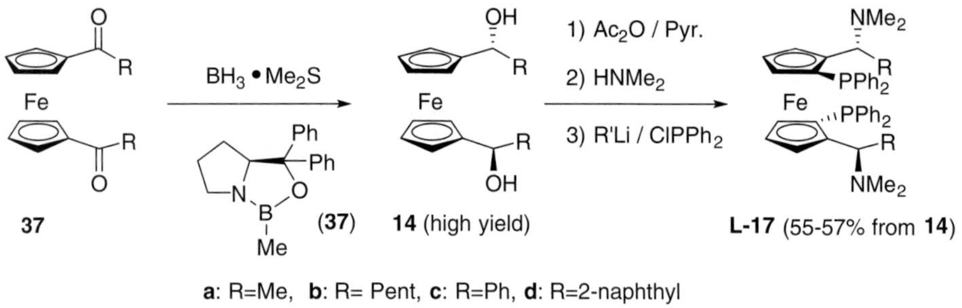

Scheme 7.23 Synthesis of diamino FERRIPHOS **L-17**.

a: R=Me, b: R= Pent, c: R=Ph, d: R=2-naphthyl

and meso products which was successfully resolved with tartaric acid. With **L-16** as a chiral ligand, up to 93% ee was achieved for the coupling reaction of vinyl bromide and 1-phenylethylzinc chloride **48**.

Knochel developed a new method to obtain Hayashi's type ligands **L-17** (diamino FERRIPHOS) via asymmetric CBS-catalyzed borane reduction of diketone as a key step to afford chiral diols **14** (Scheme 7.23) [40]. The influence of the reaction conditions upon the asymmetric reduction was investigated in detail. The chiral ligands **L-17** were then obtained, via diastereoselective lithiation of diamine, which is produced from the acetylation of **14** followed by amination, with n-BuLi (4 equiv., rt, 4 h) or t-BuLi (3 equiv., 0 °C, 4 h) followed by phosphorylation with Ph_2PCl, in 55–57% yields (Scheme 7.23).

These ligands were applied in the Pd-catalyzed asymmetric cross-coupling of 1-phenylethylmagnesium chloride or sec-butylmagnesium chloride with different vinyl bromides. The Pd-complex of **L-17c** catalyzed the asymmetric cross-coupling of 1-phenylethylmagnesium chloride and β-bromostyrene providing (S)-1,3-diphenyl-1-butene with up to 93% ee. The details are summarized in Table 7.4 [41].

Furthermore, only 25% ee was obtained in the enantioselective Rh-catalyzed transfer hydrogenation of unsymmetrical ketones with 2-propanol as the hydrogen source [42]. However, the wide family of FERRIPHOS ligands are excellent ligands for the asymmetric hydrogenation of alkyl α-acetamido acrylates (Scheme 7.24) [43].

Of the various chiral 1,1′-nitrogen-containing ferrocenyl ligands, 1,1′-bis(oxazolinyl)ferrocenes are also the most frequently utilized ones in asymmetric synthesis. Ikeda's group developed a highly diastereoselective o,o'-dilithiation of 1,1′-bis(oxazolinyl)ferrocene **51** (Scheme 7.25), leading to the synthesis of novel C_2-symmetrical 1,1′,2,2′-tetrasubstituted ferrocenyl ligands **L-18** [44a]. First, the intermediates of **51** were prepared from 1,1′-ferrocenedicarbonyl dichloride **43** in three steps. Bis(β-hydroxylamide) **49** was prepared in over 85% yield from **43** and enantiomerically pure aminoalcohols (2.2 equiv) in the presence of triethylamine (4.0 equiv). Then the amides **49** were treated with methanesulfonyl chloride (2.3 equiv) and triethylamine (4.0 equiv) at room temperature for 3 h to afford directly 1,1′-bis(oxazolinyl)ferrocenes **51** in about 90% yield without isolation of the dimesylates **50** (Scheme 7.25).

1,1′-Bis(oxazolinyl)ferrocene **51** was then treated with 2.6 equiv of BuLi at −78 °C in THF for a period of 3 h and then held for 10 min at 0 °C followed by the addition

7.3 Symmetrical 1,1',2,2'-Tetrasubstituted Ferrocenyl Ligands

Table 7.4 Asymmetric cross-coupling with diamino FERRIPHOS L-17 as chiral ligands.

Grignard reagents	Vinyl bromide	ee (%)	Grignard reagents	Vinyl bromide	ee (%)
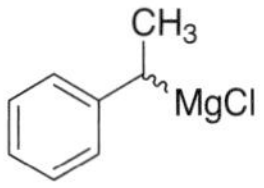 PhCH(CH₃)MgCl	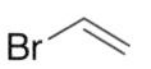 Br-CH=CH₂	82	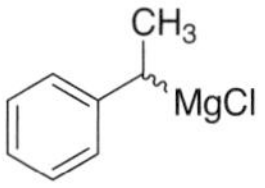 PhCH(CH₃)MgCl	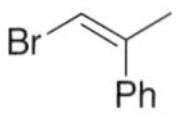 Br-C(CH₃)=CH-Ph	59
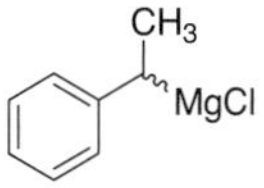 PhCH(CH₃)MgCl	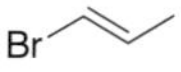 Br-CH=CH-CH₃	93	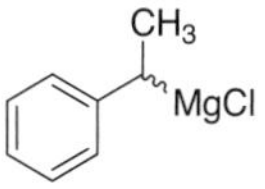 PhCH(CH₃)MgCl	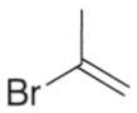 Br-C(CH₃)=CH₂	33
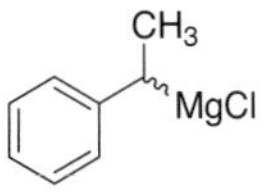 PhCH(CH₃)MgCl	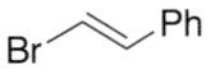 Br-CH=CH-Ph	65	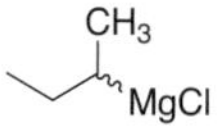 iBuCH(CH₃)MgCl	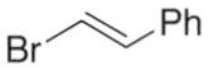 Br-CH=CH-Ph	15
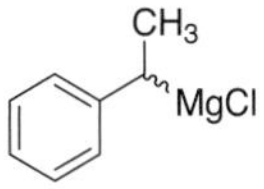 PhCH(CH₃)MgCl	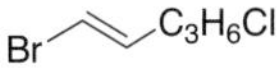 Br-CH=CH-C₃H₆Cl	11	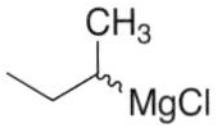 iBuCH(CH₃)MgCl	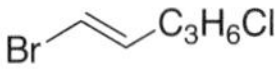 Br-CH=CH-C₃H₆Cl	75

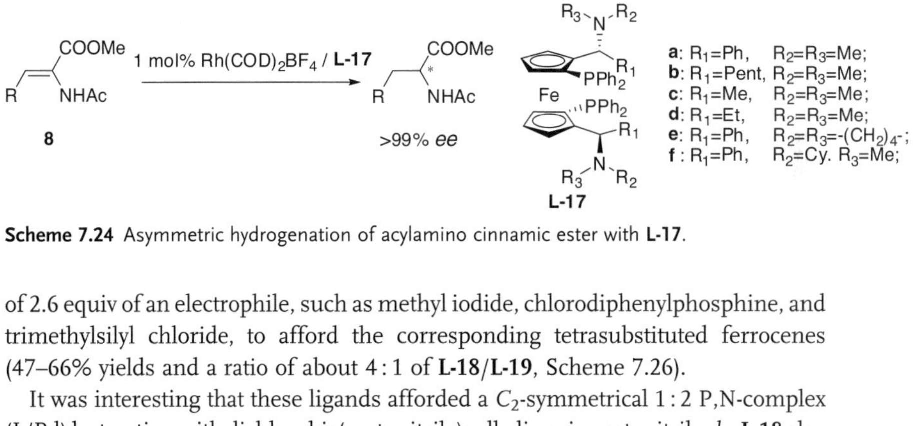

Scheme 7.24 Asymmetric hydrogenation of acylamino cinnamic ester with **L-17**.

R-CH=C(COOMe)(NHAc)   →   R-CH₂-C*H(COOMe)(NHAc)
8   1 mol% Rh(COD)₂BF₄ / **L-17**   >99% ee

L-17: Ferrocene with PPh₂ and CH(R₁)NR₂R₃ on each Cp ring

a: R₁=Ph, R₂=R₃=Me;
b: R₁=Pent, R₂=R₃=Me;
c: R₁=Me, R₂=R₃=Me;
d: R₁=Et, R₂=R₃=Me;
e: R₁=Ph, R₂=R₃=-(CH₂)₄-;
f: R₁=Ph, R₂=Cy, R₃=Me;

of 2.6 equiv of an electrophile, such as methyl iodide, chlorodiphenylphosphine, and trimethylsilyl chloride, to afford the corresponding tetrasubstituted ferrocenes (47–66% yields and a ratio of about 4 : 1 of **L-18/L-19**, Scheme 7.26).

It was interesting that these ligands afforded a C_2-symmetrical 1 : 2 P,N-complex (L/Pd) by treating with dichlorobis(acetonitrile)palladium in acetonitrile-d_6. **L-18** also showed high efficiency for the Pd-catalyzed asymmetric allylic substitution of **7a**

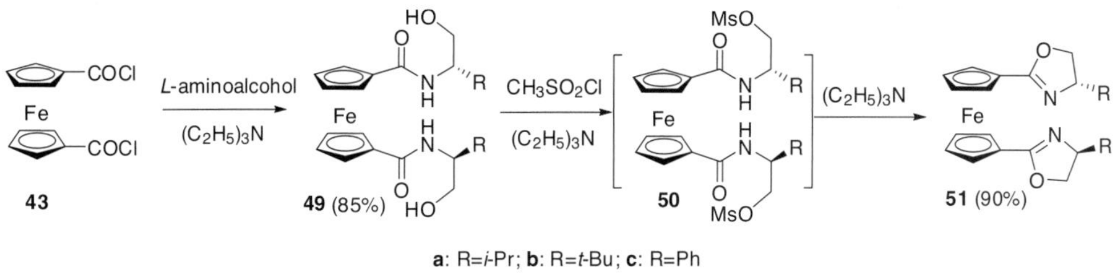

Scheme 7.25 Synthesis of **51**.

1,1'-Fc(COCl)₂ **43** → [L-aminoalcohol, (C₂H₅)₃N] → 1,1'-Fc(CONH-CH(R)-CH₂OH)₂ **49** (85%) → [CH₃SO₂Cl, (C₂H₅)₃N] → 1,1'-Fc(CONH-CH(R)-CH₂OMs)₂ **50** → [(C₂H₅)₃N] → 1,1'-Fc(oxazoline)₂ **51** (90%)

a: R=i-Pr; **b**: R=t-Bu; **c**: R=Ph

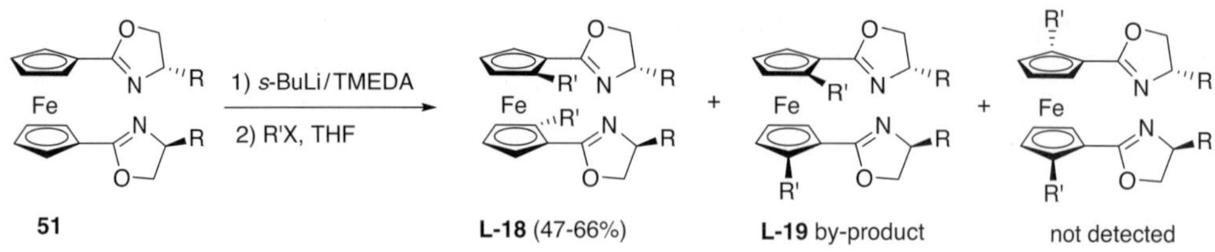

Scheme 7.26 Synthesis of L-18.

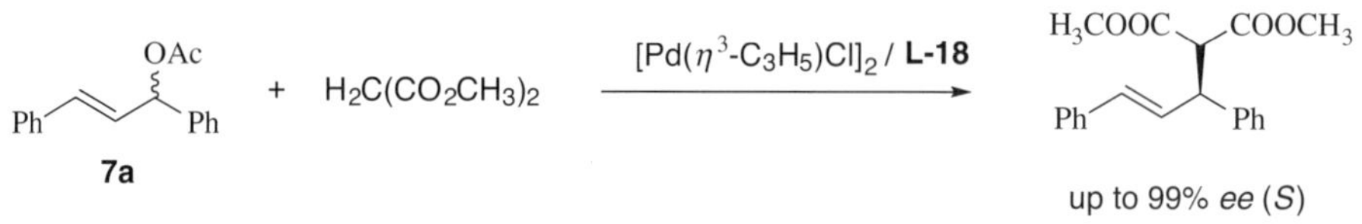

a: R=i-Pr, R'=Me; b: R=t-Bu, R'=Me; c: R=i-Pr, R'=PPh$_2$; d: R=t-Bu, R'=PPh$_2$; e: R=i-Pr, R=Si(CH$_3$)$_3$

Scheme 7.27 Asymmetric alkylation of **7a** with **L-18** as chiral ligands.

(Scheme 7.27) [44b]. Bulkier groups in the oxazoline rings afforded higher *ee* values and up to 99% *ee* was obtained with ligand **L-18d**, which was the best result for this model asymmetric catalysis at that time.

L-18 were also tested to be efficient chiral ligands in Pd-catalyzed asymmetric allylic alkylation using enamines as nucleophilic reagents, avoiding the use of unstabilized ketone enolate generated via strong bases and harsh reaction conditions (Scheme 7.28). The reaction proceeded smoothly, producing high catalytic activity and excellent enantioselectivity [45].

The complexation behavior of chiral ligands **51** and **L-18b** toward Pd(II) in acetonitrile-*d*$_3$ was investigated and 1 : 2 complexes (L/Pd) **52** were found exclusively, without the formation of expected N,N-chelating complex with Pd(II). It was interesting that the complexation behavior of both **51** and **L-18b** toward copper(I) was quite different with regard to the 1 : 2 Pd(II)-complexes. Upon complexation with copper(I) chloride, both **51** and **L-18b** afforded exclusively one of the 1:1 N,N-chelating complexes **53** and **54**. It was also interesting that ligand **51** with no planar chirality toward Cu(I) formed the complex **53** exclusively, while **L-18b** with planar chirality only afforded complex **54** which has the opposite twist direction of the two Cp rings to **53** (Scheme 7.29) [46].

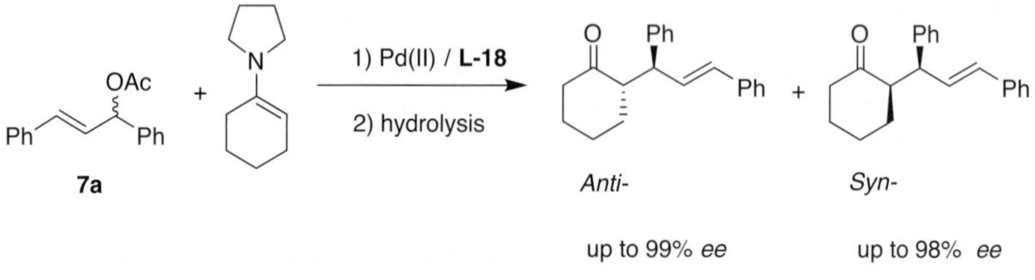

Scheme 7.28 Asymmetric allylic alkylation using enamines as nucleophilic reagents.

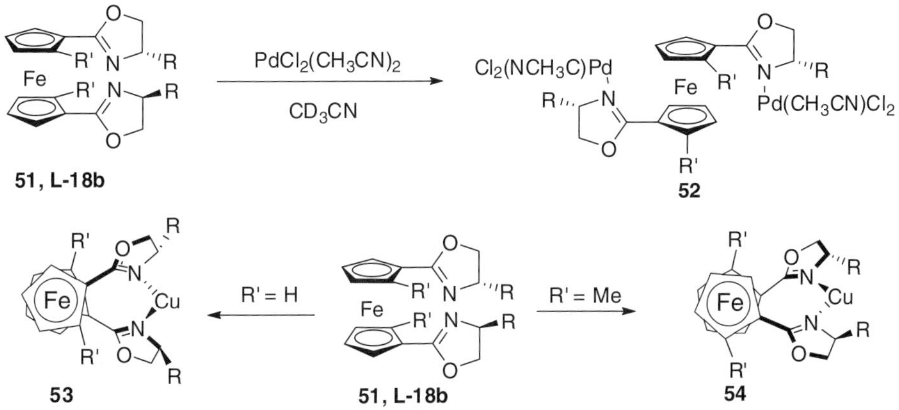

7.3 Symmetrical 1,1′,2,2′-Tetrasubstituted Ferrocenyl Ligands

Scheme 7.29 Complexation with metals.

It was also interesting that the diastereoselectivity on dilithiation of **51** was heavily dependent on the solvent. **L-19**, the by-product obtained by Zhang and Ikeda's group, was obtained as the major product by Ahn and Park's group when using diethyl ether instead of THF as solvent and 2 equiv of *t*-BuLi (or *sec*-BuLi) [47]. It was also shown that the stereoselectivity could be controlled simply by changing the temperature, solvent and/or lithiating agent [47b]. Similar to the results of Ikeda's group, up to 99% *ee* was obtained by Ahn and Park's group with ligands **L-18** for the Pd-catalyzed asymmetric allylic substitution of **7a** [48]. These ligands were also tested to be effective in the Pd-catalyzed asymmetric allylic substitution of **7c** (up to 88% *ee*) [48a]. Meanwhile, Widdowson's group also became involved in the synthesis of **L-18c**. They tried to convert ferrocenyl phosphines into the corresponding phosphine sulfides, as a means of protection and to transfer the oxazolines into other functionalities [49].

Prompted by the publications mentioned above, Knochel developed a wide family of C_2-symmetrical diphosphine ligands (the FERRIPHOS family, **L-20**) via four steps from the chiral diamine **55** (Scheme 7.30) [50a]. Dilithiation of ferrocenyl diamines **55a–c** with *t*-BuLi in diethyl ether (0 °C, 0.5 h) followed by quenching with (CCl$_2$Br)$_2$ furnished the corresponding C_2-symmetrical diaminodibromo ferrocenes **56a–c** (43–80% yields). After treatment of compounds **56a–c** with Ac$_2$O (100 °C, 2 h), diacetates **57a–c** were isolated in quantitative yield, and were then reacted with 3 equiv. of Me$_2$Zn in the presence of BF$_3$·OEt$_2$ (THF, −78 °C to rt, 1.5 h) yielding the desired dibromoferrocenes **58a–c** (92–100% yields) with high enantiomeric purity (>99% *ee*). Final bromine–lithium exchange (*n*-BuLi, THF, −78 °C, 0.25 h) followed by slow addition of Ph$_2$PCl afforded the corresponding C_2-symmetrical diphosphines FERRIPHOS **L-20** (46–68% yields).

A series of FERRIPHOS was also synthesized with a shorter route developed by Knochel. Without treating with Ac$_2$O, **56** was alkylated directly in the presence of organozinc reagents to afford **58** which can be converted to the FERRIPHOS ligands of type **L-20** in a one-pot procedure (Scheme 7.31) [50b]. **L-20** are good ligands in the Rh-catalyzed hydrogenation of both dehydroaminoacids and dehydroaminoesters as well as enol acetate (Scheme 7.31) [50].

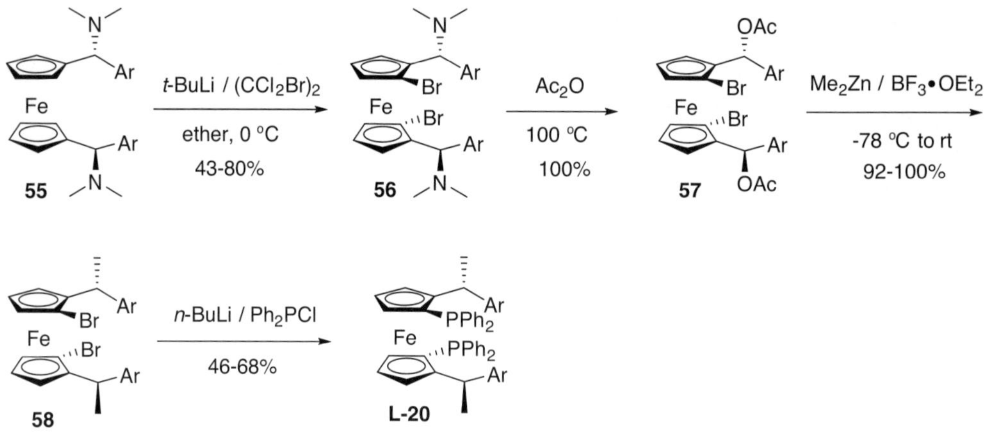

Scheme 7.30 Synthesis of L-20.

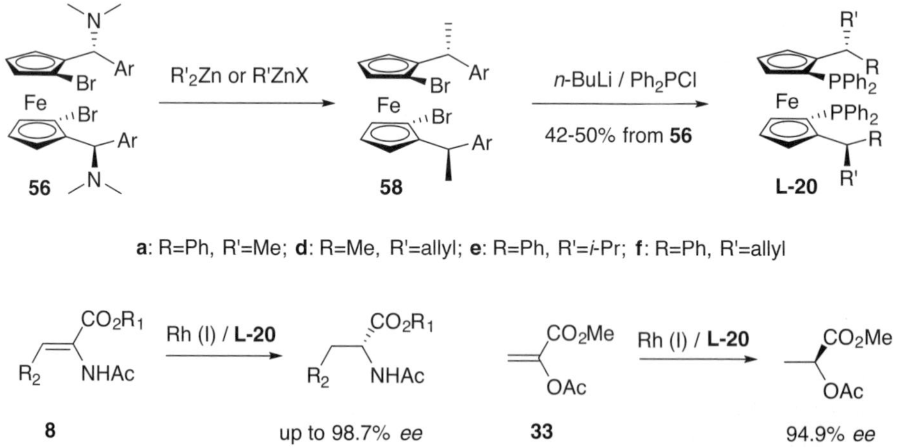

a: R=Ph, R'=Me; d: R=Me, R'=allyl; e: R=Ph, R'=i-Pr; f: R=Ph, R'=allyl

Scheme 7.31 Synthesis and application of L-20, FERRIPHOS.

7.3.2
Tetrasubstituted Ferrocenyl Ligands with Only Planar Chirality

In the above section, the tetrasubstituted diphosphine ferrocenyl ligands with multi-chiralities afforded excellent asymmetric induction. In order to further disclose the significance of the planar chiral element, many examples of C_2-symmetrical ligands with only planar chirality were synthesized and applied successfully in asymmetric synthesis.

Ikeda's group designed and synthesized first the ferrocenyl diphosphine ligands with only planar chirality via asymmetric synthesis [51]. Treatment of **L-18c** with trifluoroacetic acid in aqueous THF caused ring-opening of the oxazoline moiety to give an unstable ammonium salt. Without isolation, this salt was acetylated

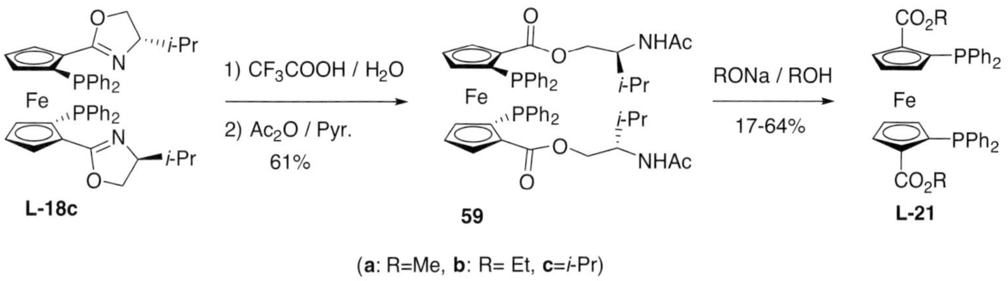

7.3 Symmetrical 1,1′,2,2′-Tetrasubstituted Ferrocenyl Ligands

Scheme 7.32 Synthetic route to L-21.

(**a**: R=Me, **b**: R= Et, **c**=*i*-Pr)

with acetic anhydride in the presence of pyridine to give ester amide **59** in 61% yield. Novel planar chiral ferrocenyl ligands **L-21** with only planar chirality were then obtained by treating **59** with alkoxide in 17–74% yields (Scheme 7.32).

L-21 were found to be very effective ligands in the Pd-catalyzed asymmetric allylic substitution (Scheme 7.33) [51]. The ligand structures have some effect on the enantioselectivity and catalytic activity. The bulkier the ester group is, the higher the enantioselectivity but the lower the catalytic activity. These results showed that the C_2-symmetrical ferrocene ligands with only planar chirality can produce an excellent asymmetric environment for metal-catalyzed asymmetric reactions.

Very recently, Zhang's group developed new chiral diphosphine ligands **L-22–24** with only planar chirality (Scheme 7.34) [52]. Thus, treating **59** with LiAlH$_4$ in THF afforded 1,1′-dihydroxymethyl-2,2′-diphenylphosphino ferrocene (**L-22**). Alcohol **L-22** could be converted to **L-23** and **L-24** via esterification and etherification, respectively. These ligands have identical chiral scaffolds but different substituents adjacent to the diphenylphosphine group. However, compared to **L-21**, they resulted in a reversal of the configuration of the products in the Pd-catalyzed asymmetric allylic amination and alkylation with excellent enantioselectivity and high catalytic activity. This phenomenon could be attributed to the different configuration of the complexes of the diphosphine ligands with Pd(II).

Kang and coworkers synthesized another type of ferrocenyl diphosphine ligands (**L-25** named as FerroPHOS) with only planar chirality (Scheme 7.35) [53a]. First, the diols **14** were converted to the corresponding diacetate, followed by treatment with excess Me$_2$NH in EtOH to afford the corresponding amino derivatives (**55**) with >90% overall yield. After diastereoselective lithiation of **55** with *n*-BuLi and trapping

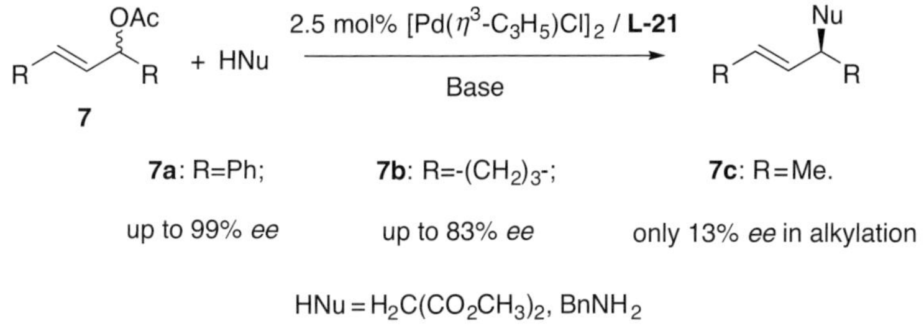

7a: R=Ph; up to 99% *ee*

7b: R=-(CH$_2$)$_3$-; up to 83% *ee*

7c: R=Me. only 13% *ee* in alkylation

HNu = H$_2$C(CO$_2$CH$_3$)$_2$, BnNH$_2$

Scheme 7.33 Asymmetric allylic alkylation using **L-21** as chiral ligands.

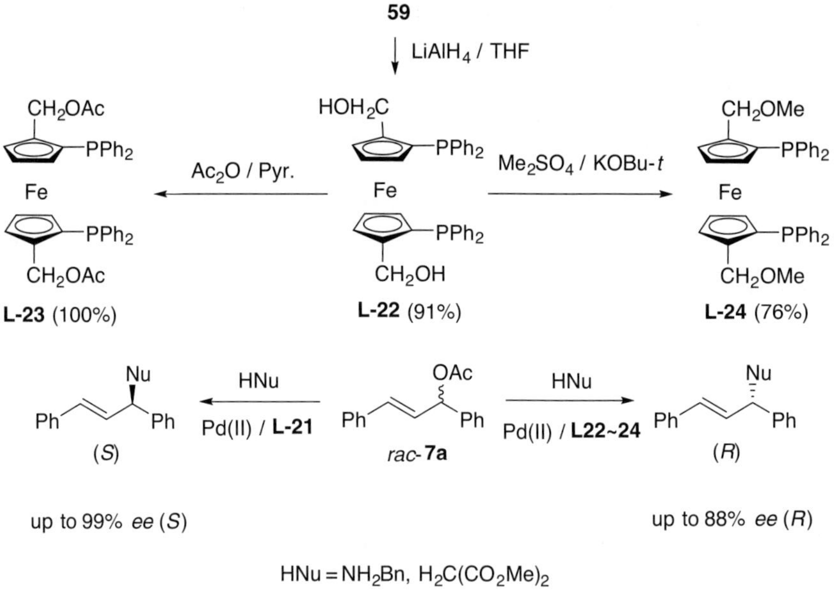

Scheme 7.34 Synthesis and application of **L-22**, **L-23** and **L-24**.

with Ph$_2$PCl, the diphosphines **L-17** were obtained (about 77% yield). The title compound FerroPHOS **L-25** was then given after removal of the central chirality from **L-17** via two steps (40–53% yields).

These ligands were successfully applied in the Rh-catalyzed enantioselective hydrogenation of α-dehydroamino acid and its derivatives with high enantioselectivity (>99% ee) [53a]. They were also effective in the asymmetric hydroboration of styrenes or indene with up to 84.6% ee (Scheme 7.36) [53b]. The observed enantioselectivity depended largely on the size of the side groups, that is, ee was significantly increased on changing the substituent R in **L-25** from methyl to phenyl and ethyl.

By (−)-sparteine-mediated *ortho*-metalation, Snieckus established a highly enantioselective synthesis of tetrasubstituted planar chiral ferrocenyl diamides (**L-26**, Scheme 7.37) [54]. The procedure contained two stages: (i) metalation of ferrocenyl diamide **61** with *n*-BuLi/(−)-sparteine **62** followed by addition of electrophilic reagents furnished, almost exclusively, the 1,1′,2-trisubstituted derivatives **63a–h**; (ii) **63** underwent the same process to afford products **L-26** with diastereoselective

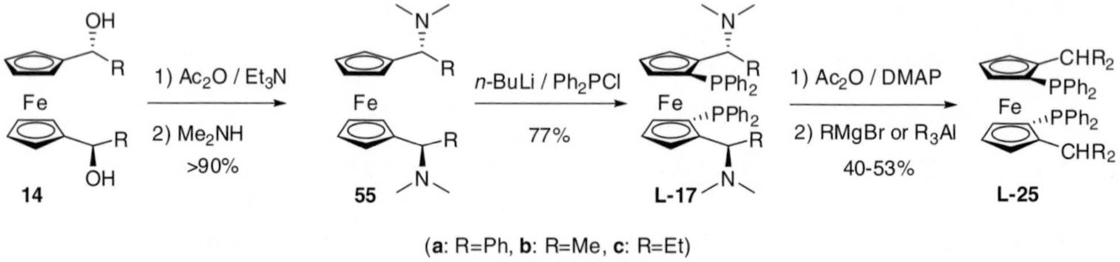

(a: R=Ph, b: R=Me, c: R=Et)

Scheme 7.35 Synthetic route to **L-25**, FerroPHOS.

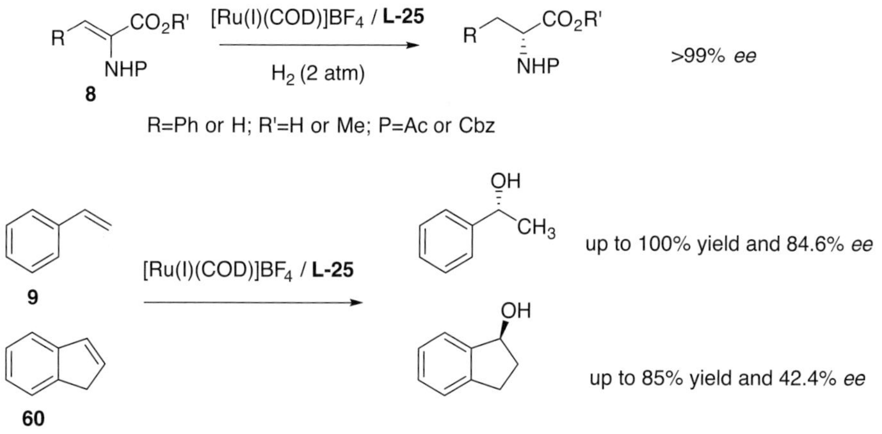

7.3 Symmetrical 1,1′,2,2′-Tetrasubstituted Ferrocenyl Ligands

Scheme 7.36 Application of L-25 in asymmetric hydrogenation.

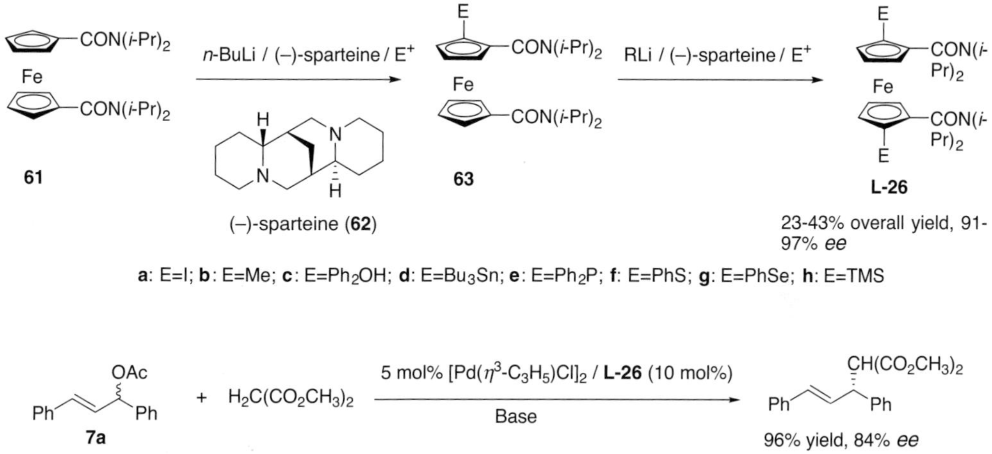

Scheme 7.37 Synthesis and application of L-26.

a: E=I; b: E=Me; c: E=Ph$_2$OH; d: E=Bu$_3$Sn; e: E=Ph$_2$P; f: E=PhS; g: E=PhSe; h: E=TMS

amplification of enantioselectivity. The identical procedure was also reported by Jendralla independently [55]. **L-26e** was tested as an efficient ligand for the enantioselective Pd-catalyzed allylic substitution of **7a**; up to 84% ee was obtained with a yield of 96% [54].

7.3.3
Other Tetrasubstituted Ferrocenyl Ligands

Ikeda's [56] and Bolm's groups [57] almost simultaneously introduced the new C_2-symmetrical N,O-chelating ligands **L-27**, independently. The synthesis of **L-27** was accomplished easily by dilithiation of 1,1′-bis(oxazolinyl)ferrocenes **51** with s-BuLi in THF at −78 °C and subsequent quenching of the dimetallated species with benzophenone (Scheme 7.38).

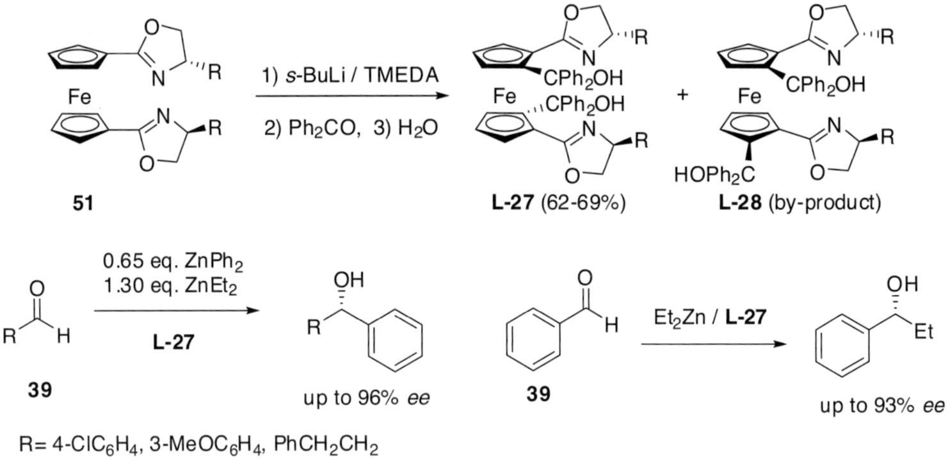

Scheme 7.38 Synthesis and application of **L-27**.

These ligands showed excellent catalytic activity and enantioselectivity in the asymmetric addition of organozinc reagents to aldehyde. Compared to the corresponding C_1-symmetrical 1,2-disubstituted ferrocenes, C_2-symmetrical **L-27** gave almost identical results with respect to both yield and enantioselectivity in the asymmetric alkylation of benzaldehyde with diethylzinc [56] and in the asymmetric catalysis of phenyl transfer from organozincs to aldehydes [57]. It is interesting that C_1-symmetrical tetrasubstituted **L-28** showed similar enantioselectivities but much higher catalytic activities than the C_2-symmetrical **L-27** [56].

C_2-symmetrical palladacycles **L-29–31** (X = I) were prepared by Kang by reacting $Pd_2(dba)_3$·$CHCl_3$ with the corresponding iodides (**66** and **68**) (Scheme 7.39) [58]. The intermediate iodides were synthesized through two approaches according to Scheme 7.39. It was noticeable that the basicity of the organolithium reagent is different in the lithiation of **51** and **64**. These ligands were applied in the asymmetric catalytic [3,3]-sigmatropic rearrangements of allylic imidates in the form of trifluoroacetate salts (X=OCOCF$_3$). The catalyst with bulkier hindrance gave higher reactivity and enantioselectivity. That is, compound **L-30c** is an excellent catalyst in the catalytic asymmetric rearrangement of allylic imidates **69**.

Ikeda's group developed the chiral C_2-symmetrical ferrocene diol **L-32**, possessing only planar chirality, which formed a 1 : 1 complex **72** by treatment with tetraisopropyl titanate (Scheme 7.40). The complex **72** was applied in the Ti-catalyzed hydrosilylation of a ketone with high yield (94%) but lack of enantioselectivity [59].

Reetz synthesized a completely different class of ferrocene-derived diphosphines having only planar chirality. A mixture containing **L-33**, **L-34** and achiral diphosphine **74** was afforded upon treating bis(tetrahydroindenyl)iron **73** with n-BuLi/TMEDA quenched by Ph$_2$PCl. **L-33** as well as **L-34** were obtained with low yield (3–17%) in analytically pure form after recrystallization from mixed solvents [60].

Both **L-33** and **L-34** can be used as ligands in the Rh-catalyzed asymmetric hydrogenation (up to 97% *ee*) and hydroboration (up to 84% *ee*) of olefins and in

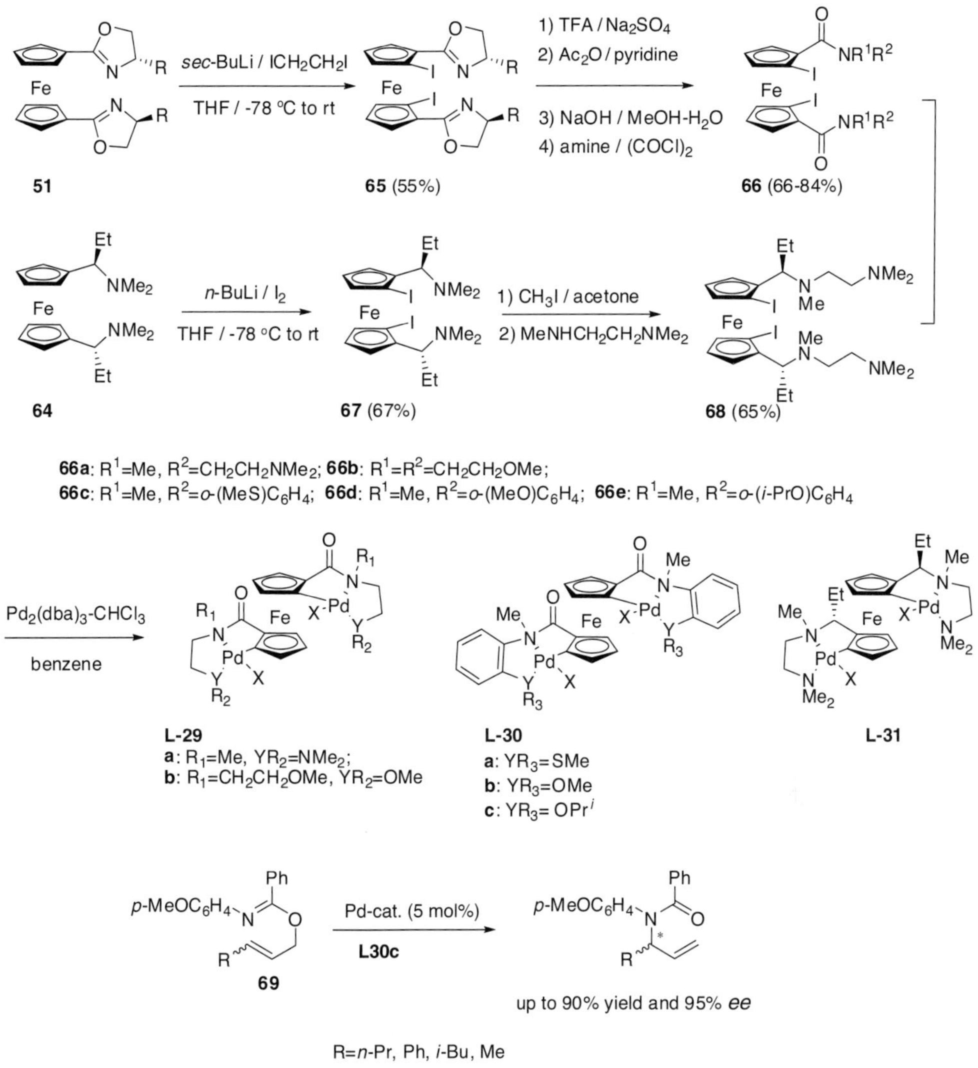

Scheme 7.39 Synthesis and application of L-29–31.

the Ir-catalyzed asymmetric hydrogenation of imines (up to 79% ee). In both cases it was shown that the C_2-symmetrical ligand **L-33** is much more stereoselective than the C_1-symmetrical analog **L-34** (Scheme 7.41).

7.4
Analogs of Ferrocenes: Symmetrical 1,1′-Bidentate Ruthenocenyl Ligands

As the analogs of ferrocenyl ligands, chiral ruthenocenes have received much less attention, especially the C_2-symmetrical chiral ruthenocenes. It is known that the

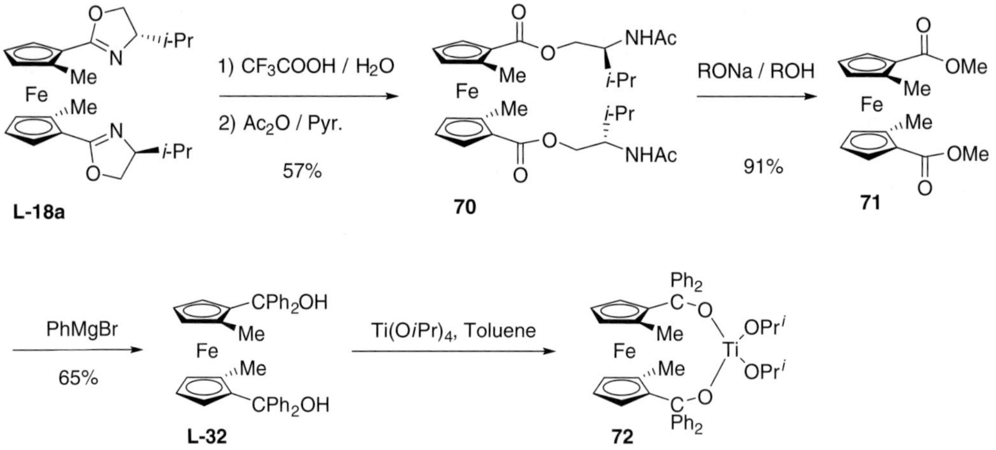

Scheme 7.40 Synthesis and complexation behavior of **L-32**.

distances between the two cyclopentadienyl rings in ferrocene and ruthenocene are 3.32 and 3.68 Å, respectively (Figure 7.4) [61]. The longer distance, by about 10%, in ruthenocenes would be expected to present some different enantioselectivity and catalytic activity in asymmetric catalysis. However, with their extraordinary similarity

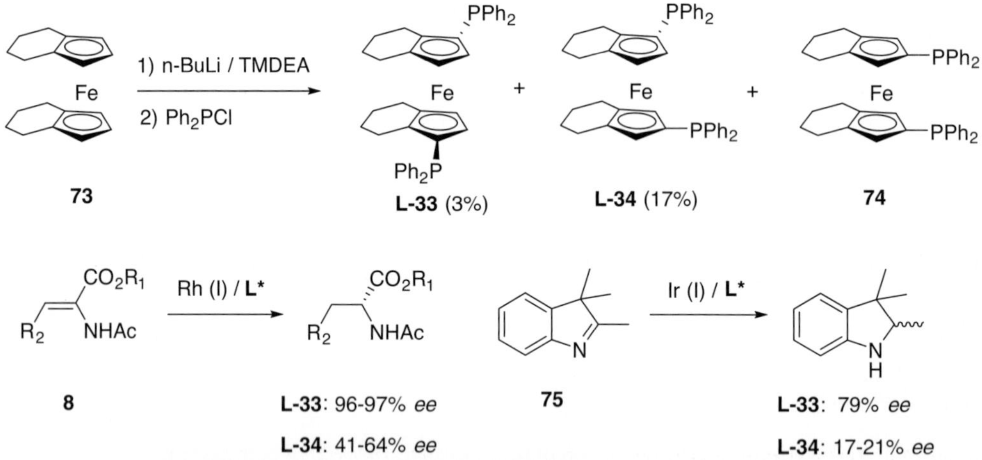

L-33: 96–97% ee
L-34: 41–64% ee

L-33: 79% ee
L-34: 17–21% ee

Scheme 7.41 Synthesis and application of **L-33** and **L-34**.

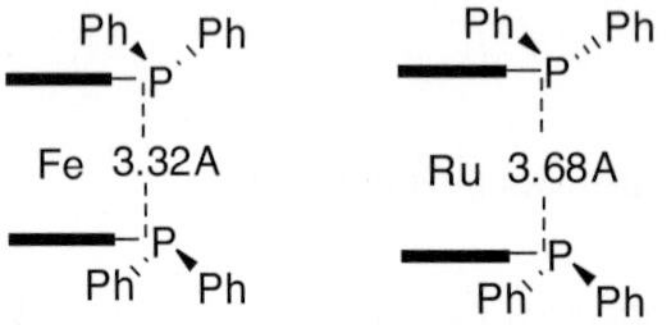

Figure 7.4 Different distance between the Cp rings in ferrocene and ruthenocene.

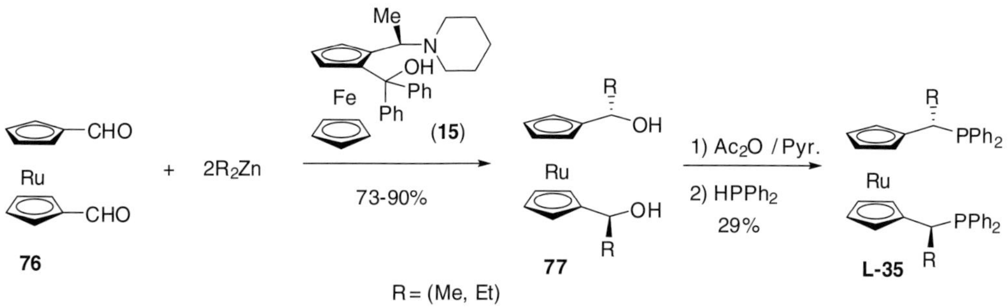

7.4 Analogs of Ferrocenes: Symmetrical 1,1′-Bidentate Ruthenocenyl Ligands | 205

Scheme 7.42 Synthetic route to **L-35**.

in structure, these two kinds of chiral ligands should provide similar functions in asymmetric catalysis. Comparing the ferrocenyl ligands with their corresponding ruthenocenyl ones can help us to further understand the chiral ferrocenes. In this section, the synthesis and application of the chiral C_2-symmetrical 1,1′-bidentate ruthenocenyl ligands are presented.

The novel chiral C_2-symmetrical ruthenocene diphosphine **L-35** was synthesized by Watanabe using the same method as the corresponding ferrocenes which gave accessibility to new ligands for asymmetric synthesis (Scheme 7.42) [17].

In the same way, Knochel synthesized the tetrasubstituted ruthenocene ligands **L-36** with moderate yield and enantioselectivity (Scheme 7.43). These ligands were applied in the Pd-catalyzed asymmetric cross-coupling of 1-phenylethylmagnesium chloride with vinyl bromide or β-bromostyrene with moderate yield (82%) and enantioselectivity (68% ee) [41, 42].

Bolm prepared the C_2-symmetrical chiral ruthenocenyl ligands **L-37** (Scheme 7.44) using the same method as for **L-27** and applied them in the metal-catalyzed

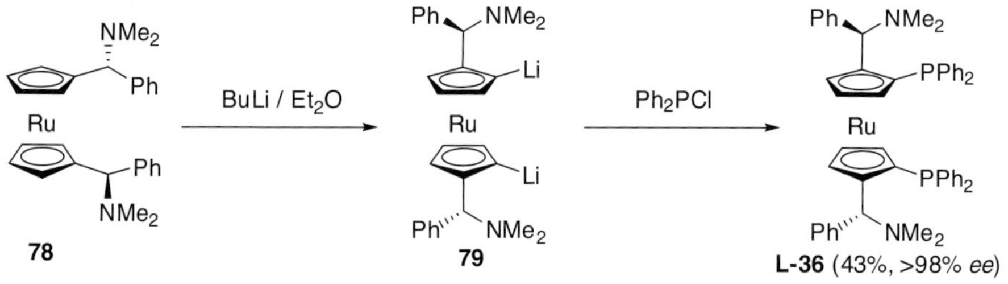

Scheme 7.43 Synthetic route to **L-36**.

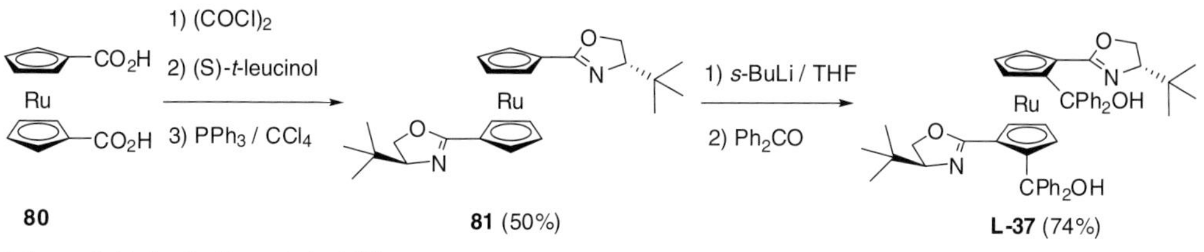

Scheme 7.44 Synthetic route to **L-37**.

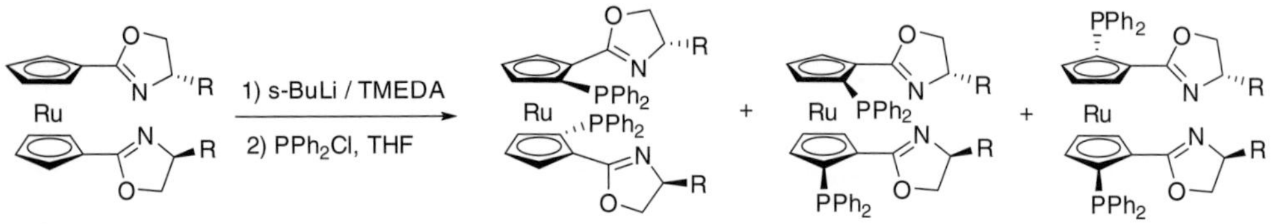

Scheme 7.45 Synthetic route to **L-38** and **L-39**.

a: R=*i*-Pr; b: R=*t*-Bu

L-38 (52-55%) **L-39** by-product not detected

asymmetric phenyl transfer from organozincs to aldehydes with high enantioselectivities (up to 96% *ee*). From the results in asymmetric catalysis, chiral ruthenocenes showed comparable selectivities and yields to their tetrasubstituted ferrocene analogs [57].

Comprehensive work on C_2-symmetrical chiral 1,1′-ruthenocenyl ligands was carried out by Zhang's group. Recently, they reported the preparation of the novel C_2-symmetrical tetrasubstituted ruthenocene P,N-chelating ligands (**L-38**) using the same synthetic route as for the corresponding chiral ferrocene analogs **L-18** (Scheme 7.45) [62]. Some procedures were optimized and the intermediates **82** were synthesized easily in high yields from the corresponding 1,1′-ruthenocenyldicarbonyl dichloride via three steps in one pot. Similar complexation behavior of **L-38** and **L-18** with palladium(II) was also observed. Compared to **L-18**, the air-stable **L-38** showed much higher catalytic activity with comparable excellent enantioselectivities in the Pd-catalyzed asymmetric allylic alkylation of **7a**. Furthermore, **L-38** was applied in the Ru-catalyzed asymmetric transfer hydrogenation of ketones **33** to their corresponding chiral alcohols using *i*-PrOH as the hydrogen donor. Up to 83.3% *ee* was observed with high conversion (98%) and catalytic activity (completed within 5 min) [63].

Furthermore, they developed the C_2-symmetrical ruthenocenyl diphosphine ligands with only planar chirality (Scheme 7.46) [64]. Thus, the ester amide **83** was prepared from **L-38a** by the same method as for **59** in 83% yield. Transesterification of **83** with methanolic sodium methoxide at room temperature for 24 h gave ligand **L-40a** in 75% yield. Ligand **L-40b** was also prepared from **84** by a similar procedure with a yield of 74%.

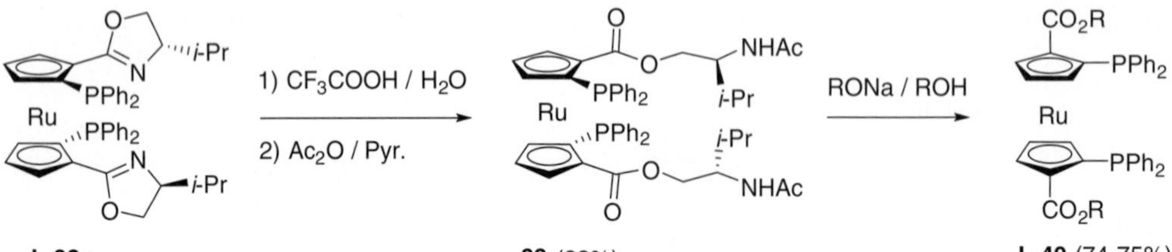

L-38a **83** (83%) **L-40** (74-75%)

(a: R=Me, b: R= Et)

Scheme 7.46 Synthetic route to **L-40**.

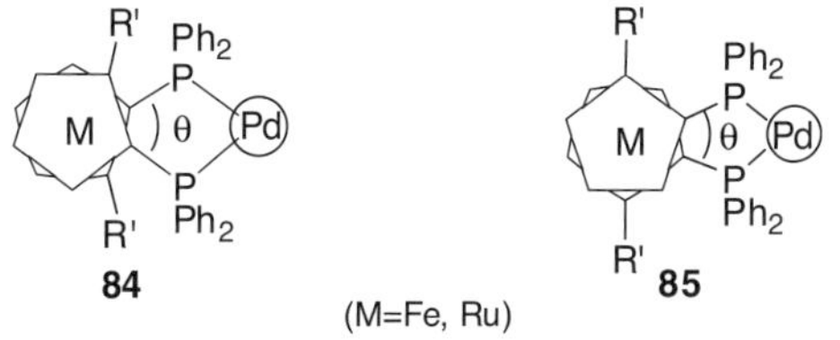

7.4 Analogs of Ferrocenes: Symmetrical 1,1'-Bidentate Ruthenocenyl Ligands

Figure 7.5 Twist angles of metal complexes.

High catalytic activity and excellent enantioselectivity were obtained in both asymmetric allylic alkylation and amination. Compared with the corresponding ferrocene ligands **L-21**, much higher catalytic activity was obtained with ruthenocene ligands **L-40** with comparable enantioselectivity. The relationship between the structure and the catalytic behavior was studied in detail. Based on the X-ray study, the Pd(II)-complexes of both **L-40** and **L-21** exist as the form of **84** rather than **85**. The Pd-complexes of **L-40** and the corresponding **L-21** with larger twist angle θ always delivered higher enantioselectivity in the asymmetric allylic substitution of 1,3-diphenyl-2-propenyl acetate. However, the opposite trend was observed for the reactions with cyclohexen-1-yl acetate as substrate. It could be deduced that the twist angle in the metallocene diphosphine ligands with only planar chirality had a pivotal influence on the asymmetric allylic substitutions (Figure 7.5).

Very recently, Zhang's group also reported the novel ruthenocene-based diphosphine ligands (**L-41–43**) with different kinds of substituents adjacent to the diphenylphosphine [52]. Ligands **L-41–43** were synthesized easily via the reduction of the ester amide **83** followed by protecting the hydroxy group, as shown in Scheme 7.47.

These ligands also showed excellent but reversed enantioselectivity in both palladium-catalyzed allylic alkylation and amination (up to 88% *ee*) compared to that with **L-40** although they have an identical chiral scaffold [52]. This is possibly because the steric interaction between the phenyl group and methylene group in R in complex **85** for ligands **L-41–43** is much smaller than that between the phenyl group and the carbonyl group in R for ligand **L-40**, which makes the formation of complex **85**

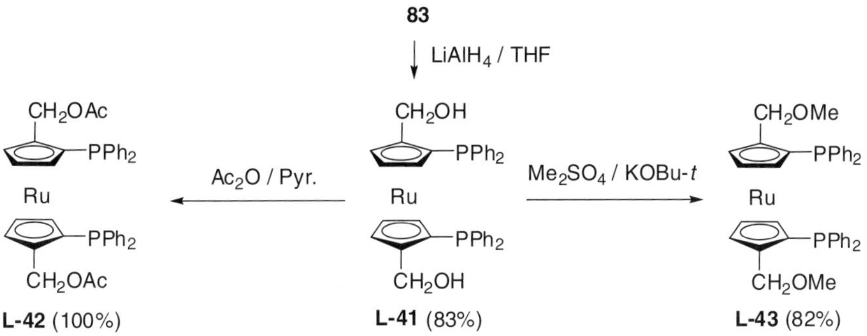

Scheme 7.47 Synthetic route to **L-41**, **L-42** and **L-43**.

easier than that of complex **84** for **L-41–43** (Figure 7.5). The opposite configuration of the complex **84** and **85** might afford a reversal of the configuration of the product.

7.5
Conclusion and Perspectives

Because of the advantages of rigidity, stability and easy availability, chiral ferrocenyl ligands are by far the most common chiral ligands employed in asymmetric catalysis. Among the various chiral ligands with a ferrocenyl backbone, 1,1'-bidentate ferrocenyl ligands have especially received the attention of scientists in the last decade for their outstanding asymmetric catalytic behavior in many kinds of metal-catalyzed reactions. It can be anticipated that studies on 1,1'-bidentate chiral ferrocenyl ligands will continue, and more and more efficient ligands of this type will be developed. Furthermore, it can be expected that 1,1'-bidentate ferrocenyl ligands will play an important role in asymmetric catalysis and will in the future be applied for the industrial production of useful chiral compounds.

7.6
Experimental: Selected Procedures

7.6.1
Typical Procedure for the Preparation of Ferrocenyldiphosphines L-2 [12c]

The respective phosphinite **5** (10 mmol) was dissolved in 10 mL of THF, degassed, and cooled to −40 °C. A suspension of 1,1'-dilithioferrocene (5 mmol) in 40 mL of diethyl ether and 5 mL of THF was cooled and slowly added to the phosphinite solution via a Teflon cannula. The reaction mixture was allowed to reach room temperature over a period of 15 h and was then quenched with water. After evaporation of solvent, the residue was extracted with CH_2Cl_2. The combined organic layers were washed with water, dried ($MgSO_4$), and concentrated. The crude borane complex **6** was suspended in 50 mL of diethylamine, degassed, and stirred at 50 °C for 5 h. The solvent was evaporated and the product diphosphine was purified by silica gel column chromatography (hexane/CH_2Cl_2 3 : 1 for **a–d**; hexane/CH_2Cl_2 2 : 1 for **e**). Minor amounts of monophosphine by-product were eluted first, followed by diphosphines **L-2a–e**, which were obtained as yellow to orange crystals. Recrystallization from CH_2Cl_2/hexane afforded analytically pure products (73–81% yields).

7.6.2
Typical Procedure for the Preparation of the (*R,R*)-ferrocenyl Diol 14 [40]

A 100 mL three-necked flask with an argon inlet was equipped with rubber septa and charged with the CBS-catalyst (82.5 mg, 0.30 mmol) in THF (4 mL) and a small fraction of a THF solution of BH_3 (0.2 mmol) and was cooled to 0 °C. A 1 M THF solution of $BH_3 \cdot Me_2S$ (0.8 mmol) and the diketone (0.5 mmol) in THF (8 mL) were added simultaneously within 15 min via syringe. After 20 min of stirring, the reaction mixture was quenched with MeOH (2 mL) and worked up as usual, affording a crude product which, after evaporation of the solvent, was purified by flash chromatography (ether) affording the desired diols **14** (74–98% yields).

7.6.3
Typical Procedure for the Preparation of ($\alpha R,\alpha' R$)-2,2'-Bis(α-N,N-dimethylaminopropyl)-(S,S)-1,1'-bis(diphenylphosphino)ferrocene L-17c [40]

Diamine (0.97 g, 2.9 mmol) was dissolved under argon in Et_2O (10 mL), cooled to 0 °C and t-BuLi ($c = 1.5$ M; 5.8 mL, 8.7 mmol) was added within 5 min. The solution was stirred at the same temperature for 0.5 h (the color changed from yellow to deep red). After quenching with Ph_2PCl (2.1 mL, 11.8 mmol) at 0 °C and stirring at rt for 3 h, the mixture was poured into a saturated solution of $NaHCO_3$ (20 mL) and extracted with CH_2Cl_2. After drying with $MgSO_4$ and filtration, the residue was purified by silica gel column chromatography (hexanes:MTBE, 3 : 1 and then 1 : 1 with 1% of Et_3N) and recrystallized in hexanes. A yellow solid **L-17c** was isolated in 40% yield (0.84 g, 1.1 mmol) as a single diastereomer (*de* 100%; *ee* >98%).

7.6.4
Typical Procedure for the Preparation of C_2-Symmetric 1,1′,2,2′-tetrasubstituted Ferrocene Derivatives L-18 [44a]

To a solution of **51** (1.0 mmol) in 20 mL of THF was added dropwise a 1.3 M solution of *sec*-BuLi in cyclohexane (2.0 mL, 2.6 mmol) at −78 °C under argon atmosphere. The reaction solution was stirred at the temperature for 3 h and then at 0 °C for 10 min to ensure complete dilithiation. Electrophiles (2.6 mmol) were added with a syringe to the solution of dilithiated species at 0 °C and then the reaction mixture was stirred at room temperature over night. The solvent was removed *in vacuo* and the residue was dissolved in 60 mL of dichloromethane. The solution was washed with water and brine, dried over Na_2SO_4, and then the solvent was evaporated *in vacuo* to give an oil (**L-18/L-19** $\approx$ 4 : 1). Enantiomerically pure **L-18** was isolated by silica gel column chromatography (ethylacetate–benzene 1 : 7) with yields of 47–66%.

7.6.5
Typical Procedure for the Preparation of the Ferrocene-Based Ester Amide 59 [51]

To a solution of compound **L-18c** (0.77 g, 0.99 mmol) in THF (20 mL) were added water (1 mL), trifluoroacetic acid (1.9 mL, 24.7 mmol), and Na_2SO_4 (9.40 g), and this suspension was stirred overnight at room temperature. After filtration and removal of the solvent under reduced pressure at below room temperature, an unstable ester ammonium salt was obtained as a brown solid. To a solution of this ester ammonium salt in dichloromethane (20 mL) were added pyridine (3.6 mL, 44.5 mmol) and acetic anhydride (6.0 mL, 38.2 mmol), and the mixture was stirred at room temperature overnight. The mixture was washed with 1 N HCl, water, and then brine and dried over Na_2SO_4. After removal of the solvent, the residue (2.01 g) obtained was purified by silica gel column chromatography (ethyl acetate) to afford pure ester amide **59** as a yellow solid (0.54 g, 61% overall yield).

7.6.6
Typical Procedure for the Preparation of (−)-(S)-(S)-1,1′-Bis(diphenylphosphino)-2,2′-bis-(methoxycarbonyl)ferrocene L-21 [51]

To a solution of ester amide **59** (0.50 g, 0.558 mmol) in THF (10 mL) was added a RONa solution prepared by the addition of RONa (22.3 mmol) in ROH (35 mL). After being stirred for 24 h, the mixture was neutralized with methanolic acetic acid, and the solvent was removed by evaporation. The residue was dissolved in dichloromethane (60 mL), and the solution was washed with water and then brine and dried over $MgSO_4$. After removal of the solvent, the residue was purified by silica gel column chromatography with ethyl acetate as an eluent to afford **L-21** as a yellow solid (17–74% yields).

7.6.7
Typical Procedure for the Preparation of 1,1′-Bis[(S)-4-isopropyloxazolin-2-yl]-ruthenocene 82 [62]

1,1′-Dicarboxylic ruthenocene (4.25 g, 13.3 mmol) was suspended in dichloromethane (70 mL) followed by addition of oxalyl chloride (11.0 mL, 106 mmol) and pyridine (0.1 mL). This mixture was refluxed for 2 h and then evaporated to dryness. The residue was washed with diethyl ether and the organic phase was evaporated to afford 1,1′-dichlorocarbonylruthenocene as a yellow–green solid. The product was used directly used in the next step without purification.

To a solution of (S)-(+)-valinol (3.20 g, 26.6 mmol) and triethylamine (11.2 mL, 58.5 mmol) in 30 mL of dichloromethane was added dropwise the above 1,1′-dichlorocarbonylruthenocence in 40 mL of dichloromethane under nitrogen atmosphere in an ice–water bath. The reaction mixture was stirred at room temperature for 24 h. To this solution was added dropwise methanesulfonyl chloride (2.80 mL, 34.6 mmol) over a period of 30 min at 0 °C, and then the solution was stirred at room temperature for 2 h. The resulting solution was washed with chilled water (5 °C) and then brine. The organic layer was dried over Na_2SO_4 and then the solvent was evaporated *in vacuo*. The residue was purified by silica gel column chromatography (petrol ether/ethyl acetate 2 : 1) to afford pure product **83** (3.5 g, 58% yield) as a light yellow solid.

Abbreviations

Ac	acetyl
BINAP	2,2′-bis(diphenylphosphino)-1,1′-binaphthyl
Bn	benzyl

Bu	butyl
Burgess reagent	(methoxycarbonylsulfamoyl)-triethylammonium hydroxide inner salt
CBS-catalyst	the Corey–Bakshi–Shibata catalyst
Cod	1,5-cyclooctadiene
Cy	cyclohexyl
Dba	dibenzylideneacetone
DIOP	2,3-O-isopropylidene-2,3-dihydroxy1,4-bis(diphenylphosphino)butane
DIPAMP	1,2-ethanediylbis(o-methoxyphenyl)phenylphosphine
DMDO	dimethyldioxirane
MOM	methoxymethyl
Ms	methylsulfonyl
MTBE	methyl tert-butyl ether
Piv	pivaloyl
PMB	p-methoxybenzyl
PNB	p-nitrobenzoyl
TBDPS	tert-butyldiphenylsilyl
TBS (TBDMS)	tert-butyldimethylsilyl
TMEDA	N,N,N',N',-tetramethylethyldiamine
TOF	turnover frequency
TON	turnover number

Acknowledgments

Financial support was provided by the National Natural Science Foundation of China and Nippon Chemical Industrial Co., Ltd.

References

1 (a) Hayashi, T. and Togni, A. (eds) (1995) *Ferrocenes*, VCH, Weinheim; (b) Togni, A. and Haltermann, R.L. (eds) (1998) *Metallocenes*, VCH, Weinheim; (c) Dai, L.X., Tu, T., You, S.L., Deng, W.P., and Hou, X.L. (2003) *Acc. Chem. Res.*, **36**, 659–667; (d) Colacot, T.J. (2003) *Chem. Rev.*, **103**, 3101–3118; (e) Richards, C.J. and Locke, A.J. (1998) *Tetrahedron: Asymmetry*, **9**, 2377–2407; (f) Arrayás, R.G., Adrio, J., and Carrtero, J.C. (2006) *Angew. Chem. Int. Ed.*, **45**, 7647–7715.

2 (a) Blaser, H.U. and Spindler, F. (1997) *Chimia*, **51**, 297–299; (b) Imwinkelried, R., Lonza, A.G., and Visp, S. (1997) *Chimia*, **51**, 300–302; (c) Blaser, H.U., Spindler, F., Jacobsen, E.N., Pfaltz, A. and Yamamoto, H. (eds) (1999) *Comprehensive Asymmetric Catalysis*, vol. 3, Springer, Berlin, p. 1427.

3 Whitesell, J.K. (1989) *Chem. Rev.*, **89**, 1581–1590.

4 (a) Kagan, H.B. and Dang, T.P. (1971) *Chem. Comm.*, 481–482; (b) Kagan, H.B. and Dang, T.P. (1972) *J. Am. Chem. Soc.*, **94**, 6429–6433; (c) Kagan, H.B., Langlois,

N., and Dang, T.P. (1975) *J. Organomet. Chem.*, **90**, 353–365; (d) Kagan, H. and Morrison, J.D. (eds) (1983) *Asymmetric Synthesis*, vol. **2**, Academic Press, New York, p. 1–39.

5 (a) Vineyard, B.D., Knowles, W.S., Sabacky, M.J., Bachman, G.L., and Weinkauff, O.J. (1977) *J. Am. Chem. Soc.*, **99**, 5946–5952; (b) Knowles, W.S. (1983) *Acc. Chem. Res.*, **16**, 106–112.

6 Knowles, W.S. (2002) *Angew. Chem.*, **114**, 2096–2107; Knowles, W.S. (2002) *Angew. Chem. Int. Ed.*, **41**, 1998–2007.

7 Miyashita, A., Yasuda, A., Takaya, H., Toriumi, K., Ito, T., Souchi, T., and Noyori, R. (1980) *J. Am. Chem. Soc.*, **102**, 7932–7934.

8 (a) Noyori, R. (2002) *Angew. Chem.*, **114**, 2108–2123; (b) Noyori, R. (2002) *Angew. Chem. Int. Ed.*, **41**, 2008–2022.

9 Brunner, H. and Zettlmeier, W. (1993) *Handbook of Enantioselective Catalysis*, VCH, Weinheim.

10 Pietrusiewicz, K.M. and Zablocka, M. (1994) *Chem. Rev.*, **94**, 1375–1411.

11 (a) Jugé, S., Stephan, M., Merdes, R., Genêt, J.-P., and Halut-Desportesc, S. (1993) *J. Chem. Soc., Chem. Comm.*, 531–533; (b) Kaloun, E.B., Merdes, R., Genêt, J.-P., Uziêl, J., and Jugé, S. (1997) *J. Organomet. Chem.*, **529**, 455–463.

12 (a) Nettekoven, U., Widhalm, M., Kamer, P.C.J., and van Leeuwen, P.W.N.M. (1997) *Tetrahedron: Asymmetry*, **8**, 3185–3188; (b) Nettekoven, U., Widhalm, M., Kamer, P.C.J., van Leeuwen, P.W.N.M., Mereiter, K., Lutz, M., and Spek, A.L. (2000) *Organometallics*, **19**, 2299–2309; (c) Nettekoven, U., Widhalm, M., Kalchhauser, H., Kamer, P.C.J., van Leeuwen, P.W.N.M., Lutz, M., and Spek, A.L. (2001) *J. Org. Chem.*, **66**, 759–770.

13 Nettekoven, U., Kamer, P.C.J., van Leeuwen, P.W.N.M., Widhalm, M., Spek, A.L., and Lutz, M. (1999) *J. Org. Chem.*, **64**, 3996–4004.

14 Nettekoven, U., Kamer, P.C.J., Widhalm, M., and van Leeuwen, P.W.N.M. (2000) *Organometallics*, **19**, 4596–4607.

15 Maienza, F., Wörle, M., Steffanut, P., and Mezzetti, A. (1999) *Organometallics*, **18**, 1041–1049.

16 (a) Togni, A., Breutel, C., Schnyder, A., Spindler, F., Landert, H., and Tijani, A. (1994) *J. Am. Chem. Soc.*, **116**, 4062–4066; (b) Herold, P., Stutz, S., Sturm, T., Weissensteiner, W., and Spindler, F. (2002) WO 02/02500; (c) Sturm, T., Weissensteiner, W., and Spindler, F. (2003) *Adv. Synth. Catal.*, **345**, 160–164; (d) de Vries, J.G. and Lefort, L. (2006) *Chem. Eur. J.*, **12**, 4722–4734; (f) Chen, W., McCormack, P.J., Mohammed, K., Mbafor, W., Roberts, S.M., and Whittall, J. (2007) *Angew. Chem. Int. Ed.*, **46**, 4141–4144.

17 Watanabe, M. (1995) *Tetrahedron Lett.*, **36**, 8991–8994.

18 (a) Longeau, A., Langer, F., and Knochel, P. (1996) *Tetrahedron Lett.*, **37**, 2209–2212; (b) Longeau, A., Durand, S., Spiegel, A., and Knochel, P. (1997) *Tetrahedron: Asymmetry*, **8**, 987–990.

19 (a) Marinetti, A., Kruger, V., and Buzin, F.-X. (1997) *Tetrahedron Lett.*, **38**, 2947–2950; (b) Marinetti, A., Genêt, J.-P., Jus, S., Blanc, D., and Ratovelomanana-Vidal, V. (1999) *Chem. Eur. J.*, **5**, 1160–1165.

20 (a) Marinetti, A., Labrue, F., and Genêt, J.-P. (1999) *Synlett*, 1975–1977; (b) Berens, U., Burk, M.J., Gerlach, A., and Hems, W. (2000) *Angew. Chem. Int. Ed.*, **39**, 1981–1984.

21 Korff, C. and Helmchen, G. (2004) *Chem. Comm.*, 530–531.

22 Chan, V.S., Bergman, R.G., and Toste, F.D. (2007) *J. Am. Chem. Soc.*, **129**, 15122–15123.

23 You, J., Drexler, H.-J., Zhang, S., Fischer, C., and Heller, D. (2003) *Angew. Chem. Int. Ed.*, **42**, 913–916.

24 (a) Lennon, I.C., Ramsden, J.A., Brear, C.J., Broady, S.D., and Muirb, J.C. (2007) *Tetrahedron Lett.*, **48**, 4623–4626; (b) Broady, S.D., Golden, M.D., Leonard, J., Muira, J.C., and Maudet, M. (2007) *Tetrahedron Lett.*, **48**, 4627–4630.

25 (a) Burk, M.J. and Harlow, R.L. (1990) *Angew. Chem., Int. Ed. Engl.*, **29**,

1462–1464; (b) Burk, M.J., Feaster, J.E., and Harlow, R.L. (1990) *Organometallics*, **9**, 2653–2655.

26 Burk, M.J. and Gross, M.F. (1994) *Tetrahedron Lett.*, **35**, 9363–9366.

27 Appleby, I., Boulton, L.T.C., Cobley, C.J., Hill, C., Hughes, M.L., de Koning, P.D., Lennon, I.C., Praquin, C., Ramsden, J.A., Samuel, H.J., and Willis, N. (2005) *Org. Lett.*, **7**, 1931–1934.

28 Liu, D., Li, W., and Zhang, X. (2002) *Org. Lett.*, **4**, 4471–4474.

29 Reetz, M.T., Gosberg, A., Goddard, R., and Kyung, S.-H. (1998) *Chem. Comm.*, 2077–2078.

30 (a) Xiao, D. and Zhang, X. (2001) *Angew. Chem. Int. Ed.*, **40**, 3425–3428; (b) Chi, Y., Zhou, Y.-G., and Zhang, X. (2003) *J. Org. Chem.*, **68**, 4120–4122.

31 Wang, Y.-Q., Yu, C.-B., Wang, D.-W., Wang, X.-B., and Zhou, Y.-G. (2008) *Org. Lett.*, **10**, 2071–2074.

32 (a) Schwink, L., Vettel, S., and Knoche, P. (1995) *Organometallics*, **14**, 5000–5001; (b) Almena, P., Juan, J., Ireland, T., and Knochel, P. (1997) *Tetrahedron Lett.*, **38**, 5961–5964.

33 Ptintener, K., Schwink, L., and Knochel, P. (1996) *Tetrahedron Lett.*, **37**, 8165–8168.

34 Kim, T.-J., Lee, H.-Y., Ryu, E.-S., Park, D.-K., Cho, C.S., Shim, S.C., and Jeong, J.H. (2002) *J. Organomet. Chem.*, **649**, 258–267.

35 (a) Bastin, S., Delebecque, N., Agbossou, F., Brocard, J., and Pélinski, L. (1999) *Tetrahedron: Asymmetry*, **10**, 1647–1651; (b) Bastin, S., Agbossou-Niedercorn, F., Brocard, J., and Pélinski, L. (2001) *Tetrahedron: Asymmetry*, **12**, 2399–2408.

36 Wang, M.-C., Hou, X.-H., Xu, C.-L., Liu, L.-T., Li, G.-L., and Wang, D.-K. (2005) *Synthesis*, **20**, 3620–3626.

37 Hua, G., Liu, D., Xie, F., and Zhang, W. (2007) *Tetrahedron Lett.*, **48**, 385–388.

38 Sawamura, M. and Ito, Y. (1992) *Chem. Rev.*, **92**, 857–871.

39 (a) Hayashi, T., Yamamoto, A., Hojo, M., and Ito, Y. (1989) *J. Chem. Soc., Chem. Comm.*, 495–496; (b) Hayashi, T., Yamamoto, A., Hojo, M., Kishi, K., and Ito, Y. (1989) *J. Organomet. Chem.*, **370**, 129–139.

40 (a) Schwink, L. and Knochel, P. (1996) *Tetrahedron Lett.*, **37**, 25–32; (b) Almena Perea, J.J., Ireland, T., and Knochel, P. (1997) *Tetrahedron Lett.*, **38**, 5961–5964.

41 Schwink, L. and Knochel, P. (1998) *Chem. Eur. J.*, **4**, 950–968.

42 Schwink, L., Ireland, T., Püntener, K., and Knochel, P. (1998) *Tetrahedron: Asymmetry*, **9**, 1143–1163.

43 Almena Perea, J.J., Lotz, M., and Knochel, P. (1999) *Tetrahedron: Asymmetry*, **10**, 375–384.

44 (a) Zhang, W., Adachi, Y., Hirao, T., and Ikeda, I. (1996) *Tetrahedron: Asymmetry*, **7**, 451–460; (b) Zhang, W., Hirao, T., and Ikeda, I. (1996) *Tetrahedron Lett.*, **37**, 4544–4548.

45 Liu, D., Xie, F., and Zhang, W. (2007) *Tetrahedron Lett.*, **48**, 7591–7594.

46 Imai, Y., Zhang, W., Kida, T., Nakatsuji, Y., and Ikea, I. (1999) *Chem. Lett.*, 243–244.

47 (a) Park, J., Lee, S., Ahn, K.H., and Cho, C.-W. (1995) *Tetrahedron Lett.*, **36**, 7263–7266; (b) Park, J., Lee, S., Ahn, K.H., and Cho, C.-W. (1996) *Tetrahedron Lett.*, **37**, 6137–6140.

48 (a) Ahn, K.H., Cho, C.-W., Park, J., and Lee, S. (1997) *Tetrahedron: Asymmetry*, **8**, 1179–1185; (b) Park, J., Quan, Z., Lee, S., Ahn, K.H., and Cho, C.-W. (1999) *J. Organomet. Chem.*, **584**, 140–146; (c) Lee, S., Koh, J.H., and Park, J. (2001) *J. Organomet. Chem.*, **637–639**, 99–106.

49 Cho, Y.-J., Carroll, M.A., White, A.J.P., Widdowson, D.A., and Williams, D.J. (1999) *Tetrahedron Lett.*, **40**, 8265–8268.

50 (a) Almena Perea, J.J., Börner, A., and Knochel, P. (1998) *Tetrahedron Lett.*, **39**, 8073–8076; (b) Lotz, M., Ireland, T., Almena Perea, J.J., and Knochel, P. (1999) *Tetrahedron: Asymmetry*, **10**, 1839–1842.

51 (a) Zhang, W., Kida, T., Nakatsuji, Y., and Ikeda, I. (1996) *Tetrahedron Lett.*, **37**, 7995–7998; (b) Zhang, W., Shimanuki, T., Kida, T., Nakatsuji, Y., and Ikeda, I. (1999) *J. Org. Chem.*, **64**, 6247–6251.

52 Xie, F., Liu, D., and Zhang, W. (2008) *Tetrahedron Lett.*, **49**, 1012–1015.

53 (a) Kang, J., Lee, J., Ahn, A., and Choi, J. (1998) *Tetrahedron Lett.*, **39**, 5523–5526; (b) Kang, J., Lee, J., Kim, J., and Kim, G. (2000) *Chirality*, **12**, 378–382.

54 (a) Laufer, R.S., Veith, U., Taylor, N.J., and Snieckus, V. (2000) *Org. Lett.*, **2**, 629–631; (b) Metallinos, C., Szillat, H., Taylor, N.J., and Snieckus, V. (2003) *Adv. Synth. Catal.*, **345**, 370–382; (c) Laufer, R., Veith, U., Taylor, N.J., and Snieckus, V. (2006) *Can. J. Chem.*, **84**, 356–369.

55 Jendralla, H. and Paulus, E. (1997) *Synlett*, 471–472.

56 Zhang, W., Yoshinaga, H., Imai, Y., Kida, Y., Nakatsuji, Y., and Ikeda, I. (2000) *Synlett*, 1512–1513.

57 Bolm, C., Hermanns, N., Kesselgruber, M., and Hildebrand, J. (2001) *J. Organomet. Chem.*, **624**, 157–161.

58 Kang, J., Yew, K.H., Kim, T.H., and Choi, D.H. (2002) *Tetrahedron Lett.*, **43**, 9509–9512.

59 Zhang, W., Youeda, Y.-I., Kida, T., Nakatsuji, Y., and Ikeda, I. (1999) *J. Organomet. Chem.*, **574**, 19–23.

60 Reetz, M.T., Beuttenmüller, E.W., Goddard, R., and Pastö, M. (1999) *Tetrahedron Lett.*, **40**, 4977–4980.

61 (a) Dunitz, J., Orgel, L., and Rich, A. (1956) *Acta Crystallogr.*, **9**, 373–375; (b) Hardgrove, G. and Templeton, D. (1959) *Acta Crystallogr.*, **12**, 28–32.

62 Liu, D., Xie, F., and Zhang, W. (2007) *Tetrahedron Lett.*, **48**, 585–588.

63 Liu, D., Xie, F., Zhao, X., and Zhang, W. (2008) *Tetrahedron*, **64**, 3561–3566.

64 Liu, D., Xie, F., and Zhang, W. (2007) *J. Org. Chem.*, **72**, 6992–6997.

8
Unsymmetrical 1,1′-Bidentate Ferrocenyl Ligands
Shu-Li You

8.1
Introduction

This chapter focuses on the preparation of enantiopure 1,1′-unsymmetrically disubstituted ferrocenes and their applications in asymmetric catalysis. Notably, Gibson, Long, and coworkers have documented a comprehensive review on the synthesis and catalytic applications of unsymmetrical ferrocene ligands [1]. Enantiopure 1,1′-bidentate ferrocenyl ligands attracted much attention soon after the discovery of ferrocene, mainly due to their unique rigid binding mode and relatively straightforward synthesis. Even today, 1,1′-bidentate ferrocenes still serve as an important scaffold for the design of new chiral ligands. It should be noted that the symmetrical 1,1′-bidentate ferrocenyl ligands, a very important class of this family, are discussed in Chapter 7. In this chapter, in addition to those having two different substituents at the 1,1′-positions of ferrocene **I**, the scope of ferrocenyl ligands will also cover unsymmetrical ferrocenyl ligands bearing identical coordination groups at the 1,1′-positions of ferrocene (**II**), for example, **L1a** [(*R*,*Sp*)-**BPPFA**] (Figure 8.1).

Due to the ready availability of **BPPFA** and its close analogs, they have been used for many catalytic asymmetric reactions. However, the corresponding work in which **BPPFA** and its close analogs were simply screened and unable to afford the optimal result will not be discussed in detail. This chapter will be divided into several parts based on the reaction type.

8.2
Palladium-Catalyzed Asymmetric Allylic Substitution Reaction

Palladium-catalyzed allylic substitution reactions have been used extensively in asymmetric carbon–carbon and carbon–heteroatom bonds formation to provide chemo-, regio-, diastereo-, and enantioselectivity [2]. The enantioselective version of this reaction has been demonstrated successfully in organic synthesis, including the

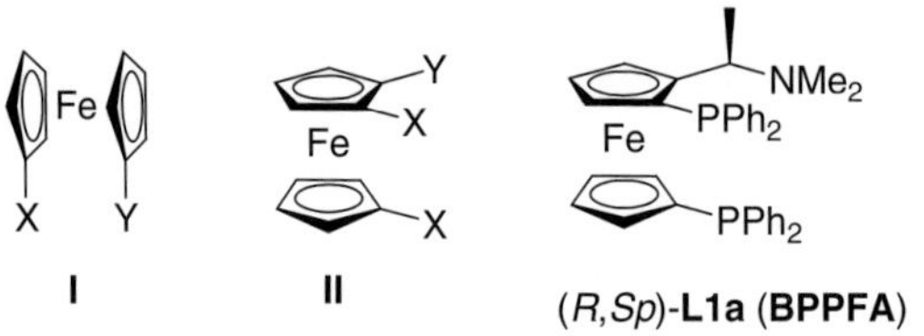
Fe
X
Y
Fe
Y
X
X
Fe
NMe2
PPh2
PPh2
I
II
(R,Sp)-L1a (BPPFA)

Figure 8.1 1,1′-Unsymmetrically disubstituted ferrocene.

total synthesis of numerous natural products [3]. Among various chiral ligands being studied in this reaction, enantiopure 1,1′-unsymmetrically disubstituted ferrocenes were found to be highly efficient and, in some cases, unique ligands.

8.2.1
A Model Reaction of Symmetrical 1,3-Disubstituted 2-Propenyl Acetates

Pioneering studies on the palladium-catalyzed allylic substitutions of *rac*-1,3-disubstituted-prop-2-enyl acetate with dimethyl malonate or benzylamine as nucleophile with the 1,1′-unsymmetrically disubstituted ferrocenyl ligands were carried out by Hayashi and coworkers. Particularly, the use of *rac*-1,3-diphenylprop-2-enyl acetate, allowing the reaction to proceed via a symmetrical π-allylpalladium intermediate to minimize the numbers of possible reaction transition states, has been demonstrated successfully with many chiral ligands.

Hayashi and coworkers have reported that ligands (*R,Sp*)-**BPPFA**, and **L1c–e** were efficient for palladium-catalyzed allylic alkylations [4]. The synthesis of these ligands is quite straightforward from (*R,Sp*)-**BPPFA**, as shown in Scheme 8.1. Compared with **BPPFA**, **L1c–e** gave higher enantioselectivities, which could be explained by the fact that the alcohol side chain might interact with the incoming nucleophile, as

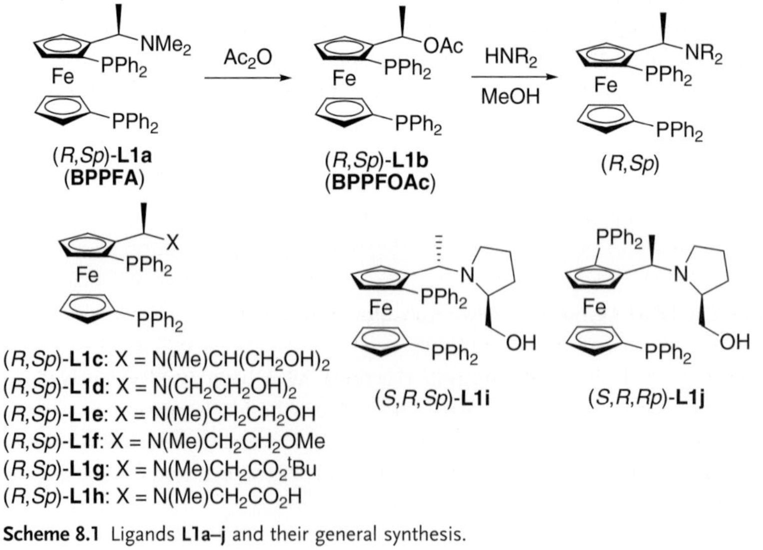
NMe2
PPh2
Fe
PPh2
Ac2O
OAc
PPh2
Fe
PPh2
HNR2
MeOH
NR2
PPh2
Fe
PPh2
(R,Sp)-L1a
(BPPFA)
(R,Sp)-L1b
(BPPFOAc)
(R,Sp)
X
PPh2
Fe
PPh2
N
PPh2
Fe
PPh2
OH
PPh2
N
Fe
PPh2
OH
(R,Sp)-L1c: X = N(Me)CH(CH2OH)2
(R,Sp)-L1d: X = N(CH2CH2OH)2
(R,Sp)-L1e: X = N(Me)CH2CH2OH
(R,Sp)-L1f: X = N(Me)CH2CH2OMe
(R,Sp)-L1g: X = N(Me)CH2CO2tBu
(R,Sp)-L1h: X = N(Me)CH2CO2H
(S,R,Sp)-L1i
(S,R,Rp)-L1j

Scheme 8.1 Ligands **L1a–j** and their general synthesis.

Table 8.1 Palladium-catalyzed allylic substitution of **1** with **L1a–j**.

$$\text{Ph}\overset{\text{OAc}}{\underset{\mathbf{1}}{\diagup\!\!\!\diagdown}}\text{Ph} \quad \xrightarrow[\substack{\text{or} \\ \text{Pd}_2(\text{dba})_3\cdot\text{CHCl}_3 \\ \text{L*}/\text{BnNH}_2/\text{THF}}]{\substack{[\text{Pd}(\text{C}_3\text{H}_5)\text{Cl}]_2/\text{L*} \\ \text{NaNu}/40\,°\text{C}/\text{THF}}} \quad \text{Ph}\overset{\text{Nu}}{\underset{}{\diagup\!\!\!\diagdown}}\text{Ph}$$

2a: Nu = CH(COMe)$_2$
2b: Nu = CH(CO$_2$Me)$_2$
2c: Nu = NHBn

Entry	Ligand	L* (mol%)	Product	Yield (%)	ee (%)	Ref.
1	(R,Sp)-**L1a**	10	2a	51	62 (S)	[4]
2	(R,Sp)-**L1c**	10	2a	97	90 (S)	[4]
3	(R,Sp)-**L1d**	10	2a	86	81 (S)	[4]
4	(R,Sp)-**L1e**	10	2a	86	71 (S)	[4]
5	(R,Sp)-**L1a**	3	2c	79	31 (R)	[5]
6	(R,Sp)-**L1c**	3	2c	84	97 (R)	[5]
7	(R,Sp)-**L1e**	3	2c	80	79 (R)	[5]
8	(R,Sp)-**L1f**	3	2c	75	27 (R)	[5]
9	(R,Sp)-**L1g**	10	2b	58	31 (S)	[6a]
10	(R,Sp)-**L1h**	10	2b	90	36 (S)	[6a]
11	(S,R,Sp)-**L1i**	1	2a	20	80 (S)	[6b]
12	(S,R,Rp)-**L1j**	1	2a	58	72 (R)	[6b]

shown in Table 8.1. Ligands **BPPFA** and **L1c–e** were also efficient for the palladium-catalyzed allylic amination [5]. The same "arm effect" was also observed in that a much lower ee (27%) of product **2c** was obtained with ligand (R,Sp)-**L1f** than with ligand (R,Sp)-**L1e**, a hydroxyl-free version of ligand (R,Sp)-**L1f**. Studies from Achiwa's and Toma's groups showed that ligands bearing carboxylic acid and prolinol could also be used for the palladium-catalyzed allylic alkylation [6].

Notably, Toma and coworkers have reported that palladium-catalyzed allylic alkylations with ligands (R,Sp)-**L1a** and (R,Sp)-**L1d** could be carried out in [bmim][BF$_4$] ionic liquid, which potentially offered easy recycle of the catalysts [7].

The 2,2'-bis(oxazolinyl)-1,1'-bis(diphenylphosphino)ferrocenes **L2** were synthesized by Park et al. and found to be efficient ligands for the palladium-catalyzed allylic alkylation to afford **2b** with up to 99% ee (Scheme 8.2) [8]. Although the X-ray diffraction of the palladium/ligand complex revealed P,P-chelation, ^{31}P NMR studies of the palladium/ligand complexes suggested that P,N-chelation is also possible.

Recently, in addition to the **BPPFA** derivatives having identical coordination groups at the 1,1'-positions of ferrocene, unsymmetrical ligands bearing two different coordination groups at 1,1'-position of ferrocene have been synthesized and applied in palladium-catalyzed allylic substitution reactions. As shown in Scheme 8.3, from 1,1'-dibromo ferrocene, (S)-**5** was synthesized through the intermediate **4**. Treatment of (S)-**5** with n-BuLi followed by trapping with Ph$_2$PCl afforded 1-diphenylphosphino-1'-oxazolinylferrocene (S)-**L3** [9]. It should be noted that an alternative synthesis of (S)-**L3** from 1,1'-bis(tributylstannyl)ferrocene was originally reported by Ikeda et al. [9a]. Starting from (S)-**5**, several planar chiral ligands **L4a–f** with different planar chiralities were synthesized conveniently.

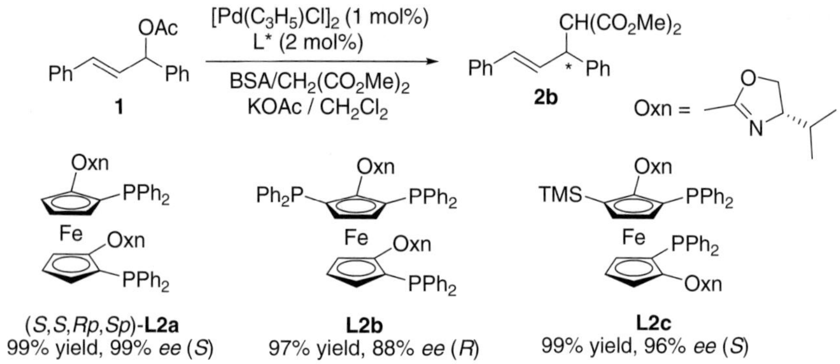

Scheme 8.2 Ligands **L2a–c** and their application in the palladium-catalyzed allylic alkylation of **1**.

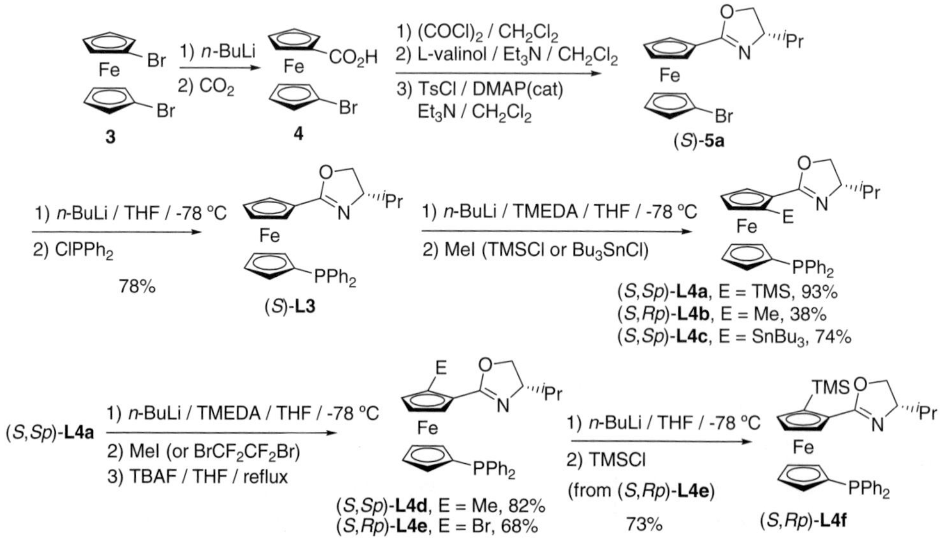

Scheme 8.3 Synthesis of ligands **L4a–f**.

An interesting feature of these ligands is that the multi-chiral elements (central, planar, and axial chirality) will be installed in one catalyst upon coordination with a metal. All the ligands worked efficiently in palladium-catalyzed allylic alkylation reactions but the different role of each chiral element and the incorporation of different chiral elements made the enantioselective outcome of the reaction intriguing. As indicated in Table 8.2, ligand (S)-**L3** gave 91% ee of **2b** in the S-configuration. By introducing planar chirality into the ligands, **L4a–c** gave **2b** in 34–83% ee but all in the R-configuration. Apparently the planar chirality plays a significant role in controlling the enantio-induction and these two elements conflict with each other in ligands **L4a–c**. These results led to the synthesis of chiral ligands containing

8.2 Palladium-Catalyzed Asymmetric Allylic Substitution Reaction

Table 8.2 Palladium-catalyzed allylic substitution reactions with **L3–7** (Figure 8.2).

$$\text{Ph}\diagup\!\!\!\diagdown\text{Ph} \;\; \xrightarrow[\substack{\text{A: BSA/CH}_2(\text{CO}_2\text{Me})_2 \\ \text{KOAc/CH}_2\text{Cl}_2 \\ \text{or} \\ \text{B: BnNH}_2/\text{ClCH}_2\text{CH}_2\text{Cl}}]{\substack{[\text{Pd}(C_3H_5)\text{Cl}]_2 (2\text{-}2.5\text{ mol\%}) \\ L^* (4\text{-}5.2\text{ mol\%})}} \;\; \text{Ph}\diagup\!\!\!\diagdown\text{Ph}$$

1 (OAc) → **2b**: Nu = CH(CO$_2$Me)$_2$; **2c**: Nu = NHBn

Entry	Ligand	Product	Yield (%)	ee (%)	Ref.
1	(S)-L3	2b	99	91 (S)	[9c]
2	(S,Sp)-L4a	2b	98	70 (R)	[9c]
3	(S,Rp)-L4b	2b	98	34 (R)	[9c]
4	(S,Sp)-L4c	2b	98	83 (R)	[9c]
5	(S,Sp)-L4d	2b	99	99 (S)	[9c]
6	(S,Rp)-L4f	2b	99	99 (S)	[9c]
7	(Sp)-L4g	2b	99	79 (R)	[9c]
8	(Sp)-L4h	2b	99	82 (R)	[9c]
9	(S,Sp)-L4d	2c	99	97 (R)	[9c]
10	(S)-L3	2c	96	90 (R)	[9c]
11	(S,Sp)-L4a	2c	25	32 (S)	[9c]
12	(Sp)-L4h	2c	99	90 (S)	[9c]
13	(S)-L5a	2b	82	84 (R)	[10]
14	(S)-L5b	2b	74	83 (R)	[10]
15	(S)-L5c	2b	62	73 (R)	[10]
16	(S)-L6a	2b	82	58 (R)	[11]
17	(S)-L6b	2b	99	56 (R)	[11]
18	(S)-L6c	2b	45	25 (R)	[11]
19	(S)-L6d	2b	96	68 (R)	[11]
20[a]	(S)-L6d	2b	98	69 (R)	[11]
21	(S,Sp)-L7a	2b	99	90 (S)	[12]
22	(S,Sp)-L7b	2b	99	75 (S)	[12]
23	(S,Sp)-L7c	2b	99	82 (S)	[12]
24	(S,Sp)-L7d	2b	98	69 (S)	[12]
25	(S,Sp)-L7e	2b	99	84 (S)	[12]
26	(S,Sp)-L7f	2b	99	84 (S)	[12]
27	(S,Sp)-L7g	2b	99	84 (S)	[12]
28	(S,Sp)-L7h	2b	100	70 (S)	[12]
29	(S,Sp)-L7i	2b	100	92 (S)	[12]

[a] Cs$_2$CO$_3$ was used as the base.

opposite planar chirality (Me or TMS group on different position), which should therefore have a combined effect with the existing central chirality. As expected, (S,Sp)-**L4d** and (S,Rp)-**L4f** afforded the alkylation product **2b** in excellent enantioselectivities (99% ee) in the S-configuration. As a direct support for the significant role of planar chirality, ligands (Sp)-**L4g–h**, containing planar chirality only, could also deliver the alkylation product in good ees (79% and 82%). In addition, the same trend was also observed for the palladium-catalyzed allylic amination reaction, the amination product **2c** was obtained in 97% ee (R), 90% ee (R), and 32% ee (S) for (S,Sp)-**L4d**,

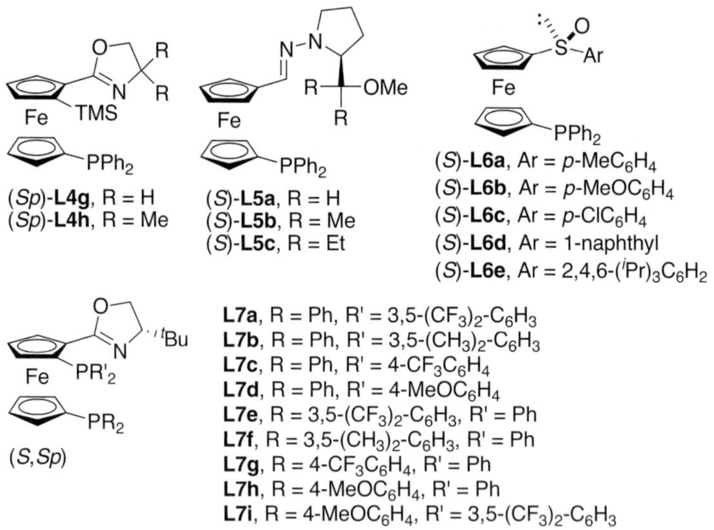

Figure 8.2 Ligands **L4g–h**, **L5–7**.

(S)-**L3**, and (S,Sp)-**L4a**, respectively. ^{1}H and ^{31}P NMR studies of the palladium complexes of these ligands showed that the existing central and planar chiralities induced different ratios during formation of the axial chirality. The high ratio of the diastereomers observed with the (S,Sp)-**L4d** might explain the excellent enantioselectivities although the kinetics was neglected here.

Phosphine-hydrazone ligands (S)-**L5a–c** were easily prepared from 1′-(diphenylphosphino)-1-ferrocenecarboxaldehyde with SAMP [(S)-1-amino-2-(methoxymethyl)-pyrrolidine] and its derivatives. These ligands were used in palladium-catalyzed asymmetric allylic alkylation to afford **2b** with good ees (up to 84%) [10].

Starting from 1,1′-bis(tributylstannyl)ferrocene, diphenylphosphino sulfinylferrocenes (S)-**L6a–e** were synthesized in high yields with excellent ees, determined by HPLC. In the palladium-catalyzed asymmetric allylic alkylation of **1** with dimethyl malonate, moderate ees (up to 69%) were obtained [11].

Interestingly, bisphosphino ferrocenyloxazolines (S,Sp)-**L7a–i** were synthesized from 1′-bromo-1-oxazolinyl ferrocene, and the procedure allowed easy installation of different disubstituted phosphino groups on the two Cp rings (Scheme 8.4) [12]. Then the electronic effects of the ligands were studied in the palladium-catalyzed asymmetric allylic alkylation reaction. Although all the ligands (S,Sp)-**L7a–i** were

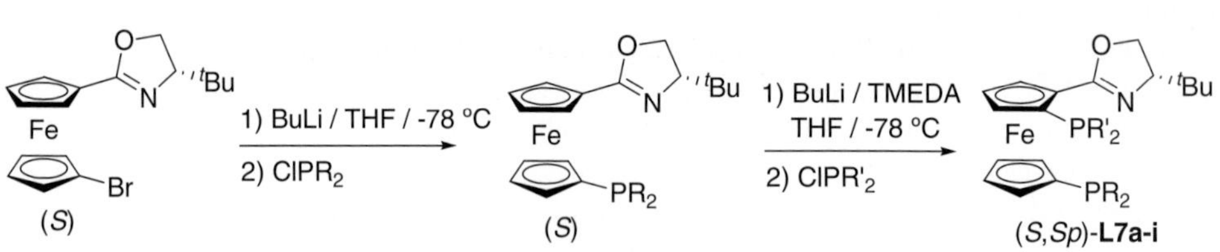

Scheme 8.4 Synthesis of ligands (S,Sp)-**L7a–i**.

effective for the reaction, higher *ee* values were observed when the electronic effect was matched with the steric effect of the ligands (entries 21–29, Table 8.2).

8.2.2
Substrate Variants of Palladium-Catalyzed Asymmetric Allylic Substitution Reaction

Although palladium-catalyzed allylic substitution of 1,3-diarylprop-2-enyl acetate can be obtained with excellent results with numerous chiral ligands, the application of the reaction is relatively limited. Allylic substitution reactions with more diversified allylic substrates and nucleophiles are highly desirable but also challenging.

By tethering the nucleophile and allylic substrate together, Yamamoto and Tsuji reported that the intramolecular allylic alkylation reaction proceeded smoothly to afford 3-vinyl-cyclohexanone in 83% yield with 48% *ee*, in the presence of Pd(OAc)$_2$ and (*S,Rp*)-**BPPFA** (Scheme 8.5) [13].

(*S,Sp*)-**L1c** was found to be efficient for the palladium-catalyzed cyclization of 2-butenylene dicarbamates **8** to afford optically active vinyloxazolidone **9** (73% *ee*) (Scheme 8.6) [14]. By utilizing a chiral amine nucleophile, palladium/(*S,Rp*)-**L1b** complex gave oxazolidinone with a moderate diastereoselectivity (66% de) [15].

Reaction of 2-butenylene dicarbonate with dimethyl malonate and methyl acetylacetate in the presence of (*R,Sp*)-**BPPFA**/palladium complex gave 2-vinylcyclopropane-1,1-dicarboxylate **12** in 26% yield with 70% *ee* (Scheme 8.7) [16].

In the presence of (*S,Sp*)-**L1c**/palladium complex, cycloaddition of ethyl 2-(benzenesulfonylmethyl)-2-propenyl carbonate with methyl acrylate and methyl vinyl ketone gave methylenecyclopentane derivatives **14** with up to 78% *ee* (Scheme 8.8) [17].

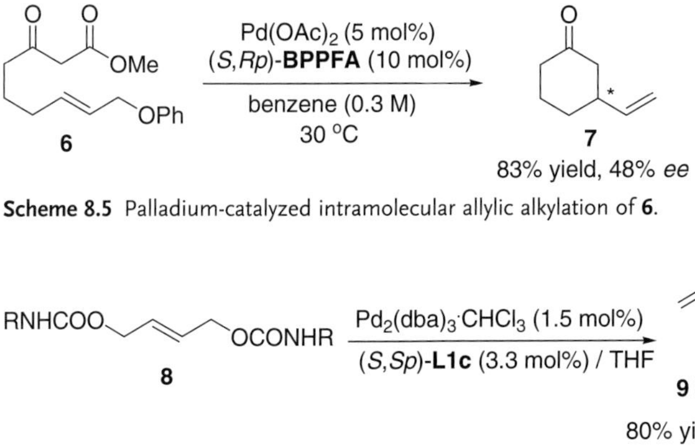

Scheme 8.5 Palladium-catalyzed intramolecular allylic alkylation of **6**.

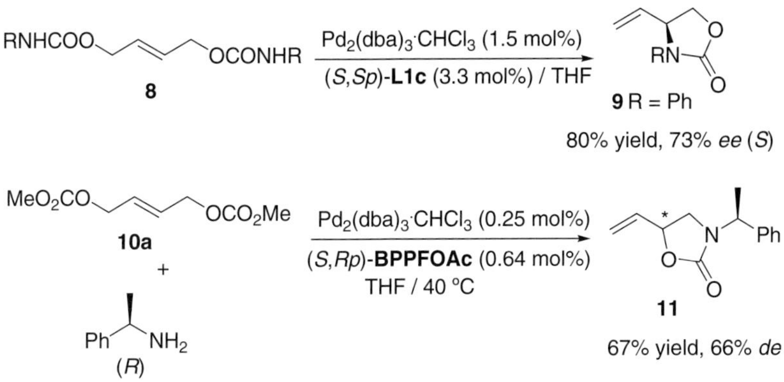

Scheme 8.6 Synthesis of optically active vinyloxazolidone.

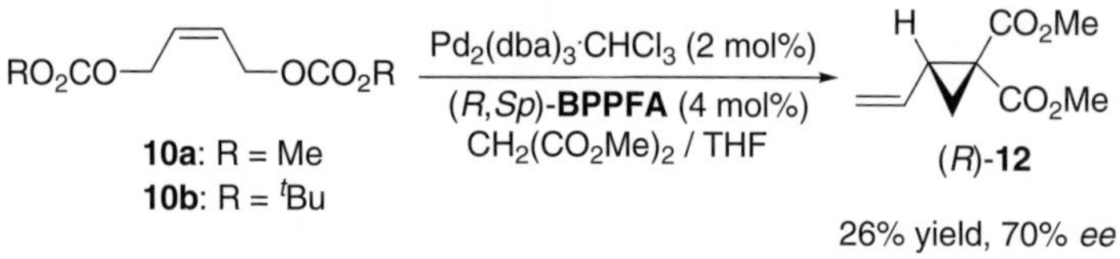

Scheme 8.7 Synthesis of optically active vinylcyclopropane.

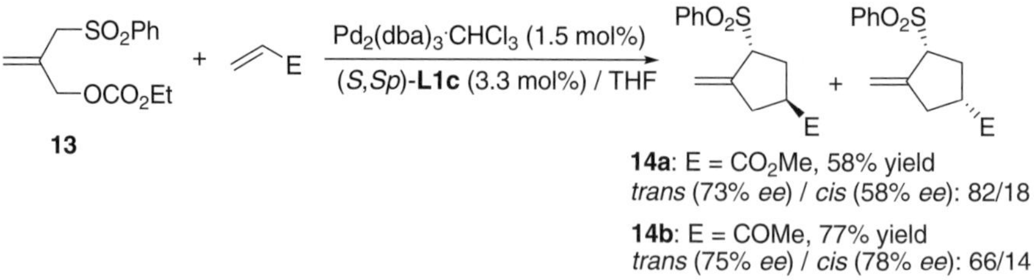

Scheme 8.8 Synthesis of optically active methylenecyclopentane derivatives.

Catalytic asymmetric allylation of sodium enolates of β-diketones with allyl carbonate in the presence of palladium/(R,Sp)-**L1e** proceeded smoothly to afford optically active ketones with a chiral quaternary carbon center. A high ee (up to 81%) was obtained for the six-membered ring substrate (Scheme 8.9) [18]. Interestingly, when the chiral ligand bearing a crown ether was used for the palladium-catalyzed allylation of 2-nitrocycloketones, a moderate ee (50%) was obtained [19].

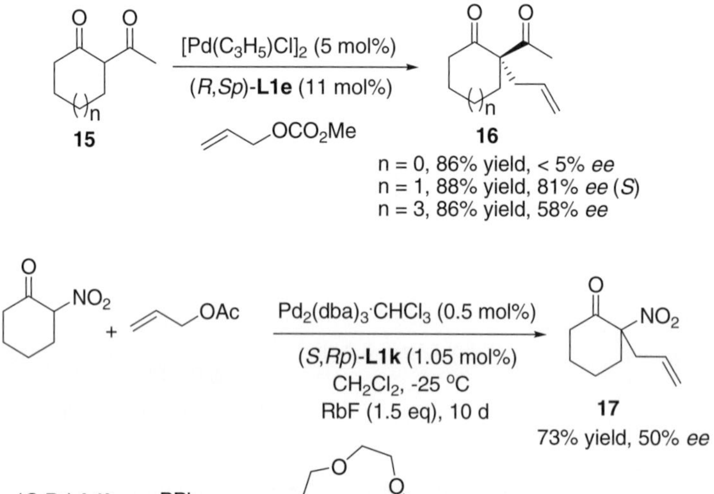

Scheme 8.9 Palladium-catalyzed allylation of prochiral nucleophiles.

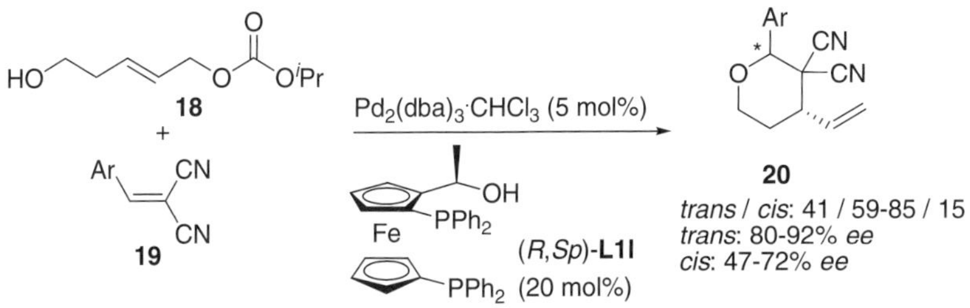

Scheme 8.10 Synthesis of optically active vinyl-tetrahydro-2H-pyran derivatives.

Reaction of allylic carbonate **18** bearing a hydroxy group with activated olefins **19** in the presence of the palladium/(*R*,*Sp*)-**L1l** complex gave vinyl-tetrahydro-2H-pyran derivatives **20**, with moderate trans/cis-ratio and excellent enantioselectivity for the trans-isomer (80–92%) (Scheme 8.10) [20].

As shown in Scheme 8.11, the planar chiral phosphino-ferrocene carboxylic acid (*Sp*)-**L7j** was synthesized from (*S*,*Sp*)-**L4a** via the hydrolysis of the oxazoline group. (*Sp*)-**L7j** was used for the palladium-catalyzed asymmetric allylic alkylations of cycloalkenyl acetates with moderate *ee*s (32–66%) [21].

Palladium-catalyzed reaction of allene **23** with iodobenzene led to the formation of the π-allyl palladium complex, which reacted with sodium malonate to afford allylic alkylation products (Scheme 8.12). With 4 mol% of Pd(dba)$_2$ and (*R*,*Sp*)-**BPPFOAc**

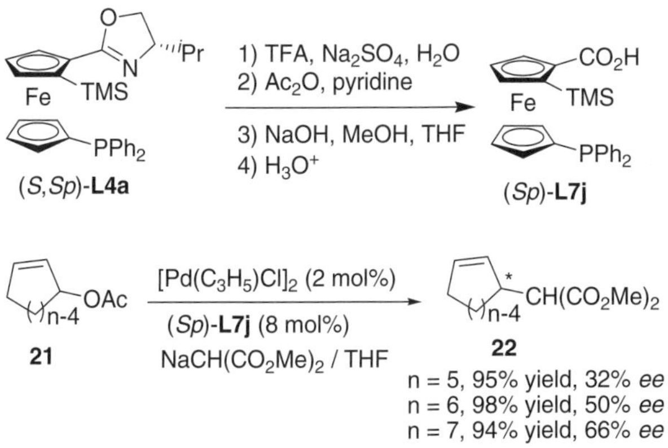

Scheme 8.11 Synthesis of (*Sp*)-**L7j** and its application in the palladium-catalyzed asymmetric allylic alkylation of **21**.

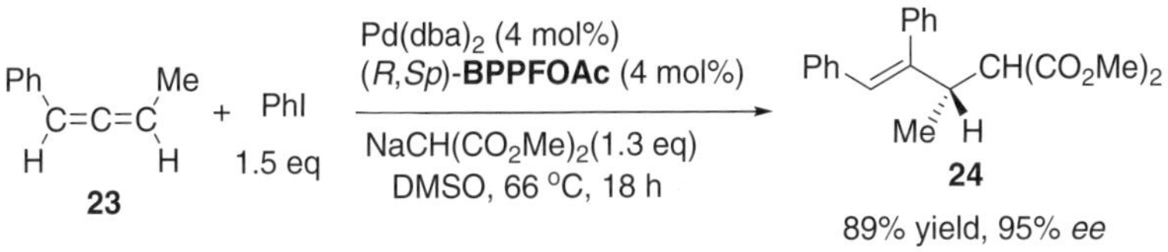

Scheme 8.12 Generating π-allyl palladium complex through the reaction of allene with iodobenzene.

(**L1b**), the asymmetric version of the above reaction occurred smoothly to give **24** in 89% yield with 95% ee [22].

8.2.3
Regioselective Control for Unsymmetrical Allylic Acetates

Regioselective control with unsymmetrically 1,3-disubstituted 2-propenyl acetate was a challenging project for Pd-catalyzed allylic substitution reactions. Initial efforts by Hayashi and Ito demonstrated that poor regioselectivities were obtained for the palladium-catalyzed allylic alkylation reaction with unsymmetrically 1,3-diarylsubstituted 2-propenyl acetate [23]. Particularly, branch selectivity for monosubstituted allylic substrates has attracted more attention since only the branch product is a chiral compound and the terminal alkene would allow versatile transformations.

Pioneering studies by Hayashi and Ito showed high regioselectivity and enantioselectivity were obtained for the palladium-catalyzed asymmetric allylic amination of 2-butenyl acetate with benzylamine (Scheme 8.13). Utilizing (*R*,*Sp*)-**L1c** as the ligand, up to 97/3 branch-linear ratio in favor of the branched product and 84% ee of the branched product were obtained [24].

Johnson *et al.* have reported that a **BPPFA** derivative, when anchored to the inner walls of the mesoporous support MCM-41, was used for the palladium-catalyzed allylic amination of cinnamyl acetate. With benzylamine as the nucleophile, the amination product was obtained with branch/linear ratio of 51/49 and 99% ee for the branched product [25].

The challenge of this reaction can be seen by the fact that only limited substrate was used in both the above-mentioned reports. In 2001, Hou, Dai and their coworkers reported the synthesis of 1,1-ferrocenyl-*P*,*N*-ligands that were highly efficient and selective for the palladium-catalyzed allylic substitution reactions of monosubstituted allylic acetates [26]. These ligands are named SiocPhox [SiocPhox is named after the Shanghai Institute of Organic Chemistry where this type of ligand is developed] and their syntheses are shown in Scheme 8.14. A new chiral center is formed on the P atom during the reaction of (*S*)-**28** with enantiopure BINOL, and both of the diastereoisomers can be easily isolated by column chromatography as orange solids. The preparation of these ligands is fairly easy, even in gram quantity, although the ligands are seemingly rather complex. Both ligands **L8** and **L9** are stable in air, and the absolute configuration at the phosphorus atom was determined by X-ray diffraction analysis.

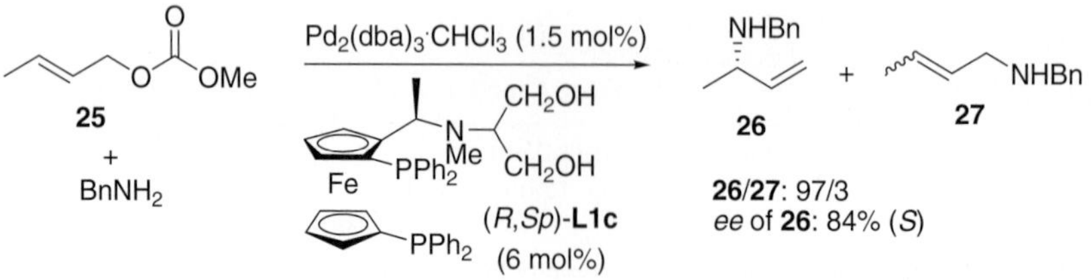

Scheme 8.13 Palladium-catalyzed regio- and enantio-selective allylic amination of **25**.

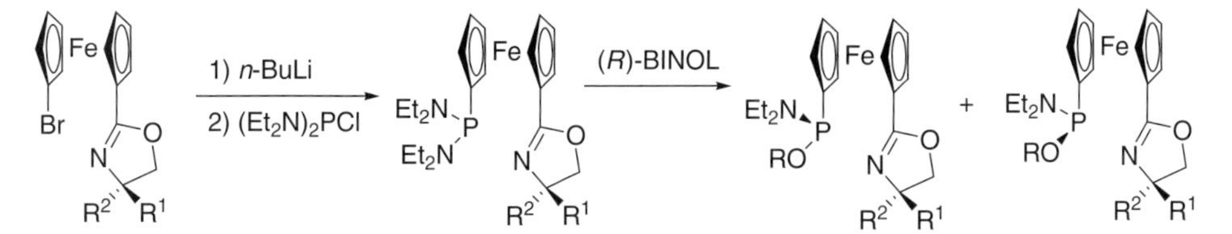

8.2 Palladium-Catalyzed Asymmetric Allylic Substitution Reaction

Scheme 8.14 Synthesis of SiocPhox **L8–9** from (S)-**5**.

(S)-**5a** $R^1 = H$, $R^2 = i$-Pr
(S)-**5b** $R^1 = H$, $R^2 = t$-Bu
(S)-**5c** $R^1 = H$, $R^2 = Ph$
(S)-**5d** $R^1 = H$, $R^2 = Bn$
(S)-**5e** $R^1 = H$, $R^2 = Me$
(S)-**5f** $R^1 = H$, $R^2 = H$
(S)-**5g** $R^1 = Me$, $R^2 = Me$

(S)-**28a–g**

(S,S$_{phos}$,R)-**L8a** $R^1 = H$, $R^2 = i$-Pr
(S,S$_{phos}$,R)-**L8b** $R^1 = H$, $R^2 = t$-Bu
(S,S$_{phos}$,R)-**L8c** $R^1 = H$, $R^2 = Ph$
(S,S$_{phos}$,R)-**L8d** $R^1 = H$, $R^2 = Bn$
(S,S$_{phos}$,R)-**L8e** $R^1 = H$, $R^2 = Me$
(S$_{phos}$,R)-**L8f** $R^1 = H$, $R^2 = H$
(S$_{phos}$,R)-**L8g** $R^1 = Me$, $R^2 = Me$

(S,R$_{phos}$,R)-**L9a**
(S,R$_{phos}$,R)-**L9b**
(S,R$_{phos}$,R)-**L9c**
(S,R$_{phos}$,R)-**L9d**
(S,R$_{phos}$,R)-**L9e**
(R$_{phos}$,R)-**L9f**
(R$_{phos}$,R)-**L9g**

ROH = (R)-BINOL

The combination of different chiralities is very important for regio- and enantioselective control for allylic substitution of monosubstituted allylic acetates. The reaction of (S)-**28** with (S)-BINOL gave the diastereoisomers of **L8** and **L9** respectively. However, they are not included since they did not provide the optimal results.

For the allylic alkylation reaction of **29**, (S,S$_{phos}$,R)-**L8d** was found to be the optimal ligand, giving excellent regio- and enantioselectivities for a wide range of substrates (Scheme 8.15). Allylic amination reaction of allyl acetates **32** with benzylamine was carried out. (S,R$_{phos}$,R)-**L9c** gave the best results, and up to >97/3 B/L ratio and 98% ee of the branched product were obtained. It should be noted that the B/L ratios were determined by ^{1}H NMR and represented the ratio of **33**/(**34** + **35**X2). Interestingly, the favored substrates and ligands for these two reactions were entirely different. The free hydroxy group might interact with the incoming nucleophile, particularly

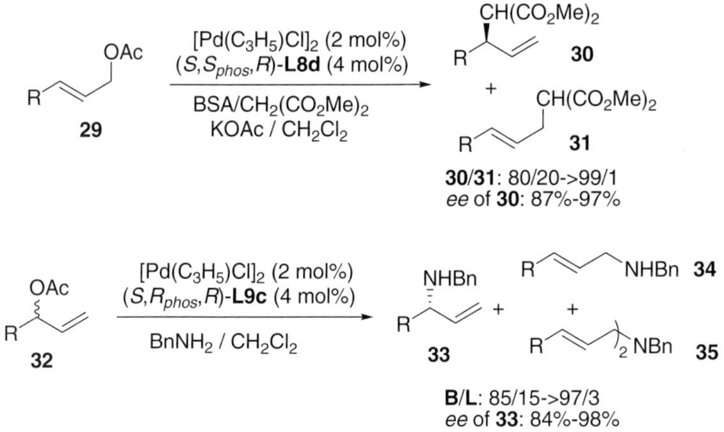

30/31: 80/20->99/1
ee of **30**: 87%–97%

B/L: 85/15->97/3
ee of **33**: 84%–98%

Scheme 8.15 Regio- and enantioselective palladium-catalyzed allylic substitution reactions with SiocPhox.

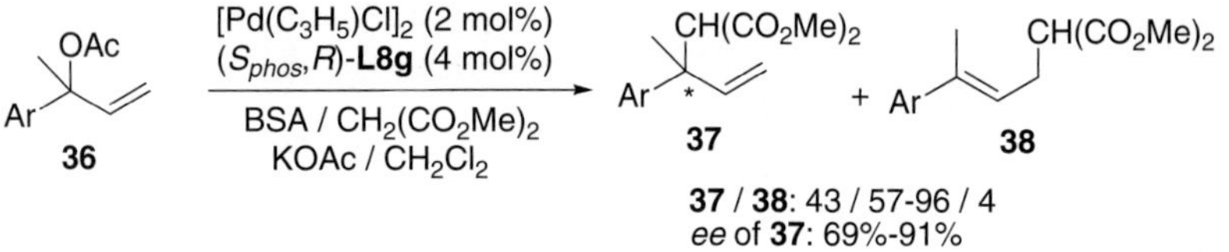

Scheme 8.16 Regio- and enantioselective palladium-catalyzed allylic alkylations to construct a quaternary carbon center with SiocPhox.

benzylamine, and the different position of the hydroxy group in the ligands would lead to different catalytic outcomes.

The regioselective allylic alkylation reaction with SiocPhox was also used for the construction of an all-carbon quaternary chiral center (Scheme 8.16). When the allylic acetates **36** were used, (S_{phos},R)-**L8g**/palladium complex catalyzed the allylic alkylation reaction of malonate methyl ester, providing the alkylation products **37** with up to 91% ee and 96/4 regioselectivity in favor of the branched product [27].

Further extending the substrate scope, the chemistry was also found to be suitable for the allylic alkylation and amination of conjugated dienyl acetates. Palladium-catalyzed asymmetric allylic alkylation of dienyl acetates **39** using (S,S_{phos},R)-**L8d** as the ligand led to the branched alkylation products **40** with excellent regio- and enantioselectivities (Scheme 8.17). Allylic amination reaction of allyl acetates **42** with benzylamine was carried out, and (S,R_{phos},R)-**L9c** gave the best results. The branched amination products were obtained with up to 98/2 branch/linear ratio and 94% ee [28].

Although ketone enolates are quite challenging nucleophiles for the enantioselective allylic substitution reaction, recently, Hou and coworkers demonstrated that ketone enolates are suitable nucleophiles for the regio- and enantioselective allylic

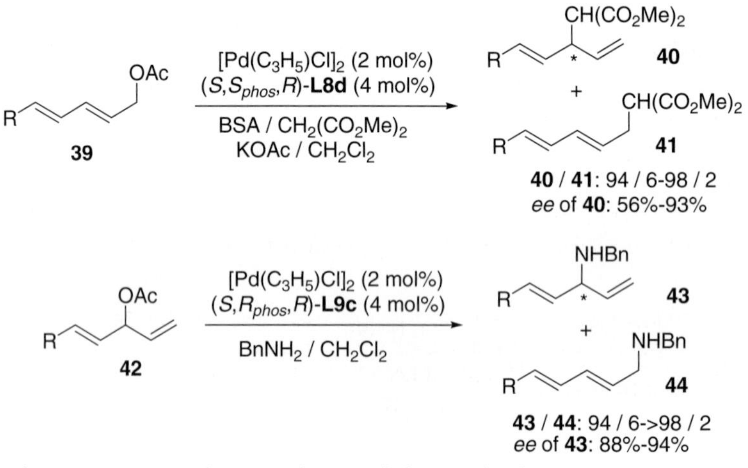

Scheme 8.17 Regio- and enantioselective palladium-catalyzed allylic substitution of the conjugated dienyl acetates with SiocPhox.

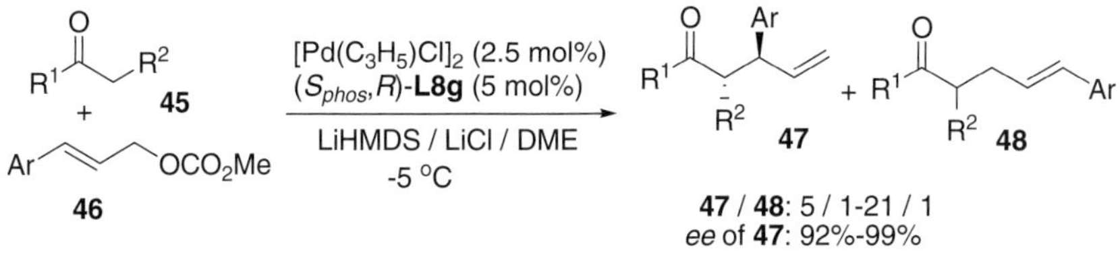

Scheme 8.18 Regio- and enantioselective palladium-catalyzed allylic substitution reaction of ketone **45** with SiocPhox.

alkylation reaction of monosubstituted allyl carbonates (Scheme 8.18). (S_{phos},R)-**L8g** was found to be the optimal ligand, and the allylic alkylation reaction of ketones **45** with **46** gave the alkylation products **47** with excellent regio- and enantioselectivities. Interestingly, one equivalent of LiCl additive was crucial for the high regioselectivity in favor of branched product. It should also be noted that two chiral centers were created in the products, offering potentially broad applications in organic synthesis [29].

SiocPhox not only displayed unique regioselectivity for the palladium-catalyzed allylic substitution reaction for monosubstituted allyl substrates but also showed interesting features in expanding the nucleophile scope. Recently, Hou and coworkers realized the highly enantioselective palladium-catalyzed allylic alkylation of acyclic amides in the presence of SiocPhox (Scheme 8.19) [30].

High enantioselectivity was observed in the palladium-catalyzed asymmetric allylic alkylation of acyclic amides in the presence of (S,S_{phos},R)-**L8d** to provide the corresponding γ,δ-unsaturated amides. Interestingly, the nature of the substituents on the nitrogen atom of the amides had a great impact on the efficiency and selectivity of the reaction, and generally N,N-diphenyl amide was required for satisfactory yields and enantioselectivities.

Instead of the branch selectivity, the linear alkylation product of cinnamyl acetates with cyclohexanone derivatives bearing an electron-withdrawing group at the α-position would also provide a useful synthetic intermediate, α,α-disubstituted cyclohexanones (Scheme 8.20). Aoyama and coworkers demonstrated that the palladium complex with (S,Rp)-**L1l** was efficient for allylic alkylation of **52** with **29**, leading to **53** bearing a quaternary carbon center with up to 90% ee [31].

A series of chiral (iminophosphoranyl)ferrocenes was synthesized from the reaction of **BPPFA** with arylazide (Scheme 8.21). These ligands have been evaluated in the Rh-catalyzed allylic alkylation of cinnamyl acetates with sodium dimethyl

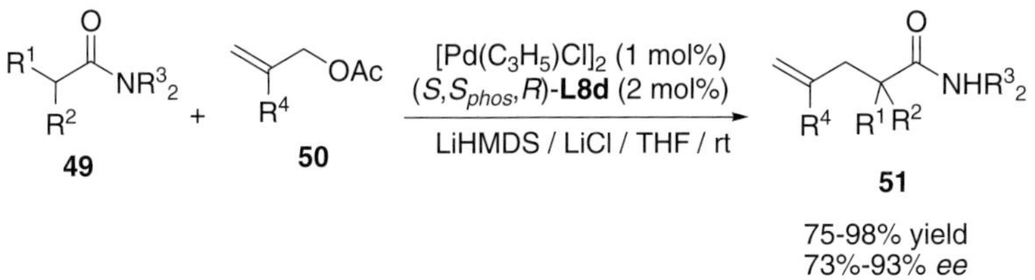

Scheme 8.19 Palladium-catalyzed allylic alkylation of acyclic amides with SiocPhox.

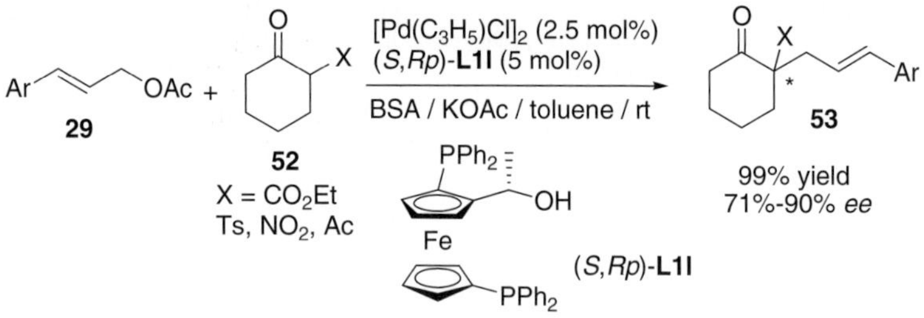

Scheme 8.20 Regio- and enantioselective palladium-catalyzed allylic alkylation of prochiral nucleophiles.

manolate. Branched alkylation products were obtained in low regioselectivities with up to 91% ee [32].

8.2.4
Applications of Palladium-Catalyzed Asymmetric Allylic Substitution Reaction

The most significant application of palladium-catalyzed asymmetric allylic substitution using unsymmetrically 1,1′-disubstituted ferrocene is the synthesis of the key intermediate of (−)-huperzine A. Terashima and coworkers examined several **BPPFA** derivatives for the allylic alkylation of β-keto ester **54** with 2-methylene-1,3-propanediol diacetate **55**. They found in the presence of (R, Sp)-**L1m** the desired double alkylation product **56** was obtained in 92% yield with 64% ee (Condition A, Scheme 8.22) [33].

By employing the same approach, He, Bai and their coworkers have elegantly modified the chiral ligand. The ligand (R,Sp)-**L1n** bearing a bulkier substituent on the

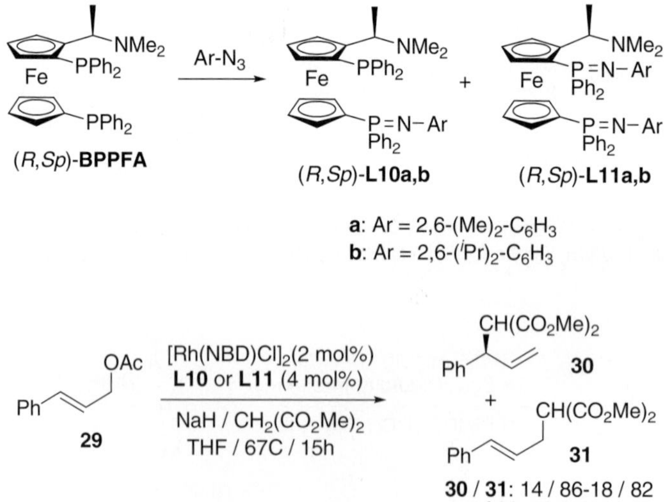

Scheme 8.21 Synthesis of **L10–11** and their applications in Rh-catalyzed allylic alkylation reactions.

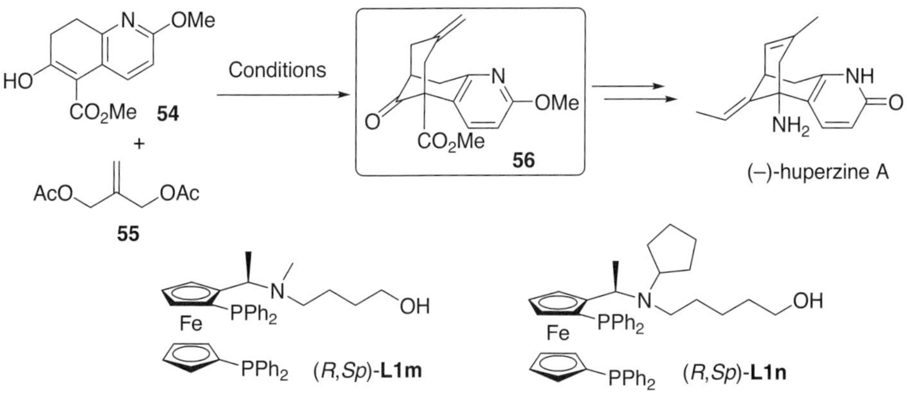

Condition A:

$Pd(OAc)_2$ (2 mol%), (*R,Sp*)-**L1m** (4 mol%)
TMG (220 mol%), DME, -30 to15 °C
6h, 92% yield, 64% *ee*

Condition B:

$[Pd(C_3H_5)Cl]_2$ (5 mol%), (*R,Sp*)-**L1n** (11 mol%)
TMG (250 mol%), toluene, -25 to -20°C
82% yield, 90.3% *ee*

Scheme 8.22 Synthesis of the key intermediate for (−)-huperzine A.

nitrogen and longer alcohol chain was efficient for the allylic alkylation, and 82% yield and 90.3% *ee* of **56** were obtained (Condition B, Scheme 8.22) [34].

By using (*R,Sp*)-**BPPFA** as the chiral ligand, Achiwa and coworkers realized an intramolecular allylic substitution of substrate **57** to give optically active vinylchroman **58** (55% yield, 54.4% *ee*) (Scheme 8.23). The product obtained here is a useful intermediate for tocopherol (vitamin E) analog drugs [35].

8.3
Gold or Silver-Catalyzed Asymmetric Aldol Reactions

8.3.1
Gold-Catalyzed Asymmetric Aldol Reactions

Asymmetric aldol reaction of an isocyanoacetate with aldehydes provides optically active 5-alkyl-4-methoxycarbonyl-2-oxazolines which are useful synthetic intermedi-

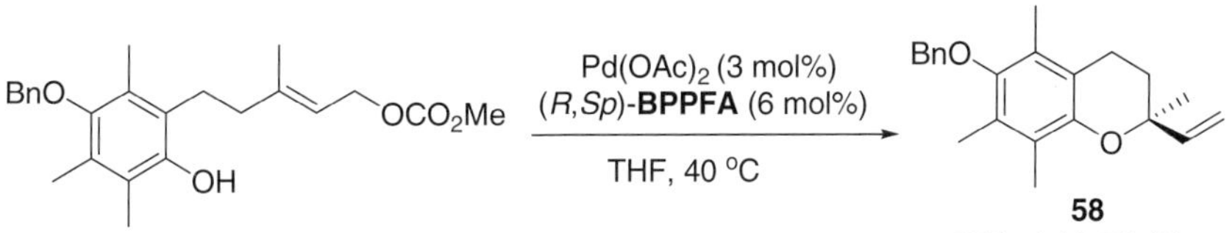

57

58
55% yield, 54.4% *ee*

Scheme 8.23 Synthesis of optically active vinylchroman.

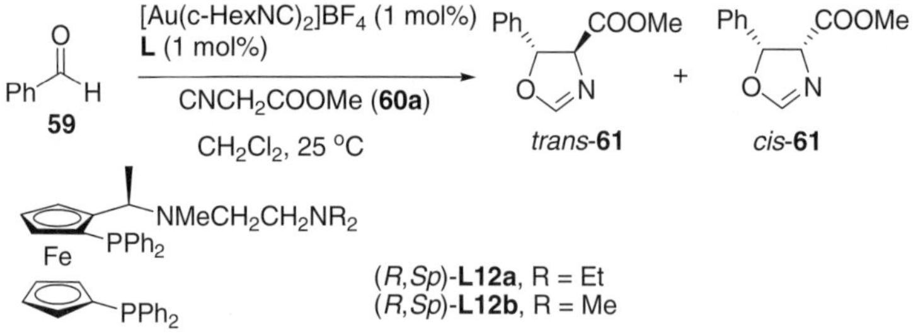

Scheme 8.24

[Au(c-HexNC)$_2$]BF$_4$ (1 mol%), L (1 mol%)

Ph-CHO (**59**) + CNCH$_2$COOMe (**60a**) → trans-**61** + cis-**61**

CH$_2$Cl$_2$, 25 °C

Ligand: Ferrocenyl with PPh$_2$ and NMeCH$_2$CH$_2$NR$_2$ substituents

(R,Sp)-**L12a**, R = Et
(R,Sp)-**L12b**, R = Me

(R,Sp)-**L12a**, 98% yield, trans/cis: 89/11 trans (4S,5R) 96% ee, cis (4R,5R) 49% ee (ref 36)
(R,Sp)-**L12b**, 91% yield, trans/cis: 90/10 trans (4S,5R) 94% ee, cis (4S,5S) 4% ee (ref 36)
(R,Sp)-**L12b**, trans/cis: 90/10 trans (4S,5R) 91% ee, cis (4S,5S) 7% ee (ref 37)
(S,Sp)-**L12b**, trans/cis: 84/16 trans (4R,5S) 41% ee, cis (4S,5S) 20% ee (ref 37)

Scheme 8.24 Gold-catalyzed aldol reaction with **L12a–b**.

ates for optically active β-hydroxyamino acids and their derivatives. In 1986, Ito, Hayashi, and coworkers reported that chiral ferrocenylphosphine-gold(I) complexes were efficient catalysts for this reaction affording 4,5-disubstituted oxazolines with high enantio- and diastereoselectivity (Scheme 8.24) [36].

The effective ligands used here are derivatives of **BPPFA** (**L1** series). The gold complex, generated *in situ* by mixing bis(cyclohexyl isocyanide)gold(I) tetrafluoroborate and (R,Sp)-**L12a–b**, catalyzed the aldol reaction of benzaldehyde and isocyanoacetate efficiently to afford 4-(methoxycarbonyl)-5-phenyl-2-oxazoline **61** with trans/cis ratio around 9/1, and 96% ee and 94% ee for the trans-isomer, respectively. To clarify the role of the planar chirality and central chirality, Pastor and Togni examined the same type of ligands having different planar and central chiralities [37]. The results showed the central chirality played an important role for controlling the configuration of the product. Changing the central chirality from R to S resulted in both a reduction of the ee of the trans-isomer and the formation of the opposite trans-oxazoline enantiomer, which also indicated the matching of two chiralities in ligand (R,Sp)-**L12b**.

Comprehensive subsequent studies by Ito, Hayashi, and their coworkers have improved the performance of the ligands by changing the amine side chain and expanded the substrate scope as well [38].

In addition to the detailed studies into the chiral cooperativity of the chiral ferrocenylamine ligands [39], Togni *et al.* have found the sulfur-containing ferrocenylphosphine ligands **L12c–f** with multi chiral centers to also be efficient for the asymmetric aldol reaction of benzaldehyde **59** with isocyanoacetate **60a** [40].

As shown in Scheme 8.25, the synthesis of the ligands started from (R,Sp)-**BPPFOAc**. Reaction of (R,Sp)-**BPPFOAc** with KSAc in AcOH, followed by treatment with LiAlH$_4$ in Et$_2$O led to the oxygen-sensitive thiol (R,Sp)-**L1o**. Treatment of the thiol with *n*-BuLi in THF and then enantiopure aziridinium iodide **62** introduced the amine side chain easily into the ferrocenyl ligands. The chiral cooperativity between different chiralities was critical to the selectivity of the reaction. These ligands were all efficient for the aldol reaction and good enantio- and diastereoselectivities were obtained.

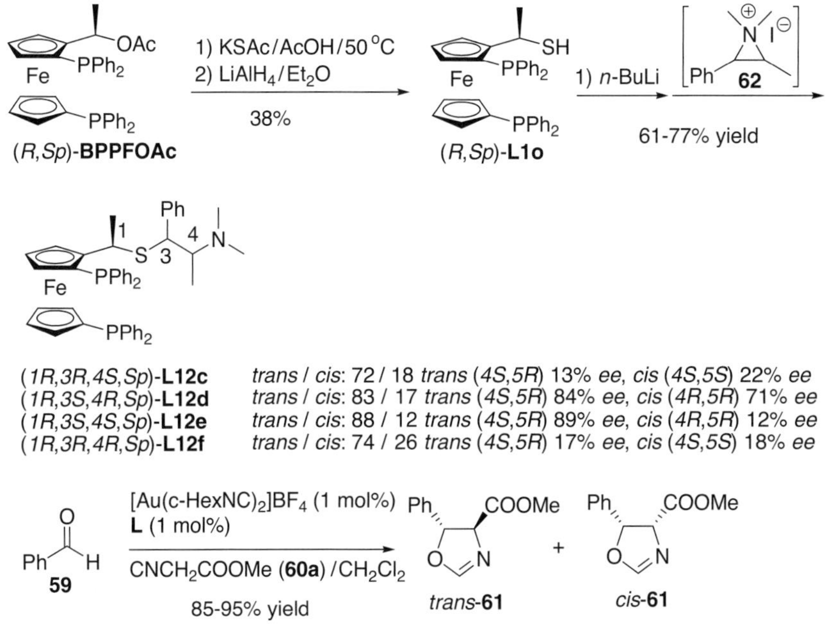

8.3 Gold or Silver-Catalyzed Asymmetric Aldol Reactions

Scheme 8.25 Synthesis of **L12c–f** and their applications in gold-catalyzed aldol reaction.

(1R,3R,4S,Sp)-**L12c**   trans / cis: 72 / 18 trans (4S,5R) 13% ee, cis (4S,5S) 22% ee
(1R,3S,4R,Sp)-**L12d**   trans / cis: 83 / 17 trans (4S,5R) 84% ee, cis (4R,5R) 71% ee
(1R,3S,4S,Sp)-**L12e**   trans / cis: 88 / 12 trans (4S,5R) 89% ee, cis (4R,5R) 12% ee
(1R,3R,4R,Sp)-**L12f**   trans / cis: 74 / 26 trans (4S,5R) 17% ee, cis (4S,5S) 18% ee

Although most of the asymmetric studies focused on the synthesis of optically active 2-oxazolines, Lin and coworkers reported a gold(I)-catalyzed enantioselective synthesis of optically active 2-imidazoline from N-sulfonylimines and isocyanoacetates (Scheme 8.26) [41]. The products are good precursors of α,β-diamino acids.

By utilizing 0.5 mol% of Me$_2$SAuCl and (R,Sp)-**L12g**, various N-sulfonylimines and ethyl isocyanoacetate **60b** underwent the cyclization smoothly to give optically pure 2-imidazolines (**64**). Interestingly, different from the formation of the trans-2-oxazolines during the aldol reaction between aldehydes and isocyanoacetates, the

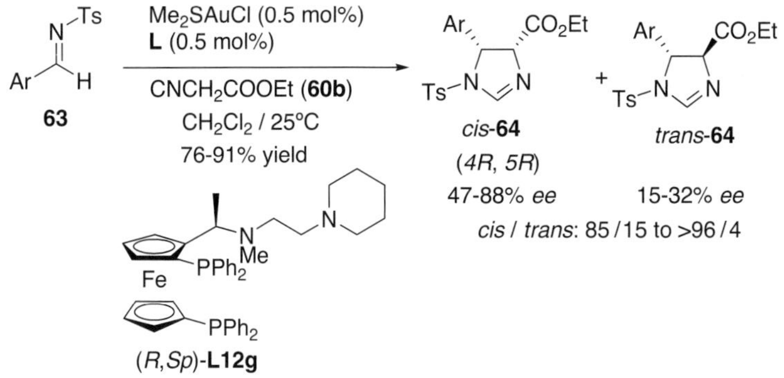

Scheme 8.26 Gold-catalyzed formation of 2-imidazoline from N-sulfonylimines and isocyanoacetates.

2-imidazoline formation from N-sulfonylimines and isocyanoacetates proceeded predominately to afford the cis-products. In general, excellent cis/trans ratios and good to excellent enantioselectivities for the cis-isomer were obtained.

8.3.2
Silver-Catalyzed Asymmetric Aldol Reactions

In 1990, Ito and coworkers reported that the silver complexes with diphenylphosphino ferrocenylamine ligands were also the chiral catalysts of choice for the stereoselective aldol reaction [42].

In the presence of 1 mol% of AgOTf and (R,Sp)-L12g or (R,Sp)-L12h, the asymmetric aldol reaction of tosylmethyl isocyanide and aldehydes gave optically active 5-substituted-4-tosyl-2-oxazolines with up to 86% ee (Scheme 8.27). It should be noted that excellent trans/cis ratios were obtained for the oxazoline products.

Interestingly, a high enantioselectivity, up to 90% could be obtained by slow addition of the isocyanoacetate over a period of 1 h to a solution of aldehyde and the silver(I) catalyst in 1,2-dichloroethane at 30 °C [43].

8.3.3
Applications of Gold or Silver-Catalyzed Asymmetric Aldol Reactions

With the established asymmetric aldol reaction, Togni et al. have used the chiral aldehyde **66** as the substrate to realize the synthesis of (2S,3R,4R,6E)-3-hydroxy-4-methyl-2-(methylamino)oct-6-enoic acid (MeBmt), an unusual amino acid in the immunosuppressive undecapeptide cyclosporine (Scheme 8.28) [44].

With the gold complex of (R,Sp)-L12b, the trans-oxazoline was obtained in 85% de. The obtained oxazoline **67** was treated with trimethyloxonium tetrafluoroborate in dichloromethane at room temperature, followed by work-up conditions with aqueous NaHCO₃ solution, affording **68** in 73% overall yield. Further hydrolysis of **68** gave MeBmt **69** in 70% yield.

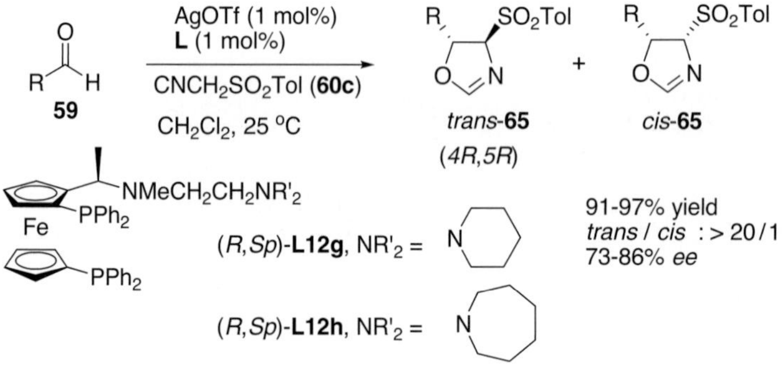

Scheme 8.27 Silver-catalyzed asymmetric aldol reaction of aldehydes with tosylmethyl isocyanoacetates.

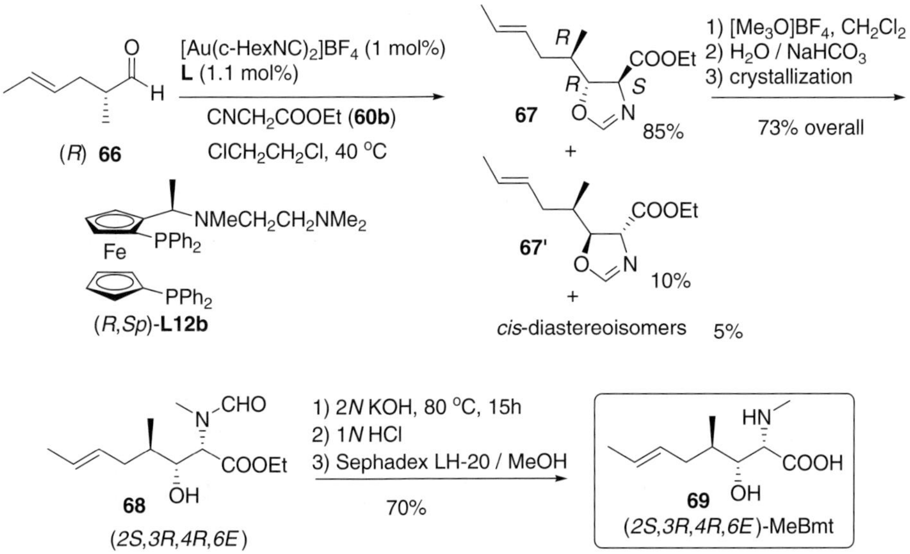

8.3 Gold or Silver-Catalyzed Asymmetric Aldol Reactions

[Scheme 8.28 diagram]

Scheme 8.28 Synthesis of MeBmt.

Asymmetric aldol reaction of methyl α-isocyanoacetate with (E)-2-hexadecenal **70** in the presence of 1 mol% of a chiral (aminoalkyl)ferrocenylphosphine-gold(I) complex gave optically active *trans*-4-(methoxycarbonyl)-5-(E)-1-pentadecenyl)-2-oxazoline (**71**) in 80% yield with 93% ee (Scheme 8.29) [45]. Treatment of **71** with concentrated HCl in methanol followed with reduction of the ester by LiAlH$_4$ led to D-*threo*-sphingosine **73** in 85% yield. The D-*erythro*-sphingosine **74** was readily accessible from the *threo* isomer by inversion of the configuration at the C-3 carbon.

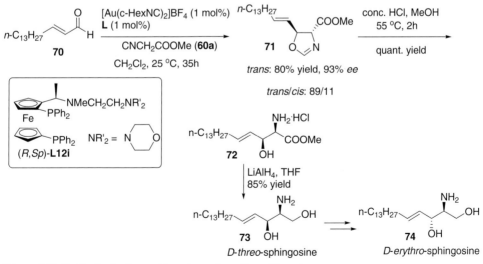

Scheme 8.29 Synthesis of D-*threo*-sphingosine and D-*erythro*-sphingosine.

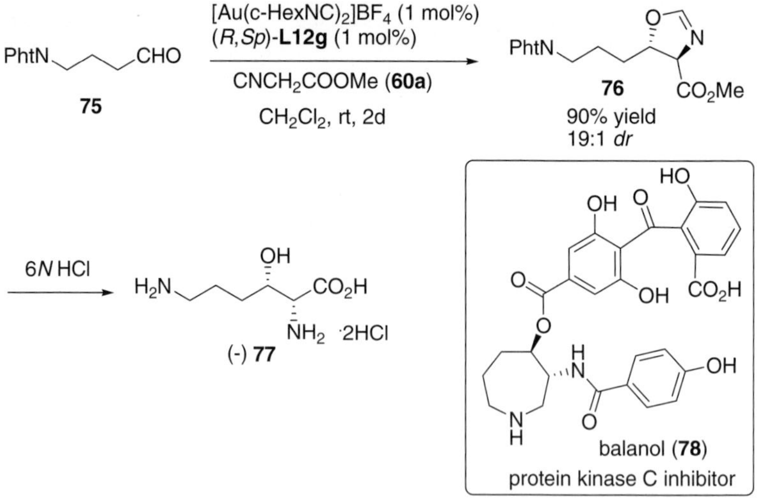

Scheme 8.30 Synthesis of key intermediate for balanol.

Hughes and coworkers utilized Hayashi's chiral ferrocenyl gold catalyst for the asymmetric aldol reaction between 4-phthalimidobutanal and methyl isocyanoacetate to synthesize oxazoline **76** in 90% yield with 19 : 1 dr ratio (Scheme 8.30). A single recrystallization from ethyl acetate/hexanes gave the solely desired isomer in an overall 73% yield. Hydrolysis of **76** with 6 N HCl gave **77**, a key intermediate for protein kinase C inhibitor balanol **78** [46].

In addition, Ito and coworkers have also demonstrated the suitability of α-isocyano Weinreb amide for the aldol reaction with different aldehydes, which allowed facile access to the synthesis of optically active α-amino aldehydes [47].

8.4
Asymmetric Hydrogenation

Catalytic asymmetric hydrosilylation and hydrogenations of prochiral unsaturated compounds, such as olefins, ketones, and imines, have been intensively studied and are considered versatile methods for the synthesis of chiral compounds. Among numerous successful chiral ligands developed in this area, enantiopure 1,1′-unsymmetrically disubstituted ferrocenes have been demonstrated as highly efficient ligands for asymmetric hydrosilylation or hydrogenation of unsaturated compounds.

8.4.1
Rh-Catalyzed Hydrosilylation

In 1974, Kumada and coworkers found that the (S,Rp)-**BPPFA** rhodium complex catalyzed asymmetric hydrosilylation of ketones in good yields but with poor enantioselectivities (25–29% ee) (Scheme 8.31) [48].

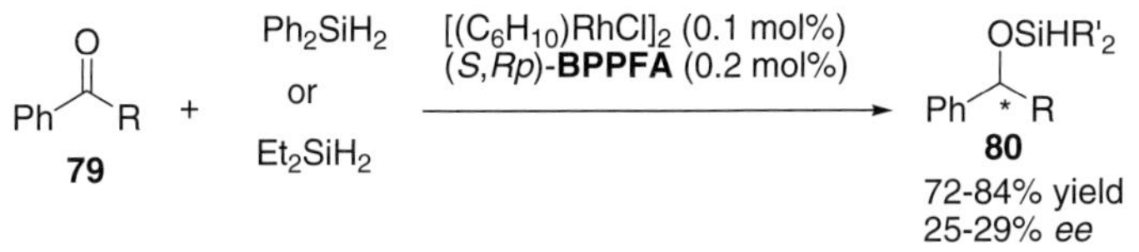

8.4 Asymmetric Hydrogenation

$$\text{Ph}\overset{\text{O}}{\underset{79}{\|}}\text{R} + \begin{array}{c}\text{Ph}_2\text{SiH}_2\\ \text{or}\\ \text{Et}_2\text{SiH}_2\end{array} \xrightarrow[(S,Rp)\text{-BPPFA (0.2 mol\%)}]{[(C_6H_{10})RhCl]_2 \text{ (0.1 mol\%)}} \text{Ph}\overset{\text{OSiHR}'_2}{\underset{\underset{80}{*}}{|}}\text{R}$$

72–84% yield
25–29% ee

Scheme 8.31 Rh-catalyzed hydrosilylation of ketones.

Uemura and coworkers synthesized chiral ferrocenylphosphines (S,Rp)-**L13a-b** possessing a sulfur or selenium moiety on the side chain of the ferrocene, which were used for rhodium-catalyzed asymmetric hydrosilylation of acetophenone (Scheme 8.32). The reaction proceeded smoothly in THF at 0 °C to give 1-phenylethanol, but unfortunately low enantioselectivity was obtained [49].

BPPFA was also used in the platinum and copper-catalyzed asymmetric hydrosilylation of acetophenone, the enantioselectivity was, however, unsatisfactory [50].

8.4.2
Rh, Ir, Ru-Catalyzed Hydrogenation

In 1976, Kumada and coworkers carried out the asymmetric hydrogenation of α-acetamidoacrylic acids **82** in the presence of [Rh(1,5-hexadiene)Cl]$_2$ and (S,Rp)-**BPPFA** in a 1 : 2.4 ratio in different solvents (Scheme 8.33). High enantioselectivities (up to 93%) were obtained [51].

In 1976, the same group used (R,Sp)-**BPPFOH** in the rhodium-catalyzed asymmetric hydrogenation of prochiral carbonyl compounds, and the corresponding alcohol was obtained with up to 83% ee (Scheme 8.34) [52].

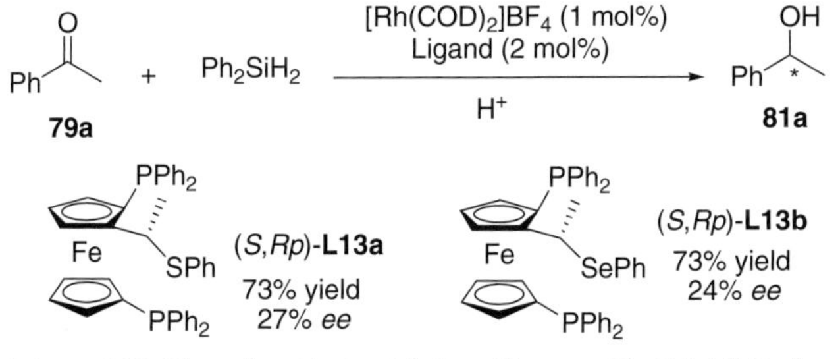

$$\text{Ph}\overset{\text{O}}{\underset{79a}{\|}}\text{Me} + \text{Ph}_2\text{SiH}_2 \xrightarrow[\text{H}^+]{\substack{[\text{Rh(COD)}_2]\text{BF}_4 \text{ (1 mol\%)} \\ \text{Ligand (2 mol\%)}}} \text{Ph}\overset{\text{OH}}{\underset{\underset{81a}{*}}{|}}\text{Me}$$

(S,Rp)-**L13a**
73% yield
27% ee

(S,Rp)-**L13b**
73% yield
24% ee

Scheme 8.32 Rh-catalyzed hydrosilylation of ketone with (S,Rp)-**L13a–b**.

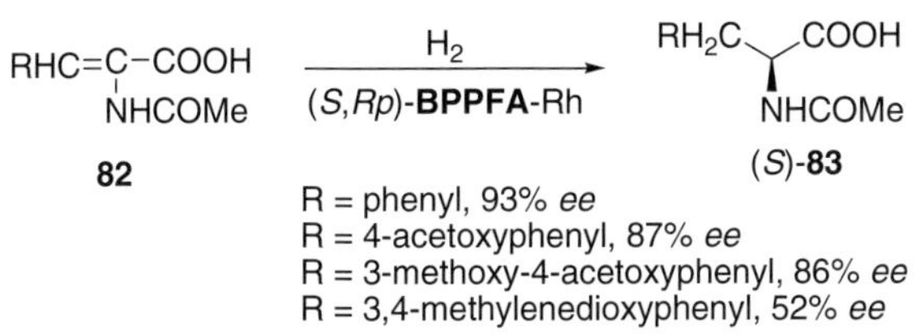

$$\underset{82}{\text{RHC=C-COOH} \atop \text{NHCOMe}} \xrightarrow[(S,Rp)\text{-}\textbf{BPPFA-Rh}]{\text{H}_2} \underset{(S)\text{-}83}{\text{RH}_2\text{C}\diagdown\text{COOH} \atop \text{NHCOMe}}$$

R = phenyl, 93% ee
R = 4-acetoxyphenyl, 87% ee
R = 3-methoxy-4-acetoxyphenyl, 86% ee
R = 3,4-methylenedioxyphenyl, 52% ee

Scheme 8.33 Rh-catalyzed asymmetric hydrogenation of α-acetamidoacrylic acids (**82**).

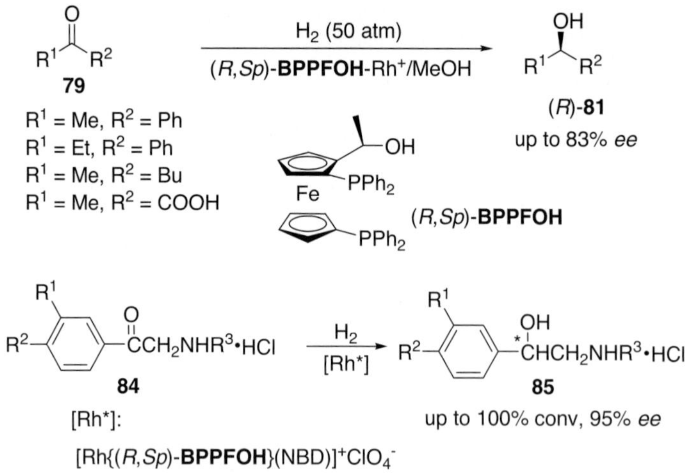

8 Unsymmetrical 1,1′-Bidentate Ferrocenyl Ligands

Scheme 8.34 Rh-catalyzed asymmetric hydrogenation of ketones.

$$R^1\text{COR}^2 \; (79) \xrightarrow[(R,Sp)\text{-BPPFOH-Rh}^+/\text{MeOH}]{H_2 \,(50\text{ atm})} R^1\text{CH(OH)}R^2 \;\; (R)\text{-81, up to 83\% ee}$$

R^1 = Me, R^2 = Ph
R^1 = Et, R^2 = Ph
R^1 = Me, R^2 = Bu
R^1 = Me, R^2 = COOH

(R,Sp)-BPPFOH: ferrocene with CH(OH)(Me) and two PPh$_2$ groups

Substrate 84: $R^2\text{-C}_6H_3(R^1)\text{-COCH}_2\text{NHR}^3\cdot\text{HCl}$ → 85: $R^2\text{-C}_6H_3(R^1)\text{-CH(OH)CH}_2\text{NHR}^3\cdot\text{HCl}$
up to 100% conv, 95% ee

[Rh*]: [Rh{(R,Sp)-BPPFOH}(NBD)]$^+$ClO$_4^-$

The **BPPFOH**-rhodium complex was also reported to catalyze the asymmetric hydrogenation of aminomethyl aryl ketones to give the corresponding 2-amino-1-arylethanol with up to 95% ee (Scheme 8.34) [53]. In 1987, Hayashi *et al.* reported the asymmetric hydrogenation of trisubstituted acrylic acid using a ferrocenylphosphine-rhodium complex (Scheme 8.35) [54]. The above asymmetric hydrogenation was found a useful application in the asymmetric synthesis of carboxylic acid

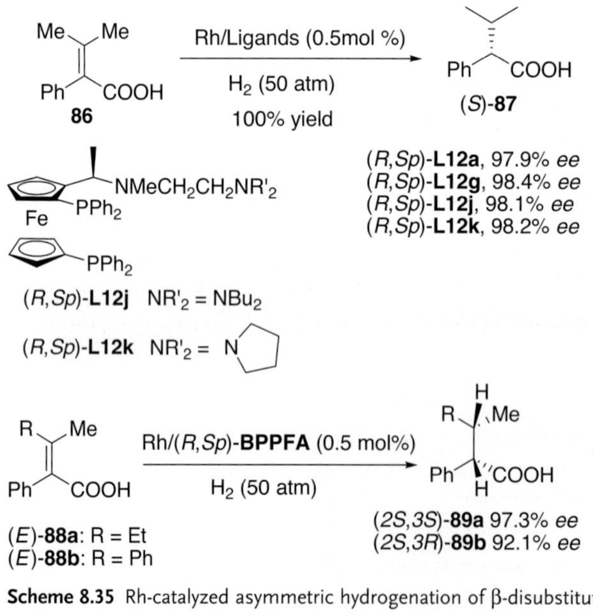

Substrate 86: Me$_2$C=C(Ph)COOH $\xrightarrow[H_2 (50\text{ atm}), 100\% \text{ yield}]{\text{Rh/Ligands (0.5 mol \%)}}$ (S)-87: Ph-CH(iPr)-COOH

(R,Sp)-L12a, 97.9% ee
(R,Sp)-L12g, 98.4% ee
(R,Sp)-L12j, 98.1% ee
(R,Sp)-L12k, 98.2% ee

Ligand: ferrocene-CH(NMeCH$_2$CH$_2$NR′$_2$)-PPh$_2$ / PPh$_2$

(R,Sp)-L12j NR′$_2$ = NBu$_2$
(R,Sp)-L12k NR′$_2$ = N(pyrrolidine)

(E)-88a: R = Et; (E)-88b: R = Ph
RC(Me)=C(Ph)COOH $\xrightarrow[H_2 (50\text{ atm})]{\text{Rh/}(R,Sp)\text{-BPPFA (0.5 mol\%)}}$ Ph-CH(Me)-CH(R)-COOH

(2S,3S)-89a 97.3% ee
(2S,3R)-89b 92.1% ee

Scheme 8.35 Rh-catalyzed asymmetric hydrogenation of β-disubstituted α-phenylacrylic acids.

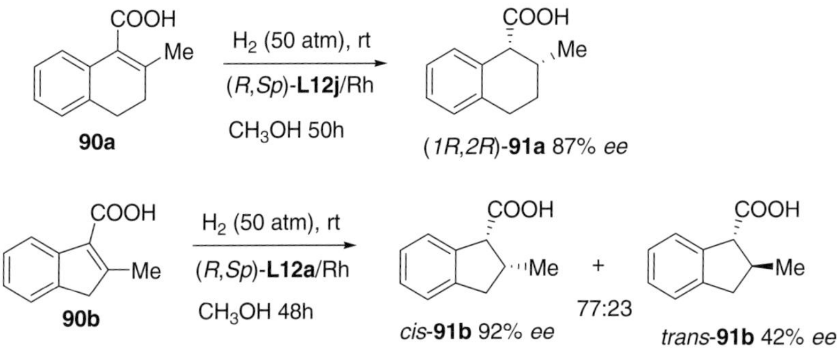

Scheme 8.36 Rh-catalyzed asymmetric hydrogenation of cyclic tetrasubstituted olefins.

containing two vicinal chiral carbon centers [55]. Cyclic tetrasubstituted olefin underwent the asymmetric hydrogenation with the rhodium catalysts to give corresponding products (Scheme 8.36).

Landis *et al.* carried out the asymmetric hydrogenation of methyl (Z)-α-acetamidocinnamate with the boron-containing ferrocenyl ligand/Rh complexes (Scheme 8.37). Relatively low *ee*s were obtained (<55%) [56].

In 1996, Shimazu and coworkers reported the asymmetric hydrogenation of itaconic esters **92** by rhodium(I)-phosphine complexes supported on Hectorite (Scheme 8.38) [57].

Reetz *et al.* developed a completely different class of ferrocene-derived diphosphines having only planar chirality. One of the compounds (−)-**L15** fell into the

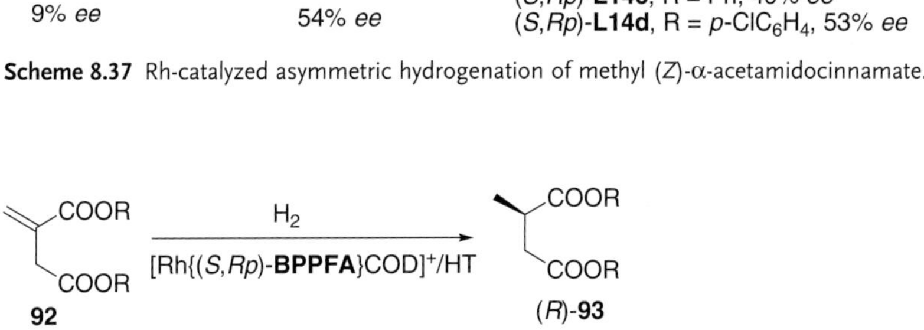

Scheme 8.37 Rh-catalyzed asymmetric hydrogenation of methyl (Z)-α-acetamidocinnamate.

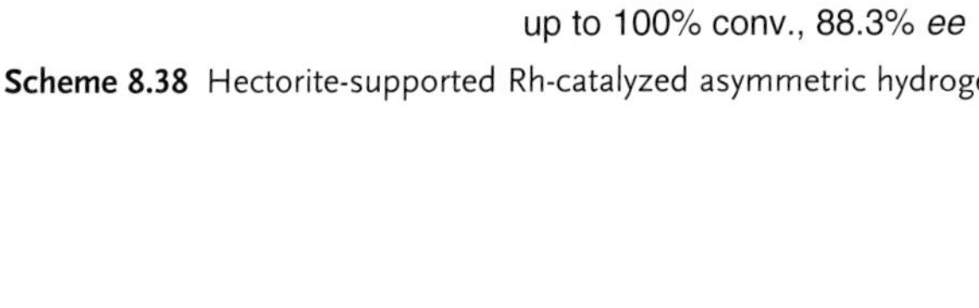

Scheme 8.38 Hectorite-supported Rh-catalyzed asymmetric hydrogenation of itaconic esters.

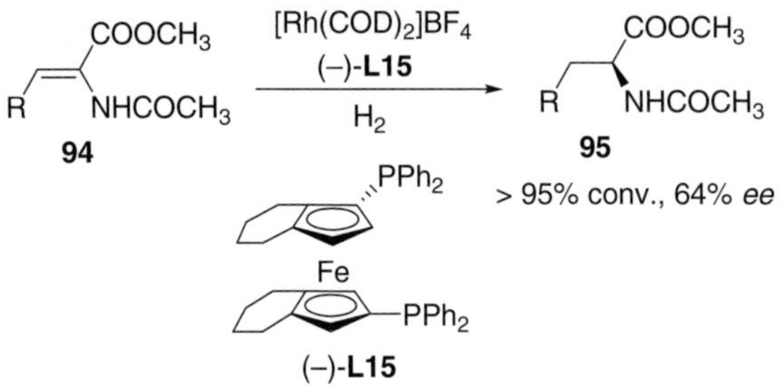

238 | *8 Unsymmetrical 1,1'-Bidentate Ferrocenyl Ligands*

Scheme 8.39 Rh-catalyzed asymmetric hydrogenation of methyl α-acetamidoacrylate derivatives.

category of 1,1'-unsymmetrical ferrocenyl ligand, and it was used in Rh-catalyzed asymmetric hydrogenation of **94** (Scheme 8.39) [58].

In 2005, Kim and coworkers synthesized a series of new chiral (iminophosphoranyl)ferrocenes. The new ligands exhibited exceptionally high enantioselectivity (up to 99%) and catalytic activity in the Rh-catalyzed asymmetric hydrogenation of (*E*)-2-methylcinnamic acid, (*Z*)-2-acetamidocinnamate and (*Z*)-2-acetamidoacrylate (Scheme 8.40) [59].

In 2006, Kim used the same ligand (*S*,*Rp*)-**L11b** for the rhodium or iridium-catalyzed enantioselective hydrogenation of unfunctionalized olefins (Scheme 8.41) [60]. In both cases, high conversion and excellent *ee*s were obtained. The alkene substrate scope covered the di, tri, and tetra-substituted alkenes.

8.4.3
Rh-Catalyzed Hydroboration

Although there are examples of the 1,1'-unsymmetrically disubstituted ferrocenyl ligands in the Rh-catalyzed hydroboration of olefins, none of them gave impressive results [58, 61].

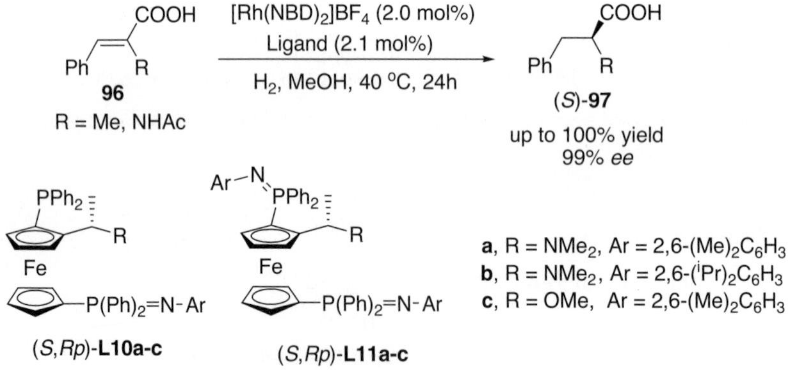

Scheme 8.40 Rh-catalyzed asymmetric hydrogenation with **L10–11**.

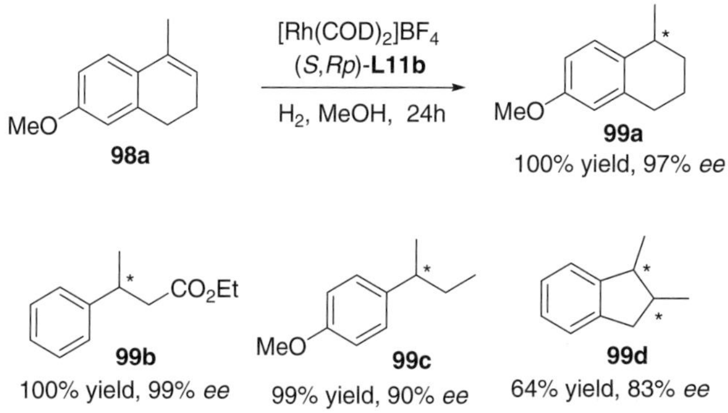

Scheme 8.41 Rh and Ir-catalyzed asymmetric hydrogenation of tri- or tetra-substituted alkenes.

8.5
Asymmetric Cross-Coupling Reaction

The transition-metal catalyzed cross-coupling reaction is one of the most important reactions to construct carbon–carbon bonds. Nickel and palladium complexes with enantiopure 1,1′-unsymmetrically disubstituted ferrocenes were found to be efficient catalysts for asymmetric cross-coupling reaction.

8.5.1
Nickel-Catalyzed Kumada Coupling Reaction

Kumada *et al.* reported the asymmetric Grignard cross-coupling reaction of 4-bromo-1-butene with phenylmagnesium bromide via isomerization catalyzed by chiral phosphine-nickel complex. (*R,Sp*)-**BPPFA** was used for the nickel-catalyzed cross-coupling reaction in 58% yield with 33.8% *ee* (Scheme 8.42) [62].

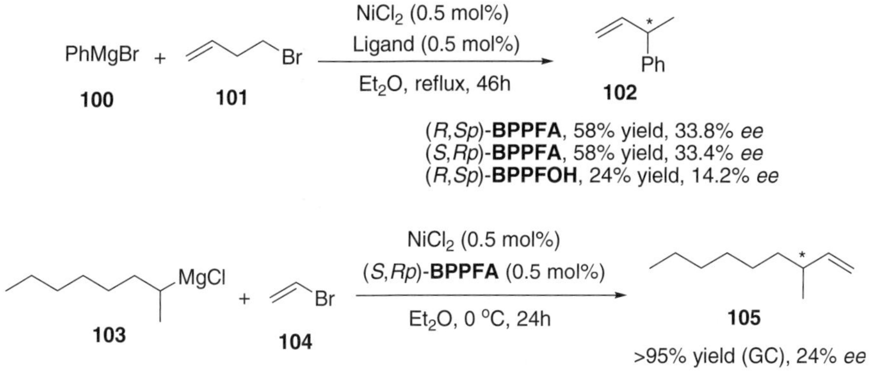

Scheme 8.42 Ni-catalyzed asymmetric Grignard cross-coupling reactions.

8.5.2
Palladium-Catalyzed Cross-Coupling Reaction

The asymmetric Grignard cross-coupling reaction of 2-octylmagnesium chloride with vinyl bromide in the presence of NiCl$_2$ and (S,Rp)-**BPPFA** gave the corresponding product with moderate optical purity (24% ee) (Scheme 8.42) [63].

Asymmetric Grignard cross-coupling reaction of 1-phenylethylmagnesium chloride with vinyl bromide in the presence of (S,Rp)-**BPPFA**/palladium complex gave the corresponding product with 61% ee (Scheme 8.43, Eq. (a)) [63].

Asymmetric Grignard cross-coupling reaction between sec-butylmagnesium bromide and bromobenzene in the presence of the (R,Sp)-**BPPFA** palladium complex gave product in 77% yield with 15.6% ee (Scheme 8.43, Eq. (b)) [64].

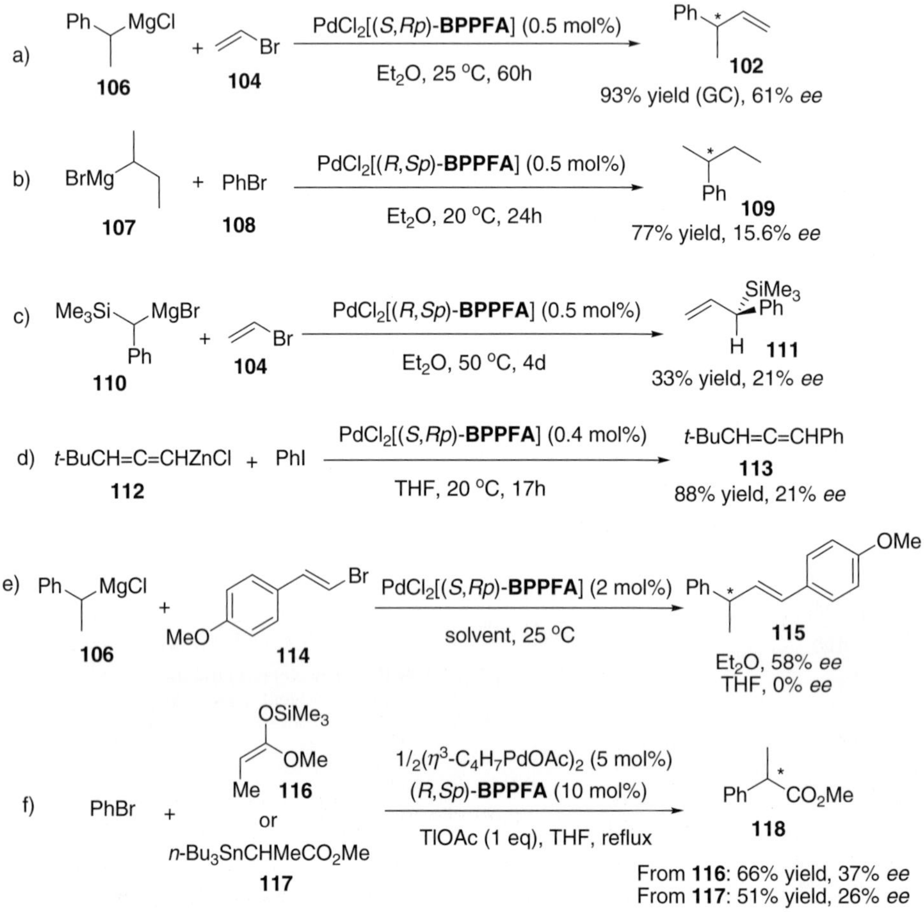

Scheme 8.43 Pd-catalyzed asymmetric cross-coupling reactions.

Asymmetric Grignard cross-coupling reaction between α-(trimethylsilyl)benzyl-Grignard reagent and vinyl bromide in the presence of the (R,Sp)-**BPPFA** palladium complex gave optically active allylsilane in 33% yield with 21% *ee* (Scheme 8.43, Eq. (c) [65].

4,4-Dimethyl-1-phenylpenta-1,2-diene was prepared by the (S,Rp)-**BPPFA** palladium complex-catalyzed asymmetric cross-coupling reaction between 4,4-dimethylpenta-1,2-dienylzinc chloride and iodobenzene in 88% yield with 21% *ee* (Scheme 8.43, Eq. (d) [66].

Baker *et al.* reported that Et$_2$O was the effective solvent in an asymmetric Grignard cross-coupling reaction. They found that the cross-coupling reaction of 1-phenylethylmagnesium chloride with (E)-1-(2-bromovinyl)-4-methoxybenzene gave 58% *ee* in Et$_2$O, but 0% *ee* in THF (Scheme 8.43, Eq. (e)) [67].

In the presence of the (R,Sp)-**BPPFA** palladium complex, asymmetric cross-coupling of bromobenzene and silyl ketene acetal or methyl 2-tributylstannylpropanoate gave methyl 2-phenylpropanoate with a moderate enantiomeric excess (Scheme 8.43, Eq. (f)) [68].

8.5.3
Palladium-Catalyzed Suzuki–Miyaura Reaction and α-Arylation of Amides

As shown in Scheme 8.44, by treatment of triflate or bromide **119** with Pd$_2$(dba)$_3$·CHCl$_3$, (S,Rp)-**BPPFOAc** and K$_2$CO$_3$, in THF at 40 °C, Cho and Shibasaki realized the asymmetric intramolecular Suzuki–Miyaura reaction, affording cyclopentane derivatives in moderate enantiomeric excess (Scheme 8.44) [69].

Chiral binaphthalenes were prepared using (S,Rp)-**BPPFA** as a ligand by asymmetric Suzuki cross-coupling reaction (Scheme 8.44) [70].

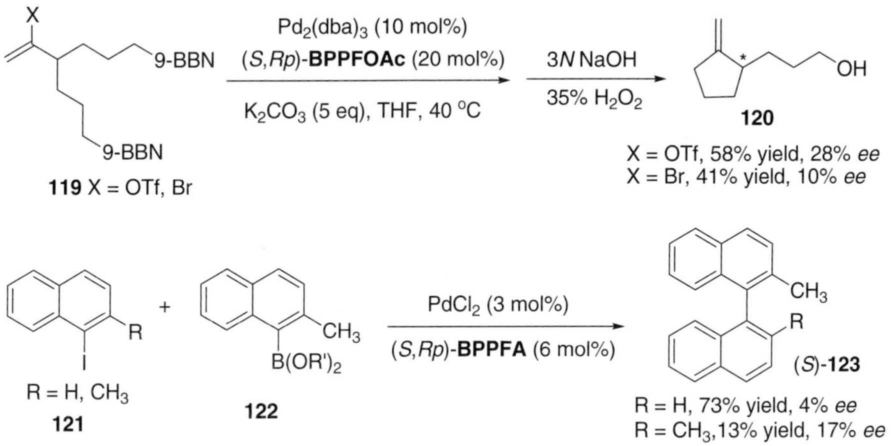

Scheme 8.44 Pd-catalyzed asymmetric cross-coupling reaction of boronic reagents.

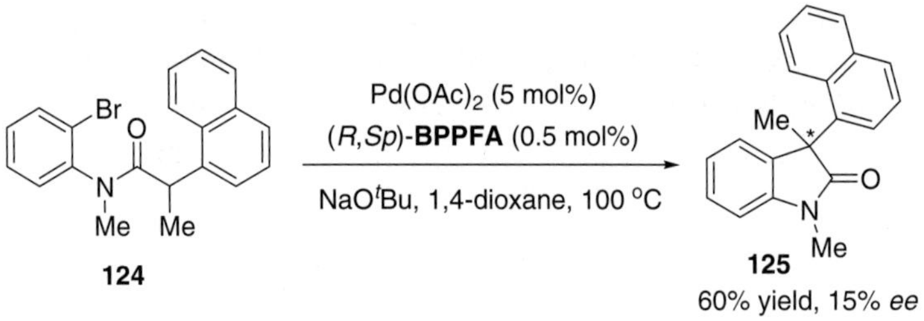

Scheme 8.45 Pd-catalyzed asymmetric α-arylation of amides.

Lee and Hartwig reported the α-arylation of an amide to synthesize substituted oxindole (Scheme 8.45). Utilization of the (R,Sp)-**BPPFA** ligand gave α,α-disubstituted oxindole in 60% yield with 15% ee [71].

Ohno et al. reported that the asymmetric allylic silylation of crotyl chloride gave 81% ee in the presence of a new chiral ligand (R,Sp)-**Et-BPPFA** (Scheme 8.46). They also found that the cycloaddition of 2-butenylene dicarbonate with methyl acetylacetate in the presence of (R,Sp)-**Et-BPPFA** gave vinyldihydrofuran in 88% yield with 77% ee [72].

8.6
Asymmetric Heck Reaction

8.6.1
Intramolecular Heck Reaction

Indolizidine derivative **131** has been synthesized with up to 86% ee by an asymmetric intramolecular Heck reaction from prochiral alkenyl iodide, where (R,Sp)-**BPPFOH** (**L1l**) was found to be an efficient ligand (Scheme 8.47) [73, 74].

Overman and coworkers found that the asymmetric intramolecular Heck reaction of **132** gave the enantioenriched spirocyclic product **133** (Scheme 8.47). With Pd(OAc)$_2$/(R,Sp)-**BPPFA** (**L1a**), the product was obtained in 39% yield with 26% ee, and it should be noted that (R)-**BINAP** gave the highest ee, up to 70% [75].

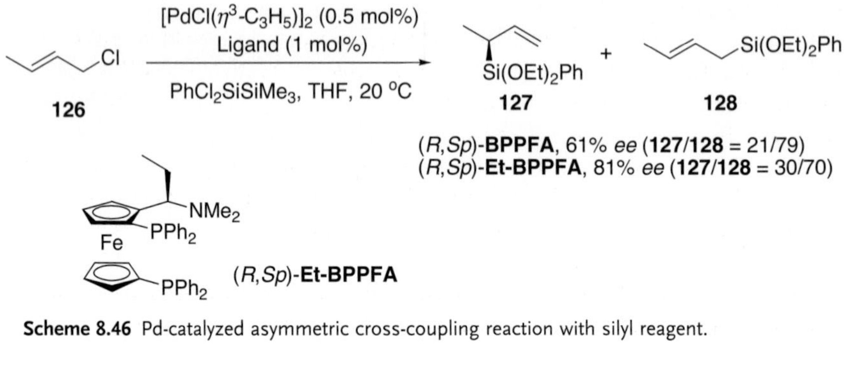

Scheme 8.46 Pd-catalyzed asymmetric cross-coupling reaction with silyl reagent.

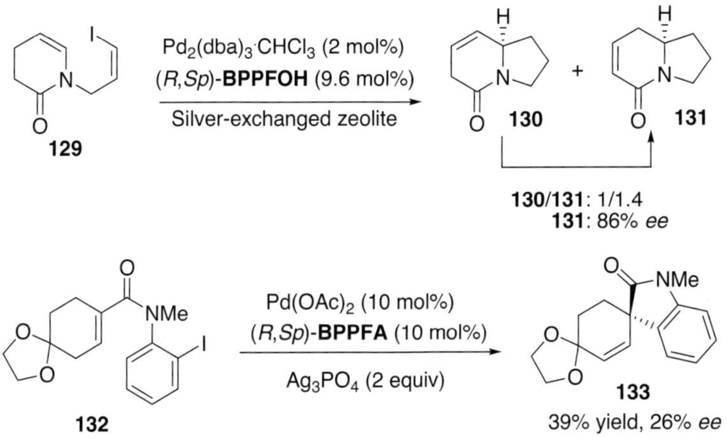

Scheme 8.47 Pd-catalyzed asymmetric intramolecular Heck reaction.

8.6.2
Intermolecular Heck Reaction

A series of planar chiral phosphine-oxazoline ferrocene derivatives was found to be efficient ligands for the palladium-catalyzed asymmetric Heck reaction of 2,3-dihydrofuran with aryl triflate, in which up to 92% *ee* was obtained (Table 8.3) [76]. Interestingly, the enantioselectivity and the absolute configuration of the product were controllable by changing the size of the planar chiral group and/or the configuration of planar chirality.

A series of planar chiral diphosphine-oxazoline ferrocene ligands was found to be efficient for the palladium-catalyzed asymmetric Heck reaction of 2,3-dihydrofuran with aryl triflate, in which up to 98% *ee* was obtained (Table 8.4) [77]. Another interesting feature about these ligands is the easily tunable electronic properties of the two diarylphosphino groups. Importantly, the electronic properties of the phosphines and the palladium precursors displayed significant impact on the selectivity of the reaction. In order to obtain high selectivity to **136a**, the re-insertion of the palladium catalyst into **136a** should be inhibited. Therefore, electron-rich ligands enhancing the electron density of the palladium center would lead to the re-insertion to give isomer **137a**, whereas electron-deficient ligands would afford a high ratio of **136a**. The experimental data are in good agreement with the hypothesis. With the ligands bearing the electron-deficient phosphine, (*S*,*Sp*)-**L7a** or **L7e**, high selectivity of **136a** is obtained. On the other hand, the relatively more isomerized product **137a** is formed with electron-rich phosphine ligands. It should be noted that Pd(OAc)$_2$ is more favorable for high selectivity of **136a** than Pd(dba)$_2$ since the acetate anion in the former case would attack the cationic palldium/**136a** complex to help the dissociation of the product.

Kang *et al*. reported the same asymmetric intermolecular Heck reaction by using a new ligand, (*Sp*,*Sp*)-**L19**. Although both relatively low reactivity and poor enantios-

Table 8.3 Palladium-catalyzed asymmetric intermolecular Heck reaction with **L3a**, and **L16–18**.

(S)-**L3a**, R = i-Pr
(S)-**L16a**, R = Bn
(S)-**L16b**, R = t-Bu
(R)-**L16c**, R = Ph

(S,Sp)-**L17a**, R¹ = TMS, R² = H
(S,Sp)-**L17b**, R¹ = H, R² = Me
(S,Rp)-**L17c**, R¹ = H, R² = TMS

(R,Rp)-**L18a**, R¹ = H, R² = TMS
(R,Rp)-**L18b**, R¹ = Me, R² = H

Entry	Ligand	Ratio (136a : 137a)	Yield (%)	ee (%)
1	(S)-**L3a**	n.d.	46	68 (R)
2	(S)-**L16a**	n.d.	80	77 (R)
3	(S)-**L16b**	n.d.	25	64 (R)
4	(S)-**L16c**	n.d.	79	42 (S)
5	(S,Sp)-**L17a**	n.d.	72	84 (S)
6	(S,Sp)-**L17b**	64 : 1	79	89 (R)
7	(S,Rp)-**L17c**	n.d.	75	92 (R)
8	(R,Sp)-**L18a**	n.d.	75	80 (R)
9	(R,Sp)-**L18b**	32 : 1	85	88 (S)

electivity were obtained, a high regioselectivity favoring the 2,5-dihydrofuran derivative over the regioisomeric 2,3-dihydrofuran derivative was observed in their study (Scheme 8.48, Eq. (a)) [78].

tert-Butyloxazoline diphosphine ligand was effectively applied in the asymmetric arylation of *N*-methoxycarbonyl-2-pyrroline, and high regio- and enantioselectivities were realized (Scheme 8.48, Eq. (b)) [79].

8.7
Miscellaneous

8.7.1
Addition of Zinc Reagent to Aldehydes

Inspired by the success of Bolm's 1,2-*N,O*-ferrocenyl ligands in the asymmetric diethylzinc addition reaction (see Chapter 6), Hou, Dai, and coworkers synthesized a series of 1,1'-*N,O*-ferrocenyl ligands and investigated their catalytic performance in the asymmetric addition of diethylzinc to aldehydes (Scheme 8.49) [80].

Table 8.4 Palladium-catalyzed asymmetric intermolecular Heck reaction with **L7**.

$$134 + PhOTf \xrightarrow[i\text{-}Pr_2NEt, 60 °C, 36h]{\text{Pd(OAc)}_2 \text{ or Pd(dba)}_2 (1.5 \text{ mol\%}) \\ \text{Ligand (3 mol\%)}} 136a + 137a$$

(S,Sp)-**L7a**, Ar = 3,5-(CF$_3$)$_2$C$_6$H$_3$
(S,Sp)-**L7b**, Ar = 3,5-(CH$_3$)$_2$C$_6$H$_3$
(S,Sp)-**L7c**, Ar = 4-CF$_3$C$_6$H$_4$
(S,Sp)-**L7d**, Ar = 4-MeOC$_6$H$_4$

(S,Sp)-**L7e**, Ar = 3,5-(CF$_3$)$_2$C$_6$H$_3$
(S,Sp)-**L7f**, Ar = 3,5-(CH$_3$)$_2$C$_6$H$_3$
(S,Sp)-**L7g**, Ar = 4-CF$_3$C$_6$H$_4$
(S,Sp)-**L7h**, Ar = 4-MeOC$_6$H$_4$
(S,Sp)-**L7j**, Ar = Ph

Entry	Ligand	Palladium	Solvent	Conv. (%)	Ratio of 136a/137a	ee (%) 136a(R)	ee (%) 137a(S)
1	L7j	Pd(OAc)$_2$	toluene	85	95:5	97	26
2	L7j	Pd(dba)$_2$	THF	98	28:72	97	29
3	L7a	Pd(OAc)$_2$	toluene	72	94:6	75	13
4	L7b	Pd(OAc)$_2$	toluene	77	90:10	98	nd
5	L7c	Pd(OAc)$_2$	toluene	98	95:5	97	29
6	L7d	Pd(OAc)$_2$	toluene	67	83:17	98	29
7	L7e	Pd(OAc)$_2$	toluene	67	99:1	92	nd
8	L7f	Pd(OAc)$_2$	toluene	73	93:7	96	nd
9	L7g	Pd(OAc)$_2$	toluene	80	64:36	83	29
10	L7h	Pd(OAc)$_2$	toluene	65	14:86	86	27
11	L7h	Pd(dba)$_2$	THF	100	15:85	94	48
12	L7h	Pd(dba)$_2$	(CH$_2$Cl)$_2$	68	8:92	nd	19

The ligands were synthesized from 1,1′-bromo-ferrocenyloxazoline by treatment with *n*-BuLi in THF at −78 °C for 30 min followed by trapping with benzophenone. After optimization of the reaction conditions, the (S)-**L21a** was found to be optimal for diethylzinc addition to aldehydes, affording optically active secondary alcohols with up to 91% *ee*. The aldehyde substrate scope included both aromatic and aliphatic aldehydes although the latter displayed relatively low enantio-selectivities.

In addition, these ferrocenyl alcohols are also capable of catalyzing the addition of an alkynylzinc reagent to aldehydes [81]. In the presence of 10 mol% of (R)-**L21b**, the reaction of alkynylzinc reagents, generated from phenylacetylene and diethyl-zinc, with aldehydes to give chiral propargyl alcohols **143** with up to 93% *ee*. The aliphatic aldehydes were suitable for the reaction with relatively low enantio-selectivities.

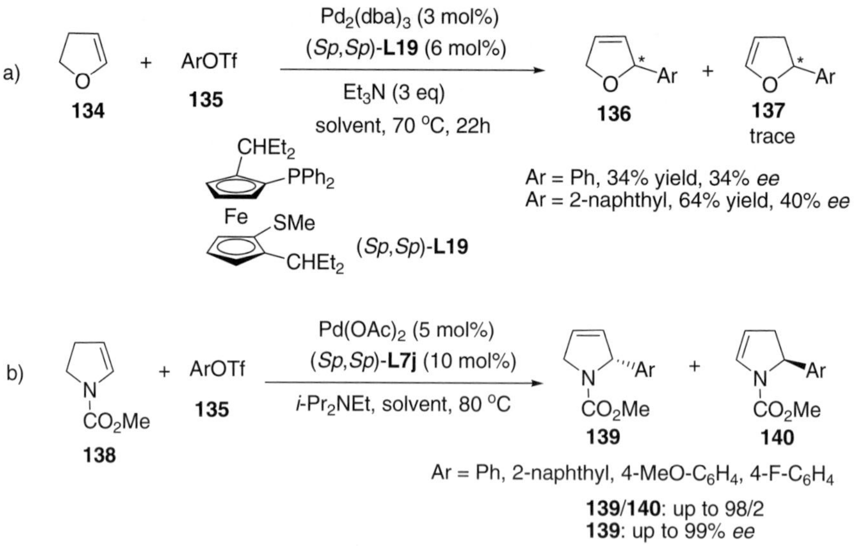

246 | 8 Unsymmetrical 1,1'-Bidentate Ferrocenyl Ligands

a) 134 + ArOTf 135 → [Pd$_2$(dba)$_3$ (3 mol%), (Sp,Sp)-L19 (6 mol%), Et$_3$N (3 eq), solvent, 70 °C, 22h] → 136 + 137 (trace)

(Sp,Sp)-L19: Fe with CHEt$_2$, PPh$_2$, SMe, CHEt$_2$ substituents

Ar = Ph, 34% yield, 34% ee
Ar = 2-naphthyl, 64% yield, 40% ee

b) 138 + ArOTf 135 → [Pd(OAc)$_2$ (5 mol%), (Sp,Sp)-L7j (10 mol%), i-Pr$_2$NEt, solvent, 80 °C] → 139 + 140

Ar = Ph, 2-naphthyl, 4-MeO-C$_6$H$_4$, 4-F-C$_6$H$_4$

139/140: up to 98/2
139: up to 99% ee

Scheme 8.48 Pd-catalyzed asymmetric Heck reactions.

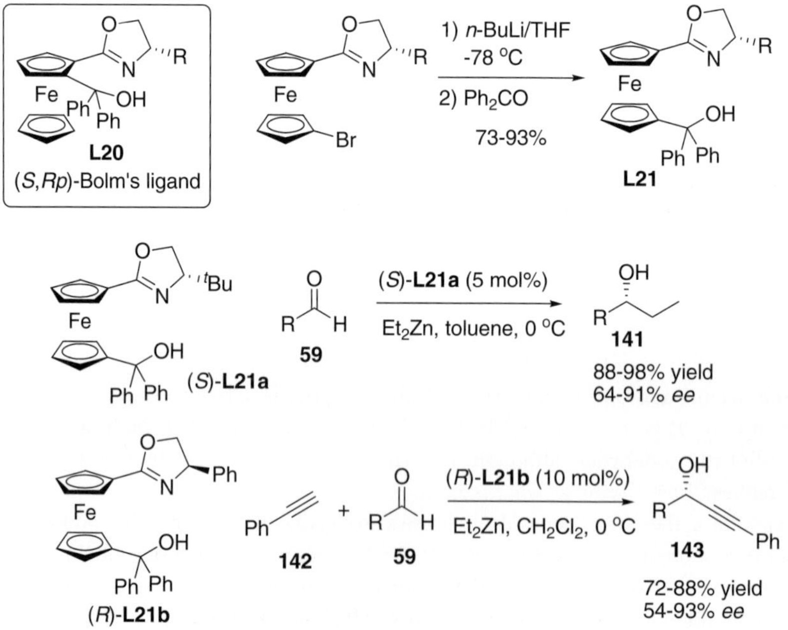

L20 (S,Rp)-Bolm's ligand

L21: via 1) n-BuLi/THF, -78 °C; 2) Ph$_2$CO, 73-93%

(S)-L21a + RCHO (59) → 141, Et$_2$Zn, toluene, 0 °C
88-98% yield
64-91% ee

(R)-L21b (10 mol%) + Ph-alkyne 142 + RCHO 59 → 143, Et$_2$Zn, CH$_2$Cl$_2$, 0 °C
72-88% yield
54-93% ee

Scheme 8.49 Asymmetric diethylzinc and alkynylzinc additions to aldehydes.

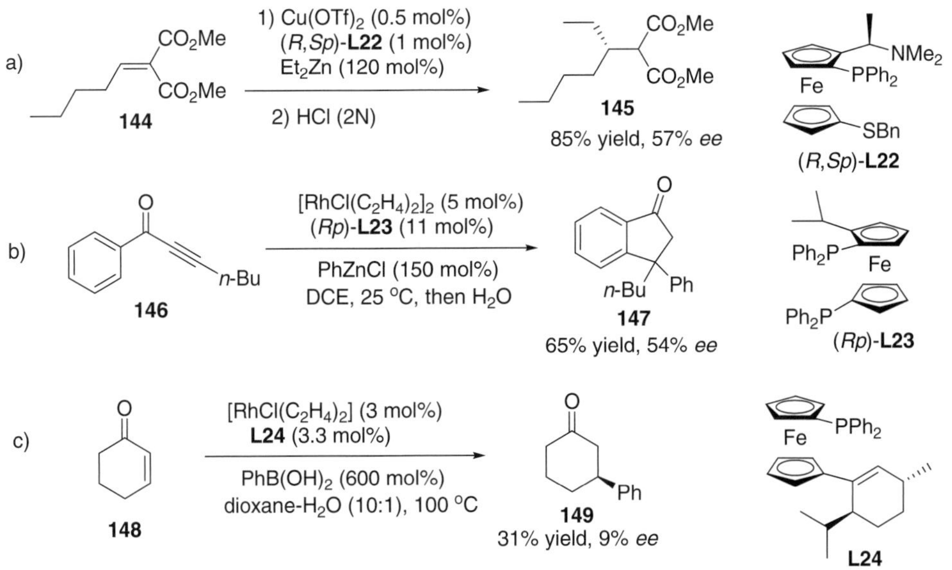

Scheme 8.50 Asymmetric conjugate additions.

8.7.2
Conjugate Addition

With (R,Sp)-L22/Cu(OTf)$_2$ as the catalyst, diethyl zinc underwent conjugate addition to alkylidiene malonates, the addition product was obtained in 85% yield with 57% ee (Scheme 8.50, Eq. (a)) [82].

Shintani and Hayashi recently reported the Rh-catalyzed addition of arylzinc reagents to aryl alkynyl ketones to give β,β-disubstituted indanones in good yield (Scheme 8.50, Eq. (b)) [83]. Interestingly, when chiral DPPF derivative (Rp)-L23 was used as the chiral ligand, the asymmetric version of this reaction was realized, affording the desired product in 65% yield with 54% ee. A novel 1,1′-chiral phosphine-olefin ferrocene ligand was recently synthesized by Bolm et al., the application of L24 in the Rh-catalyzed asymmetric addition reaction of phenylboronic acid to cyclohexenone was tested but with low yield and enantioselectivity (Scheme 8.50, Eq. (c)) [84].

8.7.3
Rh-Catalyzed Ring Opening Reaction of Oxabenzonorbornadiene

Lautens and coworkers reported the use of (R,Sp)-BPPFA for the Rh-catalyzed asymmetric alcoholysis and aminolysis of oxabenzonorbornadiene. With 2.5 mol % of [Rh(COD)Cl]$_2$ and 5 mol% of (R,Sp)- BPPFA, ring opening of **150** with oxygen

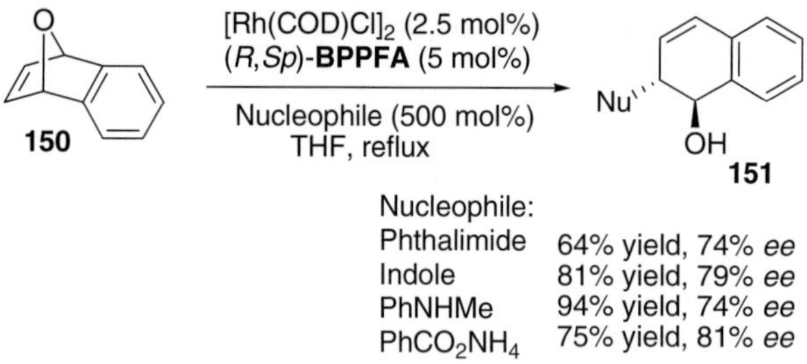

Scheme 8.51 Rh-catalyzed ring opening reaction of oxabenzonorbornadiene.

Nucleophile:
Phthalimide   64% yield, 74% ee
Indole   81% yield, 79% ee
PhNHMe   94% yield, 74% ee
PhCO$_2$NH$_4$ 75% yield, 81% ee

(PhCO$_2$NH$_4$) and nitrogen nucleophiles gave chiral diol (75% yield, 81% ee) and aminoalcohols (up to 79% ee), respectively (Scheme 8.51) [85].

8.7.4
Asymmetric Cyclopropanation of Alkenes with Diazo Esters

Starting from **BPPFA**, Kim et al. synthesized a new class of planar chiral 1,2-ferrocenediylazaphosphinines. Acetoxylation of (R,Sp)-**BPPFA** followed by amination with liquid ammonia led to (R,Sp)-**L25**. Condensation of (R,Sp)-**L25** with glyoxal, methylglyoxal, or phenylglyoxal gave (Sp)-**L26a–c**, respectively. These ligands were used for the Cu(I)-catalyzed asymmetric cyclopropanantion of alkenes with diazoesters (Scheme 8.52) [86].

In most cases, good to excellent dr ratios were obtained but with only low enantioselectivities. When bulky diazo ester **153** was used, 100% ee was achieved.

The previously described chiral(iminophosphoranyl)ferrocene (R,Sp)-**L11a** was also efficient for the Ru-catalyzed cyclopropanantion of olefins with diazoesters (Scheme 8.53)[87]. Interestingly, high diastereoselectivities, in favor of the cis-isomer, as well as excellent enantioselectivity (up to 99%) were obtained.

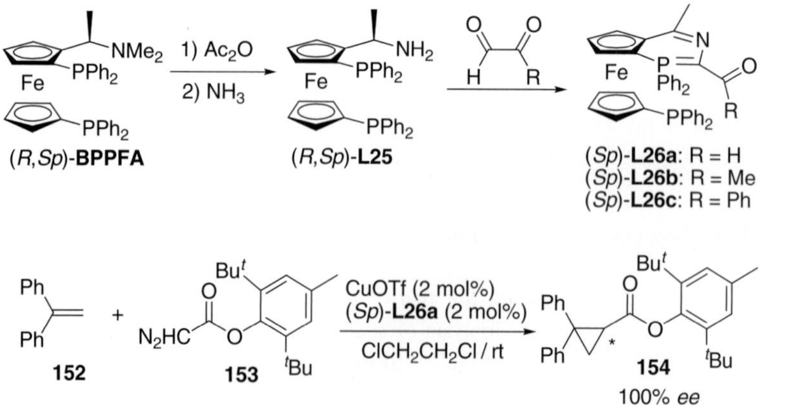

Scheme 8.52 Synthesis of (Sp)-**L26a–c** and their application in Cu-catalyzed cyclopropanation.

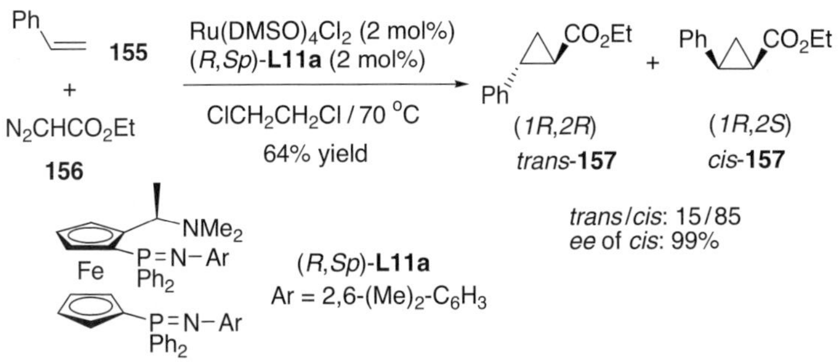

Scheme 8.53 Cu-catalyzed cyclopropanation with (R,Sp)-L11a.

8.7.5
Asymmetric Palladium-Catalyzed Hydroesterification of Styrene

Hydroesterification of olefins using CO and alcohols produces industrially valuable carboxylic acid esters, such as 2-arylpropionic acids, which are the most important class of nonsteroidal anti-inflammatory agents like ibuprofen and naproxen. Inoue et al. reported that Pd(OAc)$_2$ together with (S,Rp)-**BPPFA** catalyzed the hydroesterification of styrene to give the methyl esters in 17% yield, 44/56 branch to linear ratio, and 86% ee for the branched product **118** (Scheme 8.54) [88]. Chan, Hou, and coworkers found PdCl$_2$/(S,Rp)-**L27** was able to catalyze the hydroesterification of styrene. The hydroesterification products were obtained in only 14% yield with 40/60 branch to linear selectivity, and 64% ee for the branched product. Interestingly, although the addition of CuCl$_2$ increased the conversion, a significant decrease in enantioselectivity was observed [89].

BPPFA and its derivatives have been also utilized in many other types of asymmetric reaction, for example: the Co-catalyzed Pauson–Khand reaction (up to 46% ee) [90]; the nickel(0)-catalyzed [2 + 2 + 2] cocyclization (up to 78% ee) [91]; the intramolecular carbopalladation of allenes and subsequent intramolecular amination (up to 58% ee) [92]; and others [93].

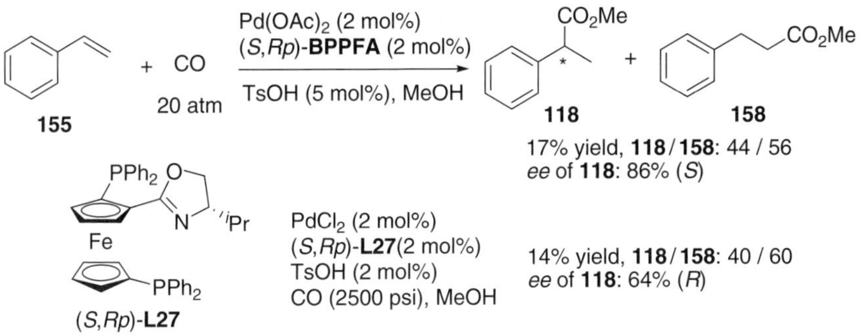

Scheme 8.54 Pd-catalyzed asymmetric hydroesterification.

8.8
Conclusion and Perspectives

As shown above, 1,1′-unsymmetrically disubstituted ferrocenes have been demonstrated to be quite useful chiral ligands for many different transition-metal catalyzed asymmetric reactions. Notably, due to the ready availability of the **BPPFA** and its close analogs, they have become one of the most popular weapons in the arsenal of chiral ligands and have been tested in many transition-metal catalyzed asymmetric reactions. However, other than the **BPPFA** and its close analogs, the development of 1,1′-unsymmetrically disubstituted ferrocene in asymmetric catalysis is still in its infancy and only limited successful examples have been reported so far. Part of the reason for this is that the facile synthetic methods of these classes of compounds have been developed only recently. For example, ready access to dibromoferrocene and selective monolithiation of dibromoferrocene have made the synthesis of 1,1′-unsymmetrically disubstituted ferrocenes facile and therefore widened their application. In addition, chiral ligands with mixed coordinating atoms have received great attention and witnessed significant progress, which has also shed light on the design of novel 1,1′-unsymmetrically disubstituted ferrocenes, such as the SiocPhox. With further progress on the derivatization of ferrocenes, we believe the 1,1′-unsymmetrically disubstituted ferrocenes will serve as efficient ligands for more transition-metal catalyzed asymmetric reactions.

8.9
Experimental: Selected Procedures

8.9.1
Synthesis of Compound (S,Sp)-L4d from (S)-5a [9c]

Compound (S)-**5a** (3.76 g, 10 mmol) was dissolved in freshly distilled THF (80 mL) under argon and cooled to −78 °C. At this temperature, n-BuLi (7.6 mL, 12 mmol, 1.6 M in n-hexane) was added, and the resulting deep red solution was stirred for 20 min. ClPPh$_2$ (2.6 mL, 1.4 mmol) was then added, and the resulting mixture was continually stirred and warmed to room temperature over 30 min. The reaction mixture was diluted with ether (400 mL), washed with saturated aqueous NaHCO$_3$, and dried over Na$_2$SO$_4$. The solvent was removed under reduced pressure, and the resulting residue was chromatographed on silica gel with ethyl acetate: petroleum ether (1 : 4) as eluent to give (S)-**L3** as an orange oil (3.70 g, 77% yield).

A solution of (S)-**L3** (3.13 g, 6.5 mmol) and TMEDA (1.2 mL, 8.0 mmol) in ether (40 mL) under argon was cooled to −78 °C. To this solution was added dropwise n-BuLi (5.0 mL, 8.0 mmol). After the resultant solution was stirred at −78 °C for 2 h, TMSCl (1.2 mL, 9.6 mmol) was added, and the dry ice bath was removed. The resulting mixture was continually stirred for 20 min, and then quenched with saturated NaHCO$_3$, diluted with ether, washed with brine, dried over MgSO$_4$, filtered, and evaporated under reduced pressure. The resulting residue was purified by column chromatography with ethyl acetate: petroleum ether (1 : 20) as an eluent to afford (S,Sp)-**L4a** as an orange oil (3.34 g, 93% yield).

Compound (S,Sp)-**L4a** (1.10 g, 2.0 mmol) was dissolved in freshly distilled THF (16 mL) under argon and cooled to −78 °C. At this temperature, n-BuLi (1.52 mL, 2.4 mmol) was added, and the resulting deep red solution was stirred for 2 h. Methyl iodide (0.4 mL, 6.0 mmol) was then added, and the resulting mixture was continually stirred and warmed to room temperature over 40 min. The reaction mixture was diluted with ether (40 mL), washed with saturated aqueous NaHCO$_3$, and dried over Na$_2$SO$_4$. The solvent was removed under reduced pressure, and the resulting residue was chromatographed on silica gel with ethyl acetate: petroleum ether (1 : 30) as eluent to give 1-diphenylphosphino-1'-[(S)-4-isopropyl-2,5-oxazolinyl]-2'(Sp)-(trimethylsilyl)-5'(Sp)-(methyl)-ferrocene as an orange oil (964 mg, 85% yield). Under an argon atmosphere, a solution of the above compound (567 mg, 1 mmol) in 1 M TBAF in THF (20 mL) was heated at reflux for 10 h. The resultant orange solution was evaporated *in vacuo* to low volume and diluted with ether (30 mL), washed with water, dried over MgSO4, filtered, and evaporated *in vacuo*. The resulting residue was column chromatographed with ethyl acetate: petroleum ether (1 : 5) as eluent to give (S,Sp)-**L4d** as an orange oil (475 mg, 96% yield). $[R]_{20}^{D} = -59.0°$ (c, 0.35, CHCl$_3$); ^{1}H NMR (400 MHz, CDCl$_3$): δ 0.90 (d, J = 6.7 Hz, 3H), 0.98 (d, J = 6.7 Hz, 3H), 1.91 (m, 1H), 2.10 (s, 3H), 3.87–4.33 (m, 9H), 4.56 (t, J = 1.55 Hz, 1H), 7.24–7.39 (m, 10H). ^{31}P NMR (161.92 MHz, CDCl$_3$): δ −17.25. MS: m/z (relative intensity) 495 (M$^+$, 100), 427 (21), 321 (33), 267 (27), 171 (25). IR (KBr): 2958, 1652, 1480, 1434, 1070, 1025, 743, 697, 505. Anal.: Calc. for C$_{29}$H$_{30}$NOPFe: C, 70.31; H, 6.10; N, 2.83. Found: C, 70.40; H, 6.09; N, 2.69.

8.9.2
Synthesis of Compound (S,R$_{phos}$,R)-L8d and (S,S$_{phos}$,R)-L8d from (S)-5d [26]

To a solution of (S)-**5d** (2.54 g, 6 mmol) in freshly distilled THF (40 mL) cooled at −78 °C, n-BuLi (4.2 mL, 6.6 mmol, 1.6 M in n-hexane) was added. The resulting dark red solution was stirred at this temperature for 20 min before adding ClP(NEt$_2$)$_2$ (1.7 mL, 8 mmol). The reaction mixture was then allowed to warm to room temperature naturally (about 30 min), then diluted with diethyl ether (20 mL), washed with water, and brine. The organic layer was dried over anhydrous Na$_2$SO$_4$, filtered and concentrated under reduced pressure. (S)-**28d** was obtained as a dark red oil and used directly for the next step.

A solution of (S)-**28d** obtained above (6 mmol scale) and (R)-BINOL (1.72 mg, 6 mmol) in freshly distilled THF (240 mL) was refluxed for 12 h (monitored by the disappearance of the starting material), and then the solvent was removed under reduced pressure. The crude product was further purified by silica gel column chromatography (petroleum ether/ethyl acetate/triethyl amine = 10 : 1 : 1) to afford (S, R$_{phos}$, R)-**L8d** (45% yield) and (S, S$_{phos}$, R)-**L8d** (36% yield), respectively.

(S, R$_{phos}$, R)-**L8d**, orange solid, m.p. 68–70 °C; $[\alpha]_D^{20} = 403°$ (c, 0.57, CHCl$_3$); ^{1}H NMR (400 MHz, CDCl$_3$) δ 0.53 (t, J = 7.0 Hz, 6H), 2.41–2.62 (m, 4H), 2.67 (dd, J = 8.9, 13.8 Hz, 1H), 3.17 (dd, J = 5.0, 13.7 Hz, 1H), 3.74 (m, 1H), 3.92 (m, 1H), 4.07

(dd, $J = 7.2$, 8.1 Hz, 1H), 4.29–4.33 (m, 3H), 4.41 (m, 1H), 4.57 (t, $J = 1.2$ Hz, 1H), 4.59–4.64 (m, 1H), 5.09 (t, $J = 1.2$ Hz, 1H), 7.11–7.41 (m, 12H), 7.79–8.03 (m, 5H), 9.25 (br, 1H); ^{31}P NMR (161.92 MHz, CDCl$_3$) δ 119.50; MS (EI) m/z (rel) 732 (M$^+$, 1), 659 (9), 541 (19), 447 (11), 315 (5), 286 (100); IR (KBr) 3055 (w), 2967 (w), 1639 (s), 1589 (m), 1504 (m), 1461 (m), 1232 (s), 1023 (s). Anal.: Calc. for C$_{44}$H$_{41}$N$_2$O$_3$PFe: C, 72.19; H, 5.60; N, 3.83. Found: C, 72.00; H, 5.66; N, 3.85.

(S, Sphos, R)-**L8d**, orange solid, m.p. 154–155 °C; $[\alpha]_D^{20} = -357°$ (c, 0.33, CHCl$_3$); ^{1}H NMR (400 MHz, CDCl$_3$) δ 0.76 (t, $J = 7.0$ Hz, 6H), 2.67 (dd, $J = 9.1$, 13.7 Hz, 1H), 2.86 (m, 4H), 3.18 (dd, $J = 4.7$, 13.7 Hz, 1H), 3.39 (m, 1H), 3.73 (m, 1H), 3.88 (m, 1H), 3.94 (m, 1H), 4.01 (t, $J = 7.8$ Hz, 1H), 4.05–4.13 (m, 2H), 4.18 (t, $J = 8.6$ Hz, 1H), 4.34–4.42 (m, 2H), 4.52 (m, 1H), 5.24 (br, 1H), 7.21–7.39 (m, 12H), 7.82–8.05 (m, 5H); ^{31}P NMR (161.92 MHz, CDCl$_3$) δ 127.86; MS (EI) m/z (rel) 732 (M$^+$, 2), 659 (42), 541 (100), 447 (20), 315 (27), 286 (74); IR (KBr) 3055 (w), 2966 (w), 1641 (s), 1588 (m), 1504 (m), 1458 (m), 1226 (s), 1023 (s); Anal.: Calc. for C$_{44}$H$_{41}$N$_2$O$_3$PFe: C, 72.19; H, 5.60; N, 3.83. Found: C, 71.73; H, 5.94; N, 3.69.

Abbreviations

Ac	acetyl
BBN	9-borabicyclo[3.3.1]nonane
BINOL	binaphthol
Bn	benzyl
Boc	*tert*-butyloxycarbonyl
BPPFA	N,N-dimethyl-1-[1′,2-bis(diphenylphosphino)ferrocenyl] ethylamine
BPPFOH	1-[1′,2-bis(diphenylphosphino)ferrocenyl] ethanol
BSA	N,O-bis(trimethylsilyl) acetamide
COD	1,5-cyclooctadiene
dba	dibenzylideneacetone
DMAP	N,N-dimethylaminopyridine
DME	1,2-dimethoxy ethane
DPPF	1,1′-bis(diphenylphosphino)ferrocene
HMDS	1,1,1,3,3,3-hexamethyldisilazane
NBD	norbornadiene
SAMP	(S)-1-amino-2-(methoxymethyl)-pyrrolidine
SiocPhox	2-{1-[1′-(2-oxazolinyl)ferrocenyl]}(diethylamino)phosphinooxy-1,1′-binaphthyl-2′-ol
TBAF	tetrabutyl ammonium fluoride
Tf	trifluoromethanesulfonyl
TFA	trifluoroacetic acid
TMEDA	N,N,N′,N′-tetramethylethylenediamine
TMG	1,1,3,3-tetramethylguanidine
TMSCl	trimethylsilyl chloride
Ts	*p*-toluenesulfonyl

Acknowledgments

Financial support was provided by the Chinese Academy of Sciences and the National Natural Science Foundation of China. I thank Mr. Ji-Bao Xia and Mr. Xue-Qiang Wang for help with the manuscript preparation.

References

1. Atkinson, R.C., Gibson, V.C., and Long, N.J. (2004) *Chem. Soc. Rev.*, **33**, 313–328.
2. For reviews: (a) Trost, B.M. and Van Vranken, D.L. (1996) *Chem. Rev.*, **96**, 395–422; (c) Trost, B.M. (2002) *Acc. Chem. Res.*, **35**, 695–705; (c) Lu, Z. and Ma, S. (2008) *Angew. Chem.*, **120**, 264–303; (2008) *Angew. Chem. Int. Ed.*, **47**, 258–297.
3. For a review: Trost, B.M. and Crawley, M.L. (2003) *Chem. Rev.*, **103**, 2921–2944.
4. Hayashi, T., Yamamoto, A., Hagihara, T., and Ito, Y. (1986) *Tetrahedron*, **27**, 191–194.
5. Hayashi, T., Yamamoto, A., Ito, Y., Nishioka, E., Miura, H., and Yanagi, K. (1989) *J. Am. Chem. Soc.*, **111**, 6301–6311.
6. (a) Yamazaki, A., Morimoto, T., and Achiwa, K. (1993) *Tetrahedron: Asymmetry*, **11**, 2287–2290; (b) Gotov, B., Toma, Š., Solčániová, E., and Cvengroš, J. (2000) *Tetrahedron*, **56**, 671–675.
7. Kmentová, I., Gotov, B., Solčániová, E., and Toma, Š. (2002) *Green Chem.*, **4**, 103–106.
8. Lee, S., Koh, J.H., and Park, J. (2001) *J. Organomet. Chem.*, **637–639**, 99–106.
9. (a) Zhang, W., Yoneda, Y.-i., Kida, T., Nakatsuji, Y., and Ikeda, I. (1998) *Tetrahedron: Asymmetry*, **9**, 3371–3380; (b) Deng, W.-P., Hou, X.-L., Dai, L.-X., Yu, Y.-H., and Xia, W. (2000) *Chem. Comm.*, 285–286; (c) Deng, W.-P., You, S.-L., Hou, X.-L., Dai, L.-X., Yu, Y.-H., Xia, W., and Sun, J. (2001) *J. Am. Chem. Soc.*, **123**, 6508–6519.
10. Mino, T., Segawa, H., and Yamashita, M. (2004) *J. Organomet. Chem.*, **689**, 2833–2836.
11. Nakamura, S., Fukuzumi, T., and Toru, T. (2004) *Chirality*, **16**, 10–12.
12. Tu, T., Hou, X.-L., and Dai, L.-X. (2004) *J. Organomet. Chem.*, **689**, 3847–3852.
13. Yamamoto, K. and Tsuji, J. (1982) *Tetrahedron Lett.*, **23**, 3089–3092.
14. Hayashi, T., Yamamoto, A., and Ito, Y. (1988) *Tetrahedron Lett.*, **29**, 99–102.
15. Tanimori, S., Inaba, U., Kato, Y., and Kirihata, M. (2003) *Tetrahedron Lett.*, **59**, 3745–3751.
16. Hayashi, T., Yamamoto, A., and Ito, Y. (1988) *Tetrahedron Lett.*, **29**, 669–672.
17. Yamamoto, A., Ito, Y., and Hayashi, T. (1989) *Tetrahedron Lett.*, **30**, 375–378.
18. Hayashi, T., Kanehira, K., Hagihara, T., and Kumada, M. (1988) *J. Org. Chem.*, **53**, 113–120.
19. (a) Swamura, M., Nagata, H., Sakamoto, H., and Ito, Y. (1992) *J. Am. Chem. Soc.*, **114**, 2586–2592; (b) Tang, W. (1995) *Youji Huaxue.*, **15**, 202–206.
20. Sekido, M., Aoyagi, K., Nakamura, H., Kabuto, C., and Yamamoto, Y. (2001) *J. Org. Chem.*, **66**, 7142–7147.
21. You, S.-L., Luo, Y.-M., Deng, W.-P., Hou, X.-L., and Dai, L.-X. (2001) *J. Organomet. Chem.*, **637–639**, 845–849.
22. Kato, F. and Hiroi, K. (2004) *Chem. Pharm. Bull.*, **52**, 95–103.
23. Hayashi, T., Yamamoto, A., and Ito, Y. (1987) *Chem. Lett.*, 177–180.
24. Hayashi, T., Kishi, K., Yamamoto, A., and Ito, Y. (1990) *Tetrahedron Lett.*, **31**, 1743–1746.
25. Johnson, B.F., Raynor, S.A., Shephard, D.S., Mashmeyer, T., Thomas, J.M., Sankar, G., Bromley, S., Oldroyd, R., Gladden, L., and Mantle, M.D. (1999) *Chem. Comm.*, 1167–1168.

26 You, S.-L., Zhu, X.-Z., Luo, Y.-M., Hou, X.-L., and Dai, L.-X. (2001) *J. Am. Chem. Soc.*, **123**, 7471–7472.

27 Hou, X.-L. and Sun, N. (2004) *Org. Lett.*, **6**, 4399–4401.

28 Zheng, W.-H., Sun, N., and Hou, X.-L. (2005) *Org. Lett.*, **7**, 5151–5154.

29 Zheng, W.-H., Zheng, B.-H., Zhang, Y., and Hou, X.-L. (2007) *J. Am. Chem. Soc.*, **129**, 7718–7719.

30 Zhang, K., Peng, Q., Hou, X.-L., and Wu, Y.-D. (2008) *Angew. Chem.*, **120**, 1765–1768; (2008) *Angew. Chem. Int. Ed.*, **47**, 1741–1744.

31 Fukuda, Y., Kondo, K., and Aoyama, T. (2007) *Tetrahedron Lett.*, **48**, 3389–3391.

32 Co, T.T., Shim, S.C., Cho, C.S., Kim, D.-U., and Kim, T.-J. (2005) *Bull. Korean Chem. Soc.*, **26**, 1359–1365.

33 (a) Kaneko, S., Yoshino, T., Katoh, T., and Terashima, S. (1997) *Tetrahedron: Asymmetry*, **8**, 829–832; (b) Kaneko, S., Yoshino, T., Katoh, T., and Terashima, S. (1998) *Tetrahedron*, **54**, 5471–5484.

34 He, X.-C., Wang, B., Yu, G., and Bai, D. (2001) *Tetrahedron: Asymmetry*, **12**, 3213–3216.

35 Mizuguchi, E. and Achiwa, K. (1997) *Chem. Pharm. Bull.*, **45**, 1209–1211.

36 Ito, Y., Sawamura, M., and Hayashi, T. (1986) *J. Am. Chem. Soc.*, **108**, 6405–6406.

37 Pastor, S.D. and Togni, A. (1989) *J. Am. Chem. Soc.*, **111**, 2333–2334.

38 (a) Ito, Y., Sawamura, M., and Hayashi, T. (1987) *Tetrahedron Lett.*, **28**, 6215–6218; (b) Ito, Y., Sawamura, M., Shirakawa, E., Hayashizaki, K., and Hayashi, T. (1988) *Tetrahedron Lett.*, **29**, 235–238; (c) Ito, Y., Sawamura, M., Kobayashi, M., and Hayashi, T. (1988) *Tetrahedron Lett.*, **29**, 6321–6324; (d) Ito, Y., Sawamura, M., Shirakawa, E., Hayashizaki, K., and Hayashi, T. (1988) *Tetrahedron*, **44**, 5253–5262; (e) Sawamura, M. and Ito, Y. (1990) *Tetrahedron Lett.*, **31**, 2723–2726; (f) Pastor, S.D. and Togni, A. (1990) *Tetrahedron Lett.*, **31**, 839–840; (g) Hayashi, T., Sawamura, M., and Ito, Y. (1992) *Tetrahedron*, **48**, 1999–2012; (h) Soloshonok, V.A. and Hayashi, T. (1994) *Tetrahedron Lett.*, **35**, 2713–2716; (i) Soloshonok, V.A. and Hayashi, T. (1994) *Tetrahedron Asymmetry*, **5**, 1091–1094; (j) Soloshonok, V.A. and Hayashi, T. (1996) *Tetrahedron*, **52**, 245–254.

39 Togni, A. and Pastor, S.D. (1990) *J. Org. Chem.*, **55**, 1649–1664.

40 (a) Togni, A. and Häusel, R. (1990) *Synlett*, 633–635; (b) Togni, A., Blumer, R.E., and Pregosin, P.S. (1991) *Helv. Chim. Acta*, **74**, 1533–1543.

41 Zhou, X.-T., Lin, Y.-R., Dai, L.-X., Sun, J., Xia, L.-J., and Tang, M.-H. (1999) *J. Org. Chem.*, **64**, 1331–1334.

42 Sawamura, M., Hamashima, H., and Ito, Y. (1990) *J. Org. Chem.*, **55**, 5935–5936.

43 Hayashi, T., Uozumi, Y., Yamazaki, A., Sawamura, M., Hamashima, H., and Ito, Y. (1991) *Tetrahedron Lett.*, **32**, 2799–2802.

44 Togni, A., Pastor, S.D., and Rihs, G. (1989) *Helv. Chim. Acta*, **72**, 1471–1478.

45 Ito, Y., Sawamura, M., and Hayashi, T. (1988) *Tetrahedron Lett.*, **29**, 239–240.

46 Hughes, P.F., Smith, S.H., and Olson, J.T. (1994) *J. Org. Chem.*, **59**, 5799–5802.

47 Sawamura, M., Nakayama, Y., Kato, T., and Ito, Y. (1995) *J. Org. Chem.*, **60**, 1727–1732.

48 Hayashi, T., Yamamoto, K., and Kumada, M. (1974) *Tetrahedron Lett.*, **15**, 4405–4408.

49 Nishibayashi, Y., Segawa, K., Singh, J.D., Fukuzawa, S.-i., Ohe, K., and Uemura, S. (1996) *Organometallics*, **15**, 370–379.

50 (a) Cullen, W.R., Evans, S.V., Han, N.F., and Trotter, J. (1987) *Inorg. Chem.*, **26**, 514–519; (b) Brunner, H. and Miehling, W. (1984) *J. Organomet. Chem.*, **275**, C17–C21.

51 Hayashi, T., Mise, T., Mitachi, S., Yamamoto, K., and Kumada, M. (1976) *Tetrahedron Lett.*, **17**, 1133–1134.

52 Hayashi, T., Mise, T., and Kumada, M. (1976) *Tetrahedron Lett.*, **17**, 4351–4354.

53 Hayashi, T., Katsumura, A., Konishi, M., and Kumada, M. (1979) *Tetrahedron Lett.*, **20**, 425–428.

54 Hayashi, T., Kawamura, N., and Ito, Y. (1987) *J. Am. Chem. Soc.*, **109**, 7876–7878.

55 Hayashi, T., Kawamura, N., and Ito, Y. (1988) *Tetrahedron Lett.*, **29**, 5969–5972.

56 Kimmich, B.F.M., Landis, C.R., and Powell, D.R. (1996) *Organometalllics*, **15**, 4141–4146.
57 Shimazu, S., Ro, K., Sento, T., Ichikuni, N., and Uematsu, T. (1996) *J. Mol. Catal. A: Chem.*, **107**, 297–303.
58 Reetz, M.T., Beuttenmüller, E.W., Goddard, R., and Pastó, M. (1999) *Tetrahedron Lett.*, **40**, 4977–4980.
59 Co, T.T., Shim, S.C., Cho, C.S., and Kim, T.-J. (2005) *Organometalllics*, **24**, 4824–4831.
60 Co, T.T. and Kim, T.-J. (2006) *Chem. Comm.*, 3537–3539.
61 (a) Sato, M., Miyaura, N., and Suzuki, A. (1990) *Tetrahedron Lett.*, **31**, 231–234; (b) Matsumoto, Y. and Hayashi, T. (1991) *Tetrahedron Lett.*, **32**, 3387–3390; (c) Hayashi, T., Matsumoto, Y., and Ito, Y. (1991) *Tetrahedron: Asymmetry*, **2**, 601–612.
62 Zembayashi, M., Tamao, K., Hayashi, T., Mise, T., and Kumada, M. (1977) *Tetrahedron Lett.*, **18**, 1799–1802.
63 Hayashi, T., Konishi, M., Fukushima, M., Mise, T., Kagotani, M., Tajika, M., and Kumada, M. (1982) *J. Am. Chem. Soc.*, **104**, 180–186.
64 Wang, Z.-X. and Chen, Z.-Y. (1993) *Youji Huaxue*, **13**, 496–500.
65 Hayashi, T., Konishi, M., Okamoto, Y., Kabeta, K., and Kumada, M. (1986) *J. Org. Chem.*, **51**, 3772–3781.
66 de Graaf, W., Boersma, J., and Van Koten, G. (1989) *J. Organomet. Chem.*, **378**, 115–124.
67 Baker, K.V., Brown, J.M., Cooley, N.A., Hughes, G.D., and Taylor, R.J. (1989) *J. Organomet. Chem.*, **370**, 397–406.
68 Galarini, R., Musco, A., Pontellini, R., and Santi, R. (1992) *J. Mol. Catal.*, **72**, L11–L13.
69 Cho, S.Y. and Shibasaki, M. (1998) *Tetrahedron: Asymmetry*, **9**, 3751–3754.
70 Cammidge, A.N. and Crépy, K.V.L. (2000) *Chem. Comm.*, 1723–1724.
71 Lee, S. and Hartwig, J.F. (2001) *J. Org. Chem.*, **66**, 3402–3415.
72 Ohno, A., Yamada, M., and Hayashi, T. (1995) *Tetrahedron: Asymmetry*, **6**, 2495–2502.
73 Nukui, S., Sodeoka, M., and Shibasaki, M. (1993) *Tetrahedron Lett.*, **33**, 4965–4968.
74 Sato, Y., Nukui, S., Sodeoka, M., and Shibasaki, M. (1994) *Tetrahedron*, **50**, 371–382.
75 Ashimori, A., Bachand, B., Overman, L.E., and Poon, D.J. (1998) *J. Am. Chem. Soc.*, **120**, 6477–6487.
76 Deng, W.-P., Hou, X.-L., Dai, L.-X., and Dong, X.-W. (2000) *Chem. Comm.*, 1483–1484.
77 Tu, T., Deng, W.-P., Hou, X.-L., Dai, L.-X., and Dong, X.-C. (2003) *Chem. Eur. J.*, **9**, 3073–3081.
78 Kang, J., Lee, J.H., and Im, K.S. (2003) *J. Mol. Catal. A*, **196**, 55–63.
79 Tu, T., Hou, X.-L., and Dai, L.-X. (2004) *Org. Lett.*, **5**, 3651–3653.
80 Deng, W.-P., Hou, X.-L., and Dai, L.-X. (1999) *Tetrahedron: Asymmetry*, **10**, 4689–4693.
81 Li, M., Zhu, X.-Z., Yuan, K., Cao, B.-X., and Hou, X.-L. (2004) *Tetrahedron: Asymmetry*, **15**, 219–222.
82 Alexakis, A. and Benhaim, C. (2001) *Tetrahedron: Asymmetry*, **12**, 1151–1157.
83 Shintani, R. and Hayashi, T. (2005) *Org. Lett.*, **7**, 2071–2073.
84 Stemmler, R.T. and Bolm, C. (2007) *Synlett*, 1365–1370.
85 (a) Lautens, M., Fagnou, K., and Rovis, T. (2000) *J. Am. Chem. Soc.*, **122**, 5650–5651; (b) Lautens, M. and Fagnou, K. (2001) *Tetrahedron*, **57**, 5067–5072.
86 Paek, S.H., Co, T.T., Lee, D.H., Park, Y.C., and Kim, T.-J. (2002) *Bull. Korean Chem. Soc.*, **23**, 1702–1708.
87 Hoang, V.D.M., Reddy, P.A.N., and Kim, T.-J. (2007) *Tetrahedron Lett.*, **48**, 8014–8017.
88 Nomura, S.O.M., Aiko, T., and Inoue, Y. (1997) *J. Mol. Catal. A: Chem.*, **115**, 289–295.
89 Wang, L., Kwok, W.H., Chan, A.S.C., Tu, T., Hou, X., and Dai, L. (2003) *Tetrahedron: Asymmetry*, **14**, 2291–2295.
90 (a) Hiroi, K., Watanabe, T., Kawagishi, R., and Abe, I. (2000) *Tetrahedron: Asymmetry*, **11**, 797–808; (b) Hiroi, K., Watanabe, T., Kawagishi, R., and Abe, I. (2000) *Tetrahedron Lett.*, **41**, 891–895.

91 Sato, Y., Nishimata, T., and Mori, M. (1994) *J. Org. Chem.*, **59**, 6133–6135.

92 Hiroi, K., Hiratsuka, Y., Watanabe, K., Abe, I., Kato, F., and Hiroi, M. (2002) *Tetrahedron: Asymmetry*, **13**, 1351–1353.

93 (a) Hayashi, T., Kishi, K., and Uozumi, Y. (1991) *Tetrahedron: Asymmetry*, **2**, 195–198; (b) Carpentier, J.-F., Pamart, L., Maciewjeski, L., Castanet, Y., Brocard, J., and Mortreux, A. (1996) *Tetrahedron Lett.*, **37**, 167–170.

9
Sulfur- and Selenium-Containing Ferrocenyl Ligands

Juan C. Carretero, Javier Adrio, and Marta Rodríguez Rivero

9.1
Introduction

Since the synthesis of ppfa (*N,N*-dimethyl-1-[2-(diphenylphosphino)ferrocenyl]ethylamine), described by Hayashi and Kumada in 1974 [1], there has been a growing interest in chiral bidentate ferrocene ligands. In particular, in the last 15 years an amazing number of ferrocene ligands, frequently combining central and planar chirality and P,P- (e.g., Josiphos, Walphos, BoPhoz, Taniaphos) or P,N-coordination (e.g., ppfa, phosphino ferrocenyl oxazolines, phosphino imine ferrocenes) have been developed, providing high reactivities and enantioselectivities in an ample variety of metal-mediated reactions [2]. However, in spite of the well known ability of thioethers and thiols to coordinate late transition metals, to the best of our knowledge, the first examples of enantiopure ferrocene sulfur ligands in asymmetric catalysis were not reported until 1996, independently, by Togni [3] and Albinati and Pregosin [4]. In both cases this pioneering work involved the use of a 1,2-disubstituted bidentate P,S-ferrocene ligand, derived from ppfa-type ligands, in standard Pd-catalyzed asymmetric allylic alkylations.

In very recent years, in parallel with the increasing interest in sulfur ligands in asymmetric catalysis [5], a variety of sulfur-containing ferrocenyl ligands (and some scattered examples of selenium related ligands) have been described by different groups and successfully employed in several asymmetric catalytic processes, although the Pd-catalyzed asymmetric alkylation continues to be the most studied reaction. In agreement with the growing interest in sulfur-containing ferrocene ligands, the first specific review on this topic has been very recently written by Bonini and Ricci [6]. In this chapter we want to give a clear view of the most significant contributions in this recent and promising field, including our own work with chiral planar 2-phosphino 1-sulfenyl ferrocenes (Fesulphos ligands), with especial focus on the results in asymmetric catalysis rather than the synthetic aspects of the preparation of the sulfur-containing ferrocene ligands.

Chiral Ferrocenes in Asymmetric Catalysis. Edited by Li-Xin Dai and Xue-Long Hou
Copyright © 2010 WILEY-VCH Verlag GmbH & Co. KGaA, Weinheim
ISBN: 978-3-527-32280-0

Table 9.1 Thio- and seleno-substituted oxazoline-containing ferrocene ligands in the asymmetric allylic alkylation of rac-1,3-diphenyl-2-propenyl acetate with dimethyl malonate.

Entry	Conditions	t	L* (amount (mol%))	Yield (%)	ee (%)	Ref.
1	BSA, KOAc, CH_2Cl_2, rt	36 h	1 (10)	98	93 (R)	[8]
2	BSA, KOAc, CH_2Cl_2, rt	6 d	2 (2)	28	75 (S)	[9]
3	BSA, LiOAc, CH_2Cl_2, rt	12 h	3a (10)	98	98 (S)	[10]
4	BSA, LiOAc, CH_2Cl_2, rt	4 d	3b (6)	71	99.3 (S)	[11]
5	BSA, KOAc, CH_2Cl_2, 36 °C	24 h	4 (2)	98	95 (R)	[12]
6	BSA, KOAc, CH_2Cl_2, rt	2 h	5 (10)	97	90 (R)	[13]

9.2
Asymmetric Allylic Substitution

9.2.1
Palladium-Catalyzed Allylic Substitution

Palladium-catalyzed allylation is a very powerful approach for the generation of carbon–carbon and carbon–heteroatom bonds [7]. Thus, asymmetric versions of this transformation have been widely explored over the last decade and it is by far the most intensively studied reaction with sulfur-containing ferrocene ligands.

Despite its low interest from a synthetic point of view, the Pd-catalyzed asymmetric allylic alkylation of rac-1,3-diphenyl-2-propenyl acetate with the enolate of dimethyl malonate (Table 9.1) has become a model test reaction to evaluate the catalytic properties of newly designed chiral ligands. This is probably because this reaction is one of the best understood catalytic asymmetric transformations and it is relatively easy to isolate intermediates, especially the palladium allylic complex.

9.2 Asymmetric Allylic Substitution

However, with few exceptions, most of the ligands afford poor results with more synthetically valuable substrates, such as substituted acyclic and cyclic allylic acetates.

9.2.1.1 Bidentate N,S-ferrocene Ligands

Among thio-substituted ferrocenes, oxazoline-containing ligands have been the most widely employed to promote the asymmetric allylation reaction. The first example was reported by Bryce et al. [8] employing a bidentate ferrocenyl system which incorporates the sulfur atom within the oxazoline framework (ligand **1**, Table 9.1, entry 1). This ligand, possessing only central chirality, led to good yields and high enantioselectivities. Modest results were obtained by Park et al. [9] employing 1-oxazolinyl-1′-(phenylthio)ferrocenes in the allylation reaction (ligand **2**, Table 9.1, entry 2). On the other hand, several 1,2-disubstituted-ferrocenyloxazolines combining both central and planar chirality have been successfully applied in this transformation. Thus, Dai, Hou et al. [10] have found that ferrocenyl-oxazolines with a phenylthioether subunit directly bonded to the ferrocene ring are highly efficient catalysts for the palladium-catalyzed alkylation (ligand **3a**, entry 3). In this case, a similar enantioselectivity and the same absolute configuration was obtained, no matter whether the planar chirality of the ligand was S or R, thus showing that the central chirality plays a major role in controlling the enantioinduction. The crystal structure of the complex formed between **3a** and the palladium precatalyst was found to be an M-type (endo-syn-syn) allylic intermediate, in which the phenyl group on the sulfur is trans to the *tert*-butyl group on the oxazoline (Scheme 9.1). The sulfur atom was proven to be a better π-acceptor than the oxazoline nitrogen atom with the Pd−C bond trans to sulfur longer than the Pd−C bond trans to nitrogen, which is in agreement with the observed asymmetric induction. NMR studies showed a 3.1 : 1 ratio of two conformers, with the major structure in the M conformation. In agreement with the greater trans effect of the sulfur atom, the very different ^{13}C NMR chemical shifts for the allyl terminus trans to sulfur ($\delta = 94.2$ ppm) and trans to nitrogen ($\delta = 76.5$ ppm) indicate that the carbon trans to the sulfur atom is more electrophilic. The same group synthesized the chiral selenide analog **3b**, which afforded a moderate reactivity but an excellent asymmetric induction (Table 9.1, entry 4) [11]. Good catalytic results were obtained by Aït-Haddou et al. [12] using ligand **4** with two stereogenic centers in the oxazoline framework (entry 5). More recently, Ferrocenyl-oxazoline-carbinol **5**, possessing an additional

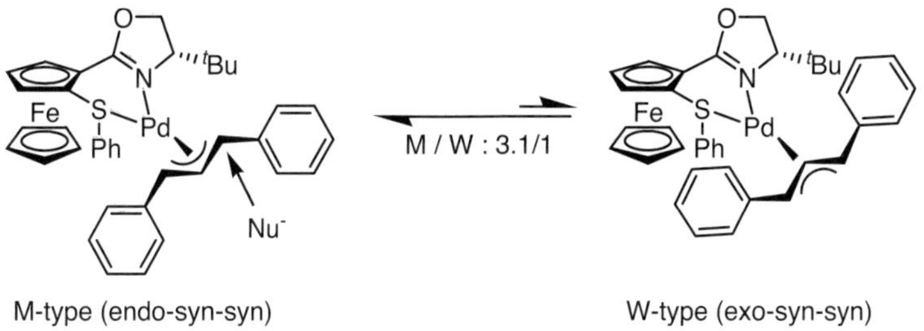

M-type (endo-syn-syn)   W-type (exo-syn-syn)

Scheme 9.1 Equilibration of π-allyl palladium-complexes of ligand **3a**.

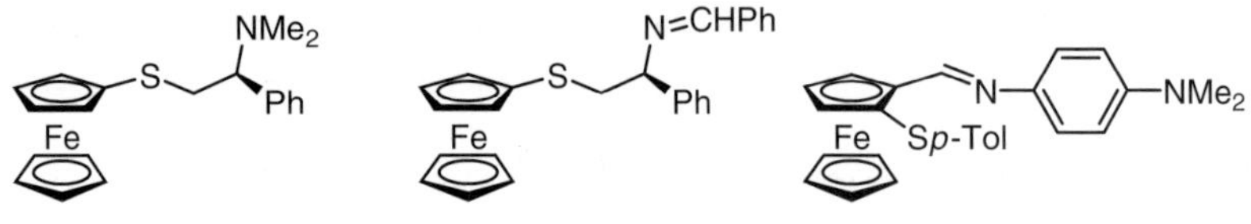

6, 96%, ee = 99% (S)   **7**, 85%, ee = 99% (R)   **8**, ee = 72% (R)

Figure 9.1 Acyclic N,S-ferrocene ligands in the asymmetric allylic alkylation of rac-1,3-diphenyl-2-propenyl acetate with dimethyl malonate.

chiral center, was described by Bonini and Zwanenburg [13]. With this ligand the alkylation product was isolated with 90% ee (entry 6).

Several acyclic N,S-bidentate ferrocene chiral ligands have also proven to be efficient in inducing enantioselectivity in the palladium-catalyzed asymmetric allylic alkylation. In particular, monosubstituted ferrocene derivatives containing chiral amino-sulfide **6** and imino-sulfide **7** backbones in the side chain have been successfully used by Bonini and coworkers in the test reaction [14] (Figure 9.1). The opposite absolute configuration obtained with these two ligands is in agreement with the attack of the nucleophile trans to the better π-acceptor, that is the sulfide for the amino alkyl derivative **6** and the imino group in the case of the ferrocenyl iminoalkyl sulfide **7**. The same research group has recently tested in this reaction a series of 1,2-disubstituted ferrocenyl-imine derivatives with planar chirality [15], affording the expected product with up to 72% ee (ligand **8**, Figure 9.1).

Another ferrocenylsulfur-imine, in this case with the chirality placed both on the carbon skeleton and the ferrocene backbone (ligand **9**, Scheme 9.2), has been evaluated by Zheng et al. [16] in the asymmetric allylic alkylation reaction of 2-cyclohexenyl acetate with dimethyl malonate. The reaction takes place with good enantioselectivity (82%), but this system was not very active, affording the desired product with only 32% yield after 7 days of reaction.

9.2.1.2 Bidentate P,S-ferrocene Ligands

Different research groups have investigated P,S-ferrocenyl chiral ligands with central and planar chirality in the test allylation reaction. The first examples employing the indenol **10** (Figure 9.2) and the thioglucose derivative **11** were reported in the mid-1990s by Togni [3] and Albinati and Pregosin [4], respectively. However, these systems led to moderate enantioselectivities. Some years later, in 2003, the simplified ligand **12** was successfully used by the groups of Hou and Dai [17].

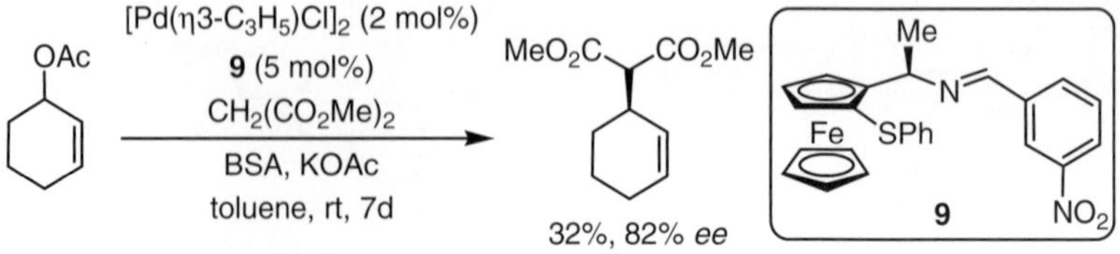

Scheme 9.2 NS-ferrocene ligand **9** in the asymmetric allylic alkylation of 2-cyclohexenyl acetate with dimethyl malonate.

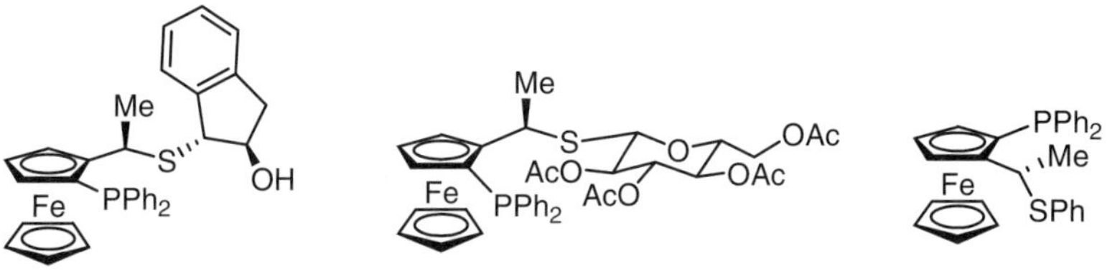

Figure 9.2 α-Sulfur-substituted P,S-ferrocene ligands in the palladium-catalyzed asymmetric allylic alkylation of *rac*-1,3-diphenyl-2-propenyl acetate with dimethyl malonate.

10, ee = 55% (*R*)   **11**, ee = 88%   **12**, ee = 93.8% (*S*)

These authors prepared a number of ferrocene ligands possessing the same basic skeleton but different donor atoms (N, S and P) to study the trans-effect of these coordinating groups. From the experimental results, as well as the X-ray and NMR analysis of the corresponding Pd complexes, the authors established the sequence of the trans-effect as C=N > P > S.

Planar chiral ferrocenyl ligands **13** (Table 9.2) bearing a stereogenic center in the β-position of the side chain were developed by Enders *et al.* [18]. At least one thioether group is present in all these ligands, being the second donor atom other than the phosphorus, selenium or another sulfur donor atom. By using S,S- (ligand **13a**, entry 1, Table 9.2) or S,Se-coordinating ligands (ligand **13b**, entry 2, Table 9.2), low reactivity and enantioselectivity were observed. On the other hand, a very different outcome was found by employing a strong π-acceptor group such as phosphorus (entries 3–5). The higher reactivity of P,S-ligands allowed the reaction to proceed

Table 9.2 β-Sulfur-substituted P,S-ferrocene ligands in asymmetric allylic alkylation.

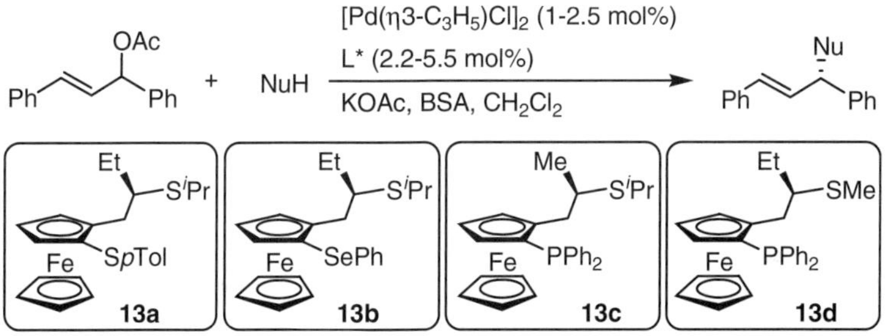

Entry	NuH	L*	T (°C)	t (h)	Yield (%)	ee (%)
1	CH$_2$(CO$_2$Me)$_2$	13a	20	100	84	20 (*R*)
2	CH$_2$(CO$_2$Me)$_2$	13b	20	100	65	44 (*R*)
3	CH$_2$(CO$_2$Me)$_2$	13c	20	1	99	80 (*R*)
4	CH$_2$(CO$_2$Me)$_2$	13d	−20	24	99	97 (*R*)
5	BnNH$_2$	13d	−20	24	50	94 (*S*)

even at low temperature in quantitative yield and up to 97% ee (ligand **13d**, entry 4). This catalytic system was successfully applied in the amination reaction, using benzylamine as nucleophile (entry 5). Employing ligand **13d**, X-ray analyses and solution NMR studies proved the formation of a seven-membered chelated palladium complex with the exo-syn-syn geometry as preferred. The ^{13}C NMR shifts for the allyl terminus ($\delta_{C\ trans\ to\ P} = 102.5$ ppm, $\delta_{C\ trans\ to\ S} = 78.2$ ppm) and the bond distances in the solid-state structure [Pd–C (trans to P) = 2.274(5) Å, Pd–C (trans to S) = 2.176(4) Å] indicate that the phosphino function is a better π-acceptor than the thioether group, thus allowing regiocontrol of the nucleophilic attack.

An important contribution to this field has been recently reported by the group of Chan who used a series of bidentate phosphinamidite-thioether ligands derived from Ugi's amine (ligands **14**, Table 9.3). These ferrocenyl P,S-ligands were successfully utilized as palladium chelates in the standard allylation reaction with malonate (entry 1, Table 9.3) and benzylamine (entry 2, Table 9.3) [19]. More interestingly, in the presence of these ligands aliphatic alcohols can be employed as nucleophiles, thus providing access to chiral ethers in high yields and enantioselectivities [20]. Benzylic (entries 3 and 4) as well as primary (entry 6) and secondary (entry 7) aliphatic alcohols underwent the reaction, which was also found to tolerate heterocycles (entry 5). The results of the asymmetric allylic etherification are remarkable, albeit the substrate scope is limited to the rac-1,3-diphenyl-2-propenyl acetate system.

Table 9.3 Phosphinamidite-thioether ferrocene ligands (S,R_p)-FerroNPS in asymmetric allylic alkylation.

Entry	NuH	L*	t (h)	Yield (%)	ee (%)
1	CH$_2$(CO$_2$Me)$_2$	14a	2	94	93.5 (R)
2	BnNH$_2$	14b	48	86	91.5 (S)
3	BnOH	14c	2	98	91.6 (S)
4	p-MeO(C$_6$H$_4$)CH$_2$OH	14c	2.5	94	94.5[a]
5	2-Py-CH$_2$OH	14c	21	71	82.9[a]
6	n-BuOH	14c	6	98	94.1[a]
7	CyOH	14c	29	58	96.2[a]

[a]Absolute configuration not determined.

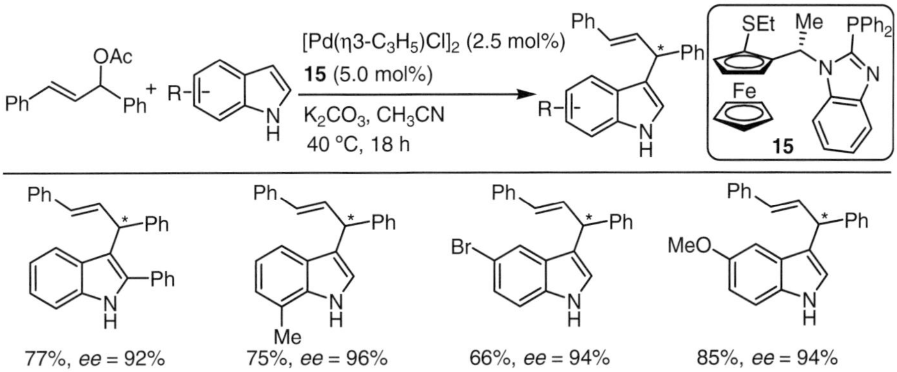

Scheme 9.3 Benzimidazole P,S-ferrocene ligand **15** in the Pd-catalyzed allylic alkylation of indoles.

As a step further, this same group found a related benzimidazole-containing ligand (**15**, Scheme 9.3) to efficiently promote the asymmetric allylic alkylation of indoles [21]. This catalytic system allowed the use of 2-substituted, 5-substituted (with both electron-rich and electron-poor groups), and 7-substituted indoles as nucleophiles. Moderate to good yields and enantioselective inductions in the range 92–96% were obtained.

A readily available family of 1-phosphino-2-sulfenylferrocenes (Fesulphos ligands, Table 9.4), having solely planar chirality, was developed by Carretero and coworkers [22]. In the Pd-catalyzed allylic substitution with 1,3-diphenyl-2-propanol acetate very good results were obtained in terms of both activity and enantioselectivity using either carbon or nitrogen nucleophiles. The steric and electronic properties of this ligand system could be easily fine-tuned by modifying the groups attached to the sulfur and phosphorus atoms. In this sense, sterically demanding thioethers (*tert*-butyl thioethers) and electronically poor phosphines provided the best results (ligand **16b**, Table 9.4). X-ray crystal structure analyses of the complex (**16a**)PdCl$_2$

Table 9.4 Fesulphos ligands in palladium-catalyzed asymmetric allylic alkylation.

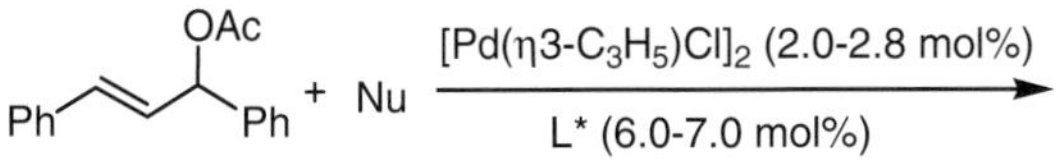

Entry	Nu	Conditions	L*	Yield (%)	ee (%)
1	CH$_2$(CO$_2$Me)$_2$	Bu$_4$NCl, BSA, CHCl$_3$, -20 °C	16a	92	96 (R)
2	CH$_2$(CO$_2$Me)$_2$	Bu$_4$NCl, BSA, CHCl$_3$, -20 °C	16b	94	97 (R)
3	BnNH$_2$	THF, rt	16b	89	98 (S)
4	KPhth	CHCl$_3$, -20 °C	16b	90	96 (S)
5	CH$_2$(CO$_2$Me)$_2$	Bu$_4$NBr, BSA, CHCl$_3$, rt	16c	82	>98 (R)
6	BnNH$_2$	THF, rt	16c	79	91 (S)

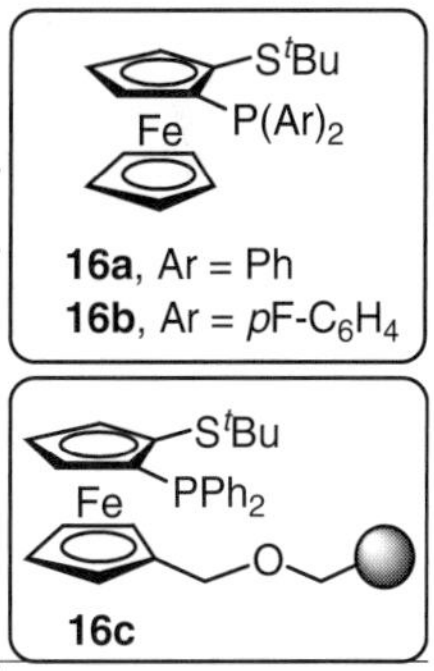

16a, Ar = Ph
16b, Ar = pF-C$_6$H$_4$

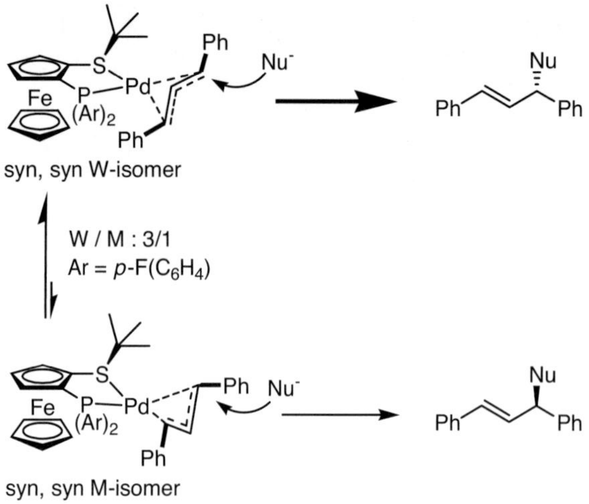

Scheme 9.4 Proposed transition state model for ligands **16**.

showed the formation of a planar five-membered palladacycle with a P,S-coordination. In the stereogenic sulfur center generated after S-Pd coordination, the *tert*-butyl group is oriented trans to the iron atom to minimize the steric hindrance. The stronger trans-effect of the phosphorus atom is reflected in the longer Pd–Cl bond distance trans to the phosphorus (2.35 Å) than that trans to the sulfur atom (2.30 Å). NMR studies of the (π-allyl)palladium complex of **16b** showed a mixture of two diastereomers in a ratio 3 : 1 in which the major isomer has W-type configuration (Scheme 9.4). Once again, the trans-effect of the heteroatom donors was reflected in the ^{13}C NMR chemical shifts of the allylic termini, with the carbon trans to the phosphorus more deshielded than that trans to the sulfur. The authors thus proposed a mechanistic course that is in accordance with the observed product configuration: the nucleophile attacks the allyl substrate trans to the best π-acceptor (the phosphorus atom) on the preferred conformation (W-type). More recently, this group has reported the synthesis of polymer-supported Fesulphos-type ligands [23]. Slightly lower yields but excellent enantioselectivities were achieved with these heterogeneous chiral ligands (entries 5–6, Table 9.4).

The use of P,S and S,S-ligands with planar chirality as the only source of chirality, was later explored by other research groups. Manoury *et al.* [24] evaluated the catalytic potential of several phosphine-thioethers and phosphine sulfide-thioethers that led to high activities and asymmetric inductions of up to 93% *ee* in the standard test allylation reaction (ligands **17** and **18**, Figure 9.3). The novel pseudo-C_2-symmetric ligand **19** (Figure 9.3) containing two planar chiralities (S_p,S_p) with two different donor atoms (P and S), developed by the group of Kang [25], was found to be highly active in the model transformation with dimethyl malonate. However, very low enantioselectivity values were attained by this ligand.

The use of planar chirality was also explored by Zhang *et al.* [26] with the application of the bis-sulfone phosphine ligand **20** in Pd-catalyzed allylic alkylations

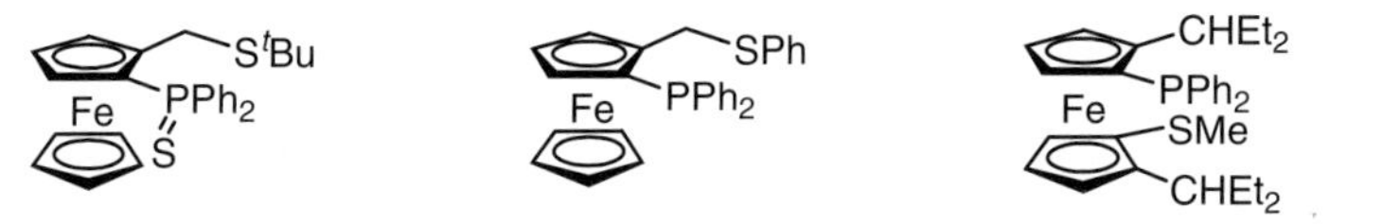

17, 90%, ee = 93% (S)   **18**, 97%, ee = 93% (R)   **19**, 99%, ee = 38.5% (S)

Figure 9.3 Other P,S and S,S-ferrocene ligands with only planar chirality in the palladium-catalyzed asymmetric allylation of rac-1,3-diphenyl-2-propenyl acetate with dimethyl malonate.

(Scheme 9.5). Different additives were evaluated in the transformation and a dramatic increase in the enantioselectivity was observed by changing the alkali metal ions from Li$^+$ to Cs$^+$. High yields and up to 98% ee were obtained in the standard test reaction by using CsOAc as the additive.

An attempt to use in this reaction ferrocenyl ligands bearing a chiral sulfoxide as the sole chirality was reported by Toru and coworkers in 2004 (Scheme 9.6) [27]. Albeit good yields and moderate enantioselectivities were obtained with the 1,3-diphenylallyl system, very poor results were achieved in the allylic substitution of 2-cyclohexenyl acetate.

9.2.2
Copper-Catalyzed Allylic Substitution

In contrast to palladium, that generally requires soft nucleophiles (that is malonate anions or amines), copper catalysis allows the use of harder nucleophiles, such as

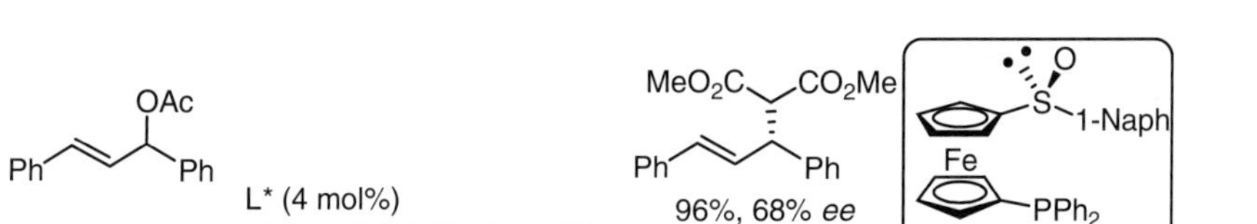

Scheme 9.5 Bis(sulfonyl)phosphino ferrocene ligands in palladium-catalyzed allylation.

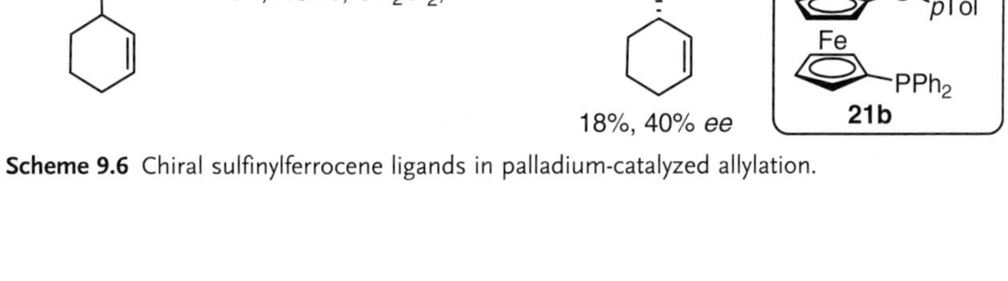

Scheme 9.6 Chiral sulfinylferrocene ligands in palladium-catalyzed allylation.

Table 9.5 Thiolate and selenide ferrocene ligands in copper-catalyzed asymmetric allylic alkylation.

Entry	R^1	R^2	L*	γ:α Ratio	Yield (%)	ee (%)
1	Cy	Me	22a	93:7	42	44 (S)
2	Cy	Et	22a	98:2	55	62 (S)
3	Cy	n-Pr	22a	96:4	77	54 (S)
4	Cy	n-Bu	22a	98:2	88	64 (S)
5	Ph	n-Bu	22a	94:6	78	42 (S)
6	Cy	n-Bu	22b	93:7	67	42 (S)
7	Ph	n-Bu	22b	94:6	78	41 (S)
8	Cy	n-Bu	22c	n.d.	65	10

simple alkyl groups, and usually affords high regioselectivities. Although copper-catalyzed substitution reactions are far behind palladium catalysis, remarkable progress in the development of copper-catalyzed enantioselective allylic alkylations has been made in the last decade [28].

Bäckvall and coworkers reported the use of chiral ferrocenyl thiolate ligands in copper-catalyzed reactions of allylic acetates with Grignard reagents (Table 9.5) [29]. The best results were attained with ligand **22a** that led almost selectively to the γ-product (branched product) in moderate yields and up to 64% *ee*. The lithium selenide **22b** afforded similar results to **22a** in the asymmetric alkylation of cinnamyl acetate. Oxazoline-containing ligand **22c** gave rise to the S_N2'-product in moderate yield and with an enantiomeric excess of only 10% (entry 8). Finally, racemic products were obtained when the sulfur group was blocked as thioether, thus showing the importance of anionic coordination of the thiolate to copper.

9.3
Other Asymmetric Palladium-Catalyzed Reactions

9.3.1
Asymmetric Heck Reactions

The reaction of 2,3-dihydrofuran with aryl halides and aryl triflates has become a typical test reaction for screening new chiral ligands in intermolecular Heck reactions [30]. Independently, the groups of Hou and Guiry [31] have reported high enantioselectivities in this reaction using ferrocene phosphine oxazoline ligands. In 2003, Kang and coworkers evaluated the efficiency of their pseudo-C_2-symmetric

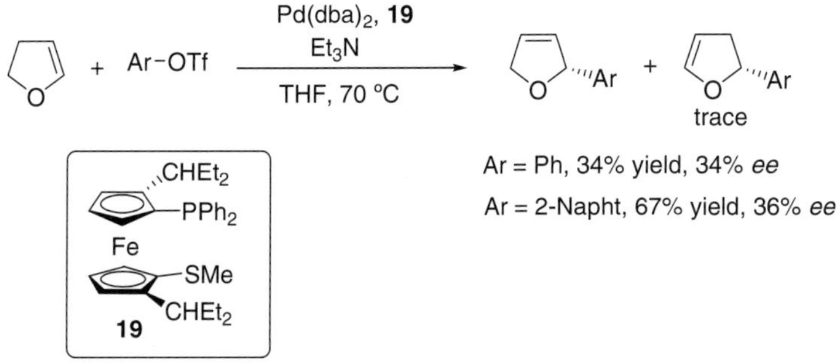

9.3 Other Asymmetric Palladium-Catalyzed Reactions

Scheme 9.7 Asymmetric intermolecular Heck reaction.

P,S-ligands **19** in the asymmetric Heck reaction of dihydrofuran with phenyl triflate [25]. Disappointingly, low yields and enantioselectivities (34% yield, 34% ee, Scheme 9.7) were obtained after screening different bases, solvents, and catalyst precursors. A positive effect on the chemical yield was observed using the sterically bulkier 2-naphthyl triflate (67% yield, 36% ee, Scheme 9.7).

9.3.2
Desymmetrization of Heterobicyclic Alkenes

The palladium-catalyzed asymmetric alkylative ring opening of meso oxabicyclic and azabicyclic alkenes with dialkyl zinc reagents, developed by Lautens et al., is among the most outstanding desymmetrization reactions [32]. Since the original work reported in 2000 using Tol-BINAP and PHOX [33] several structurally diverse chiral ligands [34], including ferrocenes [35], have been successfully employed in the process. Among them, chiral ferrocene P,S ligands developed by Carretero et al. proved to be especially reactive and stereoselective [36]. The reaction of Et_2Zn with 7-oxabenzonorbornadiene under standard reaction conditions [Et_2Zn (150 mol%), $Pd(CH_3CN)_2Cl_2$ (5 mol%), in CH_2Cl_2 at rt] using (R_p)-Fesulphos **16a** (5 mol%) as ligand provided cis-dihydronaphtol **23** in 86% yield and 80% ee. After screening the effect of the substitution at phosphorus in the Fesulphos ligand, the enantioselectivity could be improved up to 93% ee with the ligand **16d** having a dicyclohexylphosphino unit (Scheme 9.8). Mechanistic studies described by Lautens et al. [37] provided strong evidence for the participation of cationic alkyl palladium species $[L_2PdR]^+$ in the key enantioselective carbopalladation step. According to this hypothesis, the cationic catalyst generated in situ by treatment of **16d** with an equimolar amount of $NaB(Ar^F)_4$ [$Ar^F = 3,5\text{-}(CF_3)_2C_6H_3$] displayed an exceptional reactivity, affording under very mild conditions ($-25\,^\circ$C to room temperature) and low catalyst loading (0.5 mol%) the corresponding ring-opened products in good yield and excellent enantiocontrol (94–99% ee). Interestingly, the outstanding reactivity of this catalyst also allowed the highly enantioselective opening of much less reactive substrates, such as azabenzonorbornadienes and [2.2.1] oxabicyclic alkenes (Scheme 9.8).

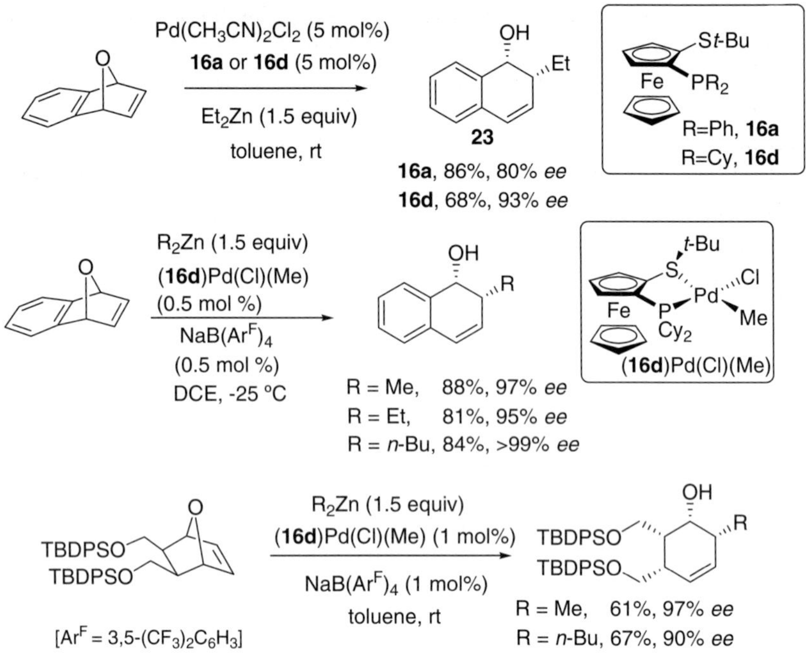

Scheme 9.8 (R_p)-Fesulphos/Pd-catalyzed asymmetric ring opening of bicyclic alkenes.

9.4
Gold-Catalyzed Reactions

Increasing attention has been devoted in recent years to homogeneous gold catalysis, mainly because of the high electronic affinity of Au(I) and Au(III) complexes for carbon–carbon triple and double bonds [38]. Furthermore, gold salts can also act as Lewis acids for the activation of carbon electrophiles. In this field, the asymmetric aldol reaction of aldehydes with isocyano esters catalyzed by a gold(I) complex of ferrocene **24**, reported by Sawamura, and Hayashi in 1986 [39], was a milestone. The high diastereo and enantioselectivity (up to 97% *ee* and trans/cis ratios from 80 : 20 to 100 : 0, Scheme 9.9) observed in the resulting oxazolines was the result of a cooperative effect of both planar and central chiralities of the ferrocene P,N- ligand. A few years later, Togni and Häusel [40] demonstrated that the replacement of the N-methyl group by sulfur (ligands type **25**) seldom affected the enantioselectivity of the reaction (Scheme 9.9).

Surprisingly, the second example of an enantioselective transformation involving a chiral gold catalyst was not reported until 2005 by Echavarren and coworkers [41]. The reaction of alkoxycyclization of 1,6-enynes in the presence of several phosphane gold complexes afforded moderate enantioselectivities. Fesulphos gold complexes of **16a**, **16d** and **16e** gave the methoxycyclization products in very good yields but low asymmetric inductions. The best enantioselectivities were obtained when (*R*)-TolBINAP was used as chiral ligand (89% yield, 53% *ee*, Scheme 9.10).

L*	trans/ cis	Yield (%)	*ee* (%) (trans)
24	90/10	91	94
25	88/12	85	89

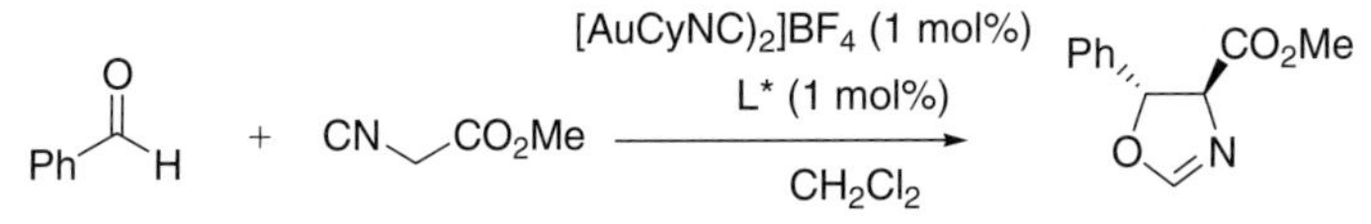

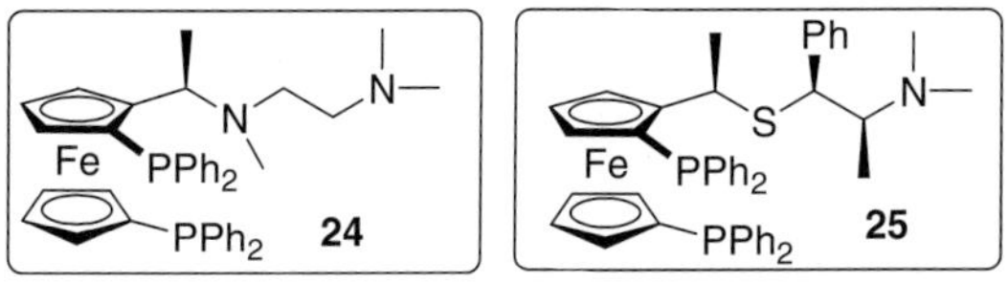

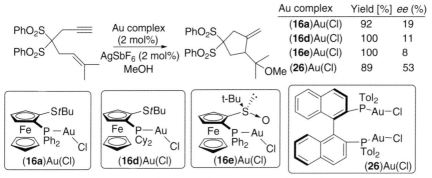

Scheme 9.9 Gold-catalyzed asymmetric aldol reaction of aldehydes with isocyano esters.

Au complex	Yield [%]	*ee* (%)
(**16a**)Au(Cl)	92	19
(**16d**)Au(Cl)	100	11
(**16e**)Au(Cl)	100	8
(**26**)Au(Cl)	89	53

Scheme 9.10 Phosphine-gold complexes in catalytic asymmetric alkoxycyclization of 1,6-enynes.

9.5
Asymmetric Reductions

The metal-catalyzed asymmetric hydrogenation of carbon–carbon and carbon–oxygen double bonds is likely the most developed type of reaction in asymmetric catalysis [42], and several processes have found industrial application [43]. In spite of the great number of new ferrocene ligands successfully tested in these reactions [2], sulfur-containing ferrocene ligands have been little explored in these reactions. Zhang *et al.* [26] evaluated the effectiveness of bis-sulfone, sulfoxide and sulfide ferrocenyl bis-phosphine ligands in the Rh-catalyzed asymmetric hydrogenation of commercially available α-(N-acetamido)acrylate. The planar chiral sulfone ligand **20** afforded the best results (93% *ee* and 99% conversion, Scheme 9.11).

99.7% conv., 93% *ee*

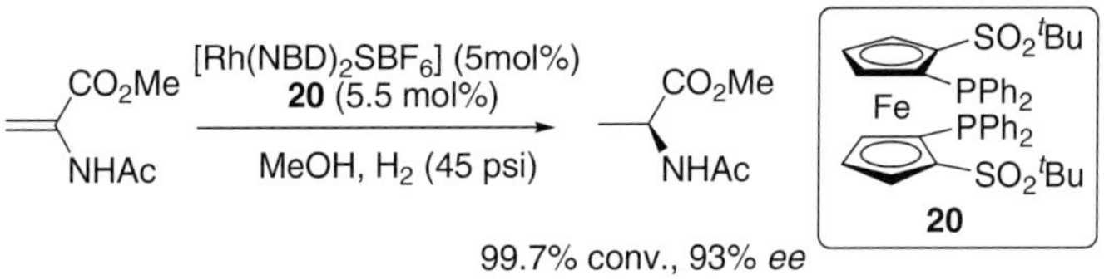

Scheme 9.11 Bis(Sulfonyl)phosphino ferrocene ligands in rhodium-catalyzed hydrogenation.

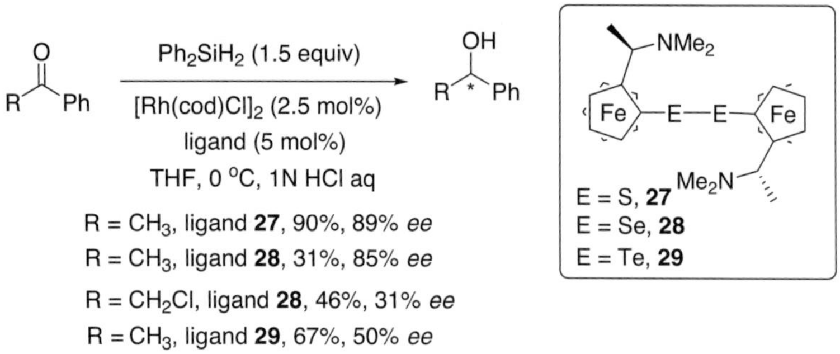

Scheme 9.12 Diferrocenyl dichalcogenide ligands in the Rh-catalyzed asymmetric hydrosilylation of ketones.

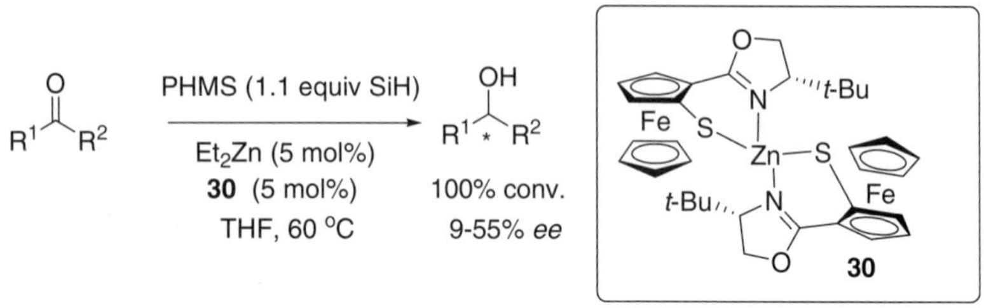

Scheme 9.13 NS-ferrocene oxazoline-Zn catalyzed asymmetric hydrosilylation of ketones.

The asymmetric hydrosilylation of carbon–carbon and carbon–heteroatom bonds is an interesting alternative to the asymmetric hydrogenation because of the mild conditions and operational simplicity [44]. In 1996, Unemura et al. reported the use of a chiral collection of diferrocenyl dichalcogenide ligands in the Rh-catalyzed asymmetric hydrosilylation of ketones [45]. Using standard conditions, ligands **27–29** afforded the corresponding chiral alcohols with good yields and asymmetric inductions up to 89% ee (Scheme 9.12). The asymmetric hydrosilylation of imines and the hydrogenation of α-acetamidocinnamic acid with the ligand **28** were also tested, albeit the enantioselectivities were moderate (up to 69%).

An N,S-chelated zinc catalyst derived from a ferrocene oxazoline has also been recently tested in the hydrosilylation of ketones using the safe and cheap PMHS (poly methylhydrosiloxane) as reducing silylating agent [46]. Unfortunately, low stereoselectivities were obtained (9–55% ee, Scheme 9.13).

9.6
Asymmetric 1,2- and 1,4-Nucleophilic Addition

9.6.1
Asymmetric 1,2 Addition to Aldehydes and Imines

The enantioselective addition of diethyl zinc to aldehydes catalyzed by chiral ligands is one of the most studied processes in asymmetric catalysis [47]. A huge number of chiral ligands, especially aminoalcohols, diamines and diols, have been studied. Ferrocene ligands have also been tested in the catalytic asymmetric addition of dialkyl zinc to aldehydes [48]. In this context, Carretero *et al.* have synthesized a sterically and electronically tunable family of enantiopure ferrocene ligands based on a 1-sulfinyl 2-sulfonamido substitution pattern [49]. The enantioselectivity in the addition of diethyl zinc to benzaldehyde proved to be highly dependent on the substitution at the nitrogen atom. The best results were achieved in the case of sulfonamide ligands **31** and **32**, which provided the alcohol (*R*)-**33** in up to 82% *ee* (Scheme 9.14). Interestingly, when the sulfoxides **31** and **32** were reduced to the corresponding thioethers **34** and **35** the enantioselectivity was very similar, showing that the planar chirality of ferrocene is the main structural factor involved in the asymmetric induction. Moderate to high enantioselectivities were also achieved with a variety of substituted aromatic aldehydes (up to 86% *ee*).

In contrast to the widely studied organometallic addition to aldehydes, the reaction with imines has not reached a similar degree of development, probably due to the generally lower chemical stability (i.e., sensitivity to hydrolysis) and poorer electrophilic character of imines in comparison with aldehydes. In spite of that, in recent years, the Lewis acid catalyzed Mannich reaction has emerged as a very efficient tool for the asymmetric synthesis of β-amino carbonyl derivatives [50]. In this field, cationic Cu(I) complexes of Fesulphos ligands have proven to be very effective catalysts in the Mannich-type addition of silyl enol ethers of ketones, esters and thioesters to *N*-sulfonyl aldimines [51]. The utilization of the dimeric Cu(I)-bromo complex of the bulky Fesulphos dinaphthyl phosphine **16f** as catalyst and 2-thienylsulfonyl imines as electrophiles provided the highest levels of enantioselectivity (86–93% *ee*; Scheme 9.15). In the case of ester and thioester Mannich products, the sulfonamide moiety was readily deprotected to the free amine by simple treatment with Mg in methanol.

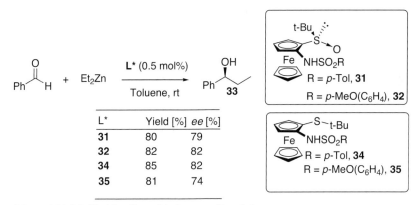

Scheme 9.14 1-Sulfinyl 2-sulfonamido ferrocene ligands in enantioselective addition of diethyl zinc to aldehydes.

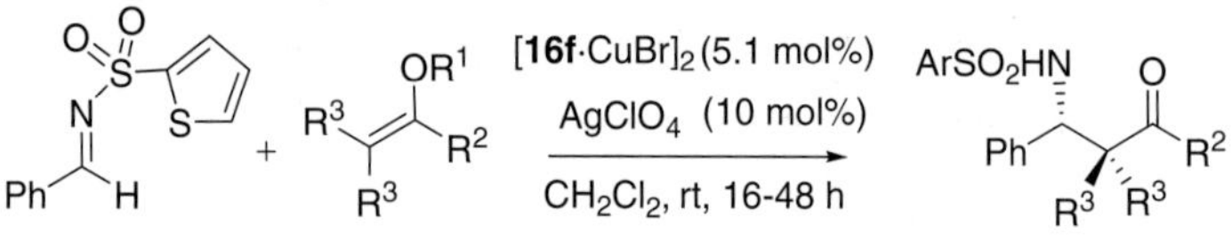

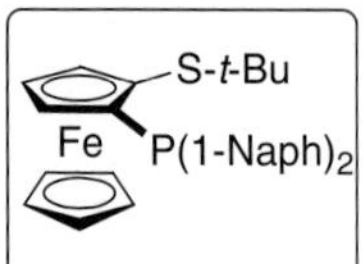

Scheme 9.15 (R_p)-Fesulphos/Cu-catalyzed asymmetric Mannich-type reaction of N-sulfonyl imines.

R^1	R^2	R^3	yield [%]	ee (%)
TBDMS	S-t-Bu	H	80	91 (>99)[a]
TBDMS	OMe	Me	78	86
TMS	Ph	H	80	93

[a] After a single recrystallization

9.6.2
Asymmetric 1,4-Conjugate Addition

The key relevance of the conjugate addition as one of the most powerful and widely used synthetic tools for C—C bond formation has encouraged the development of efficient chiral catalyst systems for the asymmetric 1,4-addition of nucleophiles to a variety of Michael-type acceptors [52]. In 2001, the group of Alexakis reported the first Cu(II)-catalyzed asymmetric conjugate addition of dialkyl zinc reagents to alkylidene malonates [53]. Among the tested ligands TADDOL-type **36** was the most effective, affording the addition products with asymmetric inductions up to 73% *ee*. A slightly lower enantioselectivity was achieved using the sulfur-containing ferrocene **37**, albeit somewhat higher than with the corresponding ferrocene ligand lacking the sulfur substitution (ligand **38**, Scheme 9.16).

9.7
Asymmetric Cycloaddition Reactions

9.7.1
[4 + 2] Cycloadditions

The Diels–Alder cycloaddition is one of the most powerful reactions in organic synthesis, and a good number of useful catalytic asymmetric variants have been developed to date [54]. The cycloaddition between bidentate 3-alkenoyl-1,3-oxazolidin-2-ones and cyclopentadiene has been frequently used as a test reaction for the evaluation of new chiral Lewis acid catalysts. However, only a small number of ferrocene-based chiral Lewis acids have provided good results in this transformation [55], and, to the best of our knowledge, the Fesulphos family has been the only sulfur-containing ferrocene chiral ligands used in this standard reaction [56]. Among a set of Fesulphos palladium complexes, the best results with regard to both reactivity and enantioselectivity were found with the palladium dichloro complex

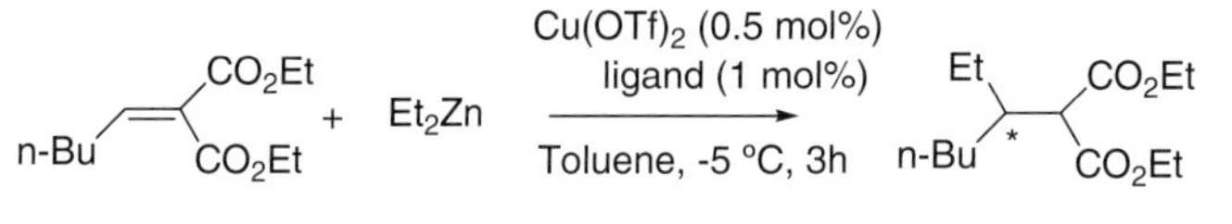

ligand	Yield (%)	*ee* (%)
36	89	65
37	85	57
38	63	53

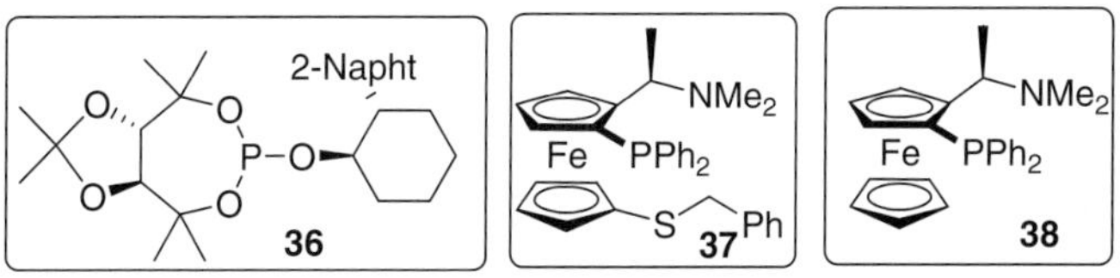

Scheme 9.16 Cu(II)-catalyzed asymmetric conjugate addition of dialkyl zinc to alkylidene malonates.

of *o*-Tol-Fesulphos **16g**, which afforded the endo cycloadduct (*R*)-**39** in excellent chemical yield (90%) and enantioselectivity (95%). Interestingly, the Cu(I)-complex of the same ligand [{**16g**CuBr}$_2$] gave rise to the opposite enantiomer (*S*)-**39** in 54% *ee* (Scheme 9.17). This reverse enantioselectivity displayed by Pd and Cu Fesulphos complexes has been explained on the basis of the different geometry of the palladium (square planar) and copper (distorted tetrahedral) complexes, as deduced by X-ray crystallographic studies on dichloropalladium and chlorocopper(I) Fesulphos complexes.

Interestingly, the Cu(I)-Fesulphos complexes also provided excellent results in the formal asymmetric aza Diels–Alder reaction between *N*-sulfonyl imines and Danishefsky-type dienes [57]. The reaction leads first to the Mannich-type addition product which, without isolation, is transformed *in situ* into the Diels–Alder adduct after addition of several drops of trifluoroacetic acid to the reaction mixture. In this transformation the copper bromide complex of the bulky 1-naphthylphosphine **16f** provided the best results (60–90% yield and enantioselectivities usually in the range 80–97% *ee*). A wide variety of aromatic, alkenyl, and even alkyl-substituted *N*-sulfonyl imines proved to be efficient substrates for this reaction (Scheme 9.18).

The synthetic potential of this methodology, which provides a convergent access to highly valuable optically active 2,3-dihydro-4-(1*H*)-dihydropyridones, has been recently illustrated by the stereodivergent synthesis of (+)-lasubines I and II [58].

39

catalyst	endo/ exo	Yield (%)	*ee* (%)
16g·$PdCl_2$	98:2	90	95 (*R*)
[**16g**·CuBr]$_2$	99:1	87	54 (*S*)

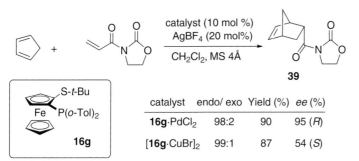

Scheme 9.17 (R_p)-Fesulphos metal complex catalyzed asymmetric Diels-Alder reaction.

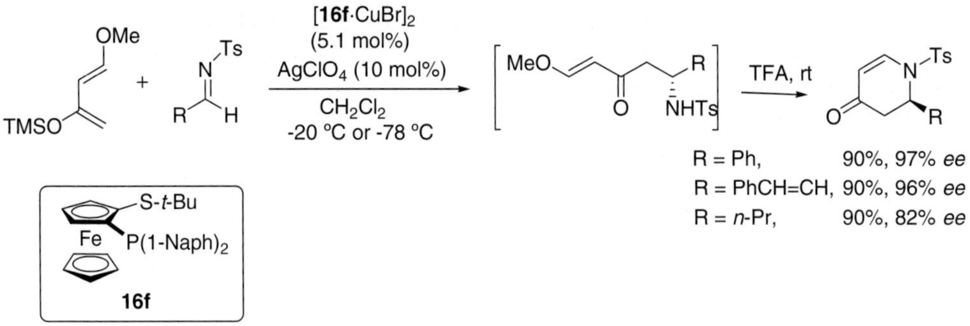

Scheme 9.18 Cu-Fesulphos catalyzed formal Aza Diels–Alder reactions of N-sulfonyl imines.

The key common intermediate N-tosyl 2,3-dihydropyridone **41** was obtained with high enantioselectivity applying the Fesulphos-Cu-catalyzed asymmetric aza Diels–Alder reaction between N-tosyl imine **40** and Danishefsky's diene (Scheme 9.19).

9.7.2
[3 + 2] Cycloadditions

The catalytic asymmetric 1,3-dipolar reaction is one of the most powerful methods for the enantioselective preparation of five-membered heterocycles [59]. In particular,

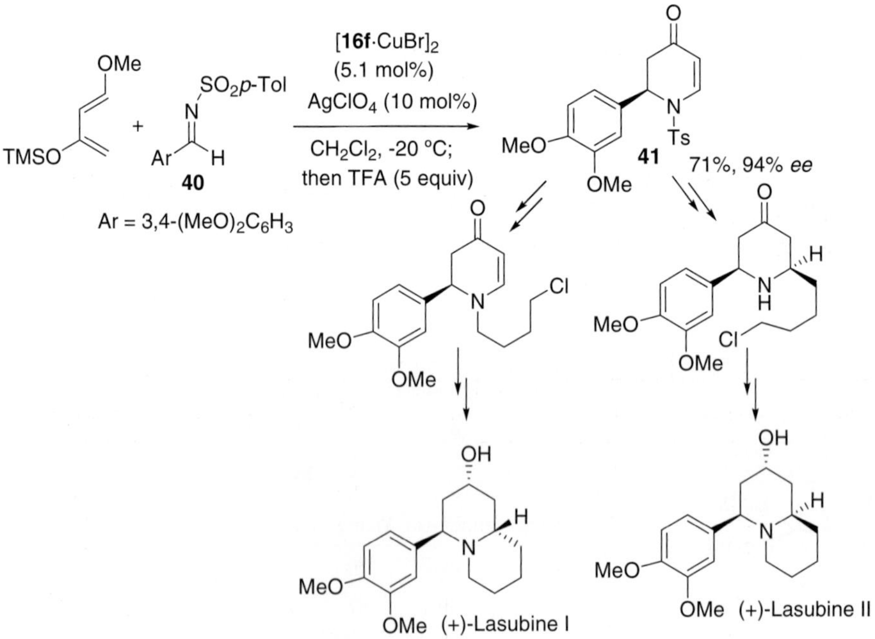

Scheme 9.19 Synthesis of (+)-Lasubines I and II.

9.7 Asymmetric Cycloaddition Reactions

the catalytic asymmetric 1,3-dipolar cycloaddition of azomethine ylides with electron-deficient alkenes constitutes an exceptionally reliable tool for the synthesis of highly substituted pyrrolidines [60]. Since the first catalytic enantioselective examples reported in 2002 [61], intense research in this field has resulted in the development of some very efficient procedures, including catalyst systems based on chiral metal complexes of bidentate P,P-, P,N-, and P,S-ferrocene ligands. The Cu(I)-Fesulphos catalyst system developed by our group has been successfully used in the cycloaddition of azomethine ylides [62], providing nearly complete endo selectivities and enantioselectivies (>99% *ee*) in the reaction of aryl imines of methyl glycinate with *N*-phenylmaleimide. The reaction has a broad scope, tolerating α-substituted azomethine ylides ($R^3 \neq H$) and ketimine ylide precursors ($R^2 \neq H$). In both cases, pyrrolidines with either a quaternary stereocenter at C-2 (Table 9.6, entries 2 and 3) or at C-5 (entries 4 and 5) were obtained with complete endo enantioselectivity and high enantiocontrol. Other dipolarophiles of varied nature, such as methyl acrylate, dimethyl maleate, dimethyl fumarate, methyl acrolein, fumarodinitrile, and nitrostyrene also participated in this 1,3-dipolar cycloaddition, providing the corresponding pyrrolidines with high enantiocontrol (76–99% *ee*), albeit the endo/exo selectivity was very dependent on the stereochemistry and substitution at the dipolarophile. In addition, the polystyrene-supported Fesulphos ligand **16c** proved to be as efficient a catalyst as the homogeneous ligand in the reaction of azomethine ylides with *N*-phenylmaleimide (entries 6 and 7). Interestingly, this chiral heterogeneous catalyst can be recycled and reused at least three times without significant loss of efficiency [23].

Table 9.6 Fesulphos/Cu-catalyzed asymmetric 1,3-dipolar cycloaddition of azomethine ylides with *N*-phenyl maleimide.

Entry	Ligand	R¹	R²	R³	Yield (%)	*ee* (%)
1	16a	Ph	H	H	81	>99
2	16a	Ph	H	Me	50	80
3	16a	2-Naph	H	Me	78	92
4	16a	Ph	Me	H	78	94
5	16a	4-ClC₆H₄	Me	H	80	>99
6	16c	Ph	H	H	95	>99
7	16c	Ph	Ph	H	95	>99

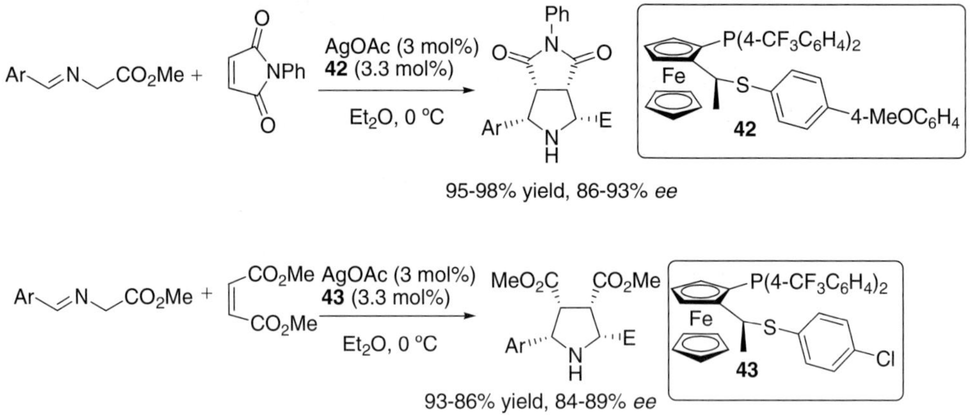

Ar = Ph, 2-Naph, 4-F-C_6H_4, 4-MeOC_6H_4, 2-Me-C_6H_4

Scheme 9.20 P,S-ferrocene ligands in Ag(I)-catalyzed asymmetric dipolar cycloaddition of azomethine ylides.

Recently Zhou and co-workers have explored other P,S-ferrocene ligands, such as **42** and **43**, in the Ag(I)-catalyzed asymmetric dipolar cycloaddition of azomethine ylides with *N*-phenylmaleimide and dimethyl maleate [63]. High enantioselectivities were achieved by tuning the substitution at both phosphorus- and sulfur-coordinating atoms. Thus, ligand **42** afforded the best results in the reaction with *N*-phenylmaleimide (86–93% *ee*, Scheme 9.20), while ligand **43** was the most enantioselective in the cycloaddition with dimethyl maleate (84–89% *ee*).

9.8 Asymmetric Nucleophilic Catalysis

The development of chiral Lewis bases for nucleophilic catalysis has attracted great attention in recent years [64]. In this field a limited number of, in some cases very efficient, catalysts having a ferrocene scaffold have been reported [65]. In 2003 Kobayashi and coworkers reported that chiral alkyl aryl sulfoxides can promote the enantioselective reaction of *N*-acylhydrazones with allyltrichlorosilanes giving rise to the corresponding homoallylic amines with high diastereo- and enantioselectivities [66]. Later, Fernández, Khiar *et al.* reported the use of ferrocenyl sulfoxides as Lewis bases in this reaction [67] (Scheme 9.21). Among several tested ferrocenyl

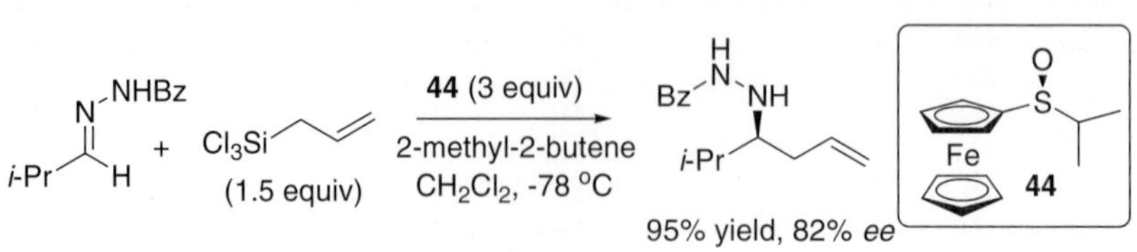

Scheme 9.21 Asymmetric allylation of hydrazones mediated by chiral sulfinyl ferrocenes.

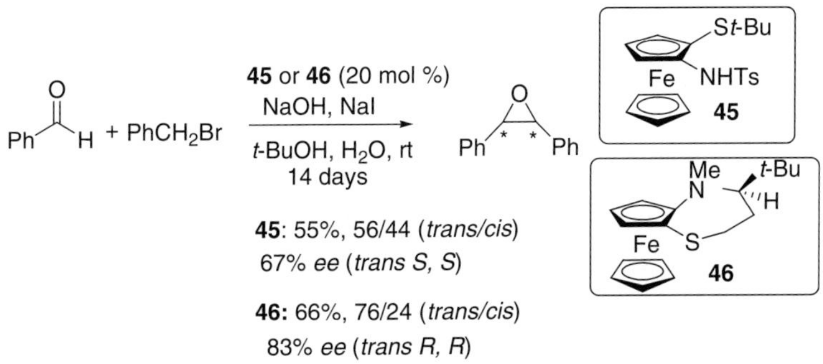

Scheme 9.22 Asymmetric epoxidation of aromatic aldehydes via chiral ferrocenyl sulfur ylides.

Reaction conditions: PhCHO + PhCH$_2$Br with 45 or 46 (20 mol %), NaOH, NaI, t-BuOH, H$_2$O, rt, 14 days → stilbene oxide (Ph-epoxide-Ph).

45: 55%, 56/44 (trans/cis), 67% ee (trans S, S)
46: 66%, 76/24 (trans/cis), 83% ee (trans R, R)

Ligand 45: ferrocene with S-t-Bu and NHTs substituents.
Ligand 46: ferrocene with cyclic N(Me)–CH(t-Bu)–S unit.

sulfoxides the isopropyl derivative (44) afforded the highest enantioselectivity (82% ee), albeit 3 equiv of the sulfoxide are required.

Metzner and coworkers have reported that chiral ferrocenyl sulfides catalyze the asymmetric Johnson–Corey epoxidation of aromatic aldehydes via sulfonium ylides. In the epoxidation of benzaldehyde to *trans*-stilbene oxide a 67% ee was obtained with the planar chiral *tert*-butylsulfenylferrocene 45 [68] (Scheme 9.22). The enantioselectivity was later improved to 83% ee using the more rigid cyclic sulfide 46 [69].

9.9
Conclusion and Perspectives

This chapter describes the most significant results on the application of sulfur-containing ferrocene ligands in asymmetric catalysis. In this recent and growing field a variety of chiral ligands have been prepared, many of them having chiral planar 1,2-disubstitution at the ferrocene moiety and S,N- or P,S-bidentate coordination. This kind of ligand has been tested in diverse metal-mediated asymmetric catalytic processes (mainly Pd-, Cu- and Rh-catalyzed reactions), providing excellent reactivities and enantioselectivities in some cases. Nevertheless, likely because this topic of research is in its infancy, the standard Pd-catalyzed allylic substitution of 1,3-diphenylpropenyl acetate is still the most studied process.

Due to the very interesting electronic and steric properties of sulfur ligands, compared with phosphorus and nitrogen ligands, and the facile synthesis and high stability of sulfur-containing ferrocene ligands, there is a great room for improvement in ferrocene ligand design, application to a much more varied set of metal-mediated reactions, and mechanistic studies. Significant progress in all these fields, as well as studies on the seldom explored ability of sulfur-containing ferrocenes in organocatalysis, is expected in the coming years.

9.10
Experimental: Selected Procedures

9.10.1
Palladium-Catalyzed Allylic Alkylation Reaction with Ligand 3a [10]

[Pd(η^3-C$_3$H$_5$)Cl]$_2$ (0.01 mmol) and ligand **3a** (0.03 mmol) were dissolved in dry CH$_2$Cl$_2$ (2 mL) and the mixture was stirred for 30 min at rt under argon atmosphere. To this solution were successively added 1,3-diphenyl-2-propenyl acetate (0.5 mmol), dimethyl malonate (1.5 mmol), N,O-bis(trimethylsilyl)acetamide (1.5 mmol) and lithium acetate (0.015 mmol). The reaction mixture was stirred at rt and monitored by TLC. After completion, the reaction mixture was diluted with CH$_2$Cl$_2$ (20 mL) and washed twice with ice-cold saturated aqueous NH$_4$Cl. The organic phase was dried over anhydrous MgSO$_4$ and then concentrated under reduced pressure. The residue was purified by preparative TLC (EtOAc/petroleum ether = 1/15) to give the pure product (98% yield, 98% *ee*).

9.10.2
Fesulphos/Palladium-Catalyzed Asymmetric Ring Opening of Bicyclic Alkenes [36]

A mixture of oxabenzonorbornadiene (0.21 mmol), palladium complex (**16**)Pd(Cl)(Me) (0.001 mmol) and NaB(ArF)$_4$ (0.001 mmol) in DCE (2 mL) was stirred at room temperature for 5 min. The resulting mixture was treated with a 1 M solution of Et$_2$Zn in hexane (0.31 mmol). Once the starting material was consumed (monitored by TLC) a few drops of brine were added and the solution was stirred at room temperature for 0.5 h. The mixture was filtered through Celite® and the filtrate was concentrated to dryness. The residue was purified by flash chromatography to give the alcohol **23** (81% yield, 95% *ee*).

9.10.3
Rhodium-Catalyzed Hydrogenation α-(N-acetamido)acrylate in the Presence of Ligand 20 [26]

A solution of [Rh-(NBD)$_2$SbF$_6$] (2.3 mg, 0.0050 mmol) and ligand **20** (4.4 mg, 0.0055 mmol) in MeOH (5 mL) was stirred in a glove box for 10 min to allow the formation of catalyst. The complex solution (1 mL) was then put in each of the hydrogenation vials. The commercially available substrate α-(N-acetamido)acrylate (14.3 mg, 0.1 mmol) was added and the mixture hydrogenated in an autoclave for 12 h. After carefully releasing the hydrogen, the reaction mixture was passed through a short silica-gel plug to remove the catalyst, providing the amino acid derivative product (99% yield, 93% *ee*).

9.10.4
Cu-Fesulphos Catalyzed Formal Aza Diels–Alder Reactions of N-Sulfonyl Imines [57]

A solution of [**16f**·CuBr]$_2$ (0.0056 mmol) and AgClO$_4$ (0.011 mmol) in CH$_2$Cl$_2$ (0.6 mL) was stirred in the dark at room temperature under argon for 2 h and treated with a solution of N-phenylsulfonyl benzenamide (0.11 mmol) in CH_2Cl_2 (1 mL). The reaction mixture was stirred at room temperature for 5 min before it was treated with Danishefsky's diene (0.15 mmol) at $-20\,°C$. Once the starting material was consumed (1 h), TFA (0.1–0.2 mL) was added and the mixture was stirred at room temperature for 30–60 min. The reaction mixture was neutralized with saturated aqueous $NaHCO_3$, extracted with CH_2Cl_2, dried over $MgSO_4$ and filtered. After evaporation of the solvent, the residue was purified by flash chromatography to provide the corresponding N-tosyl 2,3-dihydropyridone product (90% yield, 97% ee).

Abbreviations

Ac	acetyl
BINAP	2,2′-bis(diphenylphosphino)-1,1′-binaphthyl
Bn	benzyl
BSA	N,O-bis(trimethylsilyl)acetamide
Bz	benzoyl
cod	1,5-cyclooctadiene
Cy	cyclohexyl
dba	dibenzylideneacetone
DCE	1,1-dichloroethane
Naph	naphthyl
NBD	norbornadiene
Phth	phthalimide
PMHS	poly methylhydrosiloxane
Py	pyridine
TADDOL	2,2-dimethyl-$\alpha,\alpha,\alpha',\alpha'$-tetraaryl-1,3-dioxolane-4,5-dimethanol
TBDMS	tert-butyldimethylsilyl
TBDPS	tert-butyldiphenylsilyl
Tf	triflate
TFA	trifluoroacetic acid
TMS	trimethylsilyl
Tol	tolyl
Ts	p-toluensulfonyl

Acknowledgments

Financial support by the Ministerio de Educación y Ciencia (MEC, project CTQ2006-01121) and Consejería de Educación de la Comunidad de Madrid, UAM (UAM/CAM project CCG07-UAM/PPQ-1670) on our work on Fesulphos ligands is gratefully acknowledged. J. A. and M. R. R. thank the MEC for a Ramón y Cajal and a Juan de la Cierva contract, respectively.

References

1 Hayashi, T., Yamamoto, K., and Kumada, M. (1974) *Tetrahedron Lett.*, **15**, 4405.
2 (a) For recent reviews on chiral ferrocenes in asymmetric catalysis, see: Gómez Arrayás, R., Adrio, J., and Carretero, J.C. (2006) *Angew. Chem. Int. Ed.*, **45**, 7674; (b) Atkinson, R.C.J., Gibson, V.C., and Long, N.J. (2004) *Chem. Soc. Rev.*, **33**, 313; (c) Colacot, T.J. (2003) *Chem. Rev.*, **103**, 3101.
3 Spencer, J., Gramlich, V., Häusel, R., and Togni, A. (1996) *Tetrahedron: Asymmetry*, **7**, 41.
4 Albinati, A., Pregosin, P.S., and Wick, K. (1996) *Organometallics*, **15**, 2419.
5 (a) For recent reviews on chiral sulfur ligands in asymmetric catalysis, see: Mellah, M., Voituriez, A., and Schulz, E. (2007) *Chem. Rev.*, **107**, 5133; (b) Pellisier, H. (2007) *Tetrahedron*, **63**, 1297.
6 Bonini, B.F., Fochi, M., and Ricci, A. (2007) *Synlett*, 360–373.
7 (a) For recent reviews on asymmetric allylic alkylation: Trost, B.M. and Crawley, M.L. (2003) *Chem. Rev.*, **103**, 2921–2944; (b) Graening, T. and Schmalz, H.-G. (2003) *Angew. Chem. Int. Ed.*, **42**, 2580–2584; (c) Trost, B.M. (2004) *J. Org. Chem.*, **69**, 5813–5837; (d) Lu, Z. and Ma, S. (2008) *Angew. Chem. Int. Ed.*, **47**, 258–297.
8 Chesney, A., Bryce, M.R., Chubb, R.W.J., Batsanov, A.S., and Howard, J.A.K. (1997) *Tetrahedron: Asymmetry*, **8**, 2337–2346.
9 Park, J., Quan, Z., Lee, S., Ahn, K.H., and Cho, C.-W. (1999) *J. Organomet. Chem.*, **584**, 140–146.
10 (a) You, S.-L., Zhou, Y.-G., Hou, X.-L., and Dai, L.-X. (1998) *Chem. Comm.*, 2765–2766; (b) You, S.-L., Hou, X.-L., Dai, L.-X., Yu, Y.-H., and Xia, W. (2002) *J. Org. Chem.*, **67**, 4684–4695.
11 You, S.-L., Hou, X.-L., and Dai, L.-X. (2000) *Tetrahedron: Asymmetry*, **11**, 1495–1500.
12 Manoury, E., Fossey, J.S., Aït-Haddou, H., Daran, J.-C., and Balavoine, G.G.A. (2000) *Organometallics*, **19**, 3736–3739.
13 Bonini, B.F., Fochi, M., Comes-Franchini, M., Ricci, A., Thijs, L., and Zwanenburg, B. (2003) *Tetrahedron: Asymmetry*, **14**, 3321–3327.
14 Bernardi, L., Bonini, B.F., Comes-Franchini, M., Fochi, M., Mazzanti, G., Ricci, A., and Varchi, G. (2002) *Eur. J. Org. Chem.*, 2776–2784.
15 Bernardi, L., Bonini, B.F., Comes-Franchini, M., Dessole, G., Fochi, M., and Ricci, A. (2005) *Phosphorus, Sulfur, Silicon Relat. Elem.*, **180**, 1273–1277.
16 Hu, X., Bai, C., Dai, H., Chen, H., and Zheng, Z. (2004) *J. Mol. Catal. A*, **218**, 107–112.
17 Tu, T., Zhou, Y.-G., Hou, X.-L., Dai, L.-X., Dong, X.-C., Yu, Y.-H., and Sun, J. (2003) *Organometallics*, **22**, 1255–1265.
18 (a) Enders, D., Peters, R., Runsink, J., and Bats, J.W. (1999) *Org. Lett.*, **1**, 1863–1866; (b) Enders, D., Peters, R., Lochtman, R., Raabe, G., Runsink, J., and Bats, J.W. (2000) *Eur. J. Org. Chem.*, 3399–3426.
19 Lam, F.L., Au-Yeung, T.T.L., Cheung, H.Y., Kok, S.H.L., Lam, W.S., Wong, K.Y., and Chan, A.S.C. (2006) *Tetrahedron: Asymmetry*, **17**, 497–499.
20 Lam, F.L., Au-Yeung, T.T.L., Kwong, F.Y., Zhou, Z., Wong, K.Y., and Chan, A.S.C. (2008) *Angew. Chem. Int. Ed.*, **47**, 1280–1283.
21 Cheung, H.Y., Yu, W.-Y., Lam, F.L., Au-Yeung, T.T.L., Zhou, Z., Chan, T.H., and Chan, A.S.C. (2007) *Org. Lett.*, **9**, 4295–4298.
22 (a) Priego, J., García Mancheño, O., Cabrera, S., Gómez Arrayás, R., Llamas, T., and Carretero, J.C. (2002) *Chem. Comm.*, 2512–2513; (b) García Mancheño, O., Priego, J., Cabrera, S., Gómez Arrayás, R., Llamas, T., and Carretero, J.C. (2003) *J. Org. Chem.*, **68**, 3679–3686.
23 Martín-Matute, B., Pereira, S.I., Peña-Cabrera, E., Adrio, J., Silva, A.M.S.,

and Carretero, J.C. (2007) *Adv. Synth. Catal.*, **349**, 1714–1724.
24 Routaboul, L., Vincendeau, S., Daran, J.-C., and Manoury, E. (2005) *Tetrahedron: Asymmetry*, **16**, 2685–2690.
25 Kang, J., Lee, J.H., and Im, K.S. (2003) *J. Mol. Catal. A*, **196**, 55–63.
26 Raghunath, M., Gao, W., and Zhang, X. (2005) *Tetrahedron: Asymmetry*, **16**, 3676–3681.
27 Nakamura, S., Fukuzumi, T., and Toru, T. (2004) *Chirality*, **16**, 10–12.
28 (a) For recent reviews on Cu-catalyzed asymmetric allylation reaction: Yorimitsu, H. and Oshima, K. (2005) *Angew. Chem. Int. Ed.*, **44**, 4435–4439; (b) Alexakis, A., Malan, C., Lea, L., Tissot-Croset, K., Polet, D., and Falciola, C. (2006) *Chimia*, **60**, 124–130.
29 (a) Karlström, A.S.E., Huerta, F.F., Meuzelaar, G.J., and Bäckvall, J.-E. (2001) *Synlett*, 923–926; (b) Cotton, H.K., Norinder, J., and Bäckvall, J.-E. (2006) *Tetrahedron*, **62**, 5632–5640.
30 For a recent review, see: Shibasaki, M., Vogl, E.M., and Ohshima, T. (2004) *Adv. Synth. Catal.*, **346**, 1533–1552.
31 (a) Tu, T., Deng, W.-P., Hou, X.-L., Dai, L.-X., and Dong, X.-C. (2003) *Chem. Eur. J.*, **9**, 3073–3081; (b) Kilroy, T.G., Hennessy, A.J., Connolly, D.J., Malone, Y.M., Farrell, A., and Guiry, P.J. (2003) *J. Mol. Catal. A*, **196**, 65–81.
32 For an overview, see: Lautens, M., Fagnou, K., and Hiebert, S. (2003) *Acc. Chem. Res.*, **36**, 48–58.
33 Lautens, M., Renaud, J.-L., and Hiebert, S. (2000) *J. Am. Chem. Soc.*, **122**, 1804–1805.
34 (a) Dotta, P., Kuwar, P.G.A., Pregrosin, P.S., Albinati, A., and Rizzato, S. (2004) *Organometallics*, **23**, 2295–2304; (b) Imamoto, T., Sugita, K., and Yoshida, K. (2005) *J. Am. Chem. Soc.*, **127**, 11934–11935.
35 Li, M., Yan, X.-X., Hong, W., Zhu, X.-Z., Cao, B.-X., Sun, J., and Hou, X.-L. (2004) *Org. Lett.*, **6**, 2833–2835.
36 (a) Cabrera, S., Gómez Arrayás, R., and Carretero, J.C. (2004) *Angew. Chem. Int. Ed.*, **43**, 3944–3947; (b) Cabrera, S., Gómez Arrayás, R., Alonso, I., and Carretero, J.C. (2005) *J. Am. Chem. Soc.*, **127**, 17938–17947.
37 Lautens, M. and Hiebert, S. (2004) *J. Am. Chem. Soc.*, **126**, 1437–1447.
38 (a) Jiménez-Núñez, E. and Echavarren, A.M. (2007) *Chem. Comm.*, **4**, 333–346; (b) Gorin, D.J. and Toste, F.D. (2007) *Nature*, **446**, 395–403; (c) Hashmi, A.S.K. (2007) *Chem. Rev.*, **107**, 3180–3211; (d) Fürstner, A. and Davies, P.W. (2007) *Angew Chem. Int. Ed.*, **46**, 3410–3449; (e) Bongers, N. and Krause, N. (2008) *Angew. Chem. Int. Ed.*, **47**, 2–6.
39 (a) Ito, Y., Sawamura, M., and Hayashi, T. (1986) *J. Am. Chem. Soc.*, **108**, 6405–6406; (b) Hayashi, T. (1988) *Pure Appl. Chem.*, **60**, 7–12.
40 Togni, A. and Haüsel, R. (1990) *Synlett*, 633–635.
41 Muñoz, M.P., Adrio, J., Carretero, J.C., and Echavarren, A.M. (2005) *Organometallics*, **24**, 1293–1300.
42 Breuer, M., Ditrich, K., Habicher, T., Hauer, B., Kebeler, M., Stürmer, R., and Zelinski, T. (2004) *Angew. Chem. Int. Ed.*, **43**, 788–824.
43 Blaser, H.-U. and Schmidt, E. (2004) *Asymmetric Catalysis on Industrial Scale*, 1st edn, Wiley & Sons, New York.
44 Nishiyama, H. (2004) *Transition Metals for Organic Synthesis*, 2nd edn (eds M. S Beller and C. S Bolm), Wiley–VCH Verlag GmbH, Weinheim, pp. 182–191.
45 Nishibayashi, Y., Segawa, K., Singh, J.D., Fukuzawa, S.-i., Ohe, K., and Uemura, S. (1996) *Organometallics*, **15**, 370–379.
46 Gérard, S., Pressel, Y., and Riant, O. (2005) *Tetrahedron: Asymmetry*, **16**, 1889–1891.
47 (a) For reviews, see: Pu, L. and Yu, H.-B. (2001) *Chem. Rev.*, **101**, 757–824; (b) Pu, L. (2003) *Tetrahedron*, **59**, 9873–9886.
48 (a) For selected recent references, see: Bolm, C. and Rudolph, J. (2002) *J. Am. Chem. Soc.*, **124**, 14850–14851; (b) Bonini, B.F., Fochi, M., Comes-Franchini, M., Ricci, A., Thijs, L., and Zwanenburg, B. (2003) *Tetrahedron: Asymmetry*, **14**,

3321–3327; (c) Wang, M.-C., Wang, D.-K., Zhu, Y., Liu, L.-T., and Guo, Y.-F. (2004) *Tetrahedron: Asymmetry*, **15**, 1289–1294; (d) Wang, M.-C., Liu, L.-T., Zhang, J.-S., Shi, Y.-Y., and Wang, D.-K. (2004) *Tetrahedron: Asymmetry*, **15**, 3853–3859; (e) Li, M., Zhu, X.-Z., Yuan, K., Cao, B.-X., and Hou, X.-L. (2004) *Tetrahedron: Asymmetry*, **15**, 219–222; (f) Rudolph, J., Hermanns, N., and Bolm, C. (2004) *J. Org. Chem.*, **69**, 3997–4000; (g) Faux, N., Razafimahefa, D., Picart-Goetgheluck, S., and Brocard, J. (2005) *Tetrahedron: Asymmetry*, **16**, 1189–1197; (h) Wang, M.-C., Hou, X.-H., Xu, C.-L., Liu, L.-T., Li, G.-L., and Wang, D.-K. (2005) *Synthesis*, 3620–3626; (i) Rudolph, J., Schmidt, F., and Bolm, C. (2005) *Synthesis*, 840–842.

49 Priego, J., García Mancheño, O., Cabrera, S., and Carretero, J.C. (2002) *J. Org. Chem.*, **67**, 1346–1353.

50 (a) For reviews, see: Córdoba, A. (2004) *Acc. Chem. Res.*, **37**, 102–112; (b) Marques, M.M.B. (2006) *Angew Chem. Int. Ed.*, **45**, 348–352.

51 Salvador González, A., Gómez Arrayás, R., and Carretero, J.C. (2006) *Org. Lett.*, **8**, 2977–2980.

52 (a) Dialkyl zinc reagents: Alexakis, A. and Benhaim, C., (2002) *Eur. J. Org. Chem.*, 3221–3236; (b) Boronic acids: Hayashi, T. and Yamasaki, K., (2003) *Chem. Rev.*, **103**, 2829–2844; (c) Hayashi, T. (2004) *Bull. Chem. Soc. Jpn.*, **77**, 13–21; (d) Grignard reagents: Feringa, B.L., Badorrey, R., Peña, D., Harutyunyan, S.R., and Minnaard, A.J., (2004) *Proc. Natl. Acad. Sci. USA*, **101**, 5834–5838.

53 Alexakis, A. and Benhaim, C. (2001) *Tetrahedron: Asymmetry*, **12**, 1151–1157.

54 For a review, see: Carmona, D., Lamata, M.P., and Oro, L.A. (2000) *Coord. Chem. Rev.*, **200–202**, 717–772.

55 Fukuzawa, S.-i., Yahara, Y., Kamiyama, A., Hara, M., and Kikuchi, S. (2005) *Org. Lett.*, **7**, 5809–5812.

56 García Mancheño, O., Gómez Arrayás, R., and Carretero, J.C. (2005) *Organometallics*, **24**, 557–561.

57 García Mancheño, O., Gómez Arrayás, R., and Carretero, J.C. (2004) *J. Am. Chem. Soc.*, **126**, 456–457.

58 García Mancheño, O., Gómez Arrayás, R., Adrio, J., and Carretero, J.C. (2007) *J. Org. Chem.*, **72**, 10294–10297.

59 For a review, see: Pellissier, H. (2007) *Tetrahedron*, **63**, 3235–3285.

60 For a review, see: Nájera, C. and Sansano, J.M. (2005) *Angew. Chem. Int. Ed.*, **44**, 6272–6276.

61 Longmire, J.M., Wang, B., and Zhang, X. (2002) *J. Am. Chem. Soc.*, **124**, 13400–13401.

62 Cabrera, S., Gómez Arrayás, R., and Carretero, J.C. (2005) *J. Am. Chem. Soc.*, **127**, 16394–16395.

63 Zeng, W. and Zhou, Y.-G. (2007) *Tetrahedron Lett.*, **48**, 4619–4622.

64 (a) For recent reviews, see: France, S., Guerin, D.J., Miller, S.J., and Lectka, T. (2003) *Chem. Rev.*, **103**, 2985–3012; (b) Denmark, S.E. and Fu, J. (2003) *Chem. Comm.*, 167–170. (c) Vedejs, E., Daugulis, O., MacKay, J.A., and Rozners, E. (2001) *Synlett*, 1499–1505.

65 (a) Fu, G.C. (2000) *Acc. Chem. Res.*, **33**, 412–420; (b) Fu, G.C. (2004) *Acc. Chem. Res.*, **37**, 542–547.

66 Kobayashi, S., Ogawa, C., Konishi, H., and Sugiura, M. (2003) *J. Am. Chem. Soc.*, **125**, 6610–6611.

67 Fernández, I., Valdivia, V., Gori, B., Alcudia, F., Álvarez, E., and Khiar, N. (2005) *Org. Lett.*, **7**, 1307–1310.

68 (a) Minière, S., Reboul, V., Gómez Arrayás, R., Metzner, P., and Carretero, J.C. (2003) *Synthesis*, 2249–2254; (b) Minière, S., Reboul, V., and Metzner, P. (2005) *Arkivoc*, **6**, 161–177.

69 Minière, S., Reboul, V., Metzner, P., Fochi, M., and Bonini, B.F. (2004) *Tetrahedron: Asymmetry*, **15**, 3275–3280.

10
Biferrocene Ligands
Ryoichi Kuwano

10.1
Introduction

A great number of chiral ligands have been developed for metal-catalyzed asymmetric reactions [1–3]. Well-considered design of the chiral ligands is a key factor for successful achievement of high enantioselectivity in asymmetric catalysis. A promising and popular access to excellent chiral ligands is the use of a C_2-symmetric chiral backbone, because the C_2-symmetry of the chiral ligands reduces the number of possible stereoisomeric interactions between a prochiral substrate and the chiral catalyst. Indeed, many C_2-chiral ligands, such as DIOP [4] and CHIRAPHOS [5], are derived from various accessible C_2-chiral diols or diamines (Figure 10.1). An alternative approach to the development of new C_2-chiral ligands is utilization of a dimeric structure of a monodentate ligand. BINAP [6], which is one of the successful chiral ligands, possesses the dimeric structure of achiral 2-(diphenylphosphino) naphthalene. Axial chirality arises around the carbon–carbon bond linking the two naphthalenes.

As described in other chapters, ferrocene-based chiral sources have often been used for designing new chiral ligands [7] since Hayashi and Kumada developed the first chiral ferrocenyl phosphines in 1974 [8, 9]. Chiral ligands of ferrocene have brought about great successes in catalytic asymmetric synthesis. Consequently, the dimeric form of ferrocene, that is, biferrocene, should be an attractive constituent for designing new chiral ligands. Atropisomerism around the bond linking each ferrocene is not fixed, unlike with the binaphthyl compounds, while planar chirality arises on the ferrocene moiety.

The first chiral ligand bearing a biferrocene skeleton was 2,2″-bis(diphenylphosphino)-1,1″-biferrocene, BIFEP, which was documented by Sawamura and Ito in 1991 [10]. They confirmed that the ligand forms a chelate with a transition metal, but did not disclose any application to asymmetric catalysis. In the same year, they reported successful biferrocene ligands, TRAPs, which can form a chelate complex with a transition metal in a trans-manner [11]. Since then, several chiral ligands have

Chiral Ferrocenes in Asymmetric Catalysis. Edited by Li-Xin Dai and Xue-Long Hou
Copyright © 2010 WILEY-VCH Verlag GmbH & Co. KGaA, Weinheim
ISBN: 978-3-527-32280-0

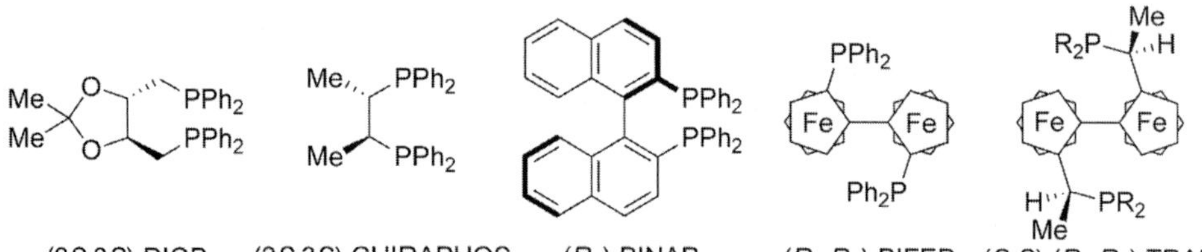

(2S,3S)-DIOP   (2S,3S)-CHIRAPHOS   (R_a)-BINAP   (R_p,R_p)-BIFEP   (S,S)-(R_p,R_p)-TRAP

Figure 10.1 Structures of representative C_2-symmetric chiral bisphosphines and biferrocene-based bisphosphines.

been developed by other researchers. This chapter treats the syntheses and applications of the chiral biferrocene ligands.

The biferrocene ligands possess planar chirality on each ferrocene unit. Two rules have been proposed for assignment of the stereochemical descriptors for the metallocene planar chirality. One procedure is based on the central chirality of the cyclopentadienyl carbon bearing the substituent with highest priority [12]. In another rule, the chirality descriptor is assigned by the arrangement of the two substituents on the planar chiral cyclopentadienyl ring [13, 14]. The latter rule is adopted in this chapter. For ferrocene derivatives possessing both central and planar chirality, the descriptor of the central chirality is written first [14] and that of the planar chirality follows in separate parentheses, as with Hayashi's ferrocenyl phosphines [8, 9]. Therefore, (R,R)-(S,S)-PhTRAP means that the PhTRAP possesses R-central chiralities and S-planar chiralities.

10.2
Trans-Chelating Chiral Bisphosphines: TRAP

The first successful chiral biferrocene ligand family was 2,2″-bis(1-phosphinoethyl)-1,1″-biferrocenes, TRAP **1** (Figure 10.2), which possess planar chiralities on the cyclopentadienyl rings and chiral centers at the α-positions of the side chains [11]. The chiral ligands are designed to coordinate to a metal atom in a bidentate trans-chelation manner. Many TRAP ligands bearing diverse aryl or alkyl groups on their phosphorus atoms have been prepared and applied to various catalytic asymmetric reactions. In naming each TRAP ligand, the abbreviation of the P-substituent is

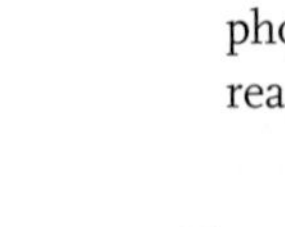

(S,S)-(R_p,R_p)-TRAP (**1**)

1a: R = Ph (PhTRAP)
1b: R = p-MeOC$_6$H$_4$ (AnisTRAP)
1c: R = p-ClC$_6$H$_4$
1d: R = p-CF$_3$C$_6$H$_4$
1e: R = 2-furyl (FurTRAP)

1f: R = Me (MeTRAP)
1g: R = Et (EtTRAP)
1h: R = Pr (PrTRAP)
1i: R = Bu (BuTRAP)
1j: R = i-Bu (i-BuTRAP)
1k: R = i-Pr (i-PrTRAP)

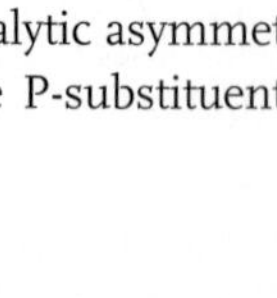

(S_p,S_p)-EtTRAP-H (**2**)

Figure 10.2 Structure of TRAP.

prefixed to "TRAP". The biferrocene ligand with no central chirality, TRAP-H **2**, has also been synthesized [15, 16]. The suffix "-H" means that the methyl groups on the chiral carbon atoms of TRAP are replaced by hydrogen atoms.

10.2.1
Synthesis of TRAP

The synthesis of (S,S)-(R_p,R_p)-TRAP **1** is outlined in Scheme 10.1 [17, 18]. The chiral ligands are synthesized from (S)-1-ferrocenyl-N,N-dimethylethylamine (S)-**3** [14], which is commercially available. The lithiation of (S)-**3** with *tert*-butyllithium occurs stereoselectively at one of its ortho-positions, inducing planar chirality on the cyclopentadienyl ring. Iodination of the lithioferrocene affords iodoferrocene (S)-(R_p)-**4** with 94% de [19]. After the quaternization of (S)-(R_p)-**4**, the resulting ammonium salt is converted into phosphine oxides (S)-(R_p)-**5a–5e** through nucleophilic substitution with a lithium diarylphosphinite. For the synthesis of TRAP bearing P-alkyl substituents, use of a lithium dialkylphosphinite as the nucleophile causes the Hofmann elimination of the ammonium group. Consequently, the amine is substituted with a dialkylphosphino group by way of the acetate obtained from the reaction of (S)-(R_p)-**4** with acetic anhydride. The acetate reacts with various dialkylphosphines in acetic acid, affording the desired trialkylphosphines in good yields. The trialkylphosphines are oxidized with hydrogen peroxide to give the phosphine oxides (S)-(R_p)-**5f–5k**. All of the above nucleophilic substitutions proceed with complete retention of the stereochemistry at the α-position of the ferrocene side chain. The biferrocene framework is constructed through the dimerization of (S)-(R_p)-**5** with copper powder. Finally, the phosphine oxide (R,R)-(S_p,S_p)-**6** is reduced with trichlorosilane and triethylamine to give (S,S)-(R_p,R_p)-TRAP (**1a–1k**). The antipode, (R,R)-(S_p,S_p)-**1**, is synthesized from (R)-**3**.

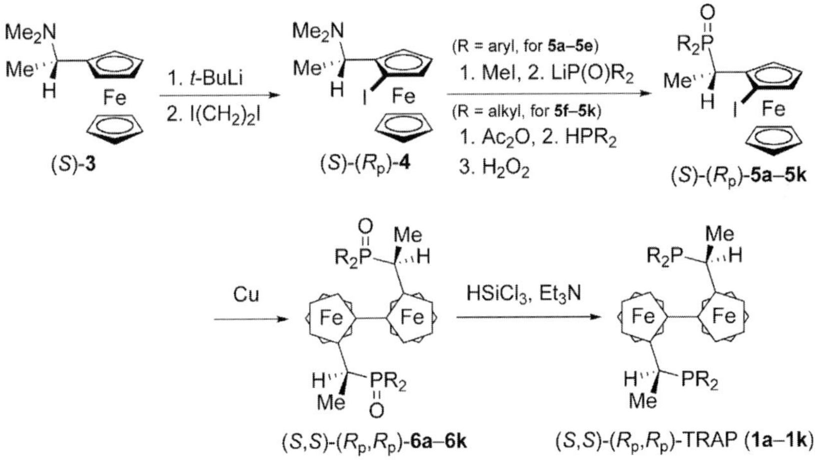

Scheme 10.1 Synthesis of TRAP.

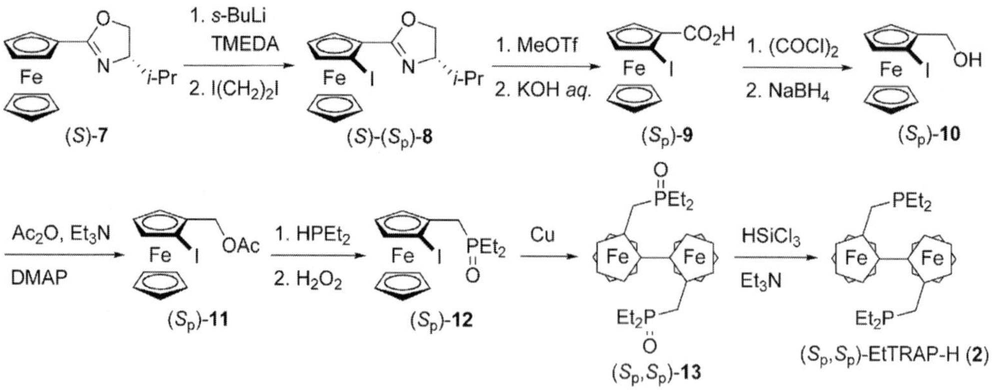

Scheme 10.2 Synthesis of TRAP-H.

Preparation of (S_p,S_p)-EtTRAP-H **2** [15, 16] starts from the diastereoselective ortholithiation of ferrocenyloxazoline (S)-**7** [20], which is readily obtained from (S)-valinol and ferrocenecarboxylic acid (Scheme 10.2) [21]. The resulting lithioferrocene reacts with 1,2-diiodoethane to give iodoferrocene (S)-(S_p)-**8** with high stereoselectivity (99 : 1). After treatment of (S)-(S_p)-**8** with methyl triflate, the oxazoline is hydrolyzed to carboxylic acid (S_p)-**9**. The carboxylic acid is transformed to the primary alcohol (S_p)-**10** by reduction with NaBH$_4$ after treatment with oxalyl chloride. The acetate (S_p)-**11**, which is obtained from the reaction of (S_p)-**10** with acetic anhydride, is subjected to nucleophilic substitution with diethylphosphine. The resulting trialkylphosphine is converted into the phosphine oxide (S_p)-**12**. The Ullmann coupling of (S_p)-**12** followed by reduction of the phosphine oxide (S_p,S_p)-**13** yields the planar chiral biferrocene ligand, (S_p,S_p)-EtTRAP-H **2**.

10.2.2
Metal Complexes of TRAP

In most cases, coordination of TRAP ligands **1** to a metal atom forms a trans-chelate, where each of the phosphorus atoms locates at the trans-position of another phosphorus [17, 18]. The biferrocene backbone of TRAP spans the metal atom.

Reaction of **1** with palladium bromide yields a square planar complex, trans-[PdBr$_2$(TRAP)], with no cis-isomer. Trans-chelation of the biferrocene ligands was demonstrated with the X-ray crystal structures of the palladium complexes (Figure 10.3). The bite angles of TRAPs range from 163–168°. As with palladium, carbonylchlororhodium(I) or iridium(I) binds the two phosphorus atoms of **1** with trans-chelation. The trans-chelation of EtTRAP-H **2** was confirmed by an NMR study of the reaction with [RhCl(CO)$_2$]$_2$ [15, 16]. Unlike the reaction of **1** with palladium, rhodium, or iridium, coordination of **1** to platinum can create a cis-chelate. Formation of trans-[PtCl$_2$(i-PrTRAP)] without the cis-isomer is observed in the reaction of PtCl$_2$(MeCN)$_2$ with **1k**, while Ph- (**1a**) or BuTRAP (**1i**) give a mixture of trans- and cis-platinum complex with ratios of 20:1 or 10:1, respectively. The ratio of trans- and

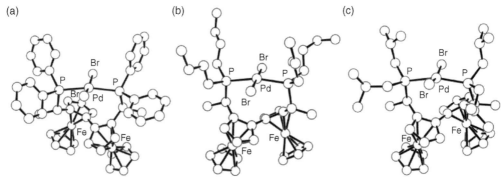

Figure 10.3 X-ray crystal structures of (a) trans-[PdBr$_2${(S,S)-(R_p,R_p)-**1a**}], (b) trans-[PdBr$_2${(R,R)-(S_p,S_p)-**1i**}], and (c) trans-[PdBr$_2${(R,R)-(S_p,S_p)-**1j**}].

cis-isomers seems to be affected by steric repulsion between the P-substituents of the bidentate ligand.

Selective formation of cis-chelating TRAP–metal complex is often observed when the metal precursor contains a ligand occupying plural vicinal coordination sites. The reaction of PhTRAP and [Pd(π-C$_3$H$_5$)(cod)]BF$_4$ yields **1a**-ligated (π-allyl)palladium, where the bisphosphine bonds to the palladium in cis-chelation [22]. Such cis-chelation of **1a** is observed in ruthenium complexes containing an η^6-arene ligand, such as [RuCl(p-cymene)(**1a**)]Cl [23]. In these cases, the multi-hapto organic ligands enforce cis-chelation on TRAP ligands.

10.2.3
Asymmetric Reactions of 2-Cyanocarboxylates

Since Murahashi et al. reported that a ruthenium complex is an effective catalyst for the Knoevenagel and Michael reaction of α-cyanocarboxylates [24, 25], some late transition metal complexes have been found to work as catalysts activating nitriles. The first successful application of TRAP to asymmetric catalysis is the Michael addition of 2-cyanopropionates **14**.

In 1992, Sawamura and Ito et al. reported that the 1,4-addition of 2-cyanopropionates **14** to vinyl ketones **15** proceeds with high enantioselectivity in the presence of the rhodium catalyst prepared from RhH(CO)(PPh$_3$)$_3$ and (S,S)-(R_p, R_p)-PhTRAP (**1a**) (Scheme 10.3) [26]. The reaction of 2-cyanopropionate **14a** with methyl vinyl ketone yields Michael adduct (R)-**16a** (R^1 = Et, R^2 = Me) with 81% ee. It is noteworthy that less than 20% ee of **16a** is afforded by using any chiral ligands other than **1**. The enantiomeric excess of the product increases to 93% ee when 2,4-dimethyl-3-pentyl ester **14b** is used as the substrate [27]. The chiral catalyst is effective for the reaction of acrolein or various vinyl ketones. Crotonaldehyde and methacrolein react with **14** to give the corresponding Michael adducts with 74–81% ee, while the products are obtained as 1:1 mixtures of diastereomers. The enantioselectivity is significantly affected by the α-substituent of the Michael

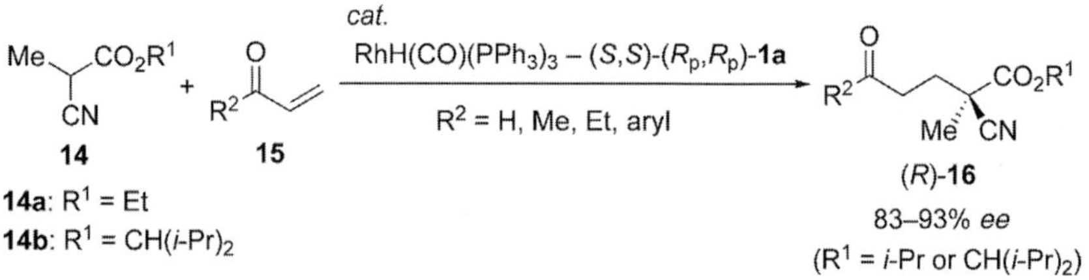

Scheme 10.3 Asymmetric Michael reaction of 2-cyanopropionate **14** with **15** catalyzed by **1a**–rhodium complex.

donor. The reaction of α-ethyl- or α-isopropylcyanoacetate proceeds with 21% or 1% ee, respectively.

The catalytic Michael addition is supposed to proceed through the deprotonated cyanoester **14** bonding to the rhodium through its cyano nitrogen [25]. The enantioselective carbon–carbon bond formation is accomplished at the α-carbon, which is distant from the metal center, as shown in Scheme 10.4. For high enantioselectivity, the chiral ligand on the rhodium needs to differentiate between the ester and α-methyl groups on the enolate of **14** as well as to select its enantioface reacting with **15**. The trans-spanning biferrocene moiety of (S,S)-(R_p,R_p)-**1a** will distinguish between the ester and methyl groups by steric repulsion. Furthermore, one of the P-phenyl substituents will obstruct the approach of **15** to the si-face of the enolate, leading to the selective formation of (R)-**16**. Such a remote stereocontrol will be achieved with high efficiency only by the concave chiral environment of TRAP. The convex chiral reaction fields created by cis-chelating bisphosphines may have no aptitude for enantioface selection.

Variants of **14**, N-methoxyl-N-methylamide **17** [28] and diethylphosphonate **21** [29], react with **15** in high enantioselectivity by using the PhTRAP–rhodium catalyst (Scheme 10.5). In these Michael reactions, Rh(acac)(CO)$_2$ is employed as the catalyst precursor. Weinreb amide (R)-**18** (R^1 = Me) is transformed into optically active aldehyde **19** or ketone **20** by way of acetal protection with ethylene glycol. The product (−)-**22** (R^1 = H) obtained from **21** and acrolein is usable for the synthesis of a protected α-amino phosphonic acid (−)-**23**, which has a quaternary chiral carbon.

The catalytic activation of α-cyanoesters by the PhTRAP–rhodium complex is applicable to the aldol reaction [30, 31]. 2-Cyanopropionate **14c** reacts with aqueous formaldehyde in the presence of Rh(acac)(CO)$_2$-(S,S)-(R_p,R_p)-**1a** catalyst, affording the aldol adduct (−)-**24a** with 93% ee (Scheme 10.6). The rhodium catalyst allows the

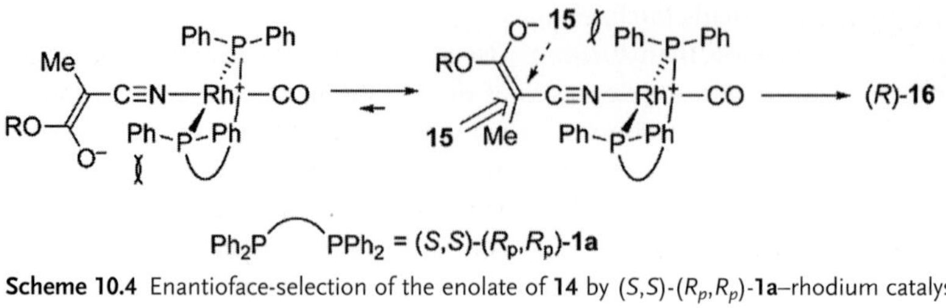

Scheme 10.4 Enantioface-selection of the enolate of **14** by (S,S)-(R_p,R_p)-**1a**–rhodium catalyst.

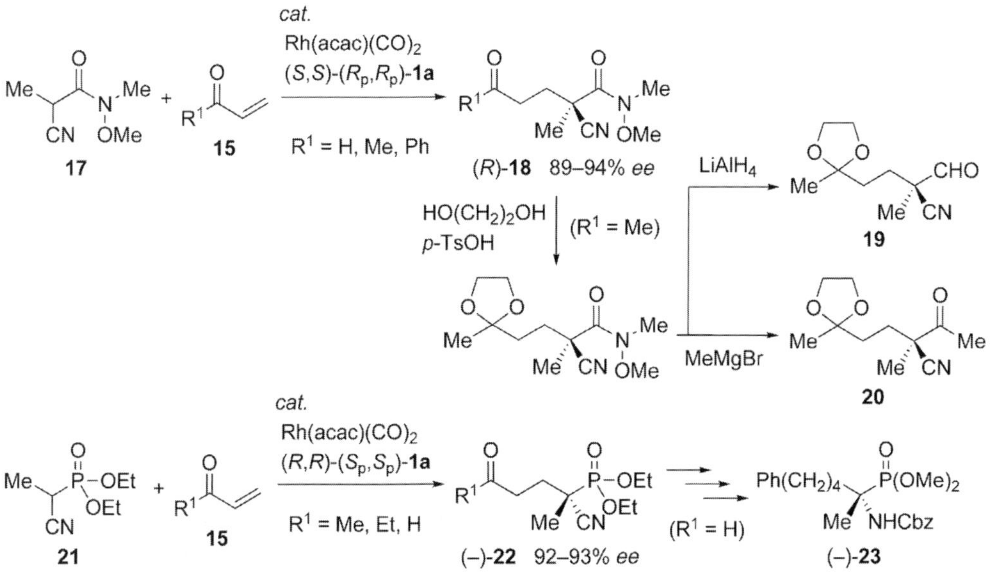

Scheme 10.5 Asymmetric Michael reactions of **17** and **21** with **15** and some transformations of the products.

aldol reaction with other aldehydes, which gives **24b** or **24c** bearing vicinal chiral centers. The size of the ester substituent in **14** significantly affects the anti/syn ratio as well as the enantioselectivity of **24**. The reaction of bulky 2,4-dimethyl-3-pentyl ester **14b** yields *anti*-**24** preferentially, with good enantiomeric excess, while ethyl ester **14a** reacts with acetaldehyde with low diastereoselectivity (anti/syn = 45 : 55).

The palladium(0) complex is known to activate an allylic ester, affording an electrophilic (π-allyl)palladium. Concerted combination of the (π-allyl)palladium chemistry with the PhTRAP–rhodium catalysis involved with cyano compounds

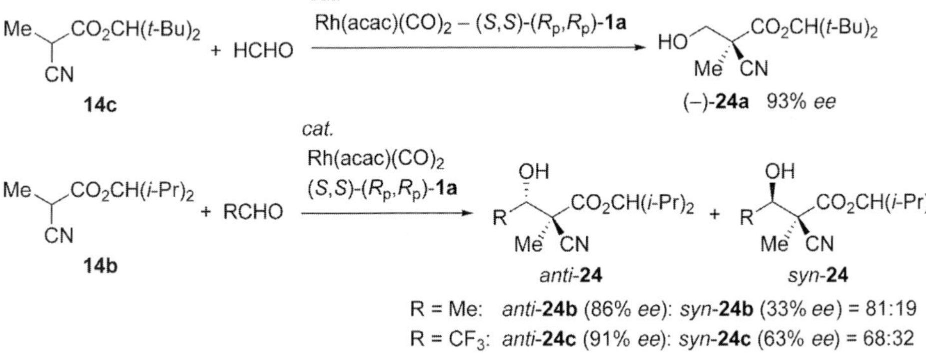

Scheme 10.6 Asymmetric aldol reaction of 2-cyanopropionate **14** catalyzed by **1a**–rhodium complex.

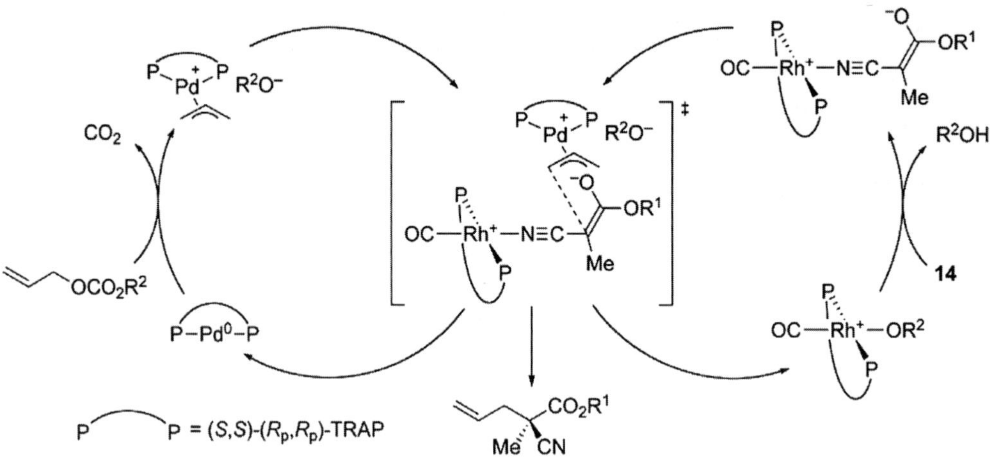

Scheme 10.7 Concept for catalytic asymmetric allylation of 2-cyanopropionate **14** using palladium–rhodium dual catalysis.

would lead to highly enantioselective allylation of **14** (Scheme 10.7). Indeed, the reaction of isopropyl 2-cyanopropionate (**14d**) with allyl hexafluoroisopropyl carbonate **25** proceeds with 93% ee (R) with the catalyst system prepared from Rh(acac)(CO)$_2$, Pd(π-C$_3$H$_5$)(Cp), and (S,S)-(R_p,R_p)-**1a** (Scheme 10.8) [22]. The enantiomeric excess of the product (R)-**26** is enhanced to 99% ee by using electron-donating AnisTRAP (**1b**) in place of **1a**. The catalyst system lacking palladium fails to produce **26**. Meanwhile, the TRAP–palladium complex catalyzes the allylation in the absence of the rhodium complex, but **26** is obtained as a racemic form. These observations suggest that the stereochemistry of the asymmetric allylation is controlled only by the chiral rhodium catalyst. The palladium complex may be merely required for the generation of electrophilic (π-allyl)palladium species and not participate in the chiral induction during the catalytic process. The chiral rhodium–palladium catalyst system is only effective for the allylation of 2-cyanopropionate **14** as with the aforementioned Michael reaction. The allylation of arylcyanoacetate occurs in low yield [32].

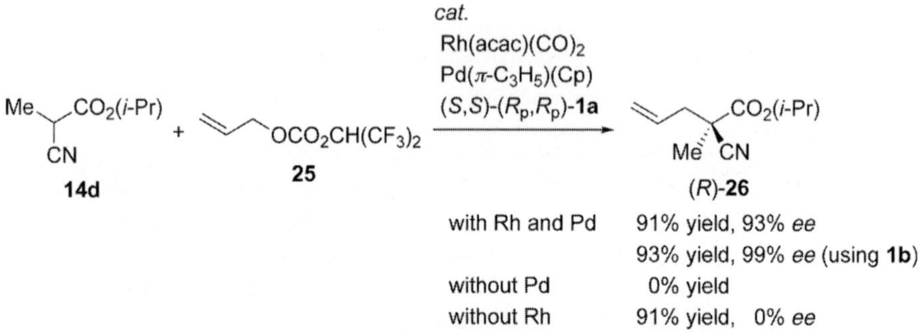

	with Rh and Pd	91% yield, 93% ee
		93% yield, 99% ee (using **1b**)
	without Pd	0% yield
	without Rh	91% yield, 0% ee

Scheme 10.8 Asymmetric allylation of 2-cyanopropionate **14** using TRAP.

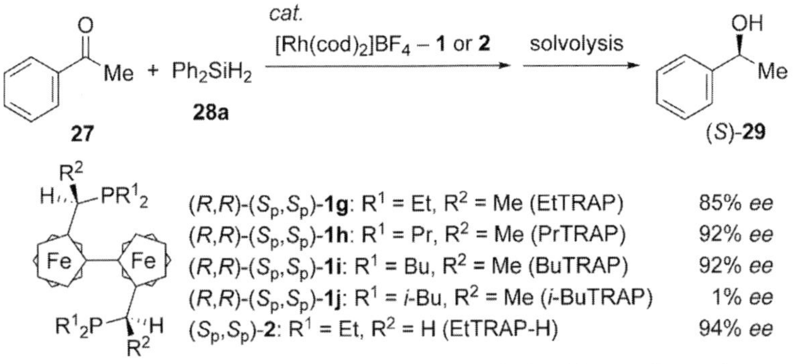
cat.
[Rh(cod)2]BF4 – 1 or 2
solvolysis
Me + Ph2SiH2
28a
27
OH
Me
(S)-29
(R,R)-(Sp,Sp)-1g: R1 = Et, R2 = Me (EtTRAP) 85% ee
(R,R)-(Sp,Sp)-1h: R1 = Pr, R2 = Me (PrTRAP) 92% ee
(R,R)-(Sp,Sp)-1i: R1 = Bu, R2 = Me (BuTRAP) 92% ee
(R,R)-(Sp,Sp)-1j: R1 = i-Bu, R2 = Me (i-BuTRAP) 1% ee
(Sp,Sp)-2: R1 = Et, R2 = H (EtTRAP-H) 94% ee

Scheme 10.9 Effect of TRAP ligand on the asymmetric hydrosilylation of acetophenone.

10.2.4 Asymmetric Hydrosilylation of Ketones

Hydrosilanes are inert to carbonyl groups with no catalyst. However, addition of hydrosilane to the carbon–oxygen double bond proceeds rapidly in the presence of a transition metal complex, affording a silyl ether [33–35]. The silyl ether is readily hydrolyzed to an alcohol. The hydrosilylation with a transition-metal catalyst had been an attractive candidate for the routine of asymmetric reduction of ketones, because it proceeds only through the reaction pathway involved with some catalyst [36].

TRAP ligands **1g**–**1i** bearing linear alkyl groups on their phosphorus atoms show high stereoselectivity for the hydrosilylation of prochiral ketones [37–40]. Treatment of acetophenone **27** with diphenylsilane **28a** affords (*S*)-**29** with 92% *ee* in the presence of the rhodium catalyst prepared from $[Rh(cod)_2]BF_4$ and (R,R)-(S_p,S_p)-PrTRAP (**1h**) or BuTRAP (**1i**) (Scheme 10.9). The choice of the P-substituents of the ligand is crucial for high stereoselectivity. Use of **1j**, whose P-substituents are β-branched primary alkyl, results in the formation of racemic **29** as well as retarding the hydrosilylation. The **1i**–rhodium catalyst converts a wide range of aryl or alkenyl methyl ketones into the corresponding chiral alcohols with over 80% *ee* (Scheme 10.10). In particular, the hydrosilylation of acetylferrocene gives 97% *ee*

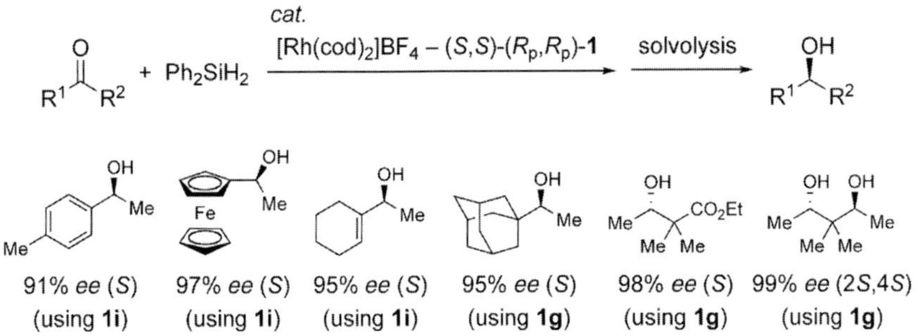
cat.
[Rh(cod)2]BF4 – (S,S)-(Rp,Rp)-1
solvolysis
+ Ph2SiH2
91% ee (S) (using 1i)
97% ee (S) (using 1i)
95% ee (S) (using 1i)
95% ee (S) (using 1g)
98% ee (S) (using 1g)
99% ee (2S,4S) (using 1g)

Scheme 10.10 Asymmetric hydrosilylation of methyl ketones catalyzed by **1i**– or **1g**–rhodium complex.

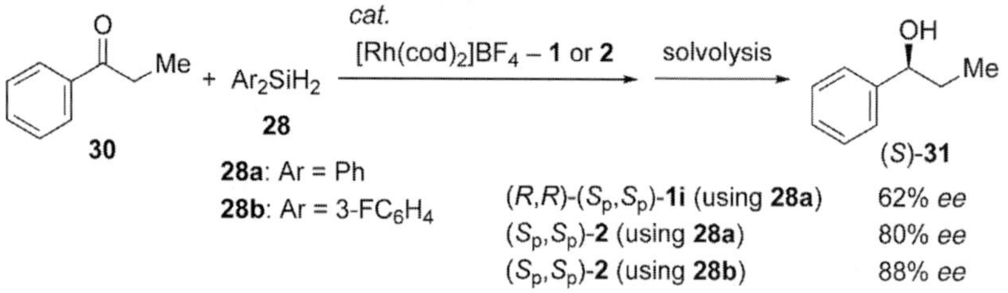

Scheme 10.11

$$\text{30} + \text{Ar}_2\text{SiH}_2 \xrightarrow[\text{solvolysis}]{\text{cat. [Rh(cod)}_2]\text{BF}_4 - \mathbf{1} \text{ or } \mathbf{2}} (S)\text{-}\mathbf{31}$$

28a: Ar = Ph
28b: Ar = 3-FC$_6$H$_4$

Catalyst	ee
(R,R)-(S_p,S_p)-**1i** (using **28a**)	62% ee
(S_p,S_p)-**2** (using **28a**)	80% ee
(S_p,S_p)-**2** (using **28b**)	88% ee

Scheme 10.11 Asymmetric hydrosilylation of propiophenone using TRAP.

of 1-ferrocenylethanol, which is useful for preparing ferrocene-based chiral ligands. EtTRAP (**1g**) exhibits higher enantioselectivity than **1i** for the enantioselective reduction of alkyl methyl ketones. Some *tert*-alkyl methyl ketones are converted into chiral alcohols with over 95% *ee* through the rhodium-catalyzed asymmetric hydrosilylation using **1g**.

The combined use of **28a** and (R,R)-(S_p,S_p)-**1i**–rhodium catalyst reduces ethyl ketone **30** to (S)-**31** with only 62% *ee* (Scheme 10.11). The stereoselectivity increases remarkably to 88% *ee* by using EtTRAP-H **2** and diarylsilane **28b** [15, 16]. The chiral rhodium catalyst exhibits 88–89% *ee* for the reaction of various primary alkyl aryl ketones. Furthermore, ligand **2** is more effective than **1g** for the hydrosilylation of acetylcyclohexane or 2-ocatanone.

10.2.5
Asymmetric Hydrogenation of Olefins

Hydrogenation of dehydroamino acid derivatives is the most-studied catalytic asymmetric reaction because it is useful for the preparation of α-amino acids [41]. Most chiral bisphosphines were originally developed for the asymmetric hydrogenation. Consequently, the hydrogenation of α-(acetamido)acrylates has frequently been used as a benchmark for new chiral ligands.

A series of TRAPs were also evaluated for the catalytic asymmetric hydrogenation of 2-(N-acetylamino)acrylate **32a** (Scheme 10.12) [42, 43]. The hydrogenation yields

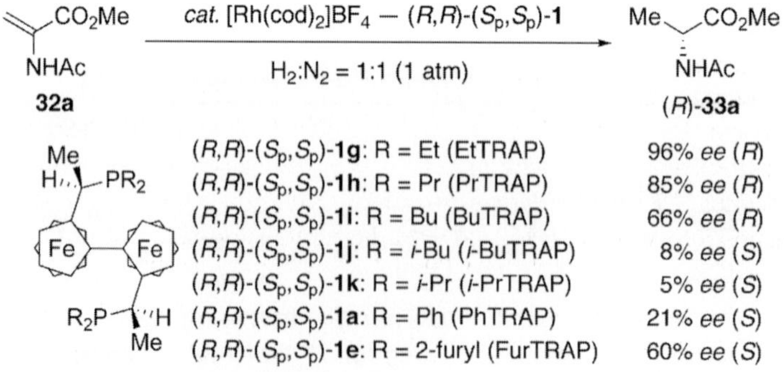

Scheme 10.12

$$\text{32a} \xrightarrow[\text{H}_2:\text{N}_2 = 1:1 \text{ (1 atm)}]{\text{cat. [Rh(cod)}_2]\text{BF}_4 - (R,R)\text{-}(S_p,S_p)\text{-}\mathbf{1}} (R)\text{-}\mathbf{33a}$$

Ligand	ee
(R,R)-(S_p,S_p)-**1g**: R = Et (EtTRAP)	96% ee (R)
(R,R)-(S_p,S_p)-**1h**: R = Pr (PrTRAP)	85% ee (R)
(R,R)-(S_p,S_p)-**1i**: R = Bu (BuTRAP)	66% ee (R)
(R,R)-(S_p,S_p)-**1j**: R = *i*-Bu (*i*-BuTRAP)	8% ee (S)
(R,R)-(S_p,S_p)-**1k**: R = *i*-Pr (*i*-PrTRAP)	5% ee (S)
(R,R)-(S_p,S_p)-**1a**: R = Ph (PhTRAP)	21% ee (S)
(R,R)-(S_p,S_p)-**1e**: R = 2-furyl (FurTRAP)	60% ee (S)

Scheme 10.12 Effect of TRAP ligand on the asymmetric hydrogenation of α-(acetamido)acrylate **32a**.

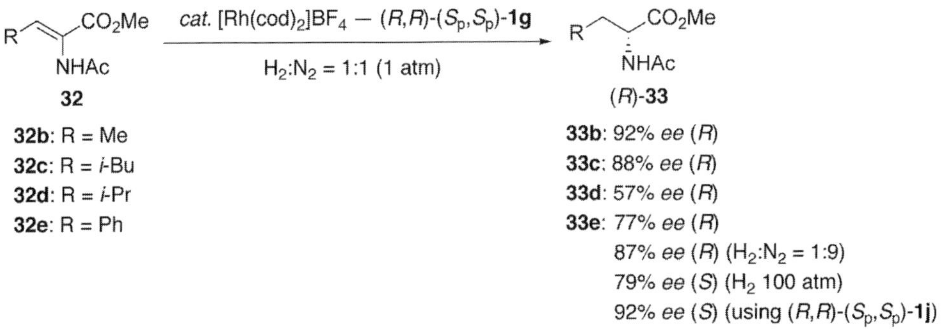

10.2 Trans-Chelating Chiral Bisphosphines: TRAP

$$\text{R} \diagup \diagdown \text{CO}_2\text{Me} \quad \xrightarrow[\text{H}_2:\text{N}_2 = 1:1 \ (1 \ \text{atm})]{\text{cat. [Rh(cod)}_2]\text{BF}_4 \ - \ (R,R)\text{-}(S_p,S_p)\text{-}\mathbf{1g}} \quad \text{R} \diagup \diagdown \text{CO}_2\text{Me}$$

NHAc   NHAc
32   (R)-**33**

32b: R = Me   **33b**: 92% ee (R)
32c: R = *i*-Bu   **33c**: 88% ee (R)
32d: R = *i*-Pr   **33d**: 57% ee (R)
32e: R = Ph   **33e**: 77% ee (R)
                                                             87% ee (R) ($H_2:N_2 = 1:9$)
                                                             79% ee (S) (H_2 100 atm)
                                                             92% ee (S) (using (R,R)-(S_p,S_p)-**1j**)

Scheme 10.13 Asymmetric hydrogenation of β-substituted α-(acetamido)acrylates catalyzed by TRAP–rhodium complex.

(R)-**33a** with the highest enantiomeric excess when (R,R)-(S_p,S_p)-EtTRAP (**1g**) is employed. The stereoselectivity deteriorates as the P-substituents of TRAP increase in size. A sterically congested rhodium catalyst bearing (R,R)-(S_p,S_p)-**1j** or **1k** produces **33a** with very low ee. The sense of chiral induction with (R,R)-(S_p,S_p)-**1a** or **1e** is opposite to that with (R,R)-(S_p,S_p)-**1g**.

A larger β-substituent of **32** is disadvantageous to enantioselective formation of (R)-**33** in the rhodium-catalyzed asymmetric hydrogenation using (R,R)-(S_p,S_p)-**1g**, as shown in Scheme 10.13. Methyl (Z)-2-(N-acetylamino)cinnamate (**32e**) is hydrogenated to (R)-**33e** with 87% ee when the reaction is carried out under 10% hydrogen in nitrogen. Surprisingly, the hydrogenation of **32e** at 100 atm of hydrogen pressure yields (S)-**33e** with 79% ee. Moreover, hydrogenation product **33e** is obtained with 92% ee (S) by using (R,R)-(S_p,S_p)-*i*-BuTRAP (**1j**), which gives **33a** with only 8% ee in the hydrogenation of **32a**. These observations are rationalized on the presumption that the catalytic asymmetric hydrogenation using the trans-chelating chiral ligands contains two competitive reaction pathways [43].

TRAPs bearing linear alkyl P-substituents are effective for asymmetric hydrogenation of β,β-disubstituted α-(acetamido)acrylates, which are converted into N,O-protected α-amino acids with two vicinal stereogenic centers (Scheme 10.14) [42, 43]. Dehydrovaline **34** is hydrogenated to the (S)-**35** with 88% ee with [Rh(cod)$_2$]BF$_4$–(R,R)-(S_p,S_p)-BuTRAP (**3i**) catalyst. The hydrogen molecule preferentially attacks the *re*-face of the prochiral enamide **34**, even at low pressure, while the *si*-face of **32** reacts with hydrogen under similar conditions. The *syn*-specific addition of hydrogen to the carbon–carbon double bond is observed in the reactions of the substrates (Z)- and (E)-**36**, which have two different substituents on their β-carbon. For the stereoselective hydrogenation of (Z)-**36**, **1g** is the most enantioselective of the TRAPs, yielding (2S,3R)-**37** with no formation of its diastereomer. In contrast, **1i** is more enantioselective than **1g** for the asymmetric transformation of (E)-**36** to (2S,3S)-**37**.

The asymmetric hydrogenation using **1** is applicable to the preparation of optically active β-hydroxy-α-amino acids (Scheme 10.15) [44]. The asymmetric synthesis of β-hydroxy-α-amino acids requires β-oxy-α-(acetamido)acrylate (Z)-**38** or (E)-**40** as the starting material. Both of the olefinic substrates are stereoselectively synthesized

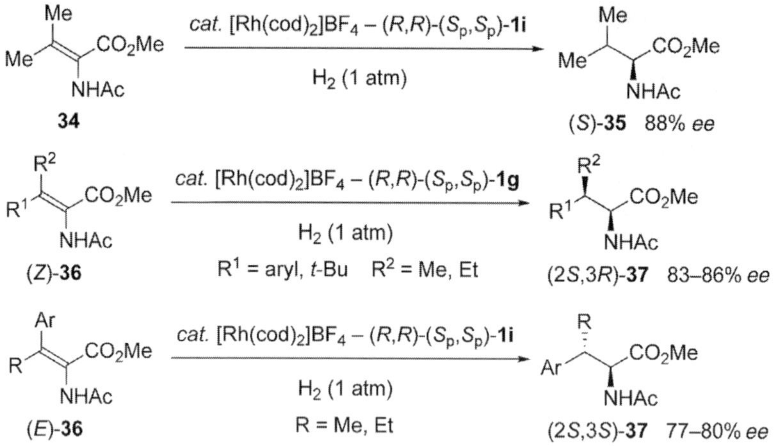

10 Biferrocene Ligands

Scheme 10.14 Asymmetric hydrogenation of β,β-disubstituted α-(acetamido)acrylates catalyzed by TRAP–rhodium complex.

from β-oxo-α-(acetamido)carboxylates with no formation of their geometrical isomer. PrTRAP (**1h**) is the ligand of choice for the hydrogenation of the β-oxo-α-(acetamido) acrylates. Silyl enolates (Z)-**38** are transformed into *erythro*-(2S,3S)-**39** with 94–97% ee by [Rh(cod)$_2$]BF$_4$–(R,R)-(S$_p$,S$_p$)-**1h** catalyst. The diastereoisomeric products, *threo*-**41**, are obtained in high stereoselectivity from the asymmetric hydrogenation of (E)-**40**. The chiral rhodium catalyst exhibits good enantioselectivity for the hydrogenation of (E)-**42**, which is useful for the asymmetric synthesis of *threo*-α,β-diamino acids [45].

TRAP–rhodium catalyst is useful for the enantioselective transformation of prochiral cyclic enamides into optically active cyclic α-amino acids, such as proline.

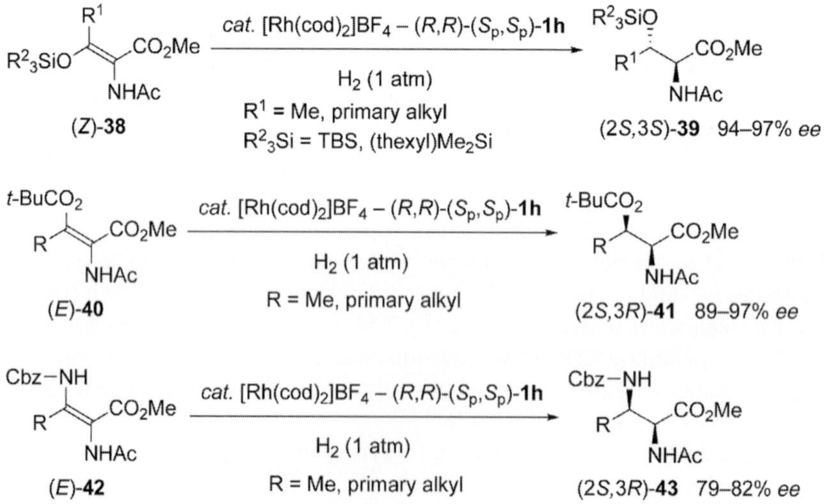

Scheme 10.15 Asymmetric hydrogenation of β-heterosubstituted α-(acetamido)acrylates catalyzed by **1h**–rhodium complex.

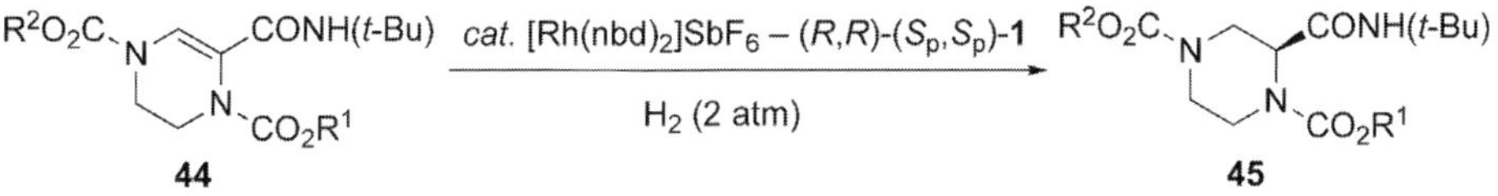

10.2 Trans-Chelating Chiral Bisphosphines: TRAP | 295

Scheme 10.16 Asymmetric hydrogenation of 1,4,5,6-tetrahydropyrazine **44** catalyzed by TRAP–rhodium complex.

44
44a: R^1 = t-Bu, R^2 = Ph
44b: R^1 = Bn, R^2 = t-Bu

45
45a: 96% ee (S) (using **1j**)
45b: 85% ee (R) (using **1f**)

The hydrogenation of 1,4,5,6-tetrahydropyrazine **44a** yields (S)-**45a** with 96% ee when (R,R)-(S_p,S_p)-i-BuTRAP (**1j**) is employed as a chiral ligand (Scheme 10.16) [46]. Interestingly, the enantiomeric product, (R)-**45b**, is obtained with 85% ee from the hydrogenation using (R,R)-(S_p,S_p)-MeTRAP (**1f**). The mechanistic study suggests that the **1j**-ligated rhodium(I) with no coordination of the olefinic substrate is the resting state in the former hydrogenation producing (S)-**45**. In the latter reaction affording the R-product, the coordination of **44** to **1f**–rhodium(I) occurs before the catalyst activates a hydrogen molecule. Furthermore, ligand **1f** chelates to the rhodium in a cis-manner during the catalytic process. The difference in reaction pathway may cause the reversal of enantioselectivity.

Asymmetric hydrogenation of cyclic dehydroamino acids **46** was attempted with the **1j**– or **1f**–rhodium catalyst, but the catalysts failed to give the product **47** with good enantiomeric excess. The stereoselectivity is remarkably improved by use of PhTRAP (**1a**) (Scheme 10.17) [47]. Tetrahydropyridine **46a** is hydrogenated to the chiral cyclic amine **47a** with 90% ee in the presence of $[Rh(nbd)_2]PF_6$–(S,S)-(R_p,R_p)-**1a** catalyst. The chiral catalyst is useful for the preparation of chiral 5- and 7-membered cyclic amines, **47b** and **47c**. It is noteworthy that benz-fused **47d** is produced with 97% ee through the asymmetric hydrogenation catalyzed by the **1a**–rhodium complex.

TRAPs work as a good chiral ligand for the rhodium-catalyzed asymmetric hydrogenation of olefins other than enamides. Itaconate **48a** undergoes the re-face addition of hydrogen by (R,R)-(S_p,S_p)-**1g**–rhodium catalyst, which gives (S)-**49a** with 96% ee (Scheme 10.18) [48]. As with the aforementioned hydrogenations of α-(acetamido)acrylates, the reverse enantioface-selection is observed in the hydrogenation of **48b**, a tetrasubstituted olefin. The hydrogenation product (S)-**49b** is

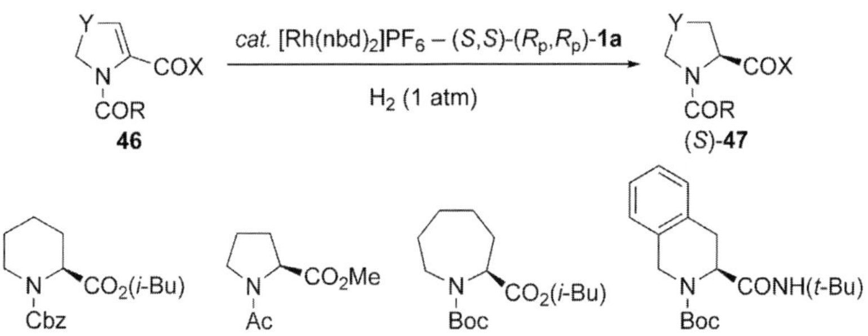

46

(S)-47

47a 92% ee (Cbz, CO_2(i-Bu))
47b 86% ee (Ac, CO_2Me)
47c 87% ee (Boc, CO_2(i-Bu))
47d 97% ee (Boc, CONH(t-Bu))

Scheme 10.17 Asymmetric hydrogenation of cyclic enamides **46** catalyzed by **1a**–rhodium complex.

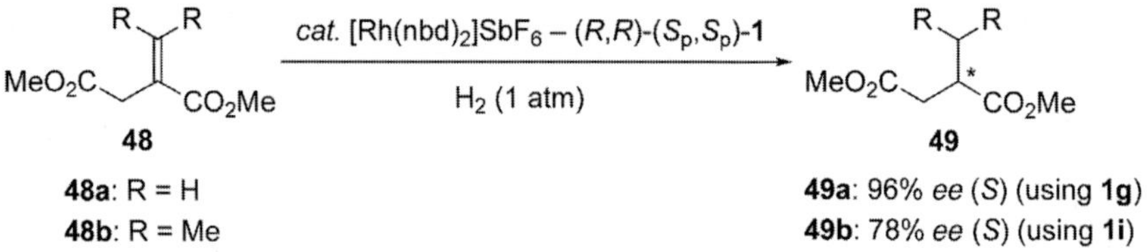

Scheme 10.18 Asymmetric hydrogenation of itaconates **48** catalyzed by TRAP–rhodium complex.

Substrate **48** (R groups with MeO$_2$C and CO$_2$Me) undergoes hydrogenation with cat. [Rh(nbd)$_2$]SbF$_6$ – (R,R)-(S$_p$,S$_p$)-**1**, H$_2$ (1 atm) to give **49**.

48a: R = H
48b: R = Me

49a: 96% ee (S) (using **1g**)
49b: 78% ee (S) (using **1i**)

produced with 78% ee by the (R,R)-(S_p,S_p)-**1i**-ligated catalyst. Only **1e** will give a reversed induction but with 6% ee.

10.2.6
Asymmetric Hydrogenations of Heteroaromatics

Nowadays, catalytic asymmetric hydrogenations of olefins and ketones are standard methods for preparing optically active compounds. Meanwhile, asymmetric catalysis for the reduction of heteroaromatics had been unexplored in the last century [49]. Highly enantioselective hydrogenation of heteroaromatics will offer a straightforward approach to optically active heterocycles, which are often seen in useful biologically active compounds. Furthermore, the methodology can create multiple chiral centers with a high degree of stereocontrol when unsymmetrical multi-substituted heteroaromatics are chosen as substrates.

In 2000, we reported that the PhTRAP (**1a**)–rhodium complex exhibits high enantioselectivity as well as catalytic activity for the hydrogenation of indoles (Scheme 10.19) [50, 51]. This is the first success in the catalytic asymmetric reduction of five-membered heteroaromatics. The chiral rhodium catalyst, which is generated *in situ* from [Rh(nbd)$_2$]SbF$_6$ and (S,S)-(R_p,R_p)-**1a** in the presence of a base, transforms various 2-substituted N-acetylindoles **50** into indolines **51** with high enantiomeric excess. Primary alkyl, phenyl, and carboxylate are used as the 2-substituent. It is noteworthy that use of any cis-chelating chiral bisphosphine results in the formation

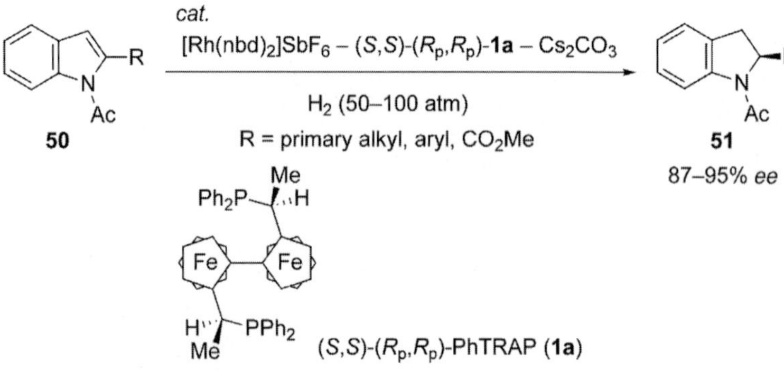

Scheme 10.19 Asymmetric hydrogenation of 2-substituted indoles **50** catalyzed by **1a**–rhodium complex.

Substrate **50** (N-Ac indole with R at 2-position) + cat. [Rh(nbd)$_2$]SbF$_6$ – (S,S)-(R_p,R_p)-**1a** – Cs$_2$CO$_3$, H$_2$ (50–100 atm), R = primary alkyl, aryl, CO$_2$Me → **51**, 87–95% ee.

(S,S)-(R_p,R_p)-PhTRAP (**1a**)

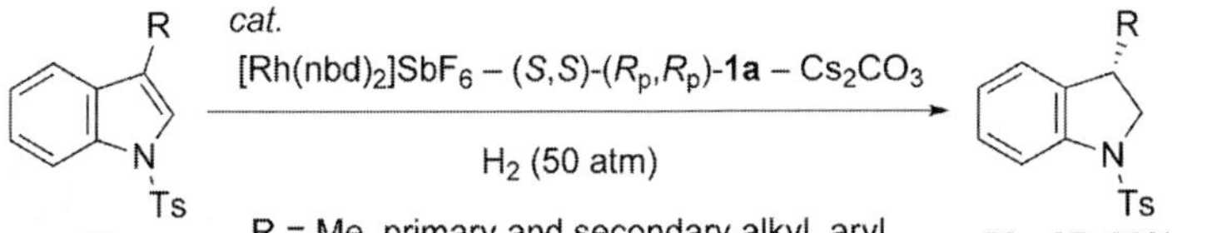

52

53 95–98% *ee*

Scheme 10.20 Asymmetric hydrogenation of 3-substituted indoles **52** catalyzed by **1a**–rhodium complex.

of the racemic product. The trans-chelating ability of the biferrocene ligand **1a** may be crucial for a high degree of enantioselection in the catalytic hydrogenation of **50**. The chiral catalyst is also effective for the enantioselective hydrogenation of 3-substituted indoles (Scheme 10.20) [51, 52]. Various *N*-tosylindoles **52** are reduced to indolines **53** with 95–98% *ee*.

The *tert*-butoxycarbonyl (Boc) group is an attractive *N*-protecting group for organic synthesis. The protective group is not only readily installed on the indole but also removed from the indoline product. The hydrogenations of *N*-Boc-indoles **54** and **56** proceed in high enantioselectivity by using ruthenium in place of rhodium (Scheme 10.21) [23]. The chiral ruthenium catalyst, [RuCl(*p*-cymene){(*S*,*S*)-(R_p, R_p)-**1a**}]Cl, converts **54a** (R = Me) into (*R*)-**55a** with 95% *ee*. A broad range of chiral indolines **55** and **57** possessing a chiral center at either the 2- or 3-position are obtained with 87–95% *ee* through the ruthenium-catalyzed asymmetric hydrogenation. Chiral **1a**–ruthenium catalyst is applicable to the asymmetric reduction of 2,3-disubstituted indole. The hydrogenation of **58** occurs with complete diastereospecificity, affording *cis*-2,3-dimethylindoline (−)-**59** with 72% *ee*.

The trans-chelating ligand is effective for the asymmetric hydrogenation of pyrroles as well as indoles [53]. The ruthenium complex generated from Ru(π-methallyl)$_2$(cod)

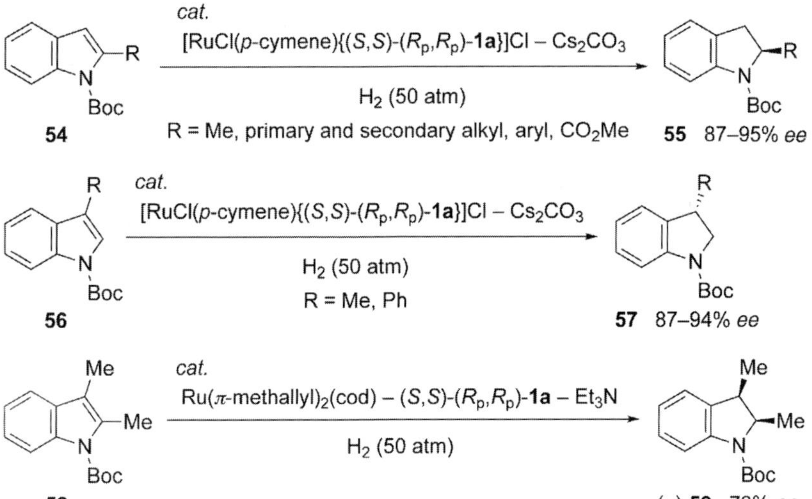

54

55 87–95% *ee*

56

57 87–94% *ee*

58

(−)-59 72% *ee*

Scheme 10.21 Asymmetric hydrogenation of *N*-Boc-indoles catalyzed by **1a**–ruthenium complex.

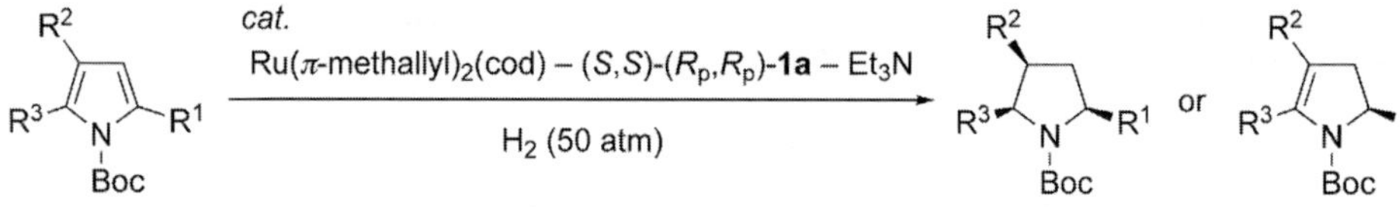

Scheme 10.22 Asymmetric hydrogenation of pyrroles **60** catalyzed by **1a**–ruthenium complex.

60a: $R^1 = CO_2Me$, $R^2 = R^3 = H$
60b: $R^1 = CO_2Me$, $R^2 = R^3 = Me$
60c: $R^1 = Ph$, $R^2 = Me$, $R^3 = Pr$
60d: R^1, R^2, $R^3 = Ph$

61a: 79% ee
61b: 96% ee
61c: 93% ee
62d: 99.7% ee

and (S,S)-(R_p,R_p)-**1a** is the most enantioselective catalyst for the asymmetric hydrogenation of **60a**, yielding protected proline (S)-**61a** with 79% ee (Scheme 10.22). The optimized catalyst system exhibits excellent performance for the asymmetric hydrogenation of 2,3,5-trisubstituted pyrroles **60b**–**60d** bearing a relatively large substituent at their 5-position. The substrates **60b** and **60c**, whose 2- and 3-substituents are methyl or primary alkyl, are fully hydrogenated into chiral pyrrolidines **61b** and **61c** with complete cis-specificity. The asymmetric hydrogenation creates three chiral centers on the pyrrolidine ring with a high level of stereocontrol in a single process. Meanwhile, optically active cyclic enamides **62** are obtained with over 98% ee from 2,3,5-triarylpyrroles such as **60d**. Further hydrogenation of **62** may be hindered by the steric repulsion between the aryl substituents.

10.2.7
Miscellaneous Reactions using TRAP

Intramolecular ene reaction of 1,6-enynes is an attractive means of access to five-membered rings bearing an exo-methylene adjacent to a chiral center. Trost et al. reported that the ene-type cycloisomerization proceeds under mild conditions in the presence of a phosphine-ligated palladium complex and a carboxylic acid [54]. Electron-deficient TRAP **1d**, which has bis{p-(trifluoromethyl)phenyl}phosphino groups, is an effective chiral ligand for the palladium-catalyzed cycloisomerization of allyl propargyl sulfonamide **63** (Scheme 10.23) [55]. 1,6-Enyne **63a** is selectively

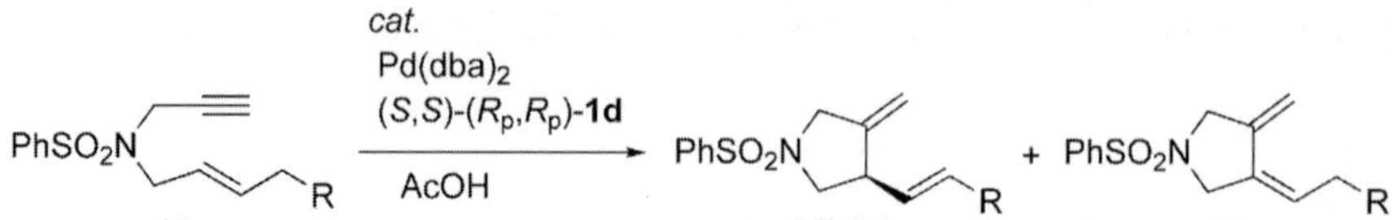

63a: R = SiMe₃
63b: R = CH₂SiCH₃
63c: R = Et (E-form)
63d: R = Et (Z-form)

64a (76% ee (R)):**65a** = >98:2
64b (95% ee (R)):**65b** = 3.5:1
64c (69% ee (R)):**65c** = 8.9:1
64d (59% ee (S)):**65d** = >15:1

Scheme 10.23 Asymmetric cycloisomerization of 1,6-enynes **63** catalyzed by **1d**–palladium complex.

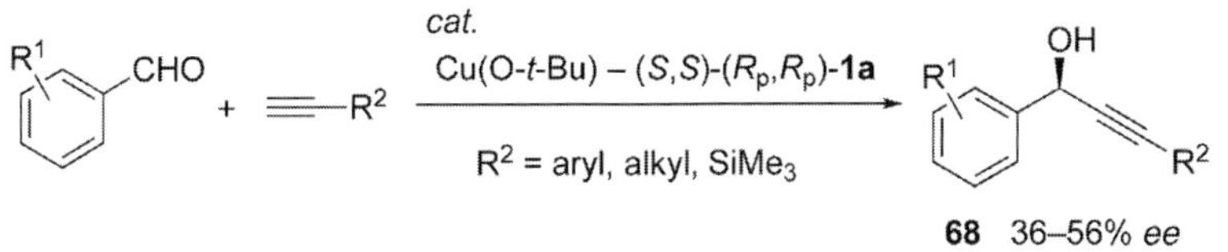

Scheme 10.24

$$R^1\text{-CHO} + \equiv\!\!-R^2 \xrightarrow[R^2 = \text{aryl, alkyl, SiMe}_3]{\text{cat. Cu(O-}t\text{-Bu)} - (S,S)\text{-}(R_p,R_p)\text{-}\mathbf{1a}} R^1\!\!-\!\!\overset{OH}{\underset{}{\text{CH}}}\!\!-\!\!\equiv\!\!-R^2$$

68 36–56% ee

Scheme 10.24 Nucleophilic addition of alkynes to aldehydes catalyzed by **1a**–copper complex.

transformed into chiral 1,4-diene **64a** with 76% ee by (S,S)-(R_p,R_p)-**1d**–palladium catalyst. No formation of achiral 1,3-diene **65a** may be caused by the α-effect of the silyl group at the allylic position of **63a**. Homoallylsilane **63b** gives a mixture of **64b** and **65b**, while the chiral 1,4-diene is obtained with 95% ee. The stereochemistry of the asymmetric cycloisomerization is reflected in the geometric isomerism of the olefinic moiety. The substrates **63c** and **63d** are converted into (R)- and (S)-enriched 1,4-diene products, respectively. Ligand **1a** was used for the palladium-catalyzed cyclization of a 1,6-diene, but the enantioselectivity is only 20% [56].

Catalytic nucleophilic addition of terminal alkynes to aldehydes is an atom-economical process forming a carbon–carbon bond. The nucleophilic addition is mediated by a copper complex, but stoichiometric copper is generally required to produce the desired propargyl alcohol in high yield. Use of PhTRAP ligand enables a catalytic amount of copper complex to produce propargyl alcohols **68** in high yield (Scheme 10.24) [57]. The large bite angle of **1a** may bring about the catalytic activity in the copper complex. However, enantiomeric excesses of the products **68** are moderate.

10.3
2,2″-Bis(diarylphosphino)-1,1″-biferrocenes: BIFEP

2,2″-Bis(diphenylphosphino)-1,1″-biferrocene (BIFEP, **69a**) was originally developed by Sawamura and Ito, but they have not reported any application of **69a** to asymmetric catalysis [10]. Since 2000, several researchers have synthesized its variants, **69b–69e** (Figure 10.4), and applied the biferrocene ligands to some catalytic asymmetric reactions.

The original synthesis of BIFEP starts from ferrocene, as shown in Scheme 10.25 [10]. Ferrocene is converted into phosphine oxide **70a** through monolithiation [58]

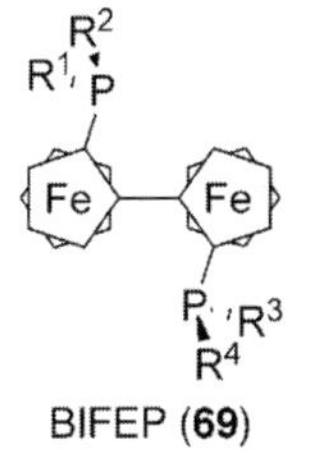

BIFEP (**69**)

(R_p,R_p)-**69a**: $R^1 = R^2 = R^3 = R^4 = $ Ph
(R_p,R_p)-**69b**: $R^1 = R^2 = R^3 = R^4 = $ 3,5-Me$_2$C$_6$H$_3$
(R_p,R_p)-**69c**: $R^1 = R^2 = $ 4-MeO-3,5-Me$_2$C$_6$H$_2$, $R^3 = R^4 = $ 3,5-(CF$_3$)$_2$C$_6$H$_3$
(S,S)-(R_p,R_p)-**69d**: $R^1 = R^3 = $ Ph, $R^2 = R^4 = $ 1-naphthyl
(S,S)-(R_p,R_p)-**69e**: $R^1 = R^3 = $ Ph, $R^2 = R^4 = $ 2-biphenylyl
(R,R)-(R_p,R_p)-**69e**: $R^1 = R^3 = $ 2-biphenylyl, $R^2 = R^4 = $ Ph

Figure 10.4 Structure of BIFEP.

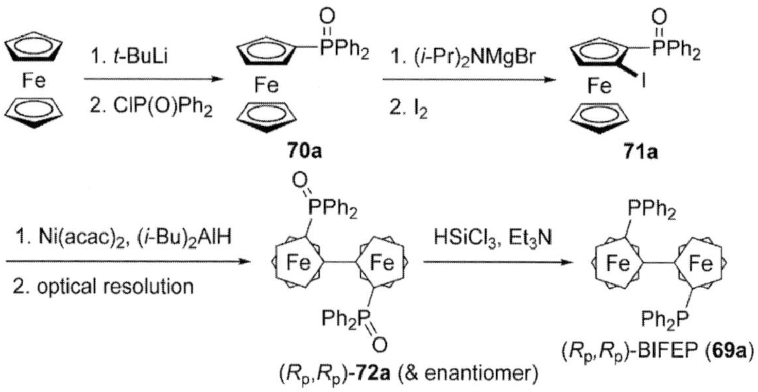

Scheme 10.25 Synthesis of BIFEP by Sawamura and Ito.

followed by treatment with diphenylphosphinyl chloride. The phosphine oxide works as a directing group, which leads to ortho-metalation with diisopropylamido magnesium bromide. The resulting Grignard reagent reacts with iodine to give racemic iodoferrocene **71a**. The nickel-mediated homocoupling of **71a** yields planar chiral biferrocene **72a** without formation of the meso-isomer. The racemic **72a** is resolved to both enantiomers by fractional recrystallization of the complexes with (−)- or (+)-dibenzoyltartaric acid. The enantiopure **72a** is reduced to **69a** by trichlorosilane–triethylamine.

Weissensteiner reported an asymmetric synthesis of the biferrocene ligands **69a–69c** (Scheme 10.26) [59]. The Ullmann-coupling of planar chiral iodoferrocene (S)-(R_p)-**73**, which is prepared by Kagan's method [60], produces biferrocene (S,S)-(S_p,S_p)-**74**. Stepwise replacement of the sulfoxides in **74** by diarylphosphino groups affords (S_p,S_p)-**69a–69c**. It is noteworthy that the sulfoxide route is adaptable to the preparation of unsymmetrical BIFEP **69c**.

Ligands **69d** and **69e**, which possess chiralities on their phosphorus, were synthesized by Widhalm and van Leeuwen as shown in Scheme 10.27 [61]. Chiral phosphine oxides (R)-**70** are prepared through the nucleophilic substitution of monolithioferrocene to the corresponding phosphinite-boranes (R)-**75**. Diastereoselective ortho-metalations of **70d** and **70e** occur with magnesium amide, affording planar chiral (R)-(S_p)-**71d** and **71e** with 94% and 50% de, respectively. The iodoferrocenes (R)-(S_p)-**71** are dimerized to give biferrocenes (R,R)-(R_p,R_p)-**72**. Although

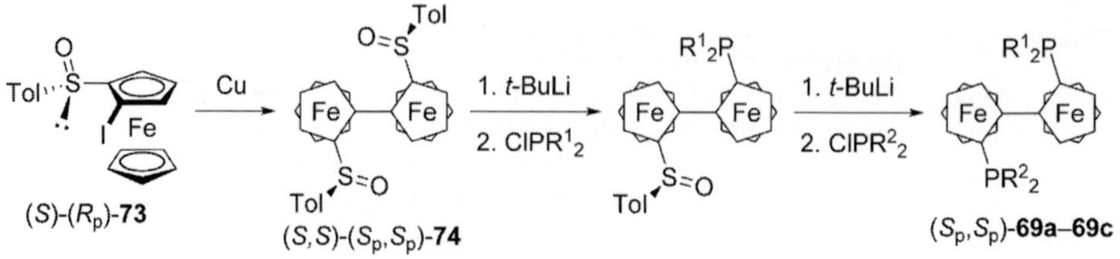

Scheme 10.26 Asymmetric synthesis of BIFEP by Weissensteiner.

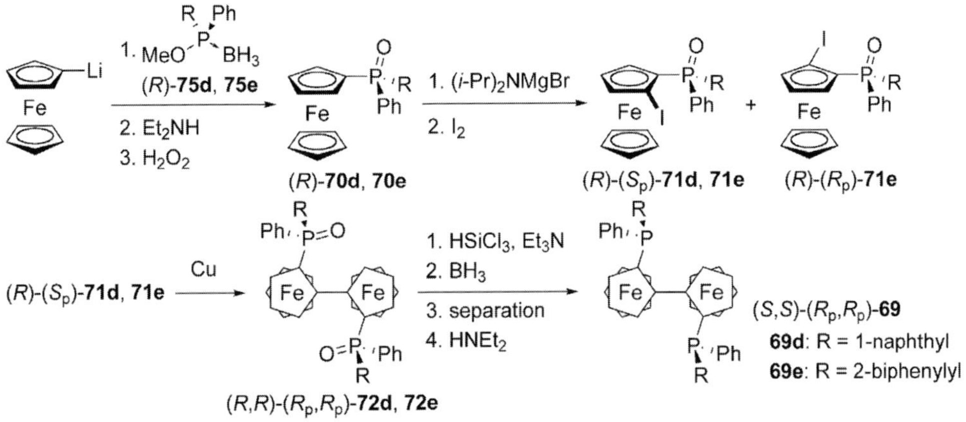

10.3 2,2″-Bis(diarylphosphino)-1,1″-biferrocenes: BIFEP

Scheme 10.27 Synthesis of P-chiral BIFEP by Widhalm and van Leeuwen.

the chiral phosphorus centers of **72** are partially racemized during the reduction of phosphine oxides, the undesirable stereoisomers can be removed through a chromatographic purification of the bis(borane) adducts of (S,S)-(R_p,R_p)-**69**. Diastereomeric (S,S)-(S_p,S_p)-**69e** is prepared from (R)-(R_p)-**71e** in a similar manner [62].

Unlike binaphthyl compounds, the biferrocenyl backbone of **69** does not inherently have axial chirality, because the cyclopentadienyl of the ferrocene cannot provide enough steric hindrance to keep the chirality. Consequently, **69**-ligated metal complexes can possess either (P)- or (M)-axial chirality. Indeed, BIFEP **69a** forms a metal complex bearing a (P)-biferrocene unit, while chelation of **69c** or **69d** to a metal induces (M)-chirality on the biferrocenyl unit. The induced axial chirality seems to be dependent on the size of the diarylphosphino groups [59, 61].

BIFEP ligands have been evaluated for several catalytic asymmetric reactions (Scheme 10.28). The ruthenium complex generated from $[RuI_2(p\text{-cymene})]_2$ and (S_p,S_p)-**69a** reduces the β-keto ester **76** to (R)-**77** with good enantiomeric excess [59]. P-Chiral ligands **69d** and **69e** have been employed for the palladium-catalyzed asymmetric allylic substitution of **78** [62]. Ligand (S,S)-(R_p,R_p)-**69d** is the most

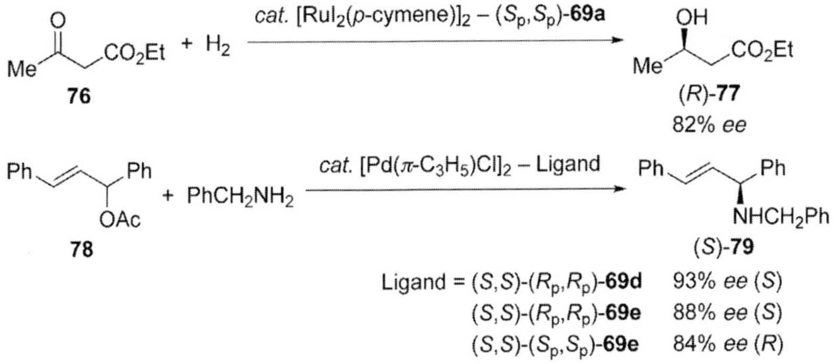

Scheme 10.28 Applications of BIFEP to asymmetric catalysis.

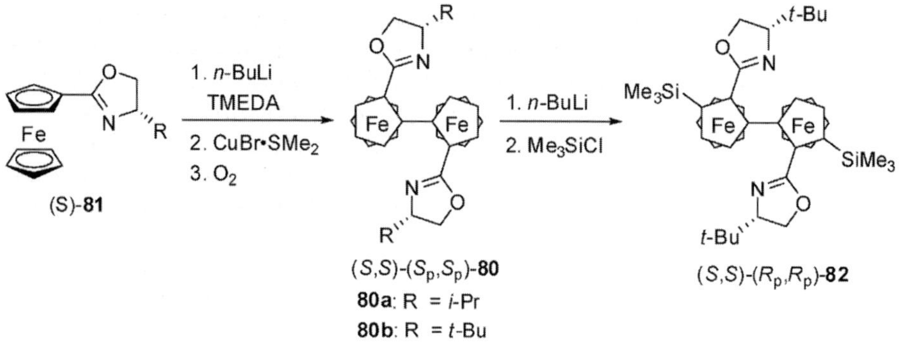

Scheme 10.29 Synthesis of biferrocene-based bis(oxazoline)s.

enantioselective of the P-chiral BIFEPs. The nucleophilic substitution with benzylamine yields (S)-79 with 93% ee when (S,S)-(R_p,R_p)-69d is used. The enantioselection in the palladium-catalyzed reactions is controlled by the planar chirality on the biferrocene unit rather than by the chiral center on phosphorus.

10.4
Miscellaneous Biferrocene-Based Chiral Ligands

Bis(oxazoline) is a powerful structural motif for designing chiral ligands. The chiral biferrocene constituent has been adopted into the bis(oxazoline) ligand by Ahn [63, 64]. The biferrocene-based bis(oxazoline)s (S,S)-(S_p,S_p)-80a and 80b are synthesized by the diastereoselective ortho-lithiation of ferrocenyloxazoline (S)-81, as shown in Scheme 10.29. After treatment of the lithioferrocene with copper(I), the ferrocenyl compound is oxidatively dimerized to give the desired 80. The trimethylsilyl group is installed at the 3- and 3″-positions of the biferrocene through the ortho-lithiation of 80b followed by treatment with chlorotrimethylsilane.

The bis(oxazoline) ligands 80 and 82 were evaluated on copper-catalyzed cyclopropanation. The silylated ligand 82 exhibits high stereoselectivity for the cyclopropanation of styrene with l-menthyl diazoacetate (Scheme 10.30) [63]. However, the ratio of trans- and cis-83 is low.

Weissensteiner and Widhalm developed a synthetic route to biferrocenoazepine, which was utilized to design a series of chiral aminophosphine ligands 84 (Scheme 10.31) [65]. The starting 1-(2-iodoferrocenyl)alkylamine is dimerized

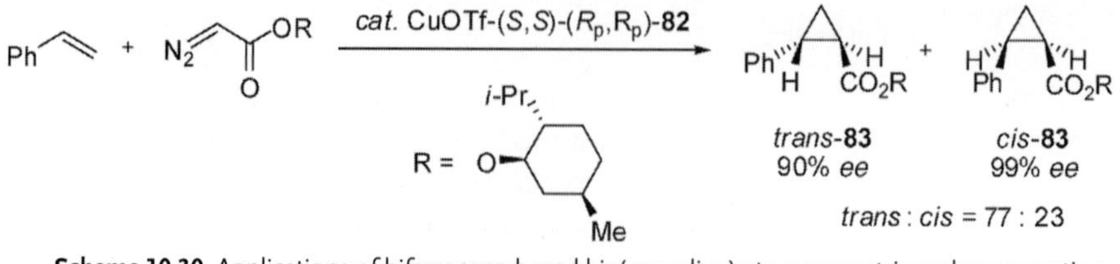

Scheme 10.30 Applications of biferrocene-based bis(oxazoline)s to asymmetric cyclopropanation.

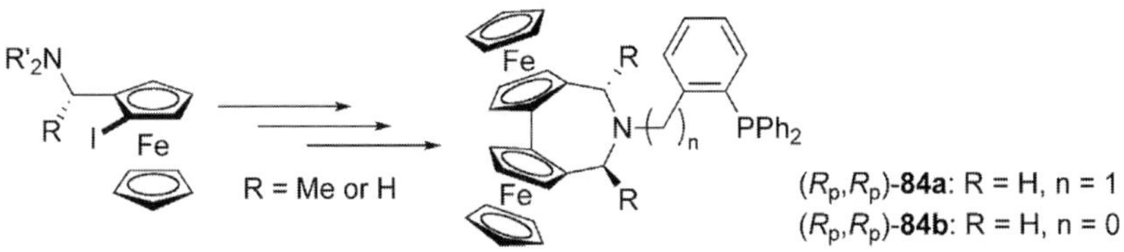

Scheme 10.31 Outline of the synthesis of biferrocene-based chiral P–N ligands **84**.

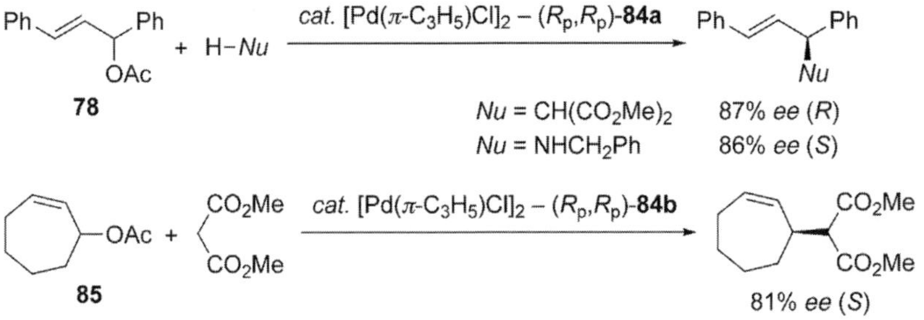

(R_p,R_p)-**84a**: R = H, n = 1
(R_p,R_p)-**84b**: R = H, n = 0

Scheme 10.32 Applications of ligands **84** to Tsuji–Trost reactions.

For substrate **78** with cat. [Pd(π-C$_3$H$_5$)Cl]$_2$ – (R_p,R_p)-**84a**:
- Nu = CH(CO$_2$Me)$_2$: 87% ee (R)
- Nu = NHCH$_2$Ph: 86% ee (S)

For substrate **85** with cat. [Pd(π-C$_3$H$_5$)Cl]$_2$ – (R_p,R_p)-**84b**: 81% ee (S)

through copper- or iron-mediated homocoupling. After the amino groups are activated with iodomethane, treatment with (2-bromoaryl)alkylamine gives biferrocenoazepine. The bromo group is replaced by the diarylphosphino group to give the desired ligands.

Tsuji–Trost reactions of various substrate combinations have been attempted by using a series of the aminophosphines **84** (Scheme 10.32). The chiral catalyst, [Pd(π-C$_3$H$_5$)Cl]$_2$–(R_p,R_p)-**84a**, exhibits good enantioselectivity for the allylic substitution of **78** with dimethyl malonate or benzylamine. The cyclic substrate **85** is alkylated with 81% ee by (R_p,R_p)-**84b**–palladium catalyst.

10.5
Conclusion

As described above, the biferrocene skeleton has been utilized to design some chiral ligands for asymmetric catalysis and proves to be useful for asymmetric catalysis. In particular, a series of TRAP **1** ligands, which can form trans-chelating metal complexes, is the most successful of the biferrocene-based chiral ligands for asymmetric catalysis. The trans-chelating ligands are useful for various catalytic asymmetric reactions, for example, reactions of 2-cyanopropionates with electrophiles, hydrosilylation of ketones, hydrogenation of carbon–carbon unsaturated bonds including heteroaromatics, and cycloisomerization of 1,6-enynes. Some other biferrocene-based ligands exhibit high enantioselectivity for the Tsuji–Trost reaction or cyclopropanation. However, it is hard to conclude that the chiral biferrocene skeleton has been leveraged for developing asymmetric catalysis. Optically active biferrocene is an unexplored structural motif for developing new chiral catalysts.

Excellent chiral ligands besides TRAP may come into existence from the utilization of biferrocene in the future.

References

1. Ojima, I. (2000) *Catalytic Asymmetric Synthesis*, Wiley-VCH, New York.
2. Tang, W. and Zhang, X. (2003) *Chem. Rev.*, **103**, 3029–3070.
3. Lia, Y.-M., Kwonga, F.-Y., Yu, W.-Y., and Chan, A.S.C. (2007) *Coord. Chem. Rev.*, **251**, 2119–2144.
4. Kagan, H.B. and Dang, T.-P. (1972) *J. Am. Chem. Soc.*, **94**, 6429–6433.
5. Fryzuk, M.D. and Bosnich, B. (1977) *J. Am. Chem. Soc.*, **99**, 6262–6267.
6. Miyashita, A., Yasuda, A., Takaya, H., Toriumi, K., Ito, T., Souchi, T., and Noyori, R. (1980) *J. Am. Chem. Soc.*, **102**, 7932–7934.
7. Arrayás, R.G., Adrio, J., and Carretero, J.C. (2006) *Angew. Chem.*, **118**, 7836–7878; (2006) *Angew. Chem. Int. Ed.*, **45**, 7674–7715.
8. Hayashi, T., Yamamoto, K., and Kumada, M. (1974) *Tetrahedron Lett.*, **15**, 4405–4408.
9. Hayashi, T., Mise, T., Fukushima, M., Kagotani, M., Nagashima, N., Hamada, Y., Matsumoto, A., Kawakami, S., Konishi, M., Yamamoto, K., and Kumada, M. (1980) *Bull. Chem. Soc. Jpn.*, **53**, 1138–1151.
10. Sawamura, M., Yamauchi, A., Takegawa, T., and Ito, Y. (1991) *J. Chem. Soc., Chem. Comm.*, 874–875.
11. Sawamura, M., Hamashima, H., and Ito, Y. (1991) *Tetrahedron: Asymmetry*, **2**, 593–596.
12. Cahn, R.S., Ingold, C., and Prelog, V. (1966) *Angew. Chem.*, **78**, 413–447; (1966) *Angew. Chem. Int. Ed. Engl.*, **5**, 333–434.
13. Schlögl, K., Fried, M., and Falk, H. (1964) *Monatsh. Chem.*, **95**, 576–597.
14. Marquarding, D., Klusacek, H., Gokel, G., Hoffmann, P., and Ugi, I. (1970) *J. Am. Chem. Soc.*, **92**, 5389–5393.
15. Kuwano, R., Uemura, T., Saitoh, M., and Ito, Y. (1999) *Tetrahedron Lett.*, **40**, 1327–1330.
16. Kuwano, R., Uemura, T., Saitoh, M., and Ito, Y. (2004) *Tetrahedron: Asymmetry*, **15**, 2263–2271.
17. Sawamura, M., Hamashima, H., Sugawara, M., Kuwano, R., and Ito, Y. (1995) *Organometallics*, **14**, 4549–4558.
18. Kuwano, R., Sawamura, M., Okuda, S., Asai, T., Ito, Y., Redon, M., and Krief, A. (1997) *Bull. Chem. Soc. Jpn.*, **70**, 2807–2822.
19. Watanabe, M., Araki, S., Butsugan, Y., and Uemura, M. (1991) *J. Org. Chem.*, **56**, 2218–2224.
20. Sammakia, T. and Latham, H.A. (1995) *J. Org. Chem.*, **60**, 6002–6003.
21. Sammakia, T., Latham, H.A., and Schaad, D.R. (1995) *J. Org. Chem.*, **60**, 10–11.
22. Sawamura, M., Sudoh, M., and Ito, Y. (1996) *J. Am. Chem. Soc.*, **118**, 3309–3310.
23. Kuwano, R. and Kashiwabara, M. (2006) *Org. Lett.*, **8**, 2653–2655.
24. Naota, T., Taki, H., Mizuno, M., and Murahashi, S. (1989) *J. Am. Chem. Soc.*, **111**, 5954–5955.
25. Murahashi, S.-I., Naota, T., Taki, H., Mizuno, M., Takaya, H., Komiya, S., Mizuho, Y., Oyasato, N., Hiraoka, M., Hirano, M., and Fukuoka, A. (1995) *J. Am. Chem. Soc.*, **117**, 12436–12451.
26. Sawamura, M., Hamashima, H., and Ito, Y. (1992) *J. Am. Chem. Soc.*, **114**, 8295–8296.
27. Sawamura, M., Hamashima, H., and Ito, Y. (1994) *Tetrahedron*, **50**, 4439–4454.
28. Sawamura, M., Hamashima, H., Shinoto, H., and Ito, Y. (1995) *Tetrahedron Lett.*, **36**, 6479–6482.
29. Sawamura, M., Hamashima, H., and Ito, Y. (2000) *Bull. Chem. Soc. Jpn.*, **73**, 2559–2562.
30. Kuwano, R., Miyazaki, H., and Ito, Y. (1998) *Chem. Comm.*, 71–72.

31 Kuwano, R., Miyazaki, H., and Ito, Y. (2000) *J. Organomet. Chem.*, **603**, 18–29.
32 Nowicki, A., Mortreux, A., and Agbossou-Niedercorn, F. (2005) *Tetrahedron: Asymmetry*, **16**, 1295–1298.
33 Ojima, I., Nihonyanagi, M., and Nagai, Y. (1972) *J. Chem. Soc., Chem. Comm.*, 938.
34 Ojima, I., Kogure, T., Nihonyanagi, M., and Nagai, Y. (1972) *Bull. Chem. Soc. Jpn.*, **45**, 3506.
35 Ojima, I., Nihonyanagi, M., Kogure, T., Kumagai, M., Horiuchi, S., Nakatsugawa, K., and Nagai, Y. (1975) *J. Organomet. Chem.*, **94**, 449–461.
36 Riant, O., Mostefaï, N., and Courmarcel, J. (2004) *Synthesis*, 2943–2958.
37 Sawamura, M., Kuwano, R., and Ito, Y. (1994) *Angew. Chem.*, **106**, 92–93; (1994) *Angew. Chem. Int. Ed. Engl.*, **33**, 111–113.
38 Sawamura, M., Kuwano, R., Shirai, J., and Ito, Y. (1995) *Synlett*, 347–348.
39 Kuwano, R., Sawamura, M., Shirai, J., Takahashi, M., and Ito, Y. (1995) *Tetrahedron Lett.*, **36**, 5239–5242.
40 Kuwano, R., Sawamura, M., Shirai, J., Takahashi, M., and Ito, Y. (2000) *Bull. Chem. Soc. Jpn.*, **73**, 485–496.
41 Chi, Y., Tang, W., and Zhang, X. (2005) *Modern Rhodium-Catalyzed Organic Reactions* (ed. P.A. Evans), Wiley-VCH Verlag GmbH, Weinheim, pp. 1–31.
42 Sawamura, M., Kuwano, R., and Ito, Y. (1995) *J. Am. Chem. Soc.*, **117**, 9602–9603.
43 Kuwano, R., Sawamura, M., and Ito, Y. (2000) *Bull. Chem. Soc. Jpn.*, **73**, 2571–2578.
44 Kuwano, R., Okuda, S., and Ito, Y. (1998) *J. Org. Chem.*, **63**, 3499–3503.
45 Kuwano, R., Okuda, S., and Ito, Y. (1998) *Tetrahedron: Asymmetry*, **9**, 2773–2775.
46 Kuwano, R. and Ito, Y. (1999) *J. Org. Chem.*, **64**, 1232–1237.
47 Kuwano, R., Karube, D., and Ito, Y. (1999) *Tetrahedron Lett.*, **40**, 9045–9049.
48 Kuwano, R., Sawamura, M., and Ito, Y. (1995) *Tetrahedron: Asymmetry*, **6**, 2521–2526.
49 Besson, M. and Pinel, C. (2003) *Top. Catal.*, **25**, 43–61.
50 Kuwano, R., Sato, K., Kurokawa, T., Karube, D., and Ito, Y. (2000) *J. Am. Chem. Soc.*, **122**, 7614–7615.
51 Kuwano, R., Kashiwabara, M., Sato, K., Ito, T., Kaneda, K., and Ito, Y. (2006) *Tetrahedron: Asymmetry*, **17**, 521–535.
52 Kuwano, R., Kaneda, K., Ito, T., Sato, K., Kurokawa, T., and Ito, Y. (2004) *Org. Lett.*, **6**, 2213–2215.
53 Kuwano, R., Kashiwabara, M., Ohsumi, M., and Kusano, H. (2008) *J. Am. Chem. Soc.*, **130**, 808–809.
54 Trost, B.M., Lee, D.C., and Rise, F. (1989) *Tetrahedron Lett.*, **30**, 651–654.
55 Goeke, A., Sawamura, M., Kuwano, R., and Ito, Y. (1996) *Angew. Chem.*, **108**, 686–687; (1996) *Angew. Chem. Int. Ed. Engl.*, **35**, 662–663.
56 Oppolzer, W. Kuo, D.L. Hutzinger, M.W. Leger, R. Durand, J.-O., and Leslie, C. (1997) *Tetrahedron Lett.*, **38**, 6213–6216.
57 Asano, Y. Hara, K. Ito, H., and Sawamura, M. (2007) *Org. Lett.*, **9**, 3901–3904.
58 Rebiere, F. Samuel, O., and Kagan, H.B. (1990) *Tetrahedron Lett.*, **31**, 3121–3124.
59 Xiao, L. Mereiter, K. Spindler, F., and Weissensteiner, W. (2001) *Tetrahedron: Asymmetry*, **12**, 1105–1108.
60 Riant, O. Argouarch, G. Guillaneux, D. Samuel, O., and Kagan, H.B. (1998) *J. Org. Chem.*, **63**, 3511–3514.
61 Nettekoven, U. Widhalm, M. Kamer, P.C.J. van Leeuwen, P.W.N.M. Mereiter, K. Lutz, M., and Spek, A.L. (2000) *Organometallics*, **19**, 2299–2309.
62 Nettekoven, U. Widhalm, M. Kalchhauser, H. Kamer, P.C.J. van Leeuwen, P.W.N.M. Lutz, M., and Spek, A.L. (2001) *J. Org. Chem.*, **66**, 759–770.
63 Kim, S.-G. Cho, C.-W., and Ahn, K.H. (1997) *Tetrahedron: Asymmetry*, **8**, 1023–1026.
64 Kim, S.-G. Cho, C.-W., and Ahn, K.H. (1999) *Tetrahedron*, **55**, 10079–10086.
65 Xiao, L. Weissensteiner, W. Mereiter, K., and Widhalm, M. (2002) *J. Org. Chem.*, **67**, 2206–2214.

11
Applications of Aza- and Phosphaferrocenes and Related Compounds in Asymmetric Catalysis
Nicolas Marion and Gregory C. Fu

11.1
Introduction

An interesting subclass of ferrocenes **1** [1] is the heteroferrocenes, where a carbon atom of the cyclopentadienyl ligand (Cp; **2**) is replaced by a heteroatom, as in **3–7** (Figure 11.1). Among these compounds, the nitrogen and phosphorus analogues have attracted the most attention [2, 3]. Thus, a wide array of aza- and phosphaferrocenes have been prepared, and they have found application in diverse fields, including coordination chemistry, materials science, physical chemistry, and biology.

Recently, the use of aza- and phosphaferrocenes in catalysis, in particular for enantioselective transformations, has become an active area of research, and this is the facet of their chemistry that is the focus of this chapter. Reactions not only of aza- and phosphaferrocenes, but also of ferrocenes that bear a fused nitrogen heterocycle, are summarized.

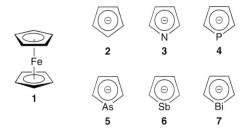

Figure 11.1 Structures of ferrocene (**1**) cyclopentadienyl (**2**) and heterocyclopentadienyls (**3–7**).

Chiral Ferrocenes in Asymmetric Catalysis. Edited by Li-Xin Dai and Xue-Long Hou
Copyright © 2010 WILEY-VCH Verlag GmbH & Co. KGaA, Weinheim
ISBN: 978-3-527-32280-0

11.2
Background on Aza- and Phosphaferrocenes

11.2.1
Azaferrocenes

The synthesis of azaferrocene **8** was first reported in 1964 independently by the groups of Pauson [4] and King [5]. King's synthesis, from $FeCl_2$, NaCp, and sodium pyrrolide, proceeded in <1% yield, whereas Pauson reacted $CpFe(CO)_2I$ with potassium pyrrolide to generate azaferrocene in 22% yield (Eq. (1)). The η^1-pyrrolide intermediate was later isolated [6], and its conversion to **8** was shown to be reversible [7]. Azaferrocene is less stable than ferrocene itself: it decomposes at 80 °C via an autodisproportionation process [8], and it is air-sensitive, especially in solution.

Extensive ^{57}Fe Mössbauer analyses of azaferrocenes have provided an understanding of their electronic structure [9]. Furthermore, electrochemical studies have established that azaferrocene is less prone to oxidize ($Fe^{II} \to Fe^{III}$) and easier to reduce ($Fe^{II} \to Fe^{I}$) than ferrocene itself [10].

Three particular aspects of the chemistry of azaferrocenes should be pointed out, due to their relevance for applications in asymmetric catalysis. First, the nitrogen of the azacyclopentadienyl ligand is nucleophilic and reacts with organic electrophiles, allowing the isolation of adducts **9** (Eq. (2) and Eq. (3)) [4] and **10** [11]. Second, the nitrogen of the azacyclopentadienyl ligand can serve as a σ donor to a transition metal; thus, a number of η^1 N-bound complexes of azaferrocenes have been reported [2]. Third, a (partial) resolution of a chiral azaferrocene derivative was described in 1969 [12, 13].

These observations establish that azaferrocenes possess key features that could allow a variety of applications in asymmetric catalysis, for example as nucleophilic catalysts and as chiral ligands for transition metals.

11.2.2
Phosphaferrocenes

Phosphaferrocene **11** was first synthesized by the group of Mathey in 1977 via reductive P–C bond cleavage of *P*-phenylphosphole to generate the phosphacyclopentadienyl ligand (Eq. (4)) [14]. They later applied a related strategy to the synthesis of a 1,1'-diphosphaferrocene [15]. Since these pioneering investigations, metal complexes that bear η^5-phospholyl ligands have been widely studied [3].

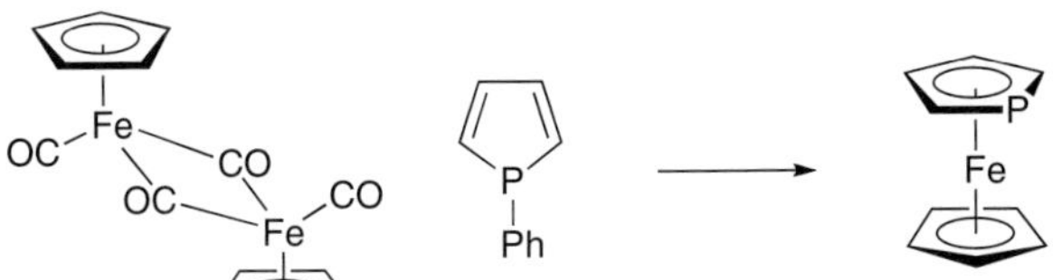

Phosphaferrocenes possess rich coordination chemistry, serving as ligands for an array of transition metals via the phosphorus lone pair. They are weaker σ-donors and stronger π-acceptors than typical sp^3-hybridized phosphines [16]. With regard to potential applications in asymmetric catalysis, the Ganter group provided a key advance in 1997 when they described a straightforward resolution of *rac*-**12** via derivatization with an enantiomerically pure diamine (Scheme 11.1) [17]. Aldehyde **12** can be converted into a variety of compounds, including **13–17**, which can be employed as bidentate ligands for metals such as ruthenium, palladium, molybdenum, and chromium [18]. For leading references to the coordination chemistry of phosphaferrocene ligands, see [3].

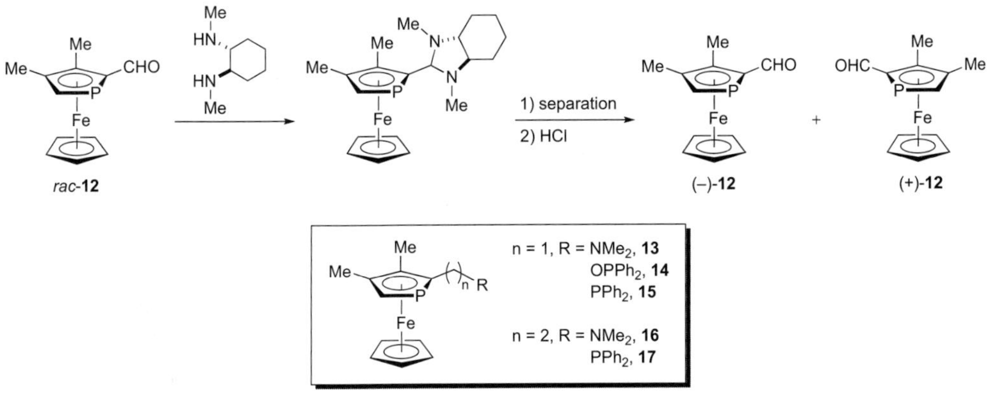

Scheme 11.1 Resolution of planar-chiral phosphaferrocene **12**.

11.3
Azaferrocenes in Catalysis

11.3.1
Nucleophilic Catalysis

In 1996, the use of azaferrocenes and related compounds (**18**, **19a**, **20**, and **21**) as nucleophilic catalysts was demonstrated for the first time [19]. The complexes can be synthesized from $FeCl_2$ (Eq. (5)–(7)), and the chiral compounds were initially resolved by chiral HPLC [20]. The N,N-dimethylaminopyridine (DMAP) derivative **21** was an especially active catalyst for processes such as the acylation of alcohols and the cyanosilylation of aldehydes [21, 22].

11.3.1.1 Acylations

O-Acylation In an initial study, the O-acylation of alcohols by diketene catalyzed by chiral azaferrocenes was investigated, specifically, the kinetic resolution of racemic secondary alcohols (Eq. (8)) [19]. This work provided proof-of-concept that an enantioselective transformation can be achieved with a planar-chiral azaferrocene, although the selectivity factor was modest.

11.3 Azaferrocenes in Catalysis

[Structure: racemic Ph-CH(OH)-Me + β-propiolactone (methyleneoxetanone) → with 10% (+)-19a, benzene, r.t., kinetic resolution → product with s = 4] (8)

$$s = \text{selectivity factor} = \frac{k \text{ (fast-reacting enantiomer)}}{k \text{ (slow-reacting enantiomer)}}$$

[Structure of (+)-19a: azaferrocene with CH₂OTES on pyrrolyl ring and Me₄ on Cp ring]

(+)-**19a**

It was subsequently determined that, through the use of a planar-chiral DMAP derivative as the catalyst, improved selectivity could be obtained and a very inexpensive acylating agent, acetic anhydride, could be employed (Scheme 11.2) [23]. Catalyst **22**, which bears a bulky C_5Ph_5 ligand, was more effective than catalyst **21**, furnishing good selectivity factors for an array of secondary aryl alkyl alcohols. Certain secondary propargylic and allylic alcohols can also be efficiently resolved by catalyst **22** (Scheme 11.2) [24]. Calculations by Zipse and coworkers support the hypothesis that the electron-donating ability of the ferrocene fragment is a key contributor to the high activity of planar-chiral DMAP derivatives such as **22** [25, 26].

Catalyst **22** was employed by McDonald for the kinetic resolution of benzylic alcohols (e.g., Eq. (9)) [27]. Similarly, Smith obtained a selectivity factor of 10 in a resolution of an allylic alcohol intermediate in his synthesis of (−)-hennoxazole A, although "the reaction was too slow to be practical" [28, 29].

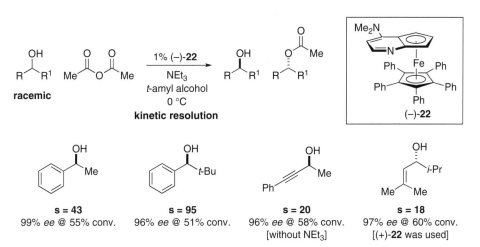

Scheme 11.2 Kinetic resolution of secondary alcohols.

A second O-acylation process to which azaferrocenes and related compounds have been applied is the addition of alcohols to ketenes. When performed with unsymmetrical disubstituted ketenes, this can be a useful method for the preparation of esters that bear a stereocenter in the α position. In 1999, it was established that azaferrocene **19b** catalyzes the addition of MeOH to an array of aryl methyl ketenes, affording arylpropionic esters in good yields and with *ee*s up to 80% (Eq. (10)) [30].

As a result of studies of the addition of pyrroles to ketenes (see next section) [31], it was discovered that planar-chiral nitrogen heterocycles can serve as enantioselective catalysts, not only through a nucleophilic pathway, but also via a Brønsted-acid pathway upon protonation of the pyridine nitrogen (Scheme 11.3).

As a consequence of these mechanistic investigations, conditions for the acylation of alcohols with ketenes were developed that employ a more basic catalyst (**23**) in combination with a more acidic alcohol (Eq. (10) vs. Eq. (11)) [32]. It was hypothesized

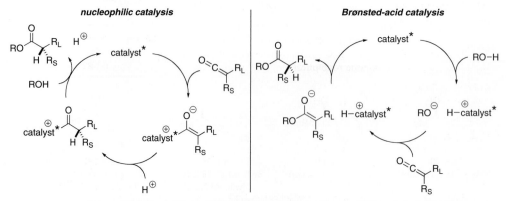

Scheme 11.3 Two of the possible pathways for the catalyzed addition of alcohols to ketenes.

that these changes would lead to a switch in mechanism from nucleophilic catalysis (Eq. (10)) to Brønsted-acid catalysis (Eq. (11)) (Scheme 11.3). By using this new method, it was possible to produce esters with significantly improved *ees* (Eq. (11)).

$$\text{ArC(=O)=CR} \xrightarrow[\text{toluene, r.t.}]{3\% \ (-)\text{-}23} \text{ester} \quad (11)$$

79-94% ee
66-97% yield

(−)-**23**

The same catalyst **23** can also achieve the coupling of an aldehyde with a ketene to generate an enol ester that bears an α stereocenter (Eq. (12)) [33]. It is likely that this reaction proceeds through chiral Brønsted-acid catalysis (Scheme 11.3), with the enolate of the aldehyde serving as the O-nucleophile. The resulting enol ester can easily be cleaved, yielding highly valuable enantioenriched carboxylic acids or alcohols.

$$\text{Ph}_2\text{CHCHO} + \text{ArC(=O)=CR} \xrightarrow[\text{CHCl}_3, \ 0\ ^\circ\text{C}]{10\% \ (-)\text{-}23} \text{enol ester} \quad (12)$$

77-98% ee
74-99% yield

N-Acylation The kinetic resolution of amines through acylation is a more difficult challenge than the corresponding reaction of alcohols, due in large part to the high nucleophilicity of amines, which leads to direct (uncatalyzed) acylation in a non-selective process [34]. However, through the careful selection of an acyl donor, the first effective non-enzymatic acylation-based method for the catalytic kinetic resolution of amines was developed, through the use of a planar-chiral DMAP derivative (Eq. (13)) [35]. Specifically, it was determined that certain O-acylated azlactones react much more rapidly with catalyst **23** than with a primary amine. Consequently, it was possible to obtain good selectivity factors for the kinetic resolution of a range of benzylic primary amines.

racemic + 2-naphthyl azlactone $\xrightarrow[\text{CHCl}_3, \ -50\ ^\circ\text{C}]{10\% \ (-)\text{-}23}$ kinetic resolution (13)

$11 \leq s \leq 27$

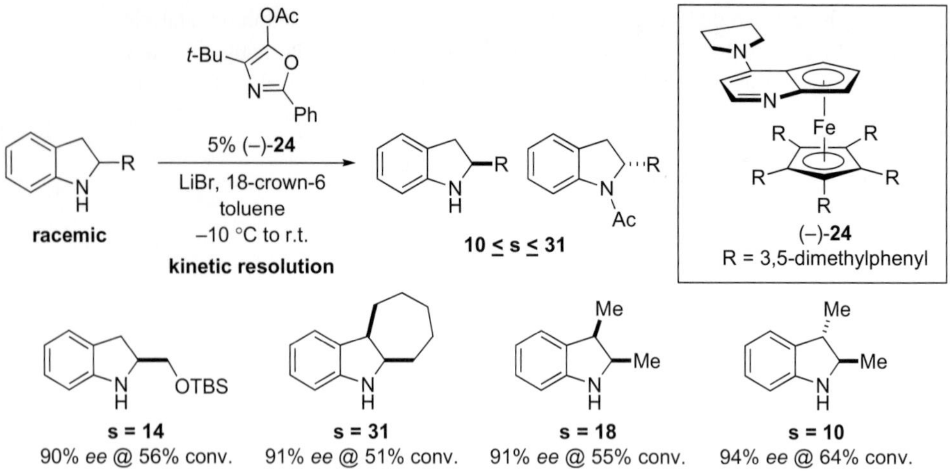

Scheme 11.4 Kinetic resolution of indolines.

Kinetic resolutions of indolines substituted in the 2-position, using a related O-acylated azlactone as the acylating agent, have also been achieved [36]. In this case, the incorporation of bulky 3,5-dimethylphenyl groups on the Cp ligand of the catalyst (**24**) led to the best enantioselectivity (Scheme 11.4).

In addition to kinetic resolutions of amines, other asymmetric N-acylations have been examined in the presence of planar-chiral DMAP derivatives. For example, complex **23** catalyzes the addition of 2-cyanopyrrole to aryl alkyl ketenes with very good enantioselectivity and yield (Eq. (14)) [31]. 2-Cyanopyrrole was an attractive choice for the nitrogen nucleophile from several standpoints, including its low reactivity toward ketenes in the absence of a catalyst, as well as the ease of derivatization of the product N-acylpyrroles. On the basis of a variety of mechanistic observations, including the formation of an ion pair upon mixing **23** with 2-cyanopyrrole, as well as the rate law, a chiral Brønsted-acid pathway was proposed for this transformation (analogous to the right-hand side of Scheme 11.3).

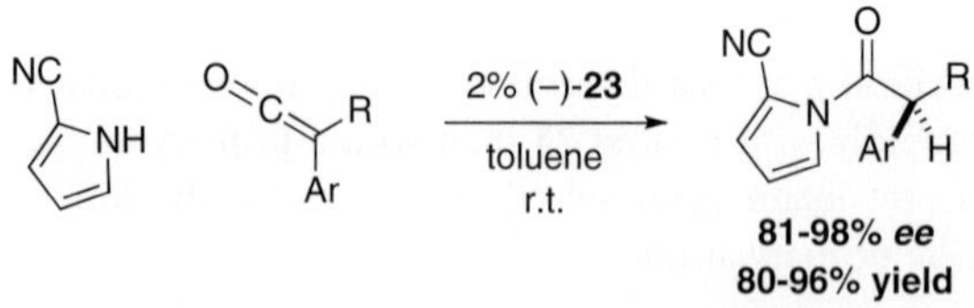

A chiral Brønsted acid-based mechanism is also believed to be operative for the catalytic asymmetric addition of hydrazoic acid to ketenes; in the same pot, the acyl azides are converted, via a Curtius rearrangement, to protected amines (Eq. (15)) [37]. In order to promote an efficient enantioselective protonation step, it was necessary to increase the Brønsted acidity of the protonated catalyst by replacing the pyrrolidino group of **23** with a methyl group (see catalyst **25**).

11.3 Azaferrocenes in Catalysis

$$N_3-H + \underset{Ar}{\overset{O=C-R}{}} \xrightarrow[\text{toluene/hexane} \\ -78 \text{ or } -90\,°C]{10\% \,(+)\text{-}25} \underset{Ar\ H}{\overset{O}{\underset{N_3}{\|}}}R \xrightarrow[\Delta]{MeOH} \underset{O\ Ar\ H}{\overset{MeO}{\underset{\|}{\overset{H}{N}}}}R \quad (15)$$

up to 97% ee
89–94% yield

(+)-25 (pentamethyl azaferrocene with methyl-substituted pyrrolyl ring)

C-Acylation Two families of C-acylation reactions have been accomplished with good enantioselectivity by using planar-chiral DMAP catalysts: rearrangement reactions and intermolecular acylations. The first success in this area, reported in 1998, employed catalyst **23** to achieve rearrangements of O-acylated azlactones to their C-acylated isomers, presumably via an ion pair (Eq 16) [38].

$$\underset{Ar}{\overset{BnO\overset{O}{\|}O}{\underset{N}{\overset{R}{\bigvee}}}} \xrightarrow[\text{t-amyl alcohol} \\ 0\,°C]{10\% \,(-)\text{-}23} \left[\underset{Ar}{\overset{BnO\overset{O}{\|}[23]^{\oplus}}{}} \ \ \underset{N}{\overset{R}{\underset{O^{\ominus}}{\bigvee}}} \right] \longrightarrow \underset{N}{\overset{BnO\overset{O}{\|}\overset{O}{\|}}{\underset{R'}{\bigvee}}}_{Ar} \quad (16)$$

88–92% ee
93–95% yield

A related rearrangement of O-acylated benzofuranones and oxindoles can be effected by catalyst **26**, again generating quaternary centers with high enantioselectivities (Eq. (17)) [39]. In this case, the ion-pair intermediate was characterized crystallographically and shown to be chemically competent.

$$\underset{X}{\overset{R^1\ O}{\bigvee}}\text{-OR} \xrightarrow[\text{CH}_2\text{Cl}_2, 35\,°C \\ X = O, NR^2 \\ R = CMe_2(CCl_3)]{5\% \,(-)\text{-}26} \underset{X}{\overset{R^1}{\bigvee}}\overset{O}{\underset{=O}{\|}}\text{OR} \quad (17)$$

88–99% ee
72–95% yield

(−)-26 (tetraphenyl azaferrocene with pyrrolidinyl-substituted pyrrolyl ring)

The second family of C-acylations involves intermolecular reactions, specifically, the enantioselective coupling of silyl ketene acetals or imines with anhydrides to generate all-carbon quaternary centers. In the presence of catalyst (−)-**26**, a wide array of silyl ketene acetals, including acyclic substrates, undergo C-acylation with

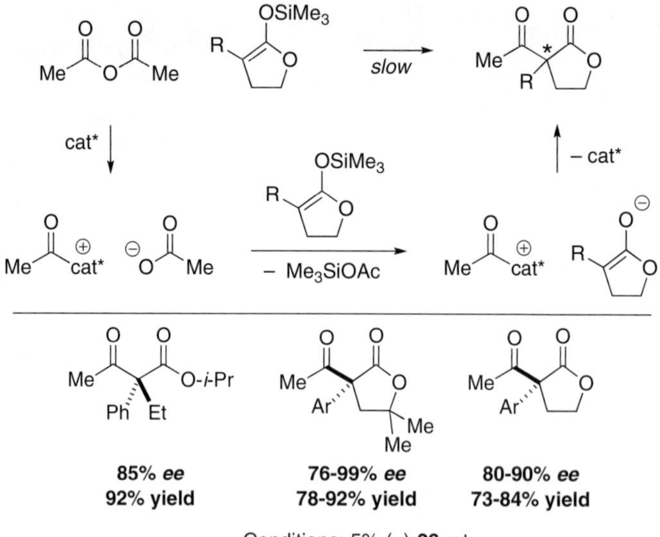

Scheme 11.5 Catalytic enantioselective C-acylation of silyl ketene acetals by anhydrides.

*ee*s up to 99% (Scheme 11.5) [40]. It was postulated that the acylation proceeds via dual activation of the electrophile (acetic anhydride → acylpyridinium ion) and the nucleophile (silyl ketene acetal → enolate).

This dual-activation strategy can also be exploited in reactions of silyl ketene imines with anhydrides, thereby providing efficient access to enantioenriched α-ketonitriles that bear a quaternary center [41]. This method has been applied to an enantioselective synthesis of (*S*)-verapamil (Eq. (18)).

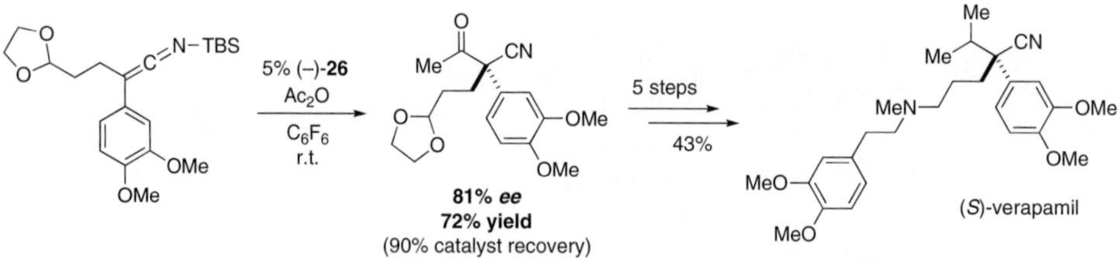

11.3.1.2 Halogenations

Because not only pyridines, but also pyridine *N*-oxides, serve as nucleophilic catalysts for a range of useful processes, the utility of planar-chiral pyridine *N*-oxides as enantioselective catalysts has been explored. Enantiopure oxygen nucleophiles had been reported to catalyze the desymmetrization of meso epoxides through ring opening with $SiCl_4$ (up to 87% *ee*) [42], so a series of planar-chiral pyridine *N*-oxides (e.g., **27–29**) were examined as potential catalysts for this transformation (Eq. (19)) [43]. When the bottom ring was either C_5Me_5 (**27**) or C_5Ph_5 (**28**), relatively modest enantioselectivities were obtained. However, a more sterically demanding analogue (**29**) served as an effective chiral catalyst. The need for a very large cyclopentadienyl group may be a consequence of the position of the nucleophilic oxygen, which is further from the FeC₅R₅ unit than is the pyridine nitrogen. If the substituents on the epoxide are alkyl rather than aryl groups, only moderate *ee* is observed.

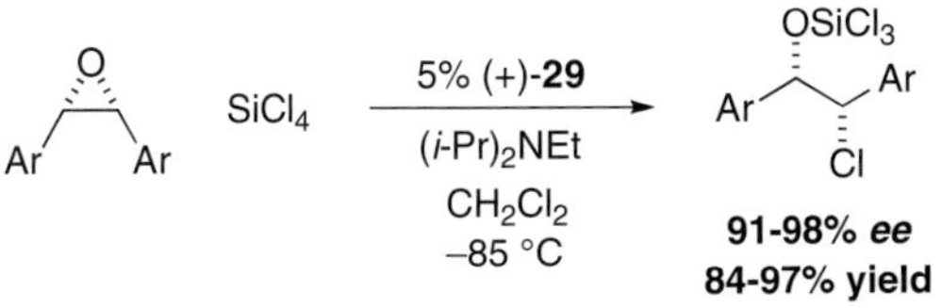

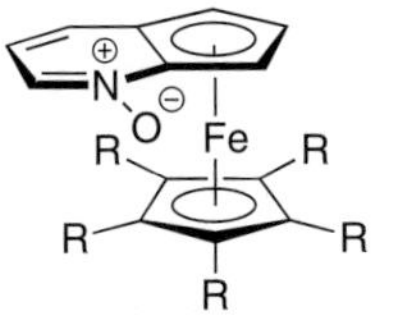

R = Me, **27**
Ph, **28**
3,5-Me₂C₆H₃, **29**

Planar-chiral heterocycles have also proved to be useful catalysts for the asymmetric chlorination of ketenes, thereby furnishing access to enantioenriched tertiary alkyl chlorides. Thus, the DMAP derivative (−)-**23** achieves enantioselective couplings of aryl alkyl ketenes with 2,2,6,6-tetrachlorocyclohexanone to provide enol esters of type **30** with good *ee* (Eq. (20)) [44]. The enol esters can be converted into highly enantioenriched α-chloro esters and β-chloro alcohols.

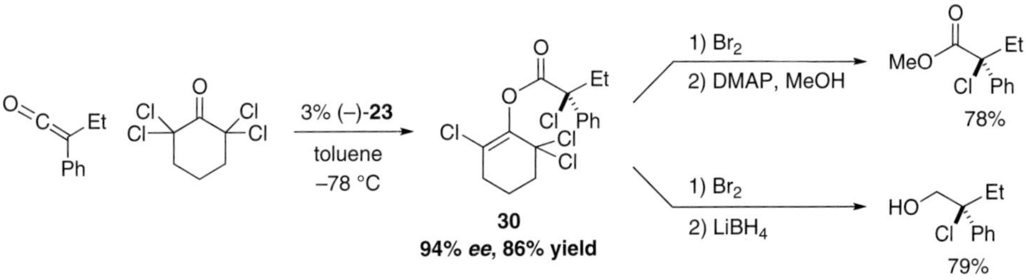

11.3.1.3 Cycloadditions

Planar-chiral DMAP derivatives can be applied as catalysts for a range of overall [2 + 2] cycloadditions of ketenes. For example, catalyst **23** couples an array of ketenes with *N*-tosyl imines to produce *cis*-β-lactams (**31**) with high stereoselectivity through a nucleophile-catalyzed Staudinger reaction (Scheme 11.6) [45]. It was subsequently discovered that, for *N*-triflyl imines, the major product of the [2 + 2] cycloaddition is the *trans*-β-lactam (Scheme 11.6) [46].

It was proposed that the divergence in cis/trans stereochemistry could be due to a change in mechanism for *N*-tosyl vs. *N*-triflyl imines (Scheme 11.7). For the less

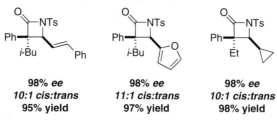

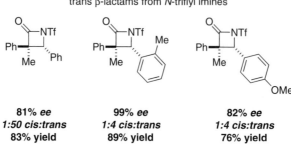

Scheme 11.6 Catalytic asymmetric Staudinger reactions.

electrophilic N-tosyl imine, the catalyst reacts initially with the ketene ("ketene-first" mechanism), whereas for the more electrophilic N-triflyl imine, the catalyst adds first to the imine ("imine-first" mechanism). Spectroscopic, isotope-effect, and other mechanistic studies are consistent with this rationale.

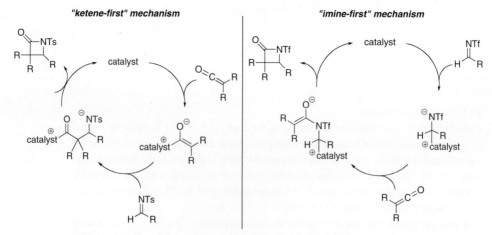

Scheme 11.7 Two of the possible pathways for nucleophile-catalyzed Staudinger reactions.

Not only imines, but also aldehydes and azo compounds, are suitable partners for ketenes in nucleophile-catalyzed asymmetric [2 + 2] cycloadditions. Thus, the planar-chiral DMAP derivative **23** produces β-lactones (Eq. (21)) [47] as well as aza-β-lactams (Eq. (22)) [48] with generally good yield and enantioselectivity.

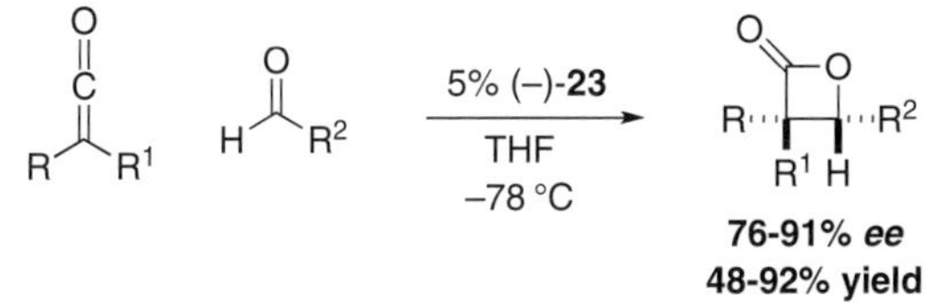

$$\underset{R}{\overset{O}{\|}}\underset{R^1}{C}=C + H\underset{R^2}{\overset{O}{\|}} \xrightarrow[\text{THF} \atop -78\,°C]{5\% \,(-)\text{-}\mathbf{23}} \quad \text{β-lactone product} \quad (21)$$

76–91% ee
48–92% yield

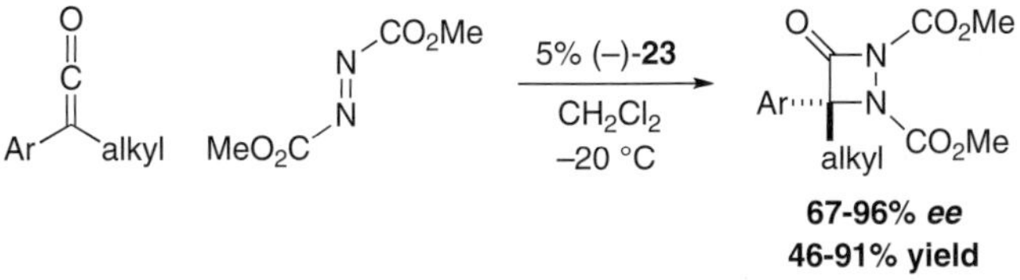

$$(22)$$

67–96% ee
46–91% yield

11.3.1.4 Nucleophilic Additions to Carbonyl Groups

It was reported in 1997 that a planar-chiral azaferrocene can catalyze the asymmetric addition of an organozinc reagent to aldehydes. Thus, **32**, which may be serving as a tridentate ligand, furnishes up to 90% ee for the reaction of diethyl zinc with benzaldehydes (Eq. (23)) [49].

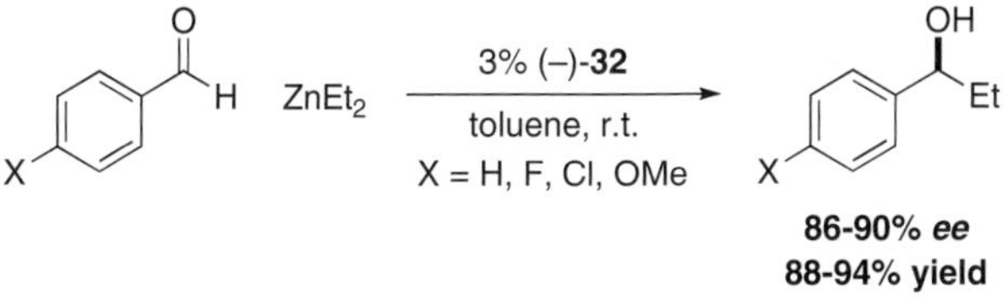

$$(23)$$

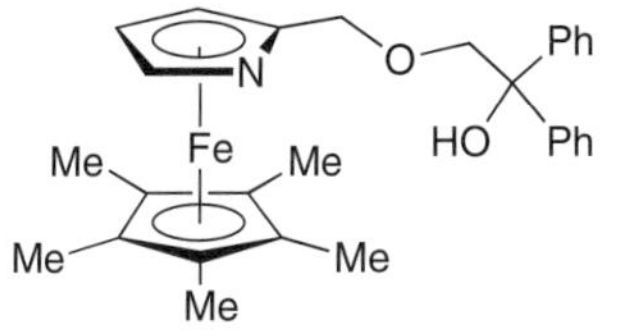

X = H, F, Cl, OMe

86–90% ee
88–94% yield

(−)-**32**

11.3.2
Transition-Metal Catalysis

11.3.2.1 Copper-Catalyzed Cyclopropanations of Olefins

C_2-Symmetric bis(azaferrocenes), such as ligand **33**, have found useful application in several transition metal-catalyzed processes. This bidentate ligand is readily accessed by the pathway outlined in Eq. (24) [50].

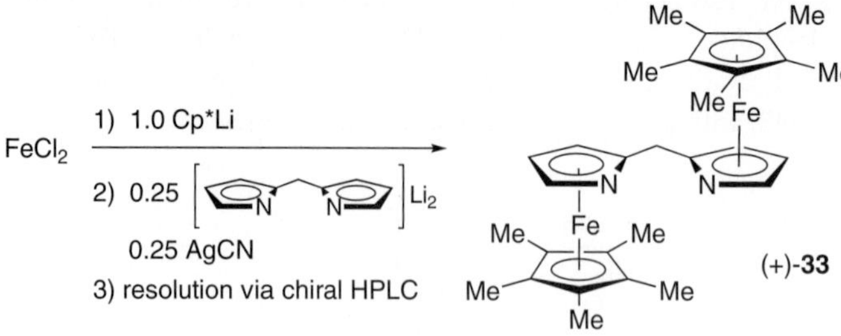

FeCl₂
1) 1.0 Cp*Li
2) 0.25 [structure]Li₂
   0.25 AgCN
3) resolution via chiral HPLC

→ (+)-33   (24)

In the presence of CuOTf, bis(azaferrocene) **33**, like structurally related ligands such as bisoxazolines and semicorrins, furnishes very good stereoselectivity in the asymmetric cyclopropanation of olefins by diazoacetates (Scheme 11.8) [50]. The best results are obtained when there is a large ester substituent, such as 2,6-di-*tert*-butyl-4-methylphenyl (BHT), on the diazo compound. High diastereo- and enantioselectivities are observed for a wide range of substrates, including aryl-, alkyl-, and silyl-substituted alkenes. It is noteworthy that, whereas a large excess of olefin is required for many cyclopropanation procedures that employ diazo compounds, in the case of CuOTf/**33**, the olefin serves as the limiting reagent [51].

Doyle and coworkers have systematically examined catalytic enantioselective intramolecular cyclopropanations to generate macrocycles [52]. They investigated a variety of chiral copper and rhodium catalysts and observed that, for certain substrates, bis(azaferrocene) **33** provided the best *ee*s and yields.

11.3.2.2 Copper-Catalyzed Insertions of Diazo Compounds

Copper/**33** has proved to be an effective catalyst for an array of insertion reactions of diazo compounds. For example, it accomplishes ring expansions of oxetanes to tetrahydrofurans with good stereoselectivity (Eq. (25)) [53]. The catalyst, not the substrate, controls the configuration of the new stereocenter.

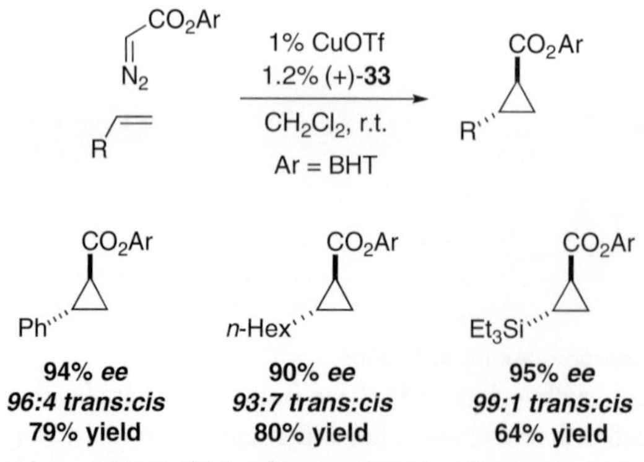

Ph'''	n-Hex'''	Et₃Si'''
94% ee	90% ee	95% ee
96:4 trans:cis	93:7 trans:cis	99:1 trans:cis
79% yield	80% yield	64% yield

Scheme 11.8 Cu/bis(azaferrocene)-catalyzed asymmetric cyclopropanations of olefins.

The versatility of ligand **33** was further confirmed by its application in copper-catalyzed enantioselective insertions of diazo compounds into O–H bonds (Eq. (26)) [54]. The structure of the alcohol plays an important role in determining both the *ee* and the yield. 2-Trimethylsilylethanol proved to be the alcohol of choice from these standpoints, as well as the facility with which the insertion product can be deprotected to reveal an α-hydroxy ester.

Copper-catalyzed enantioselective insertions of α-diazo esters into N–H bonds, which provide α-amino acid derivatives, have also been investigated with ferrocene-containing heterocycles as ligands. The method developed for O–H insertions is not highly effective for N–H insertions, but, if a C_2-symmetric bipyridine ligand (**34**) is used along with CuBr and AgSbF$_6$, good *ee*s can be achieved (Eq. (27)) [55]. The silver salt probably serves to abstract bromide from CuBr and generate a cationic copper complex. The illustrated method provides enantioselective access to *N*-Boc-protected aryl glycines; similarly, when *N*-benzyloxycarbonyl amine is employed as a substrate, *N*-Cbz-protected α-aryl amino acid derivatives are produced with high *ee*.

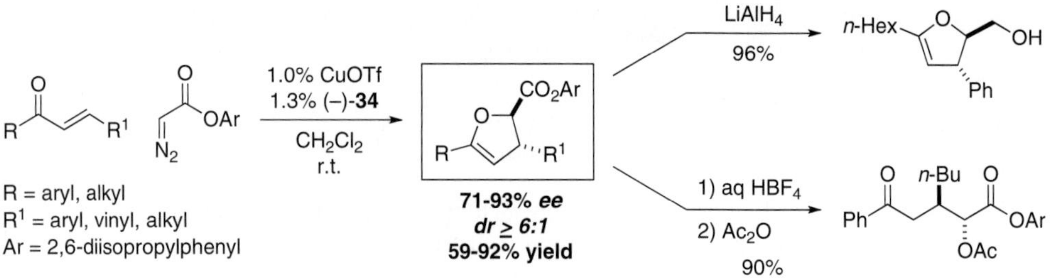

Scheme 11.9 Copper/bipyridine-catalyzed asymmetric [4 + 1] cycloadditions of enones with diazo esters.

11.3.2.3 Copper-Catalyzed Cycloadditions

Ferrocene-derived ligand **34** has also been applied to copper-catalyzed [4 + 1] cycloadditions of conjugated enones with diazo compounds (Scheme 11.9) [56]. This reaction provides access to highly substituted 2,3-dihydrofurans with good diastereo- and enantioselectivity. A range of ligands and diazo esters was examined, and the best results were obtained with bipyridine **34** and a bulky 2,6-diisopropyl-phenyl ester. The dihydrofurans can be converted into an array of cyclic and acyclic derivatives.

Other copper-catalyzed enantioselective cycloadditions can also be achieved with planar-chiral ferrocene derivatives as ligands. For example, the Kinugasa reaction, wherein a terminal alkyne couples with a nitrone to furnish a β-lactam, can be accomplished with good stereoselectivity in the presence of CuCl and bis(azaferro-cene) **35** (Scheme 11.10) [57]. Because ligand **33** provided only moderate stereo-selection, it was necessary to refine its structure through the addition of methyl groups adjacent to the nitrogen.

11.3.2.4 Olefin Polymerizations

Stimulated by reports by Brookhart that palladium and nickel complexes that bear 1,2-diimine ligands serve as highly active catalysts for olefin polymerization [58], Salo and Guan synthesized a series of bis(azaferrocene) adducts of palladium and

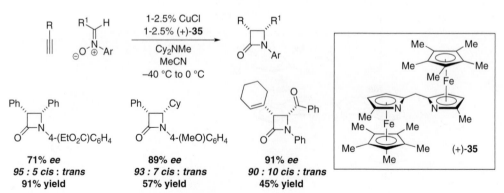

Scheme 11.10 Copper/bis(azaferrocene)-catalyzed asymmetric Kinugasa reactions.

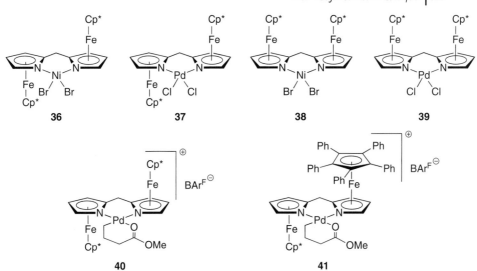

Figure 11.2 Bis(azaferrocene) catalysts of Pd and Ni for ethylene polymerization (BAr^F = tetrakis[3,5-bis(trifluoromethyl)phenyl]borate).

nickel and examined their utility in ethylene polymerization (**36–41**; Figure 11.2) [59]. Dihalide complexes **36–39** showed very low activity under standard polymerization conditions. The halide-free complexes **40** and **41** proved to be more active, but they were much less reactive than Brookhart's diimine complexes. Nevertheless, **40** and **41** did exhibit good thermal stability, forming oligomers with moderate to high branching densities at temperatures up to 120 °C.

Watanabe later reported a study of a series of nickel/iminoazaferrocene catalysts for ethylene polymerization [60]. In particular, the influence of ortho- and para-substitution on the arylimino group in compounds **42–51** was examined (Figure 11.3). With respect to the ortho substituent, a balance of steric hindrance controls the molecular weight and the branching density of the polyethylene. If there is no substituent (**42**), oligomers, rather than polymers, are observed. Polymers of increasing molecular weight are generated in going from methyl (**43**) to isopropyl (**44**). Larger groups, such as *tert*-butyl (**45**) and TMS (**46**), resulted in little or no polymerization activity. Introduction of a trifluoromethyl (**47**) or a phenyl (**48**) substituent did not improve the catalytic activity.

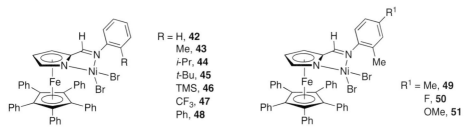

Figure 11.3 Nickel/iminoazaferrocene catalysts for ethylene polymerization.

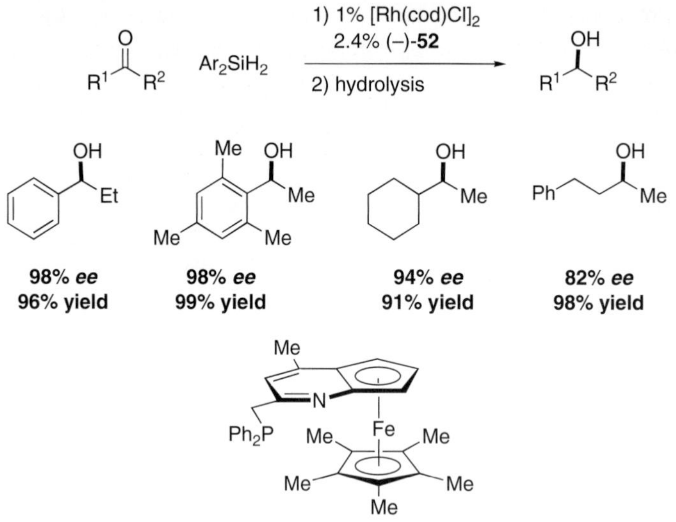

Scheme 11.11 Rhodium-catalyzed asymmetric hydrosilylations of ketones.

The nature of the group in the para-position also influences the characteristics of the product polymer. Although the addition of a methyl substituent has little effect (**49** vs. **43**), a fluoro group (**50**) leads to a highly active catalyst that produces a more branched polymer with a lower melting point. In contrast, the electron-rich methoxy-substituted compound (**51**) exhibits decreased activity when compared with catalyst **43**.

11.3.2.5 Rhodium-Catalyzed Reductions of Ketones

In addition to the N,N bidentate ligands that have been discussed earlier in this section, there has been a report of a P,N ligand that is based on a π-bound nitrogen heterocycle. Thus, when pyridine-phosphine **52** is combined with rhodium(I), hydrosilylations of ketones proceed with excellent *ee* (Scheme 11.11) [61]. The structure of the silane is crucial for obtaining good enantioselectivity, and certain hindered diarylsilanes were found to be especially effective. The scope of this process is broad, with both aryl alkyl and dialkyl ketones undergoing reduction to provide highly enantioenriched secondary alcohols.

11.4
Phosphaferrocenes in Catalysis

Unlike their aza counterparts, which have proved to be broadly useful both as nucleophilic catalysts and as ligands for transition metals, phosphaferrocenes have been employed primarily as ligands. Nevertheless, in the first application of a phosphaferrocene in a catalyzed reaction, its role was as a nucleophilic catalyst.

Specifically, a 1997 study described the use of a phosphaferrocene to catalyze ring-openings of epoxides with TMSCl to furnish chlorohydrins [62]. Beginning in 1998, a wide range of investigations have explored the utility of phosphaferrocenes as ligands for transition metals [3, 63, 64].

11.4.1
Reduction Reactions

Phosphaferrocene **53** serves as an effective ligand for rhodium-catalyzed hydrogenations of dehydroamino acids (Eq. (28)) [65]. Under mild conditions (1 atm H_2; room temperature), a range of substrates are reduced in good to excellent *ee* and yield. ^{31}P NMR studies provide strong evidence for bidentate coordination of the ligand to rhodium.

$$R\overset{O}{\underset{NHCOMe}{\diagup}}OMe \xrightarrow[\text{r.t.}]{\substack{H_2 \text{ (1 atm)} \\ 6\% \text{ (−)-}\textbf{53} \\ 5\% \text{ [Rh(cod)}_2\text{]PF}_6 \\ \text{EtOH}}} R\overset{O}{\underset{NHCOMe}{\diagup}}OMe \quad (28)$$

R = H, aryl, alkyl

79-96% *ee*
92-100% yield

(−)-**53**

More recently, a comparison of rhodium-catalyzed hydrogenations in the presence of phosphaferrocene and phospharuthenocene ligands has been provided by Carmichael and coworkers (Eq. (29)) [66]. In the case of complexes **54** and **55**, the ruthenium-based ligand furnishes better rates and enantioselectivity than the iron analogue. This study points to exciting opportunities in heteroruthenocene chemistry.

$$R\diagdown\hspace{-2pt}\text{—}\hspace{-2pt}\overset{CO_2Me}{\underset{NHAc}{\diagup}} \xrightarrow[\text{r.t.}]{\substack{H_2 \text{ (1 atm)} \\ 1\% \textbf{ 54} \text{ or } \textbf{55} \\ \text{EtOH}}} R\diagdown\hspace{-2pt}\text{—}\hspace{-2pt}\overset{CO_2Me}{\underset{NHAc}{\diagup}} \quad (29)$$

R = H, F, Cl, OMe, Me, NO_2

with **54**: 62-80% *ee*, >99% yield
with **55**: 84-96% *ee*, >99% yield

M = Fe, **54**
Ru, **55**

Rather than employing planar-chiral phosphaferrocenes, Ogasawara, Hayashi, and coworkers investigated the use of phosphaferrocene and phospharuthenocene ligands (e.g., **57–59**) generated from axially chiral binaphthyl/phosphole **56** (Eq. (30)) [67]. These intriguing monodentate ligands were applied to the palladium-catalyzed hydrosilylation of conjugated enynes, which furnishes enantioenriched allenylsilanes. As illustrated in Eq. (30), the use of a Cp*, rather than a Cp, group led to higher *ee* and yield (**57** vs. **58**), whereas the replacement of Fe with Ru resulted in a less effective process (**58** vs. **59**).

$$t\text{-Bu} \diagup\!\!\!\diagdown \quad \xrightarrow[20\,°\text{C}]{\substack{1\%\,[(\eta^3\text{-}C_3H_5)\text{PdCl}]_2 \\ 1\%\,\textbf{57-59}}} \quad t\text{-Bu} \diagdown\!\!\!\!=\!\!\!\!=\!\!\diagup\substack{H\\ \text{Me}} \quad (30)$$

with **57**: 41% *ee*, 57% yield
with **58**: 84% *ee*, 79% yield
with **59**: 75% *ee*, 59% yield

56

57, **58**, **59**

11.4.2
Palladium-Catalyzed Allylic Alkylations

Palladium-catalyzed asymmetric allylic alkylation has often been used to evaluate the potential of new chiral phosphines. Ganter's group was the first to employ phosphaferrocene ligands in this process [63a]. Enantiopure aldehyde **12** (Scheme 11.1) serves as a convenient precursor for a range of bidentate P,N and P,P ligands (Scheme 11.12). In the presence of $[(\eta^3\text{-}C_3H_5)\text{PdCl}]_2$, phosphaferrocene–pyridines **60** and **61** furnished very modest *ees* for the coupling of dimethyl malonate with 1,3-diphenylallyl acetate (Scheme 11.13, entries 1 and 2). On the other hand, Ganter and coworkers later established that bis(phosphaferrocene) **63** (Scheme 11.12) provided **64** with much higher enantioselectivity (Scheme 11.13, entry 3) [68].

Although phosphaferrocene–pyridines **60** and **61** furnished low *ees* in palladium-catalyzed allylic alkylations (Scheme 11.13, entries 1 and 2), a different family of P,N

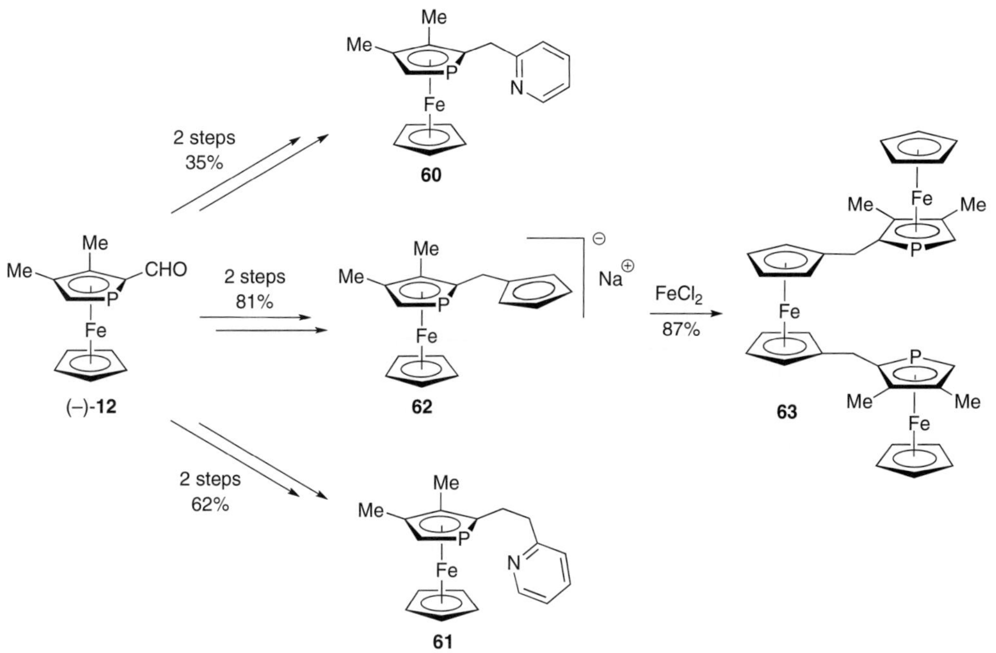

11.4 Phosphaferrocenes in Catalysis

Scheme 11.12 Synthesis of bidentate phosphaferrocenes.

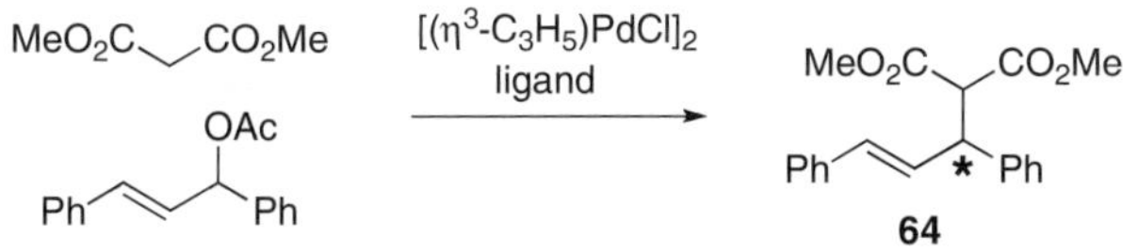

entry	ligand	ee (%)
1	60	19 (R)
2	61	11 (R)
3	63	79 (R)
4	65a	68 (R)
5	65b	79 (S)
6	66a	73 (R)
7	66b	82 (S)
8	67	99 (R)

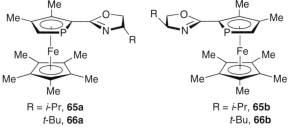

R = i-Pr, **65a**
t-Bu, **66a**

R = i-Pr, **65b**
t-Bu, **66b**

Scheme 11.13 Palladium/phosphaferrocene-catalyzed asymmetric allylic alkylations.

ligands, phosphaferrocene–oxazolines (**65** and **66**), provided moderate enantioselectivities (entries 4–7) [69]. The planar chirality of the phosphaferrocene unit, not the central chirality of the oxazoline, is the primary determinant of the stereochemical outcome of the reaction. Thus, ligands **65a** and **66a** afforded the (*R*) product, whereas **65b** and **66b** generated the (*S*)-enantiomer.

Hayashi and coworkers have reported the use of an interesting menthyl-substituted phosphaferrocene (**67**) in palladium-catalyzed allylic alkylations (Eq. (31)) [70]. Bisphosphine **67** is the most effective phosphaferrocene-based ligand reported to date for this process, furnishing product **64** in 99% isolated yield and 99% *ee* (Scheme 11.13, entry 8). Although a 1/1 mixture of **67** and [Pd] led to bidentate coordination of the ligand, use of a large excess of **67** produced a palladium complex in which two molecules of **67** are coordinated to each palladium through the diphenylphosphino group. Under the latter conditions, a lower *ee* was observed, suggesting that the chelated complex of **67** is the more enantioselective catalyst.

11.4.3
Rhodium-Catalyzed Isomerizations

Planar-chiral phosphaferrocenes have been applied to asymmetric isomerizations of allylic alcohols to aldehydes, a process for which only modest progress had previously been described (up to 53% *ee*) [71]. Thus, a rhodium/phosphaferrocene complex (**68**) efficiently isomerizes a range of allylic alcohols with generally good enantioselectivity (Scheme 11.14) [72]. Mechanistic studies established that the reaction proceeds through an intramolecular 1,3-hydrogen migration and that the chiral rhodium catalyst selectively activates one of the enantiotopic C1 hydrogens.

11.4.4
Copper-Catalyzed Conjugate Additions

Phosphaferrocene–oxazolines can be applied not only to palladium-catalyzed allylic alkylations (Scheme 11.13), but also to copper-catalyzed conjugate additions of organozinc reagents to enones (Scheme 11.15) [73]. Ligands **65a** and **65b** furnish only modest enantioselectivity for the addition of ZnEt$_2$ to chalcone (62% *ee* and 34% *ee*, respectively). Interestingly, the two phosphaferrocenes preferentially generate the same enantiomer of the product, indicating that, in contrast to palladium-catalyzed

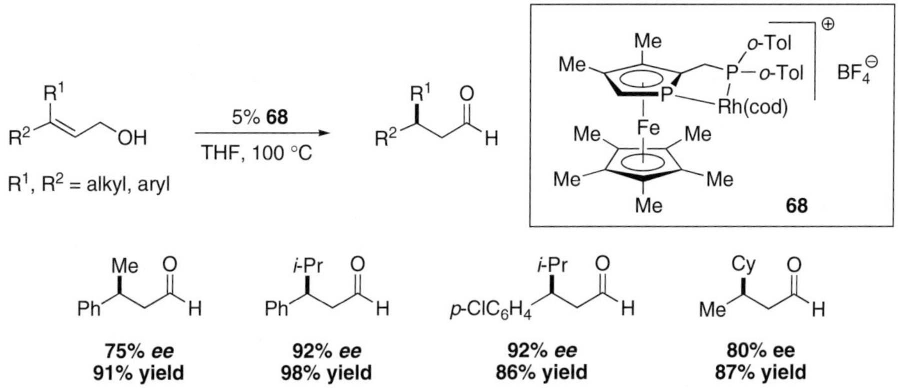

Scheme 11.14 Rhodium/phosphaferrocene-catalyzed asymmetric isomerizations of allylic alcohols.

allylic alkylations (Scheme 11.13, entries 4 and 5), the stereocenter of the oxazoline unit, not the planar chirality of the phosphaferrocene, is the dominant stereocontrol element in these conjugate additions. Better enantioselectivity is observed with phosphaferrocene–oxazoline ligands **69** and **70** (Scheme 11.15).

11.4.5
Copper-Catalyzed Cycloadditions

Phosphaferrocene–oxazolines have also proved to be useful for Kinugasa reactions. After an initial report on asymmetric intermolecular couplings catalyzed by copper/bis(azaferrocene) (Scheme 11.10), the intramolecular variant, which generates β-lactams that possess two new rings, was examined (Scheme 11.16) [74]. For substrate **71**, bis(azaferrocene) **35**, which had been effective for enantioselective intermolecular Kinugasa reactions, afforded β-lactam **72** in low ee and yield (6% ee, 30% yield). In contrast, phosphaferrocene–oxazolines **73** and **74** furnished the

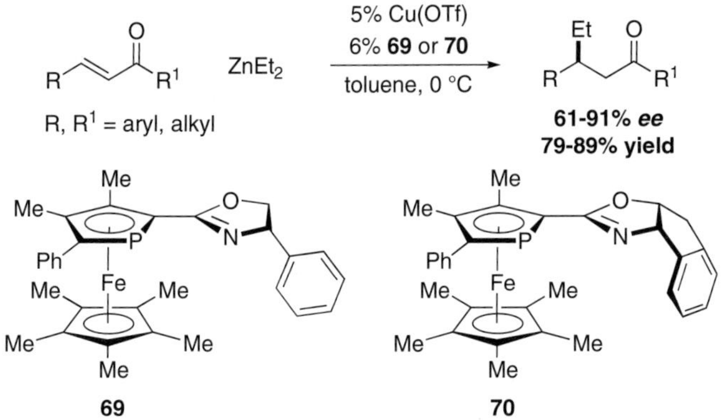

Scheme 11.15 Cu/phosphaferrocene-catalyzed asymmetric conjugate additions to enones.

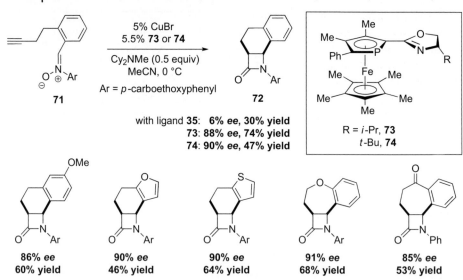

Scheme 11.16 Cu/phosphaferrocene-catalyzed asymmetric intramolecular Kinugasa reactions.

desired product with good stereocontrol (88–90% ee). In the presence of CuBr and these ligands, an array of β-lactams fused with 6- and 7-membered rings were produced with good enantioselectivity and in moderate yield. The postulated enolate intermediate **I** can be trapped by allyl iodide, thereby generating a valuable all-carbon quaternary center α to the carbonyl group (Eq. (32)).

In 2003, a mechanistically related process, the copper-catalyzed [3 + 2] cycloaddition of alkynes with azomethine imines was reported, including an asymmetric variant (Eq. (33)) [75]. Again, phosphaferrocene–oxazolines proved to be the ligands of choice. The scope of the cycloaddition was fairly broad, especially with respect to the substituents on the dipole.

R = aryl, alkenyl, alkyl
R^1 = acyl, aryl, alkyl

74–96% ee
63–100% yield

This new copper-catalyzed cycloaddition of azomethine imines can be applied to the kinetic resolution of the dipole (Eq. (34)), thereby furnishing enantio-enriched azomethine imines that can be transformed into a variety of pyrazolidinones (Eq. (35) and (36)) [76]. Phosphaferrocene–oxazoline 75 provided the highest selectivity factors among an array of chiral ligands that were examined. The enantioenriched dipoles can be derivatized with high diastereoselectivity, as exemplified in Eq. (35) and (36).

11.4.6
Palladium-Catalyzed Cross-Couplings

In 2000, Mathey, Le Floch, and coworkers reported the first use of phosphaferrocene ligands in cross-coupling reactions. Specifically, they employed a well-defined palladium(0) complex bearing two octaethyldiphosphaferrocene (76) ligands in a Suzuki reaction (Eq. (37)) [77]. Cross-couplings of aryl bromides with phenylboronic acid proceeded in essentially quantitative yield with low catalyst loadings (turnover numbers: 19 000–980 000).

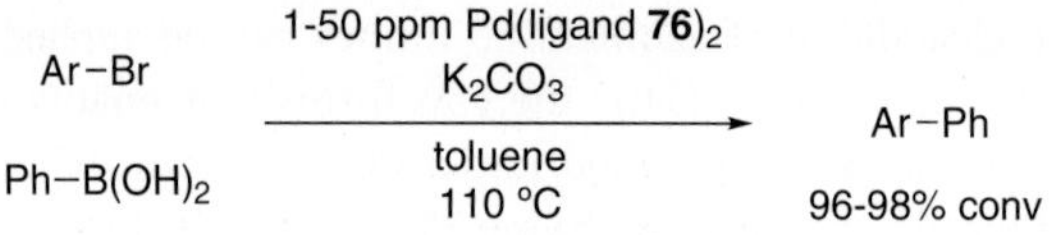

332 | *11 Applications of Aza- and Phosphaferrocenes and Related Compounds in Asymmetric Catalysis*

$$\text{Ar-Br} + \text{Ph-B(OH)}_2 \xrightarrow[\substack{\text{toluene} \\ 110\ °C}]{\substack{1\text{-}50\text{ ppm Pd(ligand 76)}_2 \\ K_2CO_3}} \text{Ar-Ph} \quad (37)$$

96–98% conv

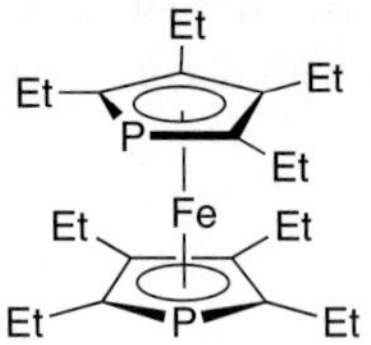

76

The same research group later expanded its investigation to the coupling of aryl halides with pinacol borane [78]. A dinuclear diphosphaferrocene complex (**77**) was the catalyst of choice among those that were examined, displaying high activity and broad versatility (Eq. (38)).

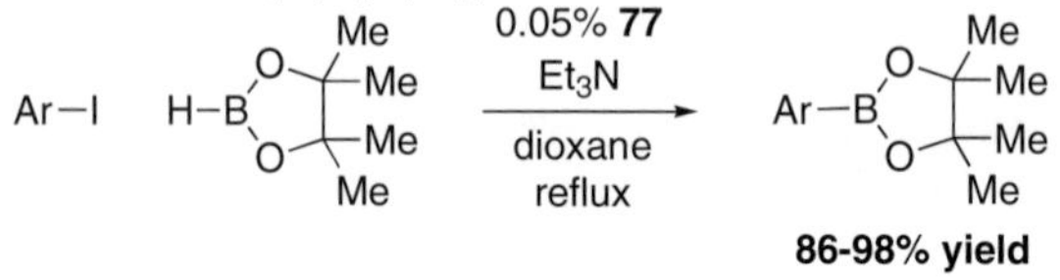

$$\text{Ar-I} + \text{H-B(OCMe}_2\text{CMe}_2\text{O)} \xrightarrow[\substack{\text{dioxane} \\ \text{reflux}}]{\substack{0.05\%\ \mathbf{77} \\ Et_3N}} \text{Ar-B(OCMe}_2\text{CMe}_2\text{O)} \quad (38)$$

86–98% yield

Ar = Ph, *p*-tol, *o*-tol, 2-thiophenyl, 2-furyl

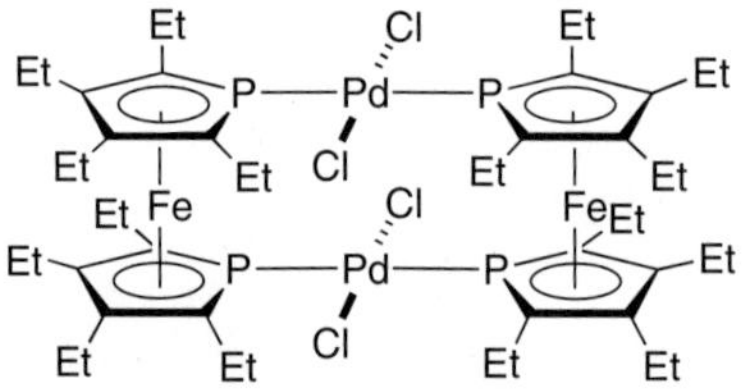

77

Ganter and coworkers have recently reported the synthesis of a series of bidentate and tridentate ligands that incorporate a phosphaferrocene and an N-heterocyclic carbene [79]. Palladium adduct **78** was examined as a catalyst for Suzuki reactions and found to generate the cross-coupling product in excellent yield even with a very low catalyst loading (Eq. (39)). As for the catalyst of Mathey and Le Floch (Eq. (37)), the nature of the active catalyst remains to be determined.

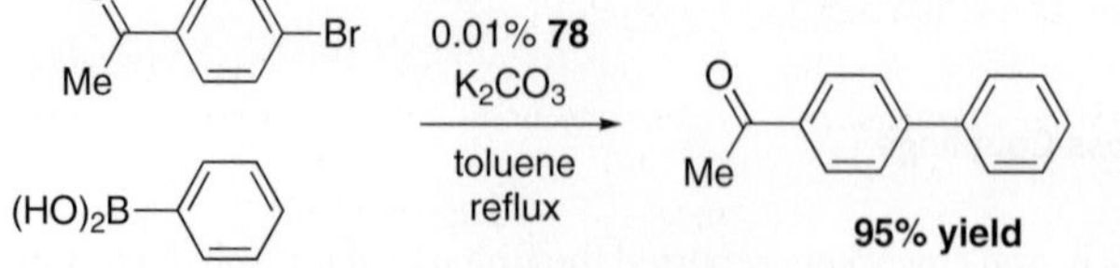

$$\text{MeC(O)-C}_6\text{H}_4\text{-Br} + (\text{HO})_2\text{B-C}_6\text{H}_5 \xrightarrow[\substack{\text{toluene} \\ \text{reflux}}]{\substack{0.01\%\ \mathbf{78} \\ K_2CO_3}} \text{MeC(O)-C}_6\text{H}_4\text{-C}_6\text{H}_5 \quad (39)$$

95% yield

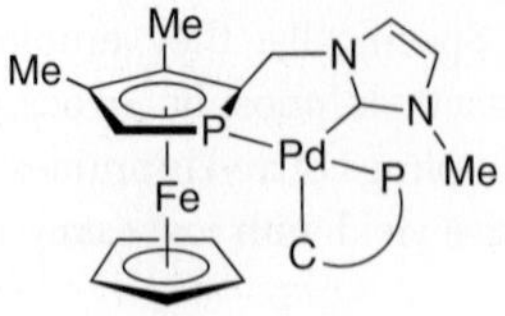

78

11.5
Conclusions

The first applications of azaferrocenes, phosphaferrocenes, and related compounds in catalysis (as nucleophilic catalysts and as ligands for transition metals) were described in the mid- and late-1990s. Since that time, they have been employed in an impressively diverse array of processes. They have been particularly useful in asymmetric catalysis where, for a number of transformations, they furnish state-of-the-art levels of enantioselectivity. There is little doubt that studies to date have only scratched the surface with regard to the potential of this family of compounds in catalysis and that further exciting applications will be described during the coming years.

References

1. For leading references, see: (a) Stepnicka, P. (ed.) (2008) *Ferrocenes: Ligands, Materials and Biomolecules*, John Wiley & Sons, New York; (b) Togni, A. and Hayashi, T. (eds) (1995) *Ferrocenes: Homogeneous Catalysis, Organic Synthesis, Materials Science*, Wiley–VCH, New York.
2. For leading references to azaferrocene chemistry, see: (a) Kuhn, N. (2000) in *Synthetic Methods of Organometallic and Inorganic Chemistry, Transition Metals Part 3*, vol. 9 (ed. W.A. Herrmann), Georg Thieme Verlag, New York, pp. 53–83; (b) Janiak, C. and Kuhn, N. (1996) in *Advances in Nitrogen Heterocycles*, vol. 2 (ed. C. Moody), JAI Press, Greenwich, pp. 179–210.
3. For leading references to phosphaferrocene chemistry, see: (a) Ganter, C. (2008) in *Phosphorus Ligands in Asymmetric Catalysis* (ed. A. Boerner), Wiley–VCH, New York, pp. 393–407; (b) Le Floch, P. (2006) *Coord. Chem. Rev.*, **250**, 627–681; (c) Carmichael, D. and Mathey, F. (2002) *Top. Curr. Chem.*, **220**, 27–51; (d) Weber, L. (2002) *Angew. Chem. Int. Ed.*, **41**, 563–572.
4. Joshi, K.K., Pauson, P.L., Qazi, A.R., and Stubbs, W.H. (1964) *J. Organomet. Chem.*, **1**, 471–475.
5. King, R.B. and Bisnette, M.B. (1964) *Inorg. Chem.*, **3**, 796–800.
6. Pauson, P.L. and Qazi, A.R. (1967) *J. Organomet. Chem.*, **7**, 321–324.
7. Efraty, A. and Jubran, N. (1980) *Inorg. Chim. Acta*, **44**, L191–L192.
8. Efraty, A., Jubran, N., and Goldman, A. (1982) *Inorg. Chem.*, **21**, 868–873.
9. For some examples, see: (a) Ernst, R.D., Wilson, D.R., and Herber, R.H. (1984) *J. Am. Chem. Soc.*, **106**, 1646–1650; (b) Cesario, M., Giannotti, C., Guilhem, J., Silver, J., and Zakrzewski, J. (1996) *J. Chem. Soc., Dalton Trans.*, 47–53.
10. For some examples, see: (a) Best, S.P., Clark, R.J.H., Deemings, A.J., McQueen, R.C.S., Powell, N.I., Acuña, C., Arce, A.J., and De Sanctis, Y. (1991) *J. Chem. Soc., Dalton Trans.*, 1111–1115; (b) Kowalski, K., Zakrzewski, J., Palusiak, M., and Domagala, S. (2006) *New J. Chem.*, **30**, 901–907.
11. Kuhn, N., Schulten, M., Zauder, E., Augart, N., and Boese, R. (1989) *Chem. Ber.*, **122**, 1891–1896.
12. Bauer, K., Falk, H., and Schlögl, K. (1969) *Angew. Chem., Int. Ed. Engl.*, **8**, 135.
13. For leading references to recent reports on the synthesis of enantioenriched azaferrocenes, see: Anderson, J.C., Osborne, J.D., and Woltering, T.J. (2008) *Org. Biomol. Chem.*, **6**, 330–339.

14 Mathey, F., Mitschler, A., and Weiss, R. (1977) *J. Am. Chem. Soc.*, **99**, 3537–3538.

15 de Lauzon, G., Deschamps, B., Fischer, J., Mathey, F., and Mitschler, A. (1980) *J. Am. Chem. Soc.*, **102**, 994–1000.

16 For example, see: Fischer, J., Mitschler, A., Ricard, L., and Mathey, F. (1980) *J. Chem. Soc., Dalton Trans.*, 2522–2525.

17 Ganter, C., Brassat, L., and Ganter, B. (1997) *Tetrahedron: Asymmetry*, **8**, 2607–2611.

18 (a) Ganter, C., Brassat, L., Glinsböckel, C., and Ganter, B. (1997) *Organometallics*, **16**, 2862–2867; (b) Ganter, C., Brassat, L., and Ganter, B. (1997) *Chem. Ber.*, **130**, 1771–1776.

19 Ruble, J.C. and Fu, G.C. (1996) *J. Org. Chem.*, **61**, 7230–7231.

20 For a resolution of DMAP derivative 21 via classical resolution, see: Wurz, R.P., Lee, E.C., Ruble, J.C., and Fu, G.C. (2007) *Adv. Synth. Catal.*, **349**, 2345–2352.

21 For overviews of the development of these families of catalysts, see: (a) Fu, G.C. (2004) *Acc. Chem. Res.*, **37**, 542–547; (b) Fu, G.C. (2000) *Acc. Chem. Res.*, **33**, 412–420.

22 For a review of the use of chiral DMAP catalysts in asymmetric synthesis, see: Wurz, R.P. (2007) *Chem. Rev.*, **107**, 5570–5595.

23 (a) Ruble, J.C., Latham, H.A., and Fu, G.C. (1997) *J. Am. Chem. Soc.*, **119**, 1492–1493; (b) Ruble, J.C., Tweddell, J., and Fu, G.C. (1998) *J. Org. Chem.*, **63**, 2794–2795.

24 Propargylic alcohols: (a) Tao, B., Ruble, J.C., Hoic, D.A., and Fu, G.C., (1999) *J. Am. Chem. Soc.*, **121**, 5091–5092; Allylic alcohols: (b) Bellemin-Laponnaz, S., Tweddell, J., Ruble, J.C., Breitling, F.M., and Fu, G.C., (2000) *Chem. Comm.*, 1009–1010.

25 Held, I., Villinger, A., and Zipse, H. (2005) *Synthesis*, 1425–1430.

26 For a study that compares the reactivity and enantioselectivity of corresponding ferrocene and ruthenocene derivatives of planar-chiral DMAPs, see: Garrett, C.E. and Fu, G.C. (1998) *J. Am. Chem. Soc.*, **120**, 7479–7483.

27 Chen, Y.-H. and McDonald, F.E. (2006) *J. Am. Chem. Soc.*, **128**, 4568–4569.

28 Smith, T.E., Kuo, W.-H., Balskus, E.P., Bock, V.D., Roizen, J.L., Theberge, A.B., Carroll, K.A., Kurihara, T., and Wessler, J.D. (2008) *J. Org. Chem.*, **73**, 142–150.

29 For another application of catalyst 22 to the kinetic resolution of a secondary alcohol, see: Harmata, M. and Kahraman, M. (1999) *J. Org. Chem.*, **64**, 4949–4952.

30 Hodous, B.L., Ruble, J.C., and Fu, G.C. (1999) *J. Am. Chem. Soc.*, **121**, 2637–2638.

31 Hodous, B.L. and Fu, G.C. (2002) *J. Am. Chem. Soc.*, **124**, 10006–10007.

32 Wiskur, S.L. and Fu, G.C. (2005) *J. Am. Chem. Soc.*, **127**, 6176–6177.

33 Schaefer, C. and Fu, G.C. (2005) *Angew. Chem. Int. Ed.*, **44**, 4606–4608.

34 For the use of an acylated planar-chiral DMAP as a stoichiometric reagent for the kinetic resolution of amines, see: Ie, Y. and Fu, G.C. (2000) *Chem. Comm.*, 119–120.

35 Arai, S., Bellemin-Laponnaz, S., and Fu, G.C. (2001) *Angew. Chem. Int. Ed.*, **40**, 234–236.

36 Arp, F.O. and Fu, G.C. (2006) *J. Am. Chem. Soc.*, **128**, 14264–14265.

37 Dai, X., Nakai, T., Romero, J.A.C., and Fu, G.C. (2007) *Angew. Chem. Int. Ed.*, **46**, 4367–4369.

38 Ruble, J.C. and Fu, G.C. (1998) *J. Am. Chem. Soc.*, **120**, 11532–11533.

39 Hills, I.D. and Fu, G.C. (2003) *Angew. Chem. Int. Ed.*, **42**, 3921–3924.

40 (a) Mermerian, A.H. and Fu, G.C. (2003) *J. Am. Chem. Soc.*, **125**, 4050–4051; (b) Mermerian, A.H. and Fu, G.C. (2005) *J. Am. Chem. Soc.*, **127**, 5604–5607.

41 Mermerian, A.H. and Fu, G.C. (2005) *Angew. Chem. Int. Ed.*, **44**, 949–952.

42 Denmark, S.E., Barsanti, P.A., Wong, K.-T., and Stavenger, R.A. (1998) *J. Org. Chem.*, **63**, 2428–2429.

43 Tao, B., Lo, M.M.-C., and Fu, G.C. (2001) *J. Am. Chem. Soc.*, **123**, 353–354.

44 Lee, E.C., McCauley, K.M., and Fu, G.C. (2007) *Angew. Chem. Int. Ed.*, **46**, 977–979.

45 Hodous, B.L. and Fu, G.C. (2002) *J. Am. Chem. Soc.*, **124**, 1578–1579.

46 Lee, E.C., Hodous, B.L., Bergin, E., Shih, C., and Fu, G.C. (2005) *J. Am. Chem. Soc.*, **127**, 11586–11587.
47 Wilson, J.E. and Fu, G.C. (2004) *Angew. Chem. Int. Ed.*, **43**, 6358–6360.
48 Berlin, J.M. and Fu, G.C. (2008) *Angew. Chem. Int. Ed.*, **47**, 7048–7050.
49 Dosa, P.I., Ruble, J.C., and Fu, G.C. (1997) *J. Org. Chem.*, **62**, 444–445.
50 Lo, M.M.-C. and Fu, G.C. (1998) *J. Am. Chem. Soc.*, **120**, 10270–10271.
51 For the application of a ferrocene-fused planar-chiral bipyridine to copper-catalyzed cyclopropanations of olefins, see: Rios, R., Liang, J., Lo, M.M.-C., and Fu, G.C. (2000) *Chem. Comm.*, 377–378.
52 Doyle, M.P., Hu, W., Chapman, B., Marnett, A.B., Peterson, C.S., Vitale, J.P., and Stanley, S.A. (2000) *J. Am. Chem. Soc.*, **122**, 5718–5728.
53 Lo, M.M.-C. and Fu, G.C. (2001) *Tetrahedron*, **57**, 2621–2634.
54 Maier, T.C. and Fu, G.C. (2006) *J. Am. Chem. Soc.*, **128**, 4594–4595.
55 Lee, E.C. and Fu, G.C. (2007) *J. Am. Chem. Soc.*, **129**, 12066–12067.
56 Son, S. and Fu, G.C. (2007) *J. Am. Chem. Soc.*, **129**, 1046–1047.
57 Lo, M.M.-C. and Fu, G.C. (2002) *J. Am. Chem. Soc.*, **124**, 4572–4573.
58 For pioneering studies, see: (a) Johnson, L.K., Killian, C.M., and Brookhart, M. (1995) *J. Am. Chem. Soc.*, **117**, 6414–6415; (b) Johnson, L.K., Mecking, S., and Brookhart, M. (1996) *J. Am. Chem. Soc.*, **118**, 267–268.
59 Salo, E.V. and Guan, Z. (2003) *Organometallics*, **22**, 5033–5046.
60 Watanabe, M. (2005) *Macromol. Rapid Commun.*, **26**, 34–39.
61 Tao, B. and Fu, G.C. (2002) *Angew. Chem. Int. Ed.*, **41**, 3892–3894.
62 Garrett, C.E. and Fu, G.C. (1997) *J. Org. Chem.*, **62**, 4534–4535. For a subsequent study, see: Wang, L.-S. and Hollis, T.K. (2000) *Org. Lett.*, **5**, 2543–2545.
63 For the first applications of phosphaferrocenes in asymmetric catalysis, see: (a) Ganter, C., Glinsböckel, C., and Ganter, B. (1998) *Eur. J. Inorg. Chem.*, 1163–1168; (b) Qiao, S. and Fu, G.C. (1998) *J. Org. Chem.*, **63**, 4168–4169.
64 For a report of an enantiopure C_2-symmetric diphosphaferrocene, see: Qiao, S., Hoic, D.A., and Fu, G.C. (1998) *Organometallics*, **17**, 773–774.
65 Qiao, S. and Fu, G.C. (1998) *J. Org. Chem.*, **63**, 4168–4169.
66 Carmichael, D., Goldet, G., Klankermayer, J., Ricard, L., Seeboth, N., and Stankevic, M. (2007) *Chem. Eur. J.*, **13**, 5492–5502.
67 Ogasawara, M., Ito, A., Yoshida, K., and Hayashi, T. (2006) *Organometallics*, **25**, 2715–2718.
68 Ganter, C., Kaulen, C., and Englert, U. (1999) *Organometallics*, **18**, 5444–5446.
69 Shintani, R., Lo, M.M.-C., and Fu, G.C. (2000) *Org. Lett.*, **2**, 3695–3697.
70 (a) Ogasawara, M., Yoshida, K., and Hayashi, T. (2001) *Organometallics*, **20**, 1014–1019; (b) Ogasawara, M., Yoshida, K., and Hayashi, T. (2001) *Organometallics*, **20**, 3913–3917.
71 Tani, K. (1985) *Pure Appl. Chem.*, **57**, 1845–1854.
72 (a) Tanaka, K., Qiao, S., Tobisu, M., Lo, M.M.-C., and Fu, G.C. (2000) *J. Am. Chem. Soc.*, **122**, 9870–9871; (b) Tanaka, K. and Fu, G.C. (2001) *J. Org. Chem.*, **66**, 8177–8186.
73 Shintani, R. and Fu, G.C. (2002) *Org. Lett.*, **4**, 3699–3702.
74 Shintani, R. and Fu, G.C. (2003) *Angew. Chem. Int. Ed.*, **42**, 4082–4085.
75 Shintani, R. and Fu, G.C. (2003) *J. Am. Chem. Soc.*, **125**, 10778–10779.
76 Suárez, A., Downey, C.W., and Fu, G.C. (2005) *J. Am. Chem. Soc.*, **127**, 11244–11245.
77 Sava, X., Ricard, L., Mathey, F., and Le Floch, P. (2000) *Organometallics*, **19**, 4899–4903.
78 Melaimi, M., Mathey, F., and Le Floch, P. (2001) *J. Organomet. Chem.*, **640**, 197–199.
79 Willms, H., Frank, W., and Ganter, C. (2008) *Chem. Eur. J.*, **14**, 2719–2729.

12
Metallocyclic Ferrocenyl Ligands
Christopher J. Richards

12.1
Introduction

This chapter is a comprehensive account of the synthesis and application of planar chiral ferrocene-based metallocycles. The first examples of the asymmetric synthesis of this class of compound date from the late 1970s, with the pioneering work of Viatcheslav Sokolov. However, it is only since the millennium that they have found widespread application, principally as catalysts for the allylic imidate rearrangement initially developed by Larry Overman.

In the vast majority of cases the metal of the metallocyclic ring is palladium. Thus this class of compound is a sub-category of palladacycles [1] and/or metallacycles [2] in which the metal is bonded to an sp^2, or less frequently, an sp^3 carbon atom. A significant feature of anionic bidentate ferrocenyl ligands is that the complexed metal M becomes a direct component of the element of planar chirality displayed by the resultant complex (Scheme 12.1). This contrasts to the majority of neutral bidentate ferrocene ligands [3] (e.g., P-P, P-N etc.), where the complexed metal is more remote from the basis of chirality [4]. Where M = Pd(II) or Pt(II), the ancillary ligands L^1 and L^2 of the resultant square-planar complex are distinguished by the larger trans-influence/effect of the anionic carbon ligand compared to the neutral heteroatom (X) ligand component. This is revealed by the longer L^1–M bond length compared to the corresponding L^2–M bond length where $L^1 = L^2$ [1)], and also by analysis of the kinetic and thermodymanic products arising from ligand substitution [5]. The coordination axis perpendicular to the square plane, as represented by diastereotopic ligands L^3 and L^4, is significantly influenced by the bulk of the metallocene. Thus associative ligand substitution reactions of square-planar complexes will most likely involve approach of the incoming ligand from the top face (L^3).

---

1) For example, from the X-ray crystal structure of *trans*-rac-**11**, the Cl–Pd bond lengths trans to carbon are 2.445(2) Å (×2) and the Cl–Pd bond lengths trans to phosphorus are 2.426(2) and 2.404(2) Å [17]. For *trans*-$(S,R_p)_2$-**47** Cl–Pd bond lengths trans to carbon and nitrogen are given as 2.485(2) and 2.339(2) Å, respectively [40a].

Chiral Ferrocenes in Asymmetric Catalysis. Edited by Li-Xin Dai and Xue-Long Hou
Copyright © 2010 WILEY-VCH Verlag GmbH & Co. KGaA, Weinheim
ISBN: 978-3-527-32280-0

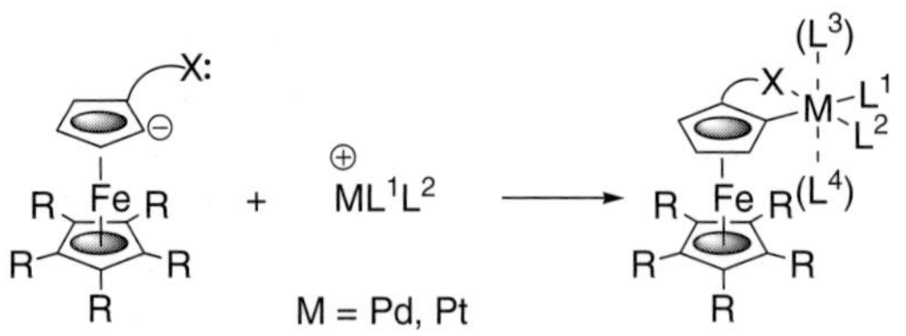

12 Metallocyclic Ferrocenyl Ligands

Scheme 12.1 The formal complexation of anionic bidentate ferrocenyl ligands.

The steric influence of the planar chiral ligand may be enhanced by introduction of non-hydrogen substituents in the bottom cyclopentadienyl ring (R = Me, Ph).

The element of planar chirality, designated either R_p or S_p, is a consequence of the absence of a plane of symmetry due to the presence of two or more different substituents on one or both of the cyclopentadienyl rings. Two methods are employed to assign the configuration of such compounds (Figure 12.1). In method 1 the substituents of the cyclopentadienyl ring are prioritized by atomic number and the sense of rotation determined by the shortest path for the arc of decreasing priority as viewed from the face of the ring not bonded to iron [6]. In method 2 the carbon atom of the cyclopentadienyl ring bearing the substituent of highest priority (*) is regarded as a virtual sp^3-hybridized carbon and the configuration assigned by application of the standard Cahn–Ingold–Prelog (CIP) system [7]. Although method 2 is a logical extension of the CIP method used for configuration assignment of saturated stereogenic centers [8], method 1 is much more frequently employed. When the atomic number of the highest priority substituent is >26 (i.e., the atomic number of iron), and the atomic number of the next substituent (typically carbon) is <26, both methods give the same outcome. This is the case with all the planar chiral metallocycles described in this chapter (M = Pd, Pt, Hg).

A variety of methods have been employed for the determination of the absolute configuration of ferrocene-based metalloacycles. Earlier studies used chemical correlation and/or circular dichroism (CD), whereas most recent reports have employed X-ray crystallography. Although the detailed appearance of the CD spectrum obviously varies from structure to structure, in all the examples examined for this review an R_p metallacycle configuration is associated with a positive value of

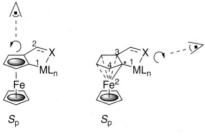

Method 1   Method 2

Figure 12.1 Assignment of configuration in planar chiral metallocycles.

molar circular dichroism (Δε) for the band arising from the major absorption in the visible region ($\lambda_{max} \sim$ 500 nm) [9]. The dominance of the planar chirality on the chiroptical properties of metallocene-based metallacycles allows a rule of thumb to be stated (which as such is not intended to be reliable for every situation) linking a positive value of the specific rotation (usually determined at 589 nm) with the R_p configuration, and a negative value with the S_p configuration. A definitive assignment of absolute configuration may be made using X-ray crystallography, either where the metallacycle contains a ligand-based stereogenic center of known configuration, or by determination of the absolute structure (Flack) parameter. Pitfalls encountered in the assignment of configuration by X-ray crystallography have recently been comprehensively reviewed [2a].

12.2
Asymmetric Synthesis of Planar Chiral Metallocyclic Complexes

The majority of metallocyclic ferrocenyl compounds are synthesized by C—H activation of a prochiral monosubstituted ferrocene containing a directing group that results in "ortho" (or α) metalation (Scheme 12.2). There are a variety of methods by which the resulting planar chiral metallocycles may be obtained enriched in one enantiomer: resolution (Section 12.2.1); diastereoselective metalation controlled by one or more stereogenic centers contained in the ferrocene substituent (Section 12.2.2); and enantioselective metalation mediated by an enantiopure chiral metallating reagent (Section 12.2.5). Following the generation of a non-racemic planar chiral metallacycle by one of these approaches, transmetalation may be employed to change the identity of the metal (Section 12.2.4). Alternatively, the metal may be introduced by oxidative addition with a scalemic halide synthesized by one of the plethora of methods now available for the asymmetric synthesis of planar chiral ferrocenes (Section 12.2.3).

Reagents for the palladation of a generalized ferrocene derivative **1** are palladium acetate, either with or without the addition of sodium acetate, or sodium tetrachloropalladate in combination with sodium acetate. If only palladium acetate is used the solvent is generally either toluene, dichloromethane or acetic acid. If sodium acetate is used with either of the two palladium sources the solvent is typically methanol. The product of these reactions is either the acetate bridged dimer **2** or the chloride bridged dimer **3**. There is the possibility that both can exist as mixtures of cis/trans-isomers with respect to the orientation of the two bidentate C—X ligands about the

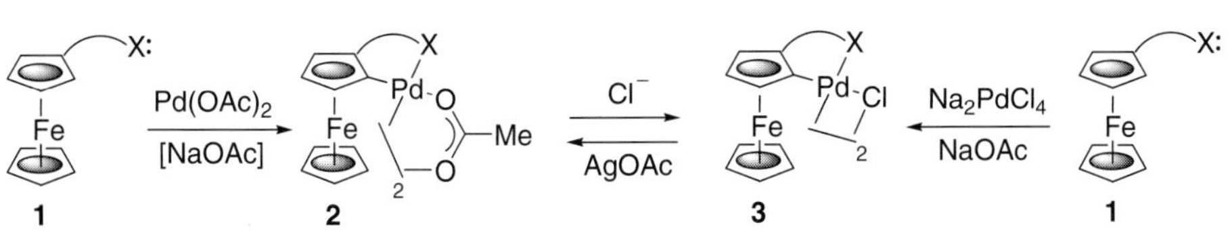

Scheme 12.2 Methods of palladation of generalized ferrocene precursor **1**.

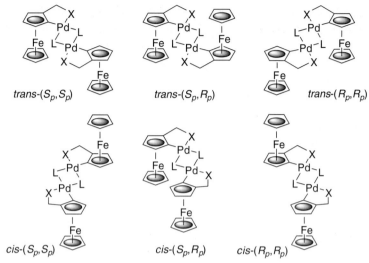

Figure 12.2 Possible stereoisomers for a bridged planar chiral palladacycle.

$Pd_2(\mu\text{-L})_2$ core. As a consequence there are six possible stereoisomers for a bridged planar chiral palladacycle that is not enantiopure (Figure 12.2). Fortunately one of the two geometrical isomers (usually the trans) is formed exclusively, simplifying the analysis of these compounds.

The acetate **2** is readily transformed into the chloride **3** on addition of lithium or sodium chloride. The reverse reaction requires the addition of silver acetate. Related platination and mercuration reactions have been performed with cis-$PtCl_2(DMSO)_2$ and $Hg(OAc)_2$/LiCl, respectively.

The mercuration of ferrocene has been experimentally determined to proceed via precomplexation of the mercurating agent to the iron atom before rate determining formation of the carbon–mercury bond (Scheme 12.3, path a) [10]. This contrasts with the Friedel–Crafts acetylation of ferrocene in which exo-attack of the acylium ion has been experimentally demonstrated (Scheme 12.3, path b) [11]. A theoretical study supports these contrasting pathways for Hg^{2+} as a representative soft electrophile,

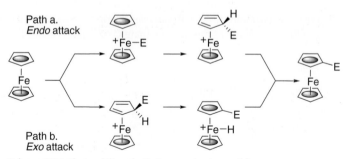

Scheme 12.3 Electrophilic substitution mechanisms of ferrocene.

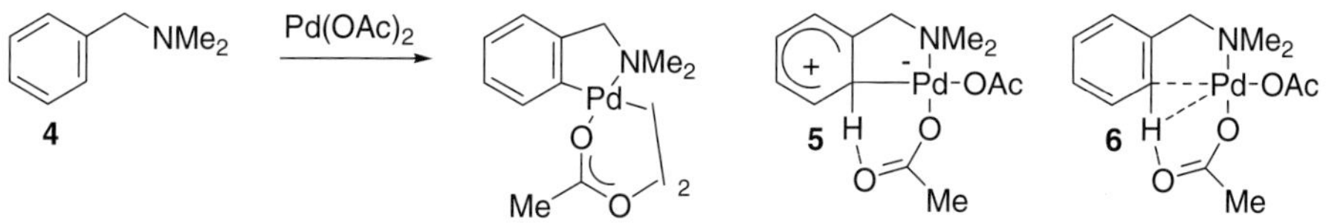

12.2 Asymmetric Synthesis of Planar Chiral Metallocyclic Complexes

Scheme 12.4

and the acylium ion as a representative hard electrophile [12]. Although the palladation of ferrocene does not appear to have been the subject of a mechanistic investigation, the corresponding reaction of benzene derivatives has been the subject of several studies. For the palladation of N,N-dimethylbenzylamine **4** an electrophilic substitution mechanism has been proposed via the Wheland intermediate **5**, in which an acetate ligand participates in proton abstraction via a six-membered transition state (Scheme 12.4) [13]. A recent theoretical study refined the mechanism, and determined that it proceeded via an agostic C–H complex **6** followed by facile hydrogen transfer to a coordinated acetate [14]. This mechanism explains the role of acetate in the palladation of **7** to give, after acetate/chloride ligand exchange, the palladacycle **8**, as in the absence of acetate only a $(FcCH_2NMe_2)_2PdCl_2$ complex was isolated (Scheme 12.5) [15].

12.2.1
Resolution

There are a number of examples of the resolution of planar chiral metallacycles by reaction with an amino acid followed by separation of the resulting diastereoisomers. This was first demonstrated with rac-**8** to which was added (S)-proline and potassium hydrogencarbonate followed by fractional crystallization to give both (S,S_p)-**9** and (S,R_p)-**9** (Scheme 12.5) [16]. In the same way palladation of **10** was followed by resolution of the resulting phosphapalladacycle **11** by combining extraction and fractional crystallization to separate (S,S_p)-**12** and (S,R_p)-**12** [17].

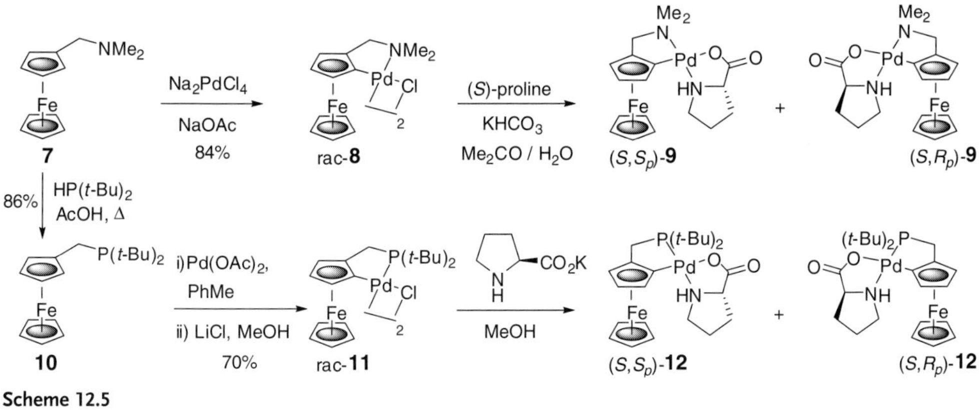

Scheme 12.5

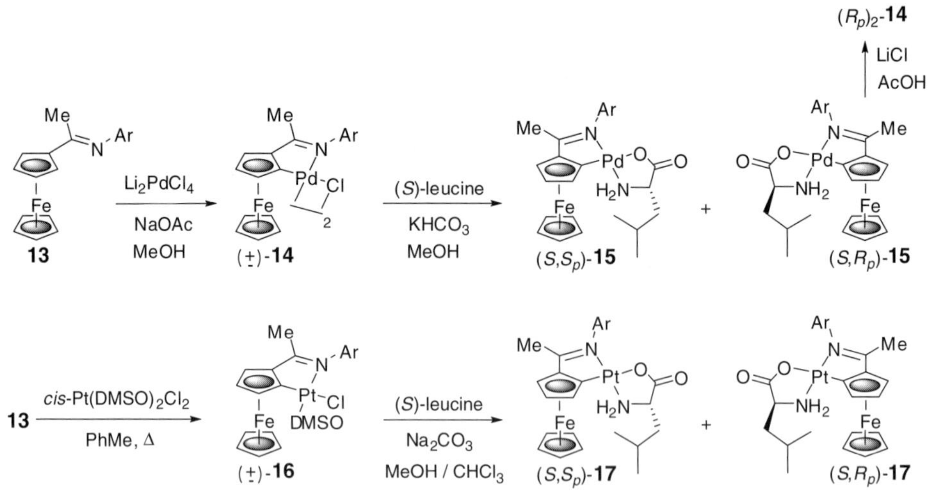

Scheme 12.6

Similarly, the reaction of the ketimine-based palladacycle **14** [18] with (S)-leucine was followed by the separation of (S,S_p)-**15** and (S, R_p)-**15** by column chromatography (Scheme 12.6) [19]. In all of these examples the amino acid ligand is readily substituted with chloride ligands to complete the formal resolution of the starting palladacycle. This methodology is also applicable to the related platinacycle **16**, where again the resulting diastereoisomers (S,S_p)-**17** and (S,R_p)-**17** were separated by chromatography [20].

12.2.2
Diastereotopic C–H Activation

N,N-Dimethyl-1-ferrocenylethylamine **18**, also known as the Ugi amine [21], is readily available in both enantiomeric forms [22]. It has been used extensively as a starting material for the synthesis of a wide variety of 1,2-disubstituted planar chiral ferrocene derivatives due to the high diastereoselectivity of metalation with butyllithium (dr = 96 : 4) [3]. The analogous palladation of **18** proceeds with the same sense of diastereoselectivity such that (R)-**18** gives $(R,S_p)_2$-**19** and $(R,R_p)_2$-**19** with a dr of 85 : 15 (Scheme 12.7) [23]. Although it was later reported that only a single diastereoisomer precipitated from the reaction mixture [24], the 85 : 15 selectivity has been independently verified [25]. Palladation of 1,1′-di(1-dimethylaminoethyl) ferrocene under the same conditions has also been reported, but in insufficient detail to permit comment on any selectivity obtained [26]. In contrast the trifluoromethyl analogue (R)-**20** gives $(R,S_p)_2$-**21** with almost complete control of diastereoselectivity [27]. Constraining the conformational freedom of the dimethylamino group by use of (R)-**22** also significantly improves the diastereoselectivity, only $(R,R_p)_2$-**23** is reported to have been isolated [28]. Finally the palladation of **24**,

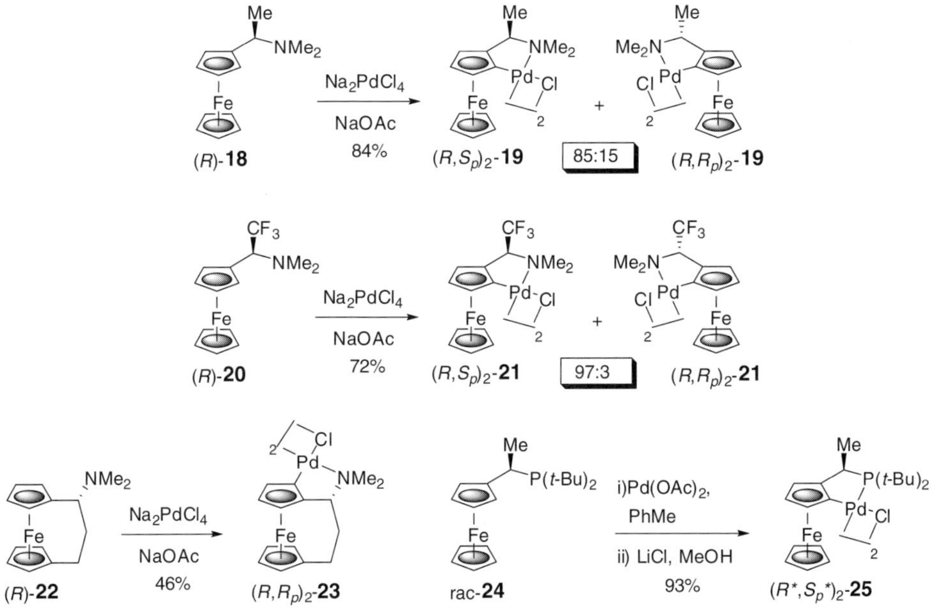

12.2 Asymmetric Synthesis of Planar Chiral Metallocyclic Complexes

Scheme 12.7

a di-*tert*-butylphosphino analogue of the Ugi amine, also results in the formation of a single diastereoisomer, $(R^*,S_p^*)_2$-**25** [29].

Another ferrocene derivative that, like the Ugi amine **18**, undergoes highly diastereoselective lithiation, is ferrocenyl oxazoline **26** [30]. Although it has been suggested that the direct cyclopalladation of 2-ferrocenyl-1,3-oxazolines is prevented by competitive substrate oxidation by palladium(II) salts [31], this reaction was subsequently reported as proceeding in 46% yield from the addition of palladium acetate to (S)-**26** to give $(S,S_p)_2$-**27** (Scheme 12.8) [32]. Again the sense of diastereoselection is the same as the corresponding lithiation reaction (see Section 12.2.3). In contrast, palladation of 4-ferrocenyl-1,3-oxazolines **28** does not result in a new element of planar chirality. Instead interannular palladation occurs to give the unprecedented trinuclear complex $(S)_2$-**29**, provided R does not contain a C—H bond in the α-position. A mixture of products was obtained where R = *iso*-propyl or ethyl [31].

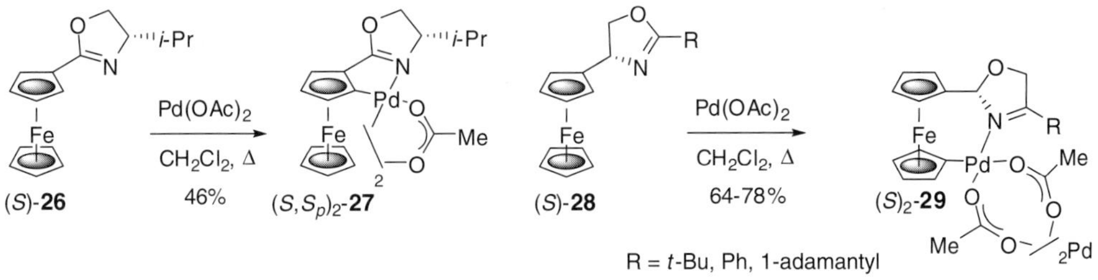

Scheme 12.8

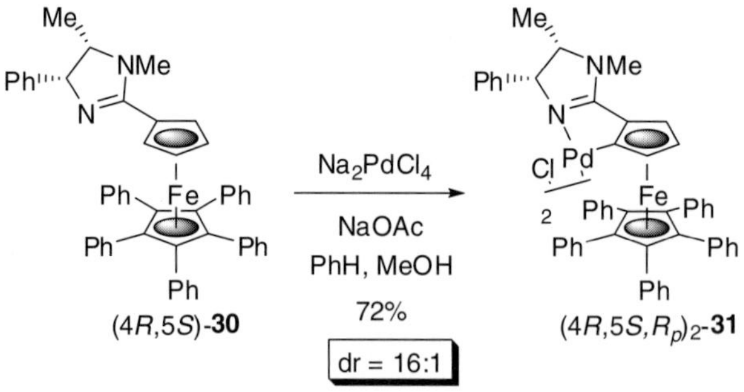

Scheme 12.9

Closely related to oxazoline **26** is the pentaphenylferrocene-derived imidazoline **30** which is resistant to lithiation but which undergoes direct palladation readily to give predominantly (4R,5S,R_p)$_2$-**31** where the 4-phenyl imidazoline substituent is endo, that is oriented towards the bulky C$_5$Ph$_5$ ligand (Scheme 12.9) [33].

The origin of diastereoselection with the related N-sulfonyl-substituted imidazolines **32** is ascribed to the operation of a relay controlling the chirality of the sulfonated pyramidal nitrogen, and thus the position of the equilibrium between **32** and **32'** [34]. Good diastereoselectivities have been reported for palladation reactions leading to ferrocenyl, pentamethylferrocenyl and pentaphenylferrocenyl derivatives (4R,5R, S_p)$_2$-**33**, and in each case the highest selectivity (20:1) was achieved with R^2 = para-tolyl (Scheme 12.10). The success of this chiral relay approach was further exploited with the bispalladation of **34** to give (S_p,S_p)$_2$-**35** as a single diastereoisomer after chromatographic purification (Scheme 12.11) [35]. This is the first diastereoselective synthesis of a ferrocenyl bispalladacycle by direct palladation. X-ray crystallography revealed the structure of **35** to be a C_2-symmetric dimer containing two (S_p,S_p)-ferrocenyl units and four chloride-bridged palladium ions in a trans, trans-geometry. It is speculated that the alternative (S_p,R_p)-bispalladated diastereoisomers polymerize rather than form stable dimers, accounting for the relatively low yield obtained in these reactions

In view of the ease with which ferrocenylimines are synthesized from either ferrocenecarboxaldehyde **36** or acetylferrocene **37**, it is not surprising that there are a number of reports on the diastereoselective palladation of ferrocenylimines **38** derived from enantiopure primary amines R*NH$_2$ (Scheme 12.12). The first

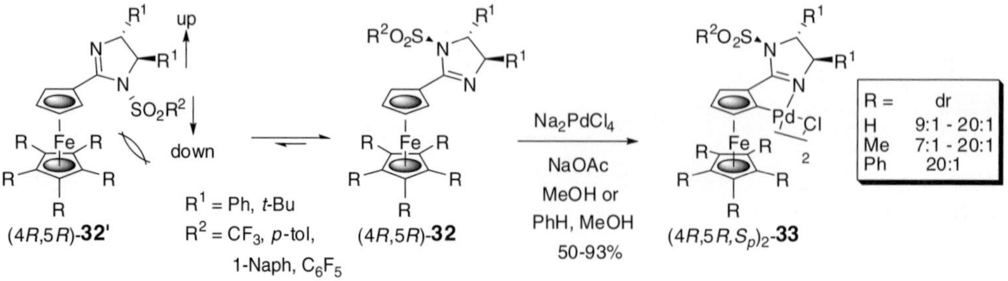

Scheme 12.10

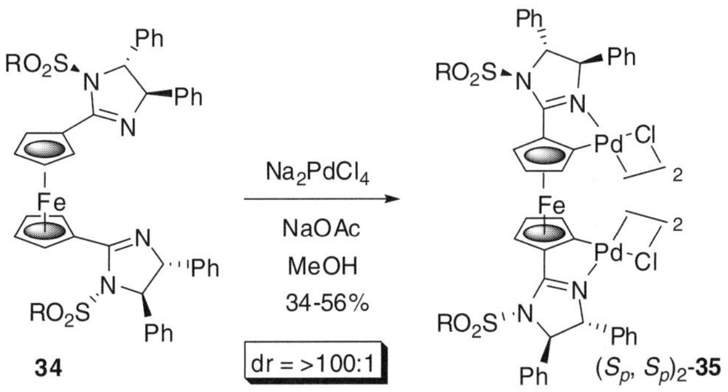

12.2 Asymmetric Synthesis of Planar Chiral Metallocyclic Complexes | 345

Scheme 12.11

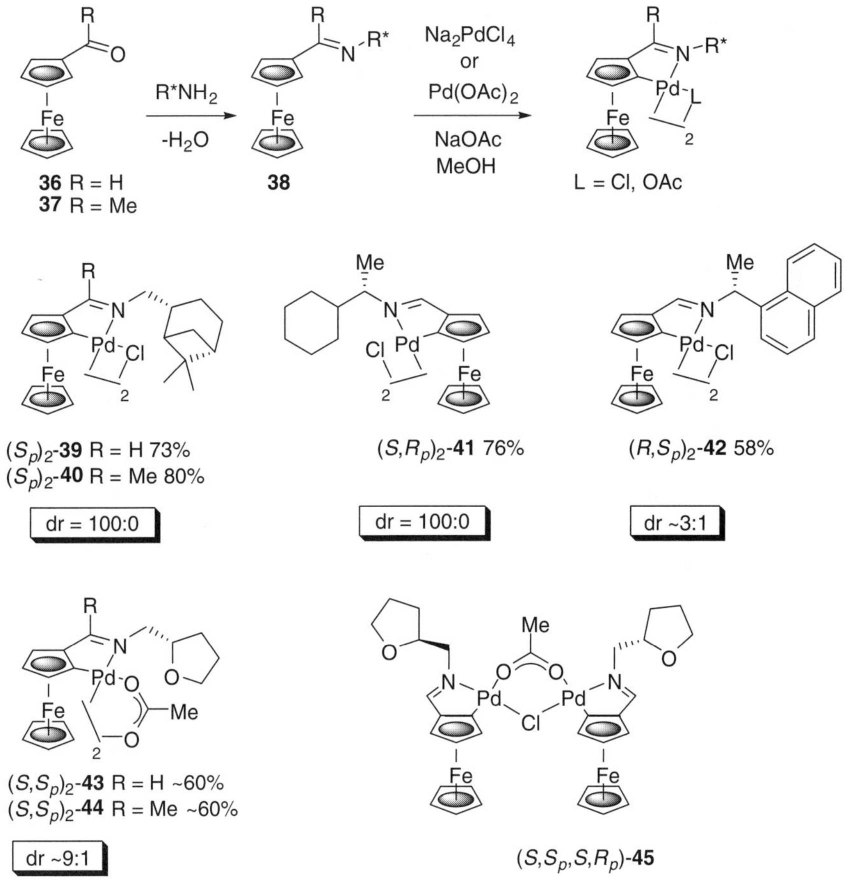

Scheme 12.12

description of this chemistry utilized (−)-*cis*-myrtanylamine as the chiral amine leading to $(S_p)_2$-**39** and $(S_p)_2$-**40** as single diastereoisomers [36]. Similarly (S)-1-cyclohexylethylamine led to $(S,R_p)_2$-**41** as a single diastereoisomer [37], and use of (R)-1-naphthylethylamine resulted in a modest preference for the product $(R,S_p)_2$-**42** displaying the same sense of diastereoselectivity [38]. Finally, both the aldimime and ketimime derived from (S)-tetrahydrofurfurylamine resulted in a 9:1 ratio of acetate bridged isomers $(S,S_p)_2$-**43** and $(S,S_p)_2$-**44**, respectively, when palladium acetate was used in the reaction [39]. Use of sodium tetrachloropalladate and sodium acetate with the aldimine derived from (S)-tetrahydrofurfurylamine gave a 6.3:1 ratio of the $(S,S_p)_2$-chloride bridged dimer and the mixed bridged dimer (S,S_p,S,R_p)-**45**. Formation of this μ-chloro μ-acetato mixed bridged dimer supports the proposed mechanism of acetate catalyzed palladation (*vide supra*) as acetate to chloride ligand exchange is normally facile, but not vice versa.

Closely related to the above examples are hydrazones derived from the hydrazine (S)-1-amino-2-(methoxymethyl)pyrrolidine. Reaction with acetylferrocene gave **46** which underwent palladation to give *trans*-$(S,R_p)_2$-**47** as the major diastereoisomer together with *trans*-$(S,S_p)_2$-**47** and *trans*-(S,R_p,S,S_p)-**47** (Scheme 12.13) [40]. This example is unusual as two minor diastereoisomers were observed containing the S_p,S_p or the R_p,S_p bridged palladacycles (see Figure 12.2). This contrasts with some of the examples described above for which incomplete diastereoselectivity resulted in only one additional product. Where the identity of this was explicitly determined [39], or may be inferred from the data provided [23, 27, 38], the configuration is always opposite to that of the major diastereoisomer formed with respect to the planar chiral components of the bridged dimer (i.e., (R_p,R_p) vs. (S_p,S_p) or vice versa).

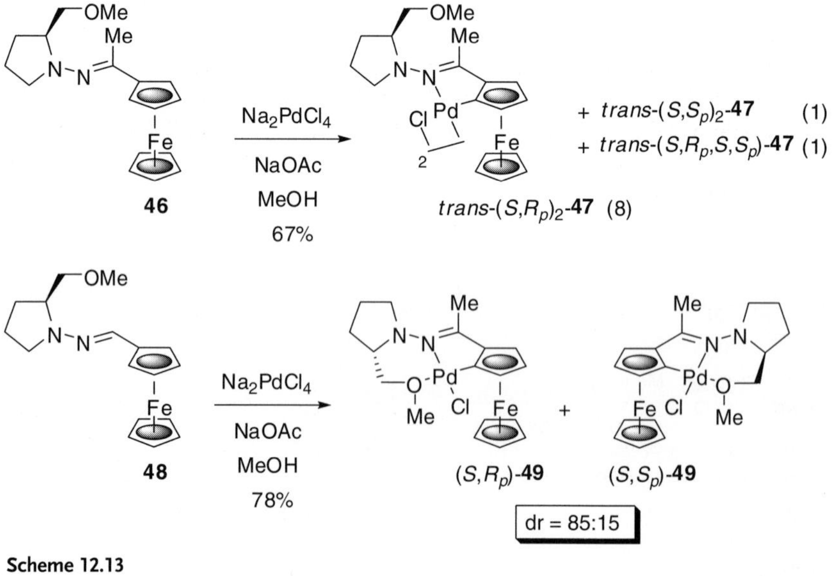

Scheme 12.13

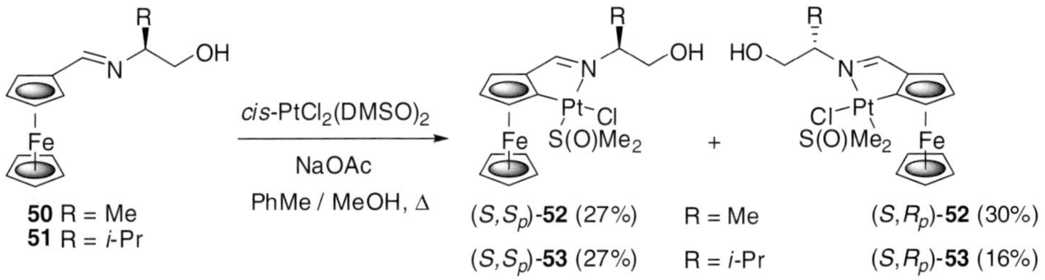

12.2 Asymmetric Synthesis of Planar Chiral Metallocyclic Complexes | 347

Scheme 12.14

50 R = Me
51 R = i-Pr

(S,S_p)-52 (27%) R = Me (S,R_p)-52 (30%)
(S,S_p)-53 (27%) R = i-Pr (S,R_p)-53 (16%)

The corresponding hydrazone **48** derived from ferrocenecarboxaldehyde behaves a little differently resulting predominantly in the ether coordinated monomer (S,R_p)-**49** [41]. Use of palladium acetate in place of sodium tetrachloropalladate with **46** and **48**, followed by the addition of lithium bromide, results in a higher diastereoselectivity for the formation of the corresponding bromide ligated complexes (9 : 1 for $(S,R_p)_2$: $(S,S_p)_2$, and 100 : 0 for $(S,R_p)_2$: $(S,S_p)_2$, respectively) [42].

There are very few examples of C—H activation with heavy metals other than palladium, and the diastereoselectivity of the reactions reported is uniformly poor. Platination of α-amino alcohol derived imines **50** and **51** requires a reaction time of 72 h to give a respectable yield of the platinacycles **52** and **53** (Scheme 12.14) [43]. It is possibly significant that the palladation of imine **51** also results in essentially no diastereoselectivity [44]. The utility of $PtCl_2(DMSO)_2$ as a cyclometallating agent, and the retention of the sulfoxide ligand in the product, led to the use of enantiomerically pure (S_s,S_s)-cis-$PtCl_2[SOMe(p-MeC_6H_4)]_2$ **54** in an attempt to control the selective formation of the new element of planar chirality (Scheme 12.15). Unfortunately, there was only a small difference in the yield of the two diastereoisomers, although the major product (S_s,R_p)-**55** selectively crystallized from the reaction solution [45]. The corresponding reaction with (S_s,S_s)-cis-$PdCl_2[SOMe(p-MeC_6H_4)]_2$ gave only a 15% yield of the palladacycle which could not be separated into diastereoisomers, or even analyzed by NMR spectroscopy, due to the low affinity for palladium of the sulfoxide ligand [46]. The chloride-bridged dimer **8** that resulted from reversible sulfoxide dissociation was also isolated and was found to be essentially racemic.

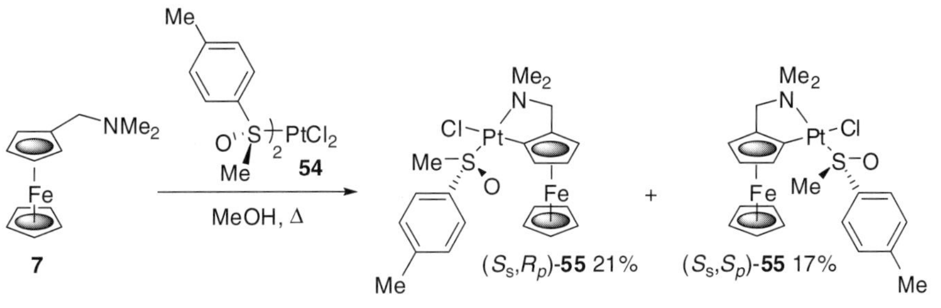

(S_s,R_p)-55 21% (S_s,S_p)-55 17%

Scheme 12.15

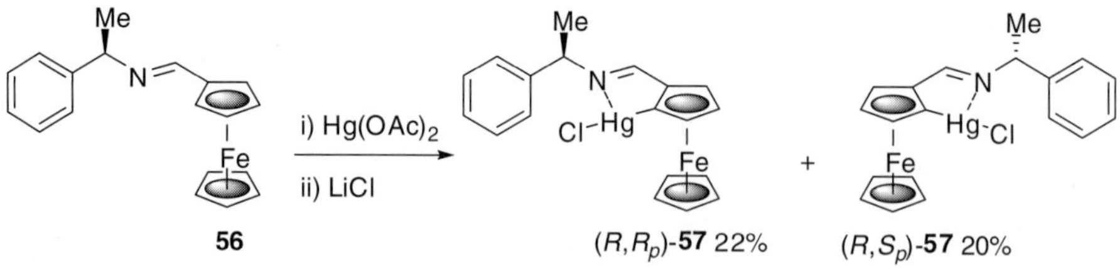

Scheme 12.16

The only reported attempt at a diastereoselective mercuration was on the α-methylbenzylamine derived imine **56**. This did result in ortho-metalation but with essentially no control over the ratio of (R,R_p)-**57** and (R,S_p)-**57** (Scheme 12.16) [47].

Finally in this section, scalemic planar chiral palladacycles have also been synthesized from a precursor where only one ferrocene "ortho" position is available for C−H activation. Employing the method of Kagan [48], acetal **58** underwent highly diastereoselective lithiation as the first step in the synthesis of enantiopure aldehydes **59** and **60**. Following reductive amination, reaction with sodium tetrachloropalladate gave palladacycles (R_p)$_2$-**61** and (R_p)$_2$-**62** due to the presence of the blocking methyl or trimethylsilyl group (Scheme 12.17) [49].

12.2.3
Oxidative Addition of Palladium(0)

The strategy of generating a scalemic planar chiral ferrocene precursor to a metallacycle has more frequently been used in conjunction with palladium(0) insertion into a carbon–halogen bond. 2-Iodoferrocenecarboxaldehyde **63**, generated using the methodology outlined in Scheme 12.17, was condensed with a series of aromatic amines and the resulting imines reacted with a source of palladium(0) to give palladacycles (S_p)$_2$-**64** (Scheme 12.18) [49].

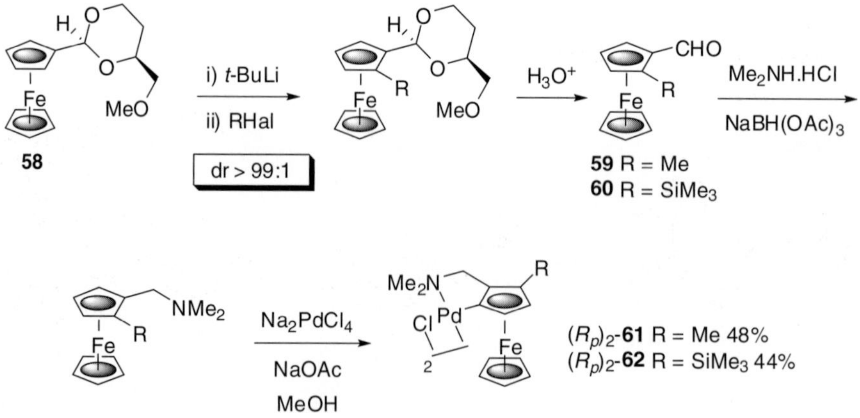

Scheme 12.17

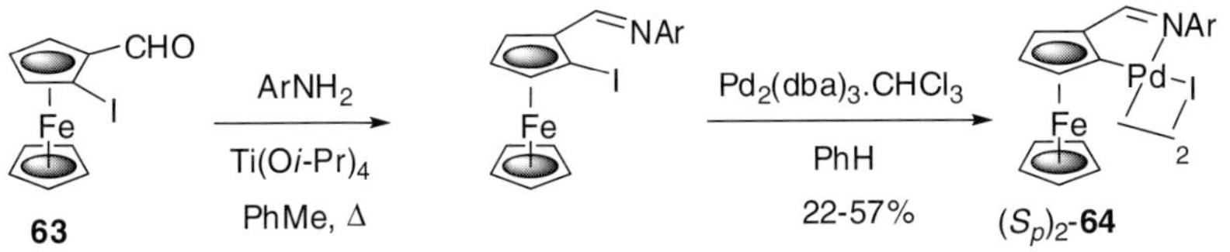

12.2 Asymmetric Synthesis of Planar Chiral Metallocyclic Complexes | 349

Scheme 12.18

An alternative to the Kagan method for the enantioselective synthesis of planar chiral ferrocenes (Scheme 12.17) is the use of ferrocenyloxazolines as the oxazoline itself may be employed as a ligand component [50]. Starting from the (S)-*tert*-leucine-derived oxazoline **65** both $(S_p)_2$- and $(R_p)_2$-palladacycles were generated using the methodology outlined in Scheme 12.19 [51]. The former was obtained by a lithiation/ iodine quench reaction to give **66** followed by the insertion of Pd(0) to give $(S,S_p)_2$-**67**. Several examples of the latter were synthesized by performing a Suzuki reaction on **66** to introduce an aryl blocking group followed by further lithiation, introduction of iodine and insertion of Pd(0) to give $(S,R_p)_2$-**68**. Alternatively, the initial lithiation was followed by introduction of a trialkylsilyl blocking group prior to the generation of $(S,R_p)_2$-**69**. The congener $(S,S_p)_2$-**70** was obtained in the same way.

In contrast to the successful direct palladation of pentaphenylferrocene derivative **30** (Scheme 12.9), the same reaction with Na_2PdCl_4 performed with the

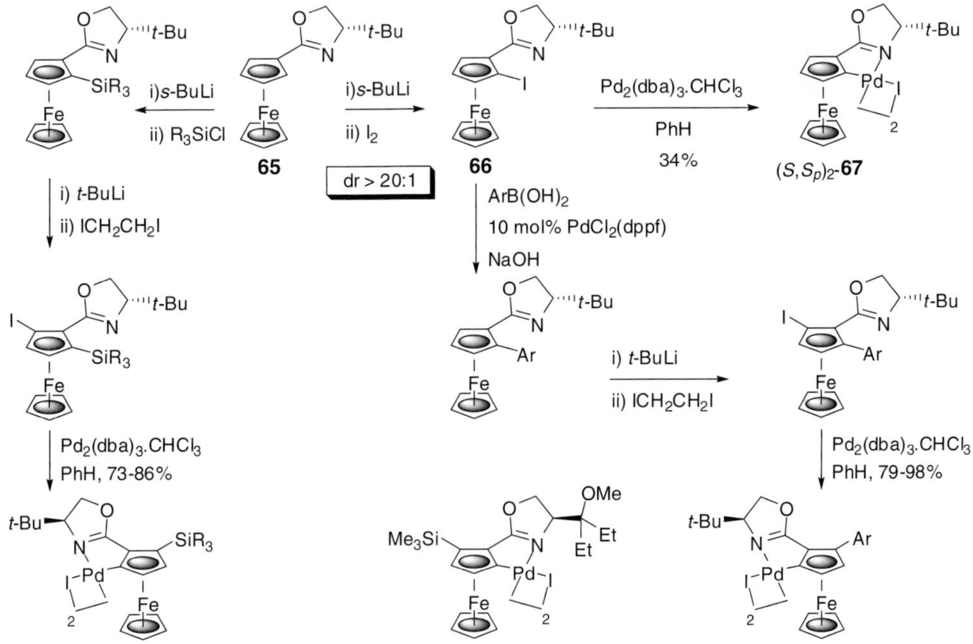

Scheme 12.19

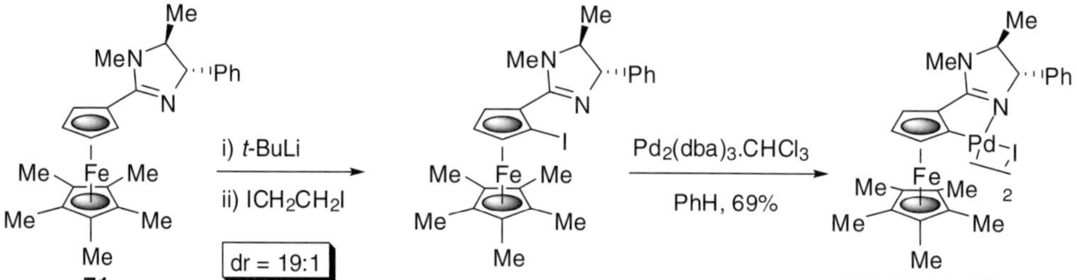

Scheme 12.20

pentamethylferrocene analogue **71** resulted in low diastereoselectivities. Thus the palladacycle (4S,5S,S_p)$_2$-**72** was selectively generated by an initial highly diastereoselective lithiation, followed by introduction of iodine and insertion of Pd(0) (Scheme 12.20). [33].

Ferrocenyl bis-palladacycles have also been generated using this oxidative addition approach. The key requirement is an efficient asymmetric synthesis of a 1,1′,2,2′-tetrasubstituted precursor for which bis-oxazoline **73** is a viable starting material. Moderately diastereoselective dilithiation [52] and introduction of two iodines was followed by purification of **74** and manipulation of the oxazoline rings to give bis-amides **75**. The corresponding palladacycles, either (S_p,S_p)-**76** or (S_p,S_p)-**77** were generated in the standard way (Scheme 12.21) [53]. Similarly, the highly diastereoselective lithiation of diamine **78** was used as the starting point for the synthesis of (S_p,S_p)-**79** (Scheme 12.21) [53].

12.2.4
Transmetalation

A further alternative to direct C—H activation for the synthesis of metallacycles is the use of transmetalation to introduce the required metal into the framework of a scalemic planar chiral ferrocene. There are, however, very few reported examples of this approach. An alternative synthesis of palladacycle (R,S_p)-**19** exploited the diastereoselective lithiation of the Ugi amine, followed by two sequential transmetalations; first using mercuric chloride to give (R,S_p)-**80**, followed by an oxidative transmetalation with a source of Pd(0) to give the palladacycle as a single diastereoisomer (Scheme 12.22) [23]. This was used to confirm the identity of the major diastereoisomer resulting from the palladation of **18** and no yields are reported. A low yield was obtained for the second transmetalation step with the chloromercuric complex (S,R_p)-**81**, resulting in (S,R_p)$_2$-**82** (Scheme 12.22) the chloride-bridged congener of complex (S,R_p)$_2$-**69** (R = Me – Scheme 12.19) [51b].

The inverse palladium to mercury transmetalation has been demonstrated with the transformation of (R_p)$_2$-**14**, obtained by resolution, as outlined in Scheme 12.6, into (R_p)-**83** by reaction with metallic mercury (Scheme 12.23) [19b]. Subsequent reaction with tellurium powder resulted in the synthesis of a series of cyclotellurides (R_p)-**84** [54].

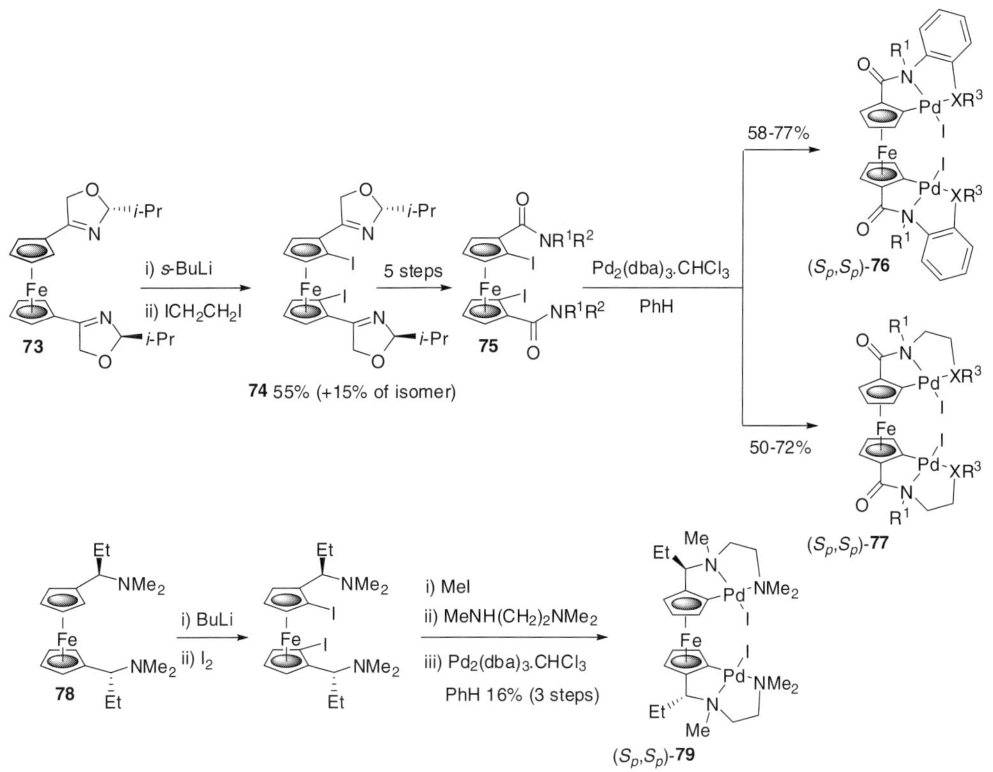

Scheme 12.21

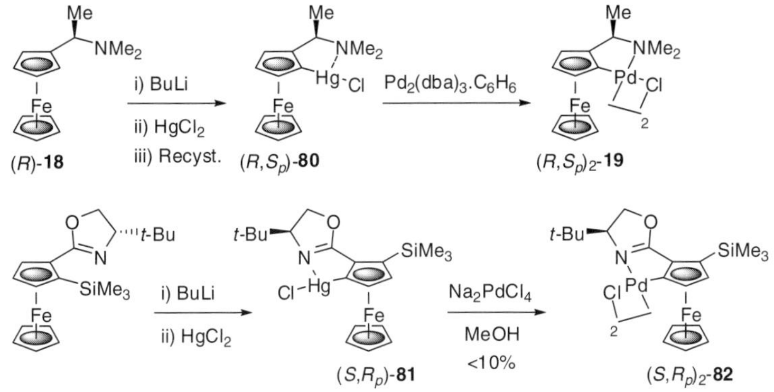

Scheme 12.22

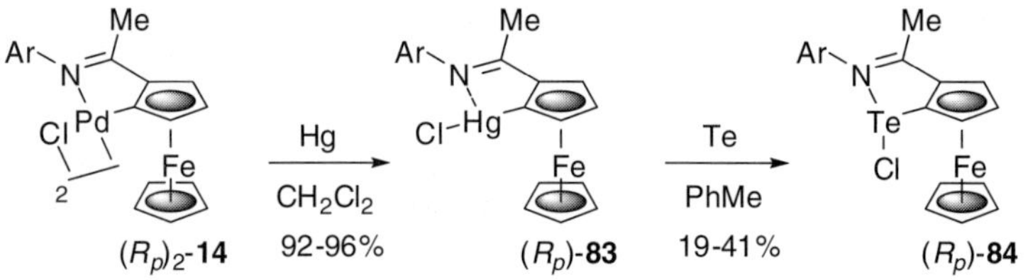

Scheme 12.23

12.2.5
Enantioselective Palladation

Rather than employing a ferrocene-substituted chiral auxiliary to control diastereoselective metalation, an attractive alternative is the use of a chiral metal reagent to effect the enantioselective synthesis of the planar chiral metallacycle. To date three examples of this strategy have been reported, all employing a stoichiometric chiral source which is not incorporated into the resultant palladacycle.

A recognition of the importance of acetate in the palladation of dimethylaminomethylferrocene 7 [15b], as discussed above, led to the use in its stead of the sodium salt of (S)-N-acetylvaline 85. This resulted in the partially enantioselective formation of $(R_p)_2$-8, the yield and ee of the product being dependent on the pH of the reaction, the optimum value for the latter being about 9–10 for maximum enantioselectivity (Scheme 12.24) [55]. In a slight modification of this procedure use of (S)- and (R)-N-acetylleucine gave $(R_p)_2$-8 and $(S_p)_2$-8, respectively, with an ee of 81% [56].

Palladacycles may be synthesized by the exchange of cyclometallated ligands [57] which typically involves the transfer of palladium from a C,N-chelate to a C,P-chelate driven by the greater strength of the phosphorus–palladium bond. The term transcyclometalation defines a subcategory of this process where the reaction proceeds without the formation of dissociated metal salts. Asymmetric transcyclometalation is thus possible, and the first example with a ferrocene substrate employed the prochiral phosphine 10. On heating with the C,N-palladacycle $(R)_2$-86 the phosphapalladacycle $(S_p)_2$-11 was formed in moderate ee together with amine 87 [58] (Scheme 12.25).

Higher enantioselectivies in asymmetric transcyclopalladation were obtained with the C,N-palladacycles $(S,R_p)_2$-89 and $(S,S_p)_2$-92. These were obtained by the highly

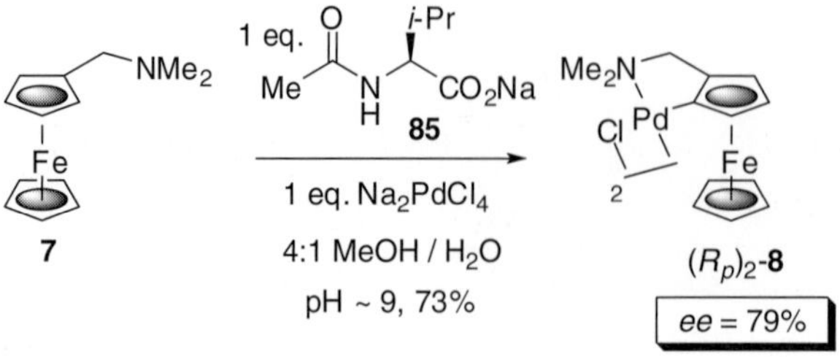

Scheme 12.24

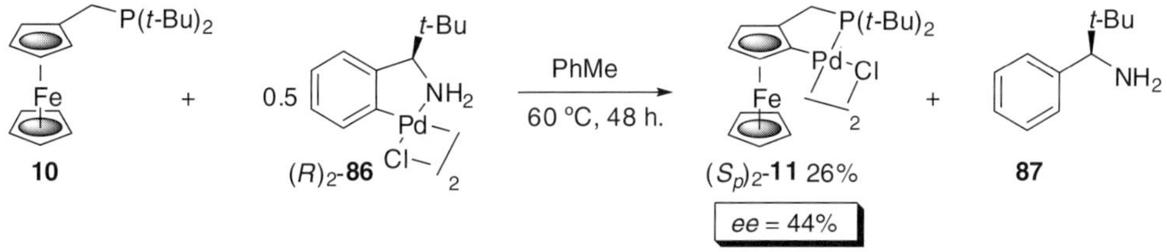

Scheme 12.25

diastereoselective palladation of the (S)-valine-derived oxazoline **88** [59] and the (S)-*tert*-leucine-derived oxazoline **91** (Scheme 12.26) [60]. That the product palladacycles have an opposite configuration of planar chirality is a consequence of the acidic reaction conditions which results in reversible palladation by Pd–C bond protonolysis. The kinetic product arising from the palladation of **88**, which has the (S,S_p)-configuration, rearranges to the more stable (S,R_p)-configuration. In contrast, palladation of **91** results in isolation of the kinetic product $(S,S_p)_2$-**92** as the bulk of the *t*-Bu group prevents formation of the thermodynamic product in which the *t*-Bu group would be oriented away from the metallocene [61].

Heating prochiral phosphine **93** with $(S,R_p)_2$-**89** resulted in clean transcyclopalladation to give highly scalemic phosphapalladacycle $(S_p)_2$-**94** and recovered oxazoline **88**. Use of the diastereoisomer $(S,S_p)_2$-**92** also resulted in high enantioselectivity, but a switching in the absolute configuration of the product phosphapalladacycle, $(R_p)_2$-**94** (Scheme 12.26) [62]. Use of the diphenylphosphino analogue of **93** with $(S,R_p)_2$-**89** and $(S,S_p)_2$-**92** gave the corresponding $(S_p)_2$ (78% *ee*) and $(R_p)_2$ (92% *ee*) phosphapalladacycles, respectively. Thus the absolute configuration of the product

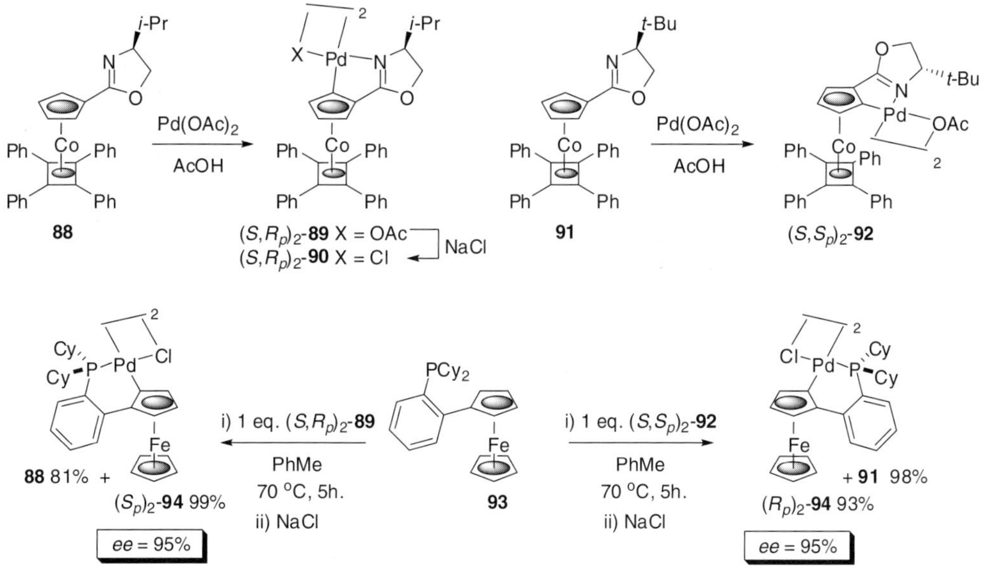

Scheme 12.26

is largely controlled by the element of planar chirality, the *i*-Pr or *t*-Bu substituents playing little part in the enantiodiscrimination.

12.3
Stoichiometric Synthetic Applications of Scalemic Planar Chiral Metallocyclic Complexes

12.3.1
Phosphine Recognition

Scalemic non-ferrocenyl metallacycles have been extensively utilized for the resolution and *ee* determination of chiral phosphines, amines and arsines [63]. Surprisingly there is only one report on the use of ferrocenyl metallacycles in this way. In this work the addition of racemic *trans*-2-(diphenylphosphino)cyclohexanol **95** to $(S,R_p)_2$-**19** was followed by recrystallization of the resulting diastereomeric phosphine adducts to give pure $(S,R_p,1R,2R)$-**96** (Scheme 12.27) [64]. The mother liquor was enriched in $(S,R_p,1S,2S)$-**97** (4 : 1), and this method was also used for the separation of the corresponding complexes formed from phosphine rac-**98**. Addition of BINAP to $(S,R_p)_2$-**19** gave two complexes that differed significantly in their reactivity; (*R*)-(+)-BINAP gave a stable bidentate adduct, and (*S*)-(−)-BINAP gave a bidentate adduct that slowly transformed into *cis*-$PdCl_2$[(*S*)-BINAP].

12.3.2
Generation of Planar Chiral Ferrocene Derivatives by M–C Bond Cleavage

Scalemic planar chiral palladacycles have been used as precursors for the synthesis of non-metallocyclic chiral ferrocene derivatives. As the synthesis and subsequent manipulation of a palladacycle requires the stoichiometric use of palladium this is a far from ideal approach to the synthesis of planar chiral ferrocene derivatives, and most of the examples predate the plethora of asymmetric methods developed over the past fifteen years [3].

The reaction of enantioenriched $(R_p)_2$-**8**, obtained by the method outlined in Scheme 12.24, with lithium diphenylphosphide gave P,N ligand **99**, ~87% *ee* (Scheme 12.28) [65]. Addition of iodine to **8** followed by lithium–halogen exchange

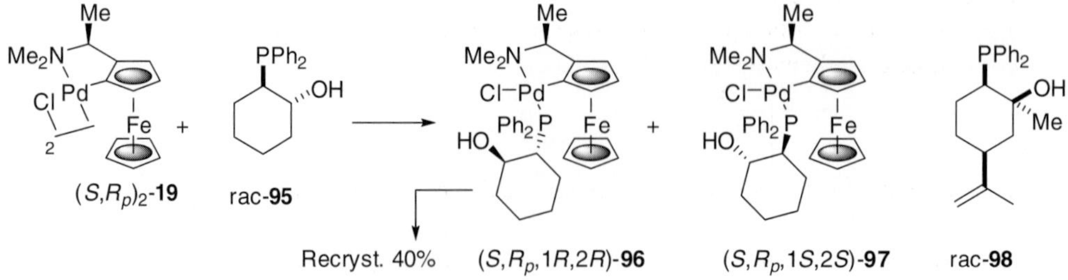

$(S,R_p)_2$-**19** rac-**95** $(S,R_p,1R,2R)$-**96** $(S,R_p,1S,2S)$-**97** rac-**98**

Scheme 12.27

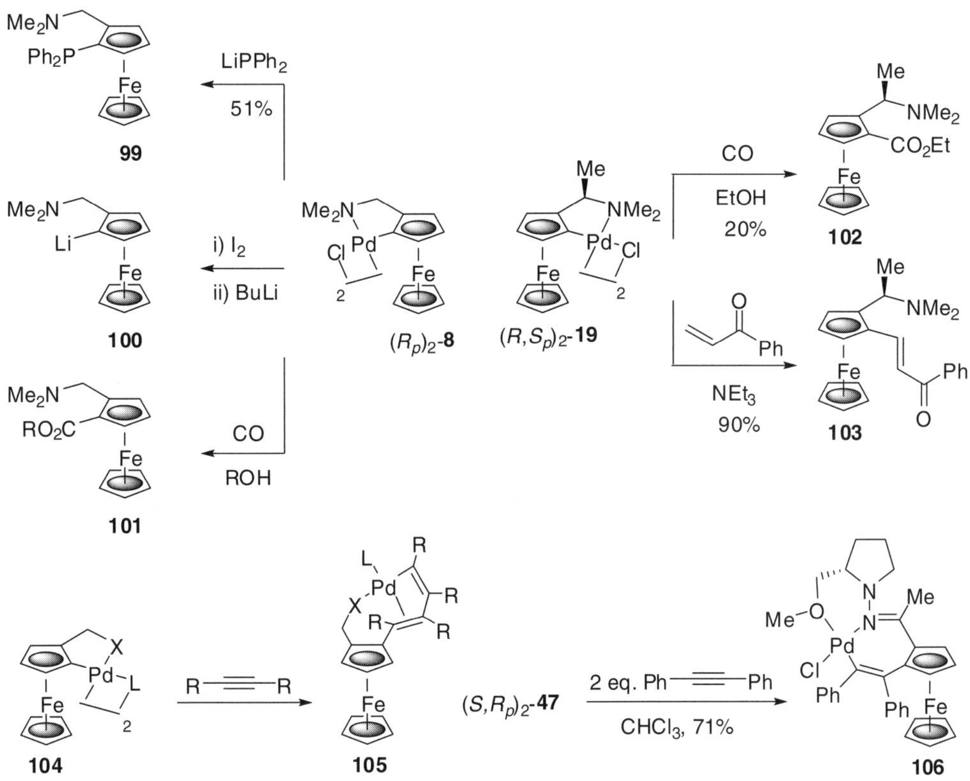

Scheme 12.28

resulted in the stereospecific synthesis of 2-lithio-N,N-dimethylaminomethylferrocene **100** [66]. Alternatively, carbonylation of **8** in the presence of an alcohol ROH gave the corresponding scalemic esters **101** [56, 67]. Similarly, the related palladacycle (R,S_p)-**19** underwent carbonylation in ethanol to give a low yield of the ethyl ester **102**, and Heck coupling with phenyl vinyl ketone to give unsaturated ketone **103** [68]. This latter method of C–C bond formation has been applied to the synthesis of scalemic planar chiral ferrocene analogues of prostaglandins [69].

A number of palladacycles **104** have been reacted with an excess of alkynes, most commonly diphenylacetylene, resulting in the generalized double-insertion product **105** (Scheme 12.28). Examples include palladacycles $(S,R_p)_2$-**19** [24], $(S_p)_2$-**39** and $(S,R_p)_2$-**41** [37], $(R,R_p)_2$-**42** [38], $(S,S_p)_2$-**43** and $(S,S_p)_2$-**44** [39], and $(S,R_p)_2$-**47** [70]. In one instance a mono-insertion product **106** was obtained from the reaction of $(S,R_p)_2$-**47** with 2 equiv of diphenylacetylene [70].

12.4
Asymmetric Catalysis with Scalemic Planar Chiral Palladocyclic Complexes

Palladium-catalyzed reactions are ubiquitous in modern organic synthesis, and many reactions have been adapted successfully to generate chiral products with excellent levels of enantioselectivity [71]. In most cases these reactions exploit kinetically accessible oxidative addition [Pd(0) to Pd(II)] and reductive elimination [Pd(II) to Pd(0)] steps in the catalytic cycle. A number of racemic planar chiral ferrocenyl palladacycles have been exploited as precatalysts for the Suzuki and Heck reactions [72], where the Pd(II) oxidation state of the palladacycle is reduced to give a Pd(0) species, the starting point of the catalytic cycle [73]. As this results in cleavage of the palladium–carbon bond, the potential benefits of the anionic bidentate ligand discussed in Section 12.1 are lost. There is the possibility of replacing the palladium substituent of a scalemic palladacycle with a non-hydrogen substituent, and exploiting the resultant planar chiral product generated *in situ* as a ligand in an asymmetric Pd(0)-catalyzed reaction. Such methodology does not appear to have been demonstrated.

Thus the potential of planar chiral palladacycles in asymmetric catalysis lies in their exploitation as chiral Lewis acids activating Lewis basic species introduced by ligand substitution. To date there are only two reactions that have successfully exploited this interaction to give scalemic organic products, and the vast majority of these reports are on the allylic imidate rearrangement.

12.4.1
Allylic Imidate Rearrangement

The allylic imidate rearrangement was first reported by Overman in 1974 as the central step in the overall conversion of allylic alcohols into allylic amines (Scheme 12.29) [74]. This concerted pericyclic [3.3]-sigmatropic rearrangement was initially performed on trichloroacetimidic esters **107**, at a temperature of 140 °C, resulting in trichloroacetamides **108**. Alternatively, addition of mercury(II) salts resulted in rate accelerations of up to 10^{12}, the reaction proceeding by a two-step iminomercuration–deoxymercuration mechanism. The discovery that palladium(II) complexes are superior catalysts [75] initiated investigations into the asymmetric allylic imidate rearrangement with cationic palladium(II) complexes containing chiral N,N- and N,P-ligands. Application as catalysts to the rearrangement of allylic imidate **109** gave **110** in moderate yield and enantioselectivity, with by-products arising from competing ionization of the imidate (Scheme 12.30) [76].

In contrast to the sluggish cationic palladium catalysts it was observed that PdCl$_2$(NCMe)$_2$ resulted in the near quantitative formation of **110** from (*E*)-**109** within minutes at room temperature. This resulted in an examination of neutral palladium(II) complexes containing two anionic ligands, and planar chiral complexes were chosen so as to project chirality perpendicular to the square plane. Palladacycle

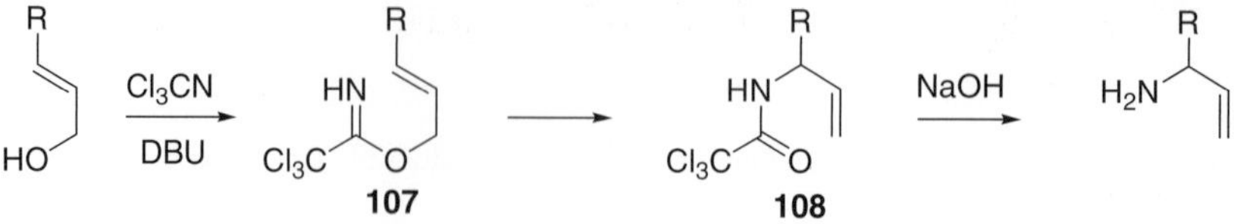

Scheme 12.29

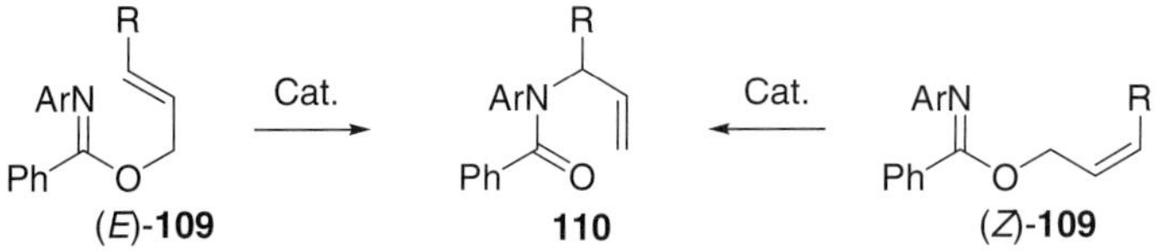

12.4 Asymmetric Catalysis with Scalemic Planar Chiral Palladocyclic Complexes

Scheme 12.30

$(R,S_p)_2$-**19** gave (R)-**110** (R = Pr, Ar = Ph) in 67% ee without any by-products arising from imidate ionization. To improve on the low yield obtained (35% after 7 days at room temperature) other bridging ligands were examined, the most effective being the trifluoroacetate derivative $(R,S_p)_2$-**111** (Scheme 12.31). A series of representative reactions with this catalyst gave (R)-**110** in good yield and moderate enantioselectivity (Table 12.1, entry 1) [25]. The diastereoisomer $(R,R_p)_2$-**111** gave (R)-**110** (R = i-Pr, Ar = Ph) with a slightly lower yield and enantioselectivity (48%).

Slightly reduced enantioselectivities were achieved with the catalyst precursors $(R_p)_2$-**61** and $(S_p)_2$-**112**, from which the active catalyst was generated *in situ* by addition of thallium trifluoroacetate and thallium triflate, respectively (entries 2 and 3). The catalyst obtained from $(S_p)_2$-**112** gave slightly higher enantioselectivities with (Z)-**109** resulting in allylic benzamide products of opposite absolute configuration (entry 4) [49]. A promising 90% ee was obtained with the trinuclear acetate bridged dimer $(S)_2$-**113** although the application of this catalyst was limited to one example (entry 5) [31]. More rigorously examined was the bis-palladacycle (S_p,S_p)-**114** which gave very good enantioselectivities with both (E)- and (Z)-**109** (entries 6 and 7) [53]. These results contrasted with the poor yield and enantioselectivity that resulted with the analogous mono-palladacycle (S_p)-**115** which was found to be unstable under

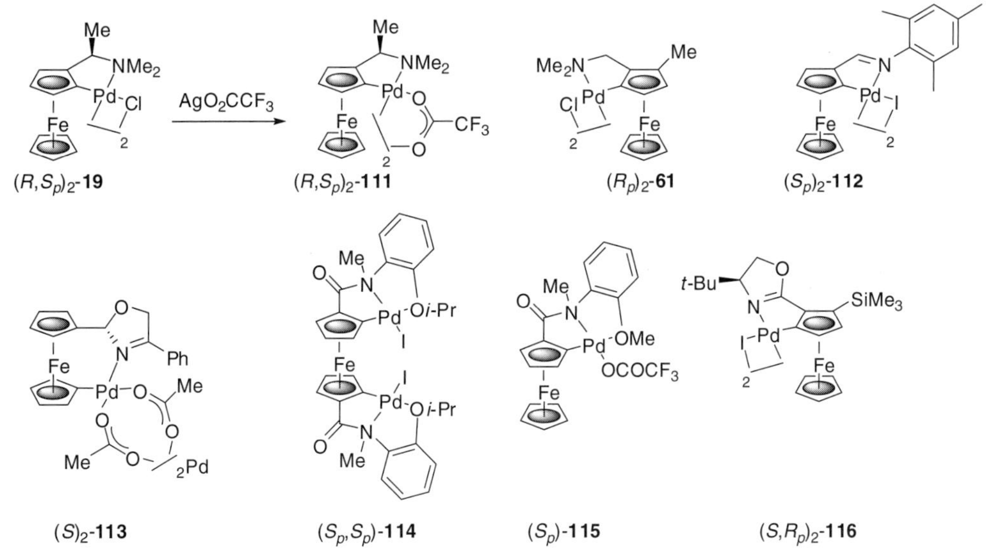

Scheme 12.31

Table 12.1 Palladacycle-catalyzed rearrangement of allylic N-arylbenzimidates **109**.

Entry	Cat. or cat. precursor	Additive	Amount of catalyst (mol%)	Substrate (Config.)	Yield (%)[a]	ee (%)	Config.	Ref.
1	$(R,S_p)_2$-**111**	—	5	(E)-**109**	47–98	47–61	R	[25]
2	$(R_p)_2$-**61**	Tl(OCOCF$_3$)	8	(E)-**109**	73	40	S	[49]
3	$(S_p)_2$-**112**	Tl(OSO$_2$CF$_3$)	11	(E)-**109**	80	46	R	[49]
4	$(S_p)_2$-**112**	Tl(OSO$_2$CF$_3$)	11	(Z)-**109**	45–78	66–73	S	[49]
5	$(S)_2$-**113**	—	5	(E)-**109**	49	90	R	[31]
6	(S_p,S_p)-**114**	Ag(OCOCF$_3$)	5	(E)-**109**	68–91	87–92	R	[53]
7	(S_p, S_p)-**114**	Ag(OCOCF$_3$)	5	(Z)-**109**	70–85	86–90	S	[53]
8	(S_p)-**115**	—	3.5	(E)-**109**	32	30	R	[53]
9	$(S,R_p)_2$-**116**	Ag(OCOCF$_3$)	5	(E)-**109**	59–97	63–84	S	[51a]
10	$(S,R_p)_2$-**116**	Ag(OCOCF$_3$)	5	(Z)-**109**	11–96	75–96	R	[51a]

[a]Reactions performed at room temperature.

the reaction conditions (entry 8) [53]. Such problems were not encountered with the planar chiral ferrocenyl oxazoline palladacycle (FOP) $(S,R_p)_2$-**116** which, following activation with silver trifluoroacetate, gave good enantioselectivities for (E)-**109** and especially (Z)-**109** (entries 9 and 10) [51a]. It was noted that activation with silver trifluoroacetate resulted in oxidation of the ferrocene to a ferrocenium cation in the catalyst resulting from $(S,R_p)_2$-**116**, and that the trimethylsilyl group of $(S,R_p)_2$-**116** is required to prevent decomposition of the resulting catalyst [51b].

Although high enantioselectivities were achieved for the rearrangement of N-arylbenzimidates, the application of this methodology to the synthesis of allylic amines is significantly hindered by the difficulty of removing the two nitrogen substituents of **110**, that is, benzoyl and typically 4-methyoxyphenyl. A solution to this problem was achieved with substrates (E)- and (Z)-**117** which, following rearrangement to **118**, were deprotected by cleavage of the trifluoroacetamide with sodium ethoxide followed by oxidative dearylation with ceric ammonium nitrate (CAN) (Scheme 12.32). Use of catalyst precursor $(S,R_p)_2$-**116** with 4 equiv of silver

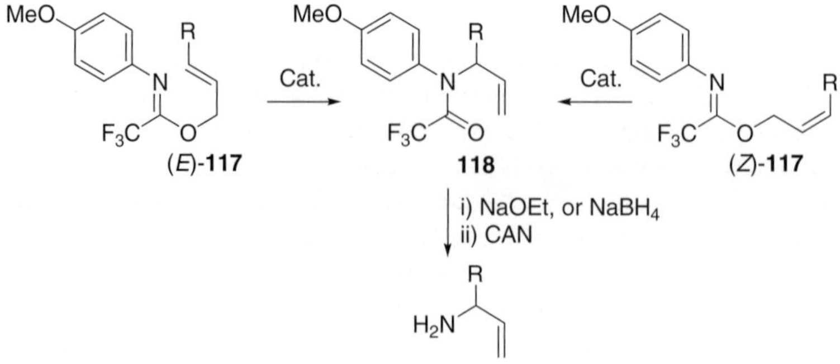

Scheme 12.32

Table 12.2 Palladacycle-catalyzed rearrangement of allylic N-(4-methoxyphenyl)trifluoroacetimidates **117**.

Entry	Cat. or cat. precursor	Additive	Amount of catalyst (mol%)	Substrate (Config.)	Yield (%)	ee (%)	Config.	Ref.
1[a]	(S,R_p)$_2$-**116**	Ag(OCOCF$_3$)	7.5[b]	(E)-**117**	46–88	45–84	S	[51b]
2[a]	(S,R_p)$_2$-**116**	Ag(OCOCF$_3$)	7.5[b]	(Z)-**117**	21	87–90	R	[51b]
3[c]	(R_p)$_2$-**31**	AgNO$_3$	5	(E)-**117**	93–98	84–88	S	[33]
4[a]	(S_p)$_2$-**119** (R = Ph)	Ag(OCOCF$_3$)	0.05[d]	(E)-**117**	95–99	92–98	R	[34]
5[c]	(S_p)$_2$-**119** (R = H)	Ag(OCOCF$_3$)	5[d]	(Z)-**117**	69–75	89–96	S	[34]
6[c]	(S_p)$_2$-**119** (R = Me)	Ag(OCOCF$_3$)	5[d]	(Z)-**117**	72–95	93–96	S	[34]
7[e]	(S_p,S_p)-**120**	AgOTs	0.1	(Z)-**117**	86–97	94–99	S	[35]
8[a]	(S,R_p)$_2$-**90**	—	5	(E)-**117**	85–92	82–95	S	[78]
9[a]	(S,R_p)$_2$-**90**	—	5	(Z)-**117**	58–87	86–97	R	[78]

[a]Reactions performed at room temperature.
[b]With 20 mol% 1,8-bis(dimethylamino)naphthalene.
[c]40 °C.
[d]With 2 equiv per catalyst dimer of 1,8-bis(dimethylamino)naphthalene.
[e]55 °C.

trifluoroacetate and the proton sponge 1,8-bis(dimethylamino)naphthalene resulted in the asymmetric transformation of both (E)-**117** (Table 12.2, entry 1) and (Z)-**117** (entry 2), although the latter rearranged only slowly [51b].

Preliminary catalytic studies with the pentaphenylferrocene palladacycle (R_p)$_2$-**31** gave very high enantioselectivities for the rearrangement of (E)-**117** (entry 3) [33]. However, interest in this catalyst was quickly superceded by the related N-tosyl FIP (ferrocenyl-imidazoline palladacycle) derivative (S_p)$_2$-**119** (R = Ph) which on activation with 4 equiv of silver trifluoroacetate gave a catalyst utilized with a loading as low as 0.05% (of a dimer, thus a Pd loading of 0.1 mol%) (entry 4). It was noted that use of 2 equiv of silver trifluoroacetate gave a new diamagnetic species that resulted in the rearrangements proceeding extremely slowly. In contrast, addition of 4 equiv gave the active catalyst as a paramagnetic species, presumably due to oxidation of the ferrocene moiety to a ferrocenium cation [34].

The catalyst precursor (S_p)$_2$-**119**, with either R = H or R = Me, was also applied successfully to the rearrangement of (Z)-**117**, although a catalyst loading of 5 mol% was required in each case (Table 12.2, entries 5 and 6) [34]. A significant development with (Z)-**117** was achieved with the ferrocenyl bispalladacycle (S_p,S_p)$_2$-**120** which, following activation with silver tosylate, gave excellent yields and enantioselectivities with just 0.1 mol% catalyst loading [35]. In this instance the sixfold excess of silver tosylate used in the activation step does not result in ferrocene oxidation, and the structure of the catalyst is given as (S_p,S_p)-**121** (Scheme 12.33) [35b]. Included in the range of substrates transformed were Z-configured trifluoroacetimidates containing an α–branched substituent (Z)-**117**, (R = i-Pr), and the methodology has been successfully extended to a give a range of highly scalemic functionalized allylic amine building blocks [35b].

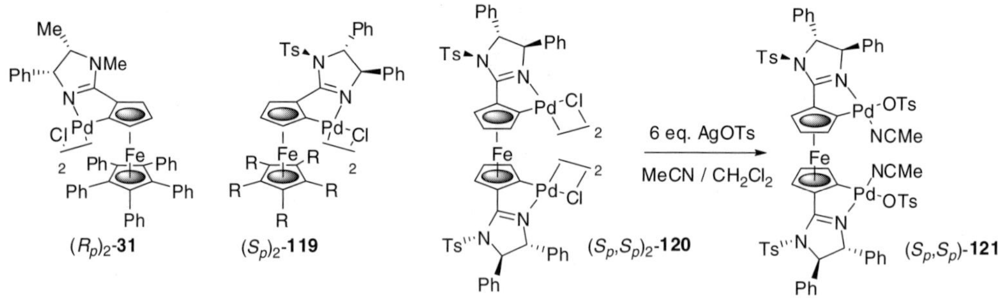

Scheme 12.33

The use of trifluoroacetimidate substrates has been extended to the asymmetric synthesis of quaternary stereogenic centers with the monopalladocyclic FIP-catalyst precursor $(S_p)_2$-**119** (Scheme 12.34) [77]. The addition of 3 equiv (with respect to $(S_p)_2$-**119**) of 1,8-bis(dimethylamino)naphthylene as an acid scavenger largely suppresses the acid-catalyzed elimination of trifluoroacetamide which is otherwise a significant problem with the 3,3-disubstituted substrates **122**. The acid may arise from the reaction of Ag(OCOCF$_3$) with trace amounts of water present in the reaction mixture [51b,77].

The application of ferrocene palladacycles as catalysts for the allylic imidate rearrangement requires comparison to the results obtained with the related cobalt oxazoline palladacycle $(S,R_p)_2$-**90** (COP-Cl). Unlike its ferrocene counterparts, this palladacycle does not require preactivation by addition of a silver salt, and as a consequence an acid scavenger is not required for the rearrangement of trifluoroacetimidates (E)-**117** and (Z)-**117**. Both of these rearrangements proceed well with the use of 5 mol% $(S,R_p)_2$-**90** (Table 12.2, entries 8 and 9) [78]. This palladacycle also catalyzes the rearrangement of (E)-trichloroacetimidates **107** to give trichloroacetamides from which allylic amines can be obtained by facile hydrolysis of the trichloroacetamide moiety (Scheme 12.29). In the first communication of this reaction a catalyst loading of 5 mol% resulted in both high yields and enantioselectivities (Scheme 12.35, Table 12.3, entry 1) [79]. This loading may be reduced to 1 mol% if the concentration is simultaneously increased [80]. The catalyst loading may be reduced 10-fold from 5 mol% of COP-Cl (a dimer) to 1 mol% by the use of the monomeric complex (S,R_p)-**124** (COP-hfacac) in acetonitrile at 50 °C (entry 2) [81]. Alternatively, just 0.25 mol% of $(S,R_p)_2$-**90** (COP-Cl) may be used in acetonitrile at 70 °C with little or no erosion of enantioselectivity (entry 3) [82].

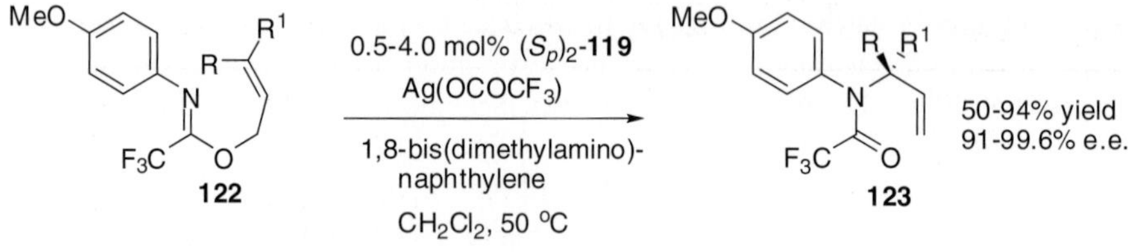

Scheme 12.34

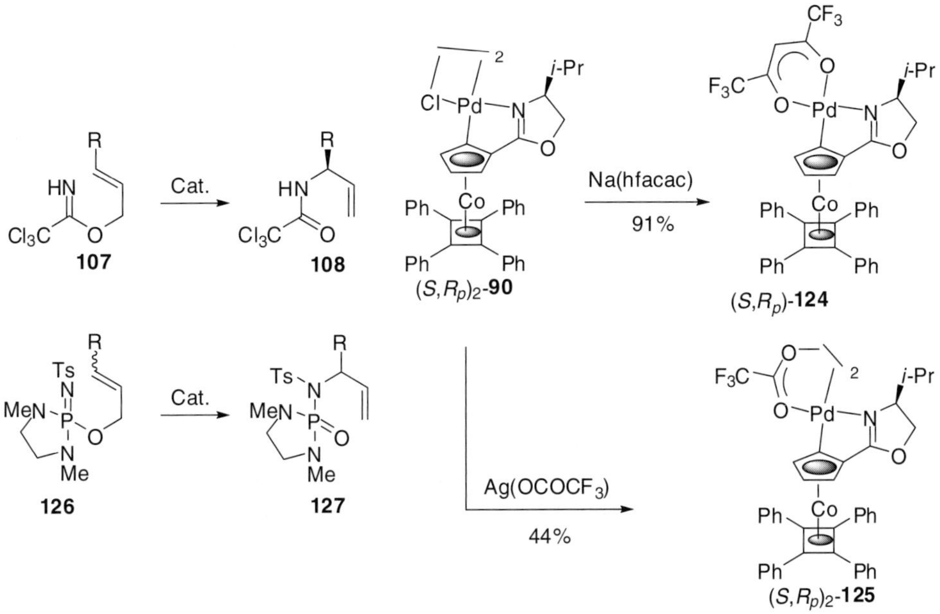

Scheme 12.35

Activation of $(S,R_p)_2$-**90** with 1 equiv per palladium of silver trifluoroacetate gives catalyst $(S,R_p)_2$-**125** [82] for the rearrangement of (allyloxy) iminodiazaphospholidines **126**. Both (E)- and (Z)-isomers of **126** react (entries 4 and 5), the latter result in a higher yield and enantiomeric excess of the product phosphoramide [83]. The same catalyst resulting from silver trifluoroacetate addition has been applied to the rearrangement of N-aryl benzimidates **109**, and again the (Z)-isomers result in significantly higher enantioselectivities [84].

The mercury(II)- or palladium(II)-catalyzed rearrangement of allylic imidates has been proposed to proceed by a stepwise cyclization-induced rearrangement (Scheme 12.36) [85], a mechanism supported by a detailed kinetic and computational

Table 12.3 Palladacycle-catalyzed rearrangement of allylic trichloroacetimidates **107** and (allyloxy) iminodiazaphospholidines **126**.

Entry	Catalyst	Temperature (°C)	Amount of catalyst (mol%)	Substrate (Config.)	Yield (%)	ee (%)	Config.	Ref.
1	$(S,R_p)_2$-**90**	38	5	(E)-**107**	96–98	92–96	S	[79]
2	$(S,R_p)_2$-**124**	50	1	(E)-**107**	82–95	91–95	S	[81]
3	$(S,R_p)_2$-**90**	70	0.25	(E)-**107**	68–82	84–94	S	[82]
4	$(S,R_p)_2$-**125**	45	5	(E)-**126**	44–60	84–86	S	[83]
5	$(S,R_p)_2$-**125**	45	5	(Z)-**126**	82–92	91–96	R	[83]

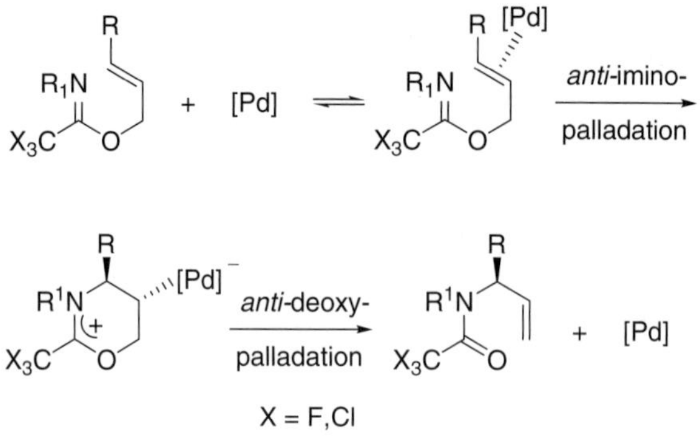

Scheme 12.36

analysis of the palladium(II)-catalyzed rearrangement of allylic trichloroacetimidates [86]. The evidence gained in this study points to C–N bond formation as the rate and enantiodetermining step, which requires that palladium–olefin coordination be reversible. This offers an explanation for the higher yields obtained with (E)-trichloro- and (E)-trifluoroacetimidates from which both the R group and palladium can adopt a pseudo-equtorial orientation in the resulting six-membered cyclic transition state. The corresponding (Z)-isomers will lead to a higher energy conformation containing a pseudo-axial R group or palladium moiety.

The lowest energy transition state structure in the reaction catalyzed by $(S,R_p)_2$-**90** is represented by **128**, and contains an olefin coordinated trans to the oxazoline ligand (Scheme 12.37). This structure is consistent with the thermodynamic addition product of all halogen-bridged palladacycles with neutral ligands such as sulfoxides, phosphines and so on (e.g., see Schemes 12.15 and 12.27), and is a consequence of the preferred trans-geometry of the π-accepting chloride ligand with respect to the strongest π-donor, the cyclopentadienyl ring. The second-lowest transition state structure **129** is destabilized by steric hindrance between the trichloroacetimidate R

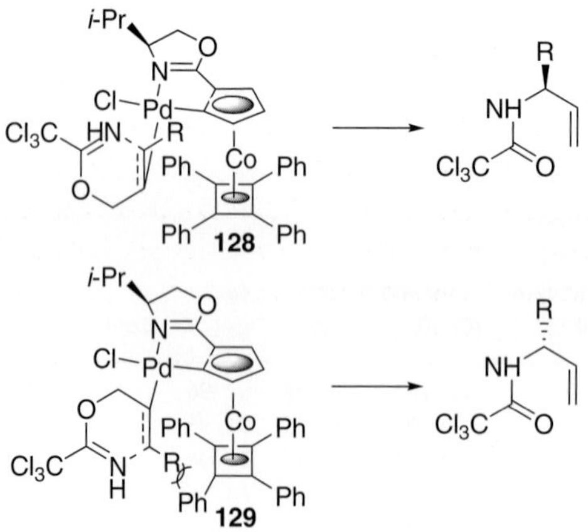

Scheme 12.37

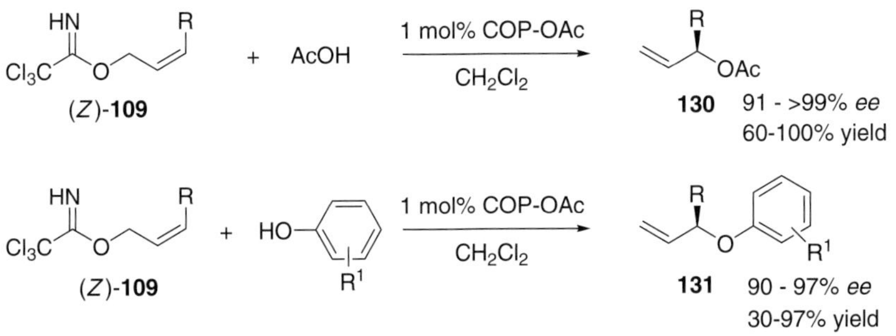

Scheme 12.38

substituent and the bulky tetraphenylcyclobutadiene group. Thus the planar chirality is predominantly responsible for the high enantioselectivity observed in the $(S,R_p)_2$-90 catalyzed rearrangement. This model of enantioinduction has also been proposed for the rearrangement of trifluoroacetimidates 117 with the catalyst derived from $(R_p)_2$-119 [34]. There is the possibility that for the more electron-rich allylic N-arylbenzimidates C—N bond formation is faster, such that palladium–olefin coordination is then rate-limiting [86].

The relative lack of reactivity of (Z)-trichloroacetimidates 109 to intramolecular rearrangement enables them to be used with added nucleophiles. In the presence of acetic acid (and other carboxylic acids) and 1 mol% of the acetate-bridged dimer $(S,R_p)_2$-89 (COP-OAc), (Z)-trichloroacetimidates 109 are converted enantioselectively into allylic esters 130 (Scheme 12.38) [87]. This esterification reaction has been adapted to give an iterative approach to the synthesis of 1,3-polyols [88]. The catalytic addition of phenols to (Z)-109 results in the enantioselective formation of allylic aryl ethers 131 [89].

12.4.2
Intramolecular Aminopalladation

Finally, the one exception to the use of a trichloroacetamide leaving group is the intramolecular aminopalladation of (Z)-allylic acetates, exemplified by substrates 132 (Scheme 12.39). Both the ferrocene-based FOP (ferrocenyl oxazoline palladacycle) catalyst generated from $(S,R_p)_2$-116 (Table 12.4, entries 1–3) [90] and the COP (cobalt oxazoline palladacycle) catalyst $(S,R_p)_2$-89 (entries 4–5) [91] have been applied successfully to this and related transformations. The exact mechanistic detail of these reactions remains to be determined but what evidence there is points to

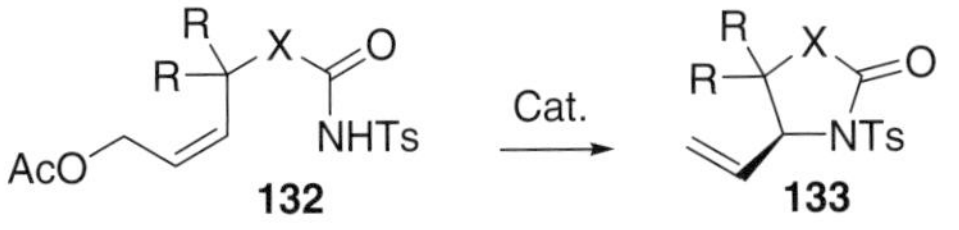

Scheme 12.39

Table 12.4 Palladacycle-catalyzed enantioselective synthesis of N-tosyl vinyl substituted 2-imidazolidinones, 2-pyrrolidinones and 2-oxazolidinones.

Entry	Cat. or cat. precursor	Additive	Amount of catalyst (mol%)	X	R	Yield (%)	ee (%)	Ref.
1	$(S,R_p)_2$-116	$Ag(OCOCF_3)$	5	NH	H	96	90	[90]
2	$(S,R_p)_2$-116	$Ag(OCOCF_3)$	5	CH_2	H	95	90	[90]
3	$(S,R_p)_2$-116	$Ag(OCOCF_3)$	5	O^a	Me	77	96	[90]
4	$(S,R_p)_2$-89	—	1	O^a	Me	89	98	[91]
5	$(S,R_p)_2$-89	—	1	O^a	H	87	95	[91]

^{a}Substrate **132** generated *in situ* from tosyl isocyanate.

palladium remaining in a +2 oxidation state, rather than the reaction proceeding via Pd(II)/Pd(IV) or Pd(0)/Pd(II) catalytic cycles [90].

12.5
Conclusion

This chapter describes the extensive methodology now available for the synthesis of scalemic planar chiral ferrocene-based palladacycles. Related non-palladium metallacycles are few in number and limited to platinacycles and mercuracycles, and in neither case has a stereoselective method been identified for the introduction of the metal by C–H activation. The stoichiometric application of all these metallacycles has so far proved limited, but very significant results have recently been achieved for the use of palladacycles as catalysts in asymmetric synthesis. This is principally, but not exclusively, the catalysis of the allylic imidate rearrangement. If the chemistry of the closely related planar chiral cobalt metallocene-based palladacycles is also considered, there are a number of examples of palladacycle catalysis operating by the activation of alkenes to both intramolecular and intermolecular nucleophilic attack. Very high enantioselectivities have been obtained in many cases, and that the palladium is coordinated directly to the cyclopentadienyl ring of the planar chiral ligand has been demonstrated to be central to the enantioselectivity of trichloro- and trifluoroacetimidate rearrangement. There would appear to be plenty of additional potential for the use of planar chiral palladacycles as Lewis acid catalysts, and for the application of related metallacycles in many other areas of asymmetric catalysis.

References

1 (a) Dupont, J., Pfeffer, M., and Spencer, J. (2001) *Eur. J. Inorg. Chem.*, 1917–1927;
(b) Dupont, J., Consorti, C.S., and Spencer, J. (2005) *Chem. Rev.*, **105**, 2527–2571;
(c) Canty, A.J. (1995) *Comprehensive Organometallic Chemistry II*, vol. 9, Pergamon, Oxford, New York, pp. 242–248.

2 (a) Djukic, J.-P., Hijazi, A., Flack, H.D., and Bernardinelli, G. (2008) *Chem. Soc. Rev.*, **37**, 406–425; (b) Morales-Morales, D., and Jensen, C.M.C (eds) (2007) *The Chemistry of Pincer Type Complexes*, Elsevier, Amsterdam; (c) Alt, H.G., Licht, E.H., Licht, A.I., and Schneider, K.J. (2006) *Coord. Chem. Rev.*, **250**, 2–17; (d) Wu, Y., Huo, S., Gong, J., Cui, X., Ding, L., Ding, K., Du, C., Liu, Y., and Song, M. (2001) *J. Organomet. Chem.*, **637–639**, 27–46.

3 (a) Richards, C.J. and Locke, A.J. (1998) *Tetrahedron: Asymmetry*, **9**, 2377–2407; (b) Colacot, T.J. (2003) *Chem. Rev.*, **103**, 3101–3118; (c) Arrayás, R.G., Adrio, J., and Carretero, J.C. (2006) *Angew. Chem. Int. Ed.*, **45**, 7674–7715.

4 The exceptions are ferrocene-like bidentate ligands where one of the heteroatoms is incorporated into a cyclopentadienyl ring. See for example (a) Shintani, R., Lo, M.M.-C., and Fu, G.C. (2000) *Org. Lett.*, **2**, 3695–3697; (b) Ganter, C., Brassat, L., Glinsböckel, C., and Ganter, B. (1997) *Organometallics*, **16**, 2862–2867.

5 Black, D.S., Deacon, G.B., and Edwards, G.L. (1994) *Aust. J. Chem.*, **47**, 217–227.

6 (a) Schlögl, K. and Fried, M. (1964) *Monatsh. Chem.*, **95**, 558–575; (b) Schlögl, K., Fried, M., and Falk, H. (1964) *Monatsh. Chem.*, **95**, 576–597.

7 Schlögl, K. (1967) *Top. Stereochem.*, **1**, 39–91.

8 Cahn, R.S., Ingold, C., and Prelog, V. (1966) *Angew. Chem., Int. Ed. Engl.*, **5**, 385–415.

9 Sokolov, V.I., Bulygina, L.A., Babievskii, K.K., and Yamskov, I.A. (2007) *Russ. Chem. Bull., Int. Ed.*, **56**, 1464–1466. Circular dichroism spectra are also reported in the following references: [16,19,20].

10 Cunningham, A.F. (1997) *Organometallics*, **16**, 1114–1122.

11 (a) Cunningham, A.F. (1991) *J. Am. Chem. Soc.*, **113**, 4864–4870; (b) Cunningham, A.F. (1994) *Organometallics*, **13**, 2480–2485.

12 Mayor-López, M.J., Weber, J., Mannfors, B., and Cunningham, A.F. (1998) *Organometallics*, **17**, 4983–4991.

13 (a) Ryabov, A.D., Sakodinakaya, I.K., and Yatsimirsky, A.K. (1985) *J. Chem. Soc., Dalton Trans.*, 2629–2638; (b) Ryabov, A.D. (1990) *Chem. Rev.*, **90**, 403–424.

14 Davies, D.L., Donald, S.M.A., and Macgregor, S. (2005) *J. Am. Chem. Soc.*, **127**, 13754–13755.

15 (a) Moynahan, E.B., Popp, F.D., and Werneke, W.F. (1969) *J. Organomet. Chem.*, **19**, 229–232; (b) Gaunt, J.C. and Shaw, B.L. (1975) *J. Organomet. Chem.*, **102**, 511–516.

16 (a) Komatsu, T., Nonoyama, M., and Fujita, J. (1981) *Bull. Chem. Soc. Jpn.*, **54**, 186–189; (b) Sokolov, V.I., Nechaeva, K.S., and Reutov, O.A. (1983) *J. Organomet. Chem.*, **253**, C55–C58.

17 Dunina, V.V., Gorunova, O.N., Livantsov, M.V., Grishin, Y.K., Kuz'mima, L.G., Kataeva, N.A., and Churakov, A.V. (2000) *Tetrahedron: Asymmetry*, **11**, 3967–3984.

18 Huo, S.Q., Wu, Y.J., Du, C.X., Zhu, Y., Yuan, H.Z., and Mao, X.A. (1994) *J. Organomet. Chem.*, **483**, 139–146.

19 (a) Wu, Y.J., Cui, X.L., Du, C.X., Wang, W.L., Guo, R.Y., and Chen, R.F. (1998) *J. Chem. Soc., Dalton Trans.*, 3727–3730; (b) Cui, X.L., Wu, Y.J., Du, C.X., Yang, L.R., and Zhu, Y. (1999) *Tetrahedron: Asymmetry*, **10**, 1255–1262.

20 Wu, Y.-J., Ding, L., Wang, W.-L., and Du, C.-X. (1998) *Tetrahedron: Asymmetry*, **9**, 4035–4041.

21 Marquarding, D., Klusacek, H., Gokel, G., Hoffmann, P., and Ugi, I. (1970) *J. Am. Chem. Soc.*, **92**, 5389–5393.

22 (a) Gokel, G.W. and Ugi, I.K. (1972) *J. Chem. Ed.*, **49**, 294–296; (b) Boaz, N.W. (1989) *Tetrahedron Lett.*, **30**, 2061–2064.

23 Sokolov, V.I., Troitskaya, L.L., and Reutov, O.A. (1977) *J. Organomet. Chem.*, **133**, C28–C30.

24 López, C., Bosque, R., Solans, X., and Font-Bardia, M. (1996) *Tetrahedron: Asymmetry*, **7**, 2527–2530.

25 Hollis, T.K. and Overman, L.E. (1997) *Tetrahedron Lett.*, **38**, 8837–8840.

26 Sokolov, V.I., Troitskaya, L.L., and Bondareva-Don, V.L. (2005) *Molecules*, **10**, 649–652.

27 Sokolov, V.I., Troitskaya, L.L., and Rozhkova, T.I. (1987) *Gazz. Chim. Ital.*, **117**, 525–527.

28 Sokolov, V.I., Troitskaya, L.L., Gautheron, B., and Tainturier, G. (1982) *J. Organomet. Chem.*, **235**, 369–373.

29 Dunina, V.V., Gorunova, O.N., Livantsov, M.V., Grishin, Y.K., Kuz'mina, L.G., Kataeva, N.A., and Churakov, A.V. (2000) *Inorg. Chem. Comm.*, **3**, 354–357.

30 (a) Richards, C.J., Damalidis, T., Hibbs, D.E., and Hursthouse, M.B. (1995) *Synlett*, 74–76; (b) Sammakia, T., Latham, H.A., and Schaad, D.R. (1995) *J. Org. Chem.*, **60**, 10–11; (c) Nishibayashi, Y. and Uemura, S. (1995) *Synlett*, 79–81.

31 Moyano, A., Rosol, M., Moreno, R.M., López, C., and Maestro, M.A. (2005) *Angew. Chem. Int. Ed.*, **41**, 1865–1869.

32 Xia, J.-B. and You, S.-L. (2007) *Organometallics*, **26**, 4869–4871.

33 Peters, R., Xin, Z., Fischer, D.F., and Schweizer, W.B. (2006) *Organometallics*, **25**, 2917–2920.

34 Weiss, M.E., Fischer, D.F., Xin, Z., Jautze, S., Schweizer, W.B., and Peters, R. (2006) *Angew. Chem. Int. Ed.*, **45**, 5694–5698.

35 (a) Jautze, S., Seiler, P., and Peters, R. (2007) *Angew. Chem. Int. Ed.*, **46**, 1260–1264; (b) Jautze, S., Seiler, P., and Peters, R. (2008) *Chem. Eur. J.*, **14**, 1430–1444.

36 Zhao, G., Xue, F., Zhang, Z.-Y., and Mak, T.C.W. (1997) *Organometallics*, **16**, 4023–4026.

37 Zhao, G., Wang, Q.-G., and Mak, T.C.W. (1998) *Tetrahedron: Asymmetry*, **9**, 2253–2257.

38 Benito, M., López, C., Solans, X., and Font-Bardía, M. (1998) *Tetrahedron: Asymmetry*, **9**, 4219–4238.

39 Zhao, G., Yang, O.-C., and Mak, T.C.W. (1999) *Organometallics*, **18**, 3623–3636.

40 (a) Zhao, G., Wang, Q.-G., and Mak, T.C.W. (1998) *Tetrahedron: Asymmetry*, **9**, 1557–1561; (b) Zhao, G., Wang, Q.-G., and Mak, T.C.W. (1998) *Polyhedron*, **17**, 1–8.

41 Zhao, G., Wang, Q.-G., and Mak, T.C.W. (1998) *Organometallics*, **17**, 3437–3441.

42 Zhao, G., Wang, Q.-G., and Mak, T.C.W. (1998) *J. Chem. Soc., Dalton Trans.*, 3785–3789.

43 (a) López, C., Caubet, A., Pérez, S., Solans, X., and Font-Bardia, M. (2004) *Chem. Comm.*, 540–541; (b) López, C., Caubet, A., Pérez, S., Solans, X., Font-Bardía, M., and Molins, E. (2006) *Eur. J. Inorg. Chem.*, 3974–3984.

44 Du, L.-Z., , Gong, J.-F., Xu, C., Zhu, Y., Wu, Y.-J., and Song, M.-P. (2006) *Inorg. Chem. Comm.*, **9**, 410–414.

45 Ryabov, A.D., Panyashkina, I.M., Polyakov, V.A., Howard, J.A.K., Kuz'mina, L.G., Datt, M.S., and Sacht, C. (1998) *Organometallics*, **17**, 3615–3618.

46 Ryabov, A.D., Panyashkina, I.M., Kazankov, G.M., Polyakov, V.A., and Kuz'mina, L.G. (2000) *J. Organomet. Chem.*, **601**, 51–56.

47 Wu, Y., Cui, X., Zhou, N., Song, M., Yun, H., Du, C., and Zhu, Y. (2000) *Tetrahedron: Asymmetry*, **11**, 4877–4883.

48 (a) Riant, O., Samuel, O., and Kagan, H.B. (1993) *J. Am. Chem. Soc.*, **115**, 5835–5836; (b) Riant, O., Samuel, O., Flessner, T., Taudien, S., and Kagan, H.B. (1997) *J. Org. Chem.*, **62**, 6733–6745.

49 Cohen, F. and Overman, L.E. (1998) *Tetrahedron: Asymmetry*, **9**, 3213–3222.

50 Sutcliffe, O.B. and Bryce, M.R. (2003) *Tetrahedron: Asymmetry*, **14**, 2297–2325.

51 (a) Donde, Y. and Overman, L.E. (1999) *J. Am. Chem. Soc.*, **121**, 2933–2934; (b) Anderson, C.E., Donde, Y., Douglas, C.J., and Overman, L.E. (2005) *J. Org. Chem.*, **70**, 648–657.

52 Use of TMEDA as an additive and ether as the reaction solvent improves the diastereoselctivity to 10:1, see: Locke, A.J., Pickett, T.E., and Richards, C.J. (2001) *Synlett*, 141–143.

53 Kang, J., Yew, K.H., Kim, T.H., and Choi, D.H. (2002) *Tetrahedron Lett.*, **43**, 9509–9512.
54 Wu, Y., Yang, L., Cui, X., Du, C., and Zhu, Y. (2003) *Tetrahedron: Asymmetry*, **14**, 1073–1077.
55 (a) Sokolov, V.I. and Troitskaya, L.L. (1978) *Chimia*, **32**, 122–123; (b) Sokolov, V.I., Troitskaya, L.L., and Reutov, O.A. (1979) *J. Organomet. Chem.*, **182**, 537–546.
56 Ryabov, A.D., Firsova, Y.N., Goral, V.N., Ryabov, E.S., Shevelkova, A.N., Troitskaya, L.L., Demeschik, T.V., and Sokolov, V.I. (1999) *Chem. Eur. J.*, **4**, 806–813.
57 (a) Ryabov, A.D. and Yatsimirsky, A.K. (1984) *Inorg. Chem.*, **23**, 789–790; (b) Ryabov, A.D. (1987) *Inorg. Chem.*, **26**, 1252–1260.
58 (a) Dunina, V.V., Razmyslova, E.D., Gorunova, O.N., Livantsov, M.V., and Grishin, Y.K. (2003) *Tetrahedron: Asymmetry*, **14**, 2331–2333; (b) Dunina, V.V., Gorunova, O.N., Kuznetsova, E.D., Turubanova, E.I., Livantsov, M.V., Grishin, Y.K., Kuz'mina, L.G., and Churakov, A.V. (2006) *Russ. Chem. Bull., Int. Ed.*, **55**, 2193–2211.
59 Stevens, A.M. and Richards, C.J. (1999) *Organometallics*, **18**, 1346–1348.
60 Prasad, R.S., Anderson, C.E., Richards, C.J., and Overman, L.E. (2005) *Organometallics*, **24**, 77–81.
61 Yeamine, M.R. and Richards, C.J. (2007) *Tetrahedron: Asymmetry*, **18**, 2613–2616.
62 Roca, F.X., Motevalli, M., and Richards, C.J. (2005) *J. Am. Chem. Soc.*, **127**, 2388–2389.
63 Wild, S.B. (1997) *Coord. Chem. Rev.*, **166**, 291–311.
64 López, C., Bosque, R., Sainz, D., Solans, X., and Font-Bardía, M. (1997) *Organometallics*, **16**, 3261–3266.
65 Sokolov, V.I., Troitskaya, L.L., and Reutov, O.A. (1980) *J. Organomet. Chem.*, **202**, C58–C60.
66 (a) Gruselle, M., Malezieux, B., Sokolov, V.I., and Troitskaya, L.L. (1994) *Inorg. Chem. Acta*, **222**, 51–61; (b) Malézieux, B., Gruselle, M., Troitskaya, L.L., and Sokolov, V.I. (1998) *Tetrahedron: Asymmetry*, **9**, 259–269.
67 Troitskaya, L.L. and Sokolov, V.I. (1985) *J. Organomet. Chem.*, **285**, 389–393.
68 Kasahara, A., Izumi, T., and Watabe, H. (1979) *Bull. Chem. Soc. Jp.*, **52**, 957–958.
69 (a) Sokolov, V.I., Troitskaya, L.L., and Khrushchova, N.S. (1983) *J. Organomet. Chem.*, **250**, 439–446; (b) Sokolov, V.I. (1983) *Pure Appl. Chem.*, **55**, 1837–1844.
70 Zhao, G., Wang, Q.-G., and Mak, T.C.W. (1999) *J. Organomet. Chem.*, **574**, 311–317.
71 Tsuji, J. (2004) *Palladium Reagents and Catalysts: New Perspectives for the 21st Century*, Wiley, New York.
72 See for example: (a) Wu, Y., Hou, J., Yun, H., Cui, X., and Yuan, R. (2001) *J. Organomet. Chem.*, **637–639**, 793–795; (b) Roca, F.X. and Richards, C.J. (2003) *Chem. Comm.*, 3002–3003; (c) Gong, J., Liu, G., Du, C., Zhu, Y., and Wu, Y. (2005) *J. Organomet. Chem.*, **690**, 3963–3969; (d) Xu, C., Gong, J.-F., Yue, S.-F., Zhu, Y., and Wu, Y.-J. (2006) *Dalton Trans.*, 4730–4739; (e) Xu, C., Gong, J.-F., and Wu, Y.-J. (2007) *Tetrahedron Lett.*, **48**, 1619–1623; (f) Li, J., Cui, M., Yu, A., and Wu, Y. (2007) *J. Organomet. Chem.*, **692**, 3732–3742; (g) Mu, B., Li, T., Xu, W., Zeng, G., Liu, P., and Wu, Y. (2007) *Tetrahedron*, **63**, 11475–11488; (h) Cui, M., Li, J., Yu, A., Zhang, J., and Wu, Y. (2008) *J. Mol. Cat. A*, **290**, 67–71.
73 Phan, N.T.S., Sluys, M.V.D., and Jones, C.W. (2006) *Adv. Synth. Catal.*, **348**, 609–679.
74 (a) Overman, L.E. (1974) *J. Am. Chem. Soc.*, **96**, 597–599; (b) Overman, L.E. (1976) *J. Am. Chem. Soc.*, **98**, 2901–2910; (c) Overman, L.E. (1980) *Acc. Chem. Res.*, **13**, 218–224.
75 (a) Ikariya, T., Ishikawa, Y., Hirai, K., and Yoshikawa, S. (1982) *Chem. Lett.*, 1815–1818; (b) Schenck, T.G. and Bosnich, B. (1985) *J. Am. Chem. Soc.*, **107**, 2058–2066; (c) Metz, P., Mues, C., and Schoop, A. (1992) *Tetrahedron*, **48**, 1071–1080; (d) Mehmandoust, M., Petit, Y., and Larchevêque, M. (1992) *Tetrahedron*

Lett., **33**, 4313–4316; (e) Gonda, J., Helland, A.-C., Ernst, B., and Belluš, D. (1993) *Synthesis*, 729–734; (f) Overman, L.E. and Zipp, G.G. (1997) *J. Org. Chem.*, **62**, 2288–2291.

76 (a) Calter, M., Hollis, T.K., Overman, L.E., Ziller, J., and Zipp, G.G. (1997) *J. Org. Chem.*, **62**, 1449–1456; (b) Uozumi, Y., Kato, K., and Hayashi, T. (1998) *Tetrahedron: Asymmetry*, **9**, 1065–1072.

77 Fischer, D.F., Xin, Z., and Peters, R. (2007) *Angew. Chem. Int. Ed.*, **46**, 7704–7707.

78 Overman, L.E., Owen, C.E., Pavan, M.M., and Richards, C.J. (2003) *Org. Lett.*, **5**, 1809–1812.

79 Anderson, C.E. and Overman, L.E. (2003) *J. Am. Chem. Soc.*, **125**, 12412–12413.

80 Anderson, C.E., Overman, L.E., and Watson, M.P. (2005) *Org. Synth.*, **82**, 134–137.

81 Kirsch, S.F., Overman, L.E., and Watson, M.P. (2004) *J. Org. Chem.*, **69**, 8101–8104.

82 Nomura, H. and Richards, C.J. (2007) *Chem. Eur. J.*, **13**, 10216–10224.

83 Lee, E.E. and Batey, R.A. (2005) *J. Am. Chem. Soc.*, **127**, 14887–14893.

84 Kang, J., Kim, T.H., Yew, K.H., and Lee, W.K. (2003) *Tetrahedron: Asymmetry*, **14**, 415–418.

85 Overman, L.E. (1984) *Angew. Chem., Int. Ed. Engl.*, **23**, 579–586.

86 Watson, M.P., Overman, L.E., and Bergman, R.G. (2007) *J. Am. Chem. Soc.*, **129**, 5031–5044.

87 Kirsch, S.F. and Overman, L.E. (2005) *J. Am. Chem. Soc.*, **127**, 2866–2867.

88 Binder, J.T. and Kirsch, S.F. (2007) *Chem. Comm.*, 4164–4166.

89 Kirsch, S.F., Overman, L.E., and White, N.S. (2007) *Org. Lett.*, **9**, 911–913.

90 Overman, L.E. and Remarchuk, T.P. (2002) *J. Am. Chem. Soc.*, **124**, 12–13.

91 Kirsch, S.F. and Overman, L.E. (2005) *J. Org. Chem.*, **70**, 2859–2861.

A
Show Case of the Most Effective Chiral Ferrocene Ligands in Various Catalytic Reactions

Note: References for each reaction show the author names, year, chapter in this book and reference number in the chapter. For example, in A.1.1 the authors are Dübner and Knochel, the year of publication was 1999 and the full reference is given in Chapter 3, Reference [14].

A.1
Asymmetric Allylic Substitution Reactions

A.1.1
Copper(I)-Catalyzed Asymmetric Allylic Alkylation Reactions

up to 89% yield
up to 98% *ee*
B/L: 97/3-99/1

Dubnër and Knochel (**1999**) [3-14]

Chiral Ferrocenes in Asymmetric Catalysis. Edited by Li-Xin Dai and Xue-Long Hou
Copyright © 2010 WILEY-VCH Verlag GmbH & Co. KGaA, Weinheim
ISBN: 978-3-527-32280-0

A.1.2
Palladium-Catalyzed Asymmetric Allylic Substitution Reactions

A.1.2.1 Alkylation

Palladium-Catalyzed Asymmetric Allylic Alkylation

Ph–CH(OAc)–CH=CH–Ph + MeOOC–CH$_2$–COOMe $\xrightarrow[\text{Ligand}]{[Pd(C_3H_5)Cl]_2}$ Ph–C*(COOMe)(COOMe)–CH=CH–Ph

Ligand:

Ligand	Results	Reference
Fe–PPh$_2$, CH(Me)–N=CH–(4-pyridyl)	up to 95% yield, up to 98% ee	Zheng (**2004**) [5-53a]
Fe–PPh$_2$, CH(Me)–N(Me)–triazine(OMe)$_2$	up to 93% yield, up to 99% ee	Zheng (**2005**) [5-53b]
Fe–PPh$_2$, CH(cyclohexyl-NMeN-Me)	up to 99% yield, up to 99.6% ee	Jin (**2006**) [5-52]
Fe–indanyl(NMe$_2$)(PPh$_2$)	up to 92% yield, up to 96% ee	Kikuchi (**2007**) [5-55c]
Fe–(PAr$_2$, Oxn-R)$_2$ (C$_2$-symmetric)	up to 99% ee	Zhang (**1996**) [7-44b]
Fe(PPh$_2$, CO$_2$R)$_2$	up to 99% ee	Zhang (**1999**) [7-51]
Fe(Oxn, PPh$_2$)$_2$	up to 99% yield, up to 99% ee	Park (**2001**) [8-8]
Fe(PPh$_2$, Oxn)$_2$ with TMS	up to 99% yield, up to 96% ee	Park (**2001**) [8-8]
Oxn = oxazoline-iPr		
Fe–Me, Oxn-iPr, PPh$_2$	up to 99% yield, up to 99% ee	
Fe–TMS, Oxn-iPr, PPh$_2$	up to 99% yield, up to 99% ee	Hou and Dai (**2000**) [8-9]
Fe–Oxn-tBu, PR'$_2$, PR$_2$	up to 99% yield, up to 92% ee	Hou and Dai (**2004**) [8-12]
Fe(SO$_2$tBu, PPh$_2$)$_2$	up to 98% ee	Zhang (**2005**) [9-26]
Fe–Oxn–SMe	up to 98% yield, up to 93% ee	Bryce (**1997**) [9-8]
Fe–SPh, Oxn-tBu	up to 98% yield, up to 98% ee	Dai and Hou (**1998**) [9-10]
Fe–SPh, Oxn–CH(Ph)(OTBDMS)	up to 98% yield, up to 95% ee	Ait-Haddou (**2000**) [9-12]
Fe–S–CH(Ph)(NMe$_2$)	up to 96% yield, up to 99% ee	Bonini (**2002**) [9-14]

Palladium-Catalyzed Asymmetric Allylic Alkylation with Cyclic-Substrate

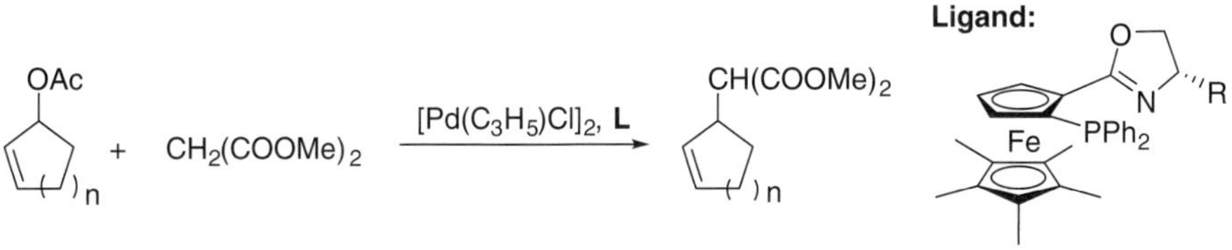

up to 93% yield
up to 94% ee
Helmchen (**2006**) [5-65]

Asymmetric Allylic Alkylation Using Enamines as Nucleophilic Reagents

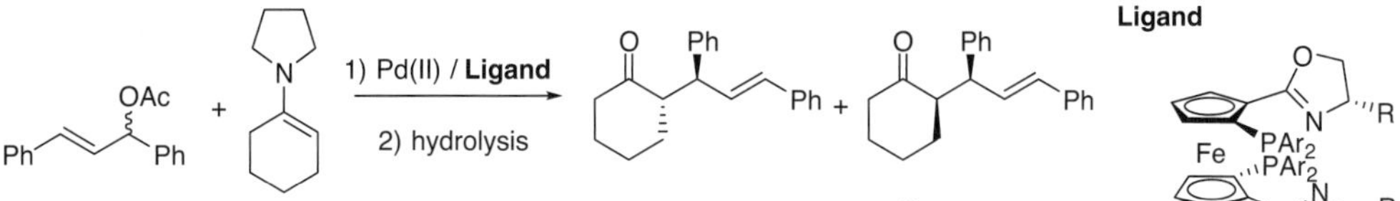

Anti-
up to 99% ee

Syn-
up to 98% ee

Zhang (**2007**) [7-45]

A.1.2.2 Amination

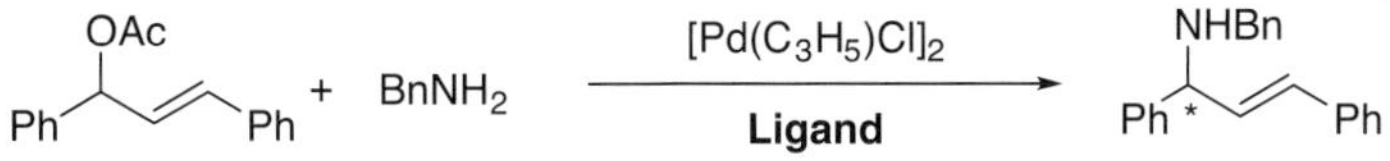

Ligand:

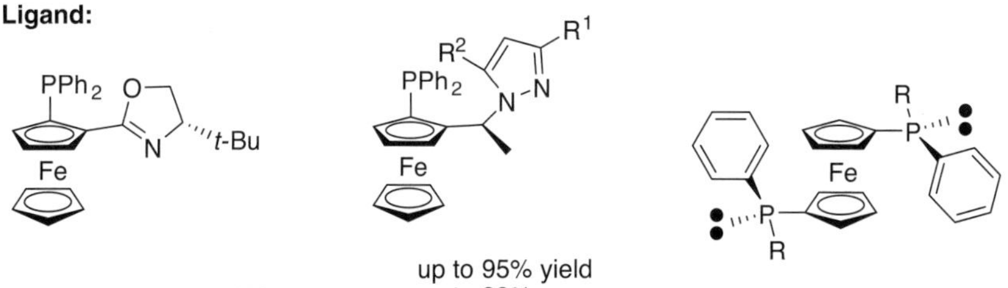

up to 97.2% ee (R)
Hou and Dai (**2002**) [5-60]

up to 95% yield
up to 99% ee
Togni (**1996**) [5-57]

up to 99% ee
van Leeuwen (**1997**) [7-12]

A.1.3
Generating π-Allyl Palladium Complex Through the Reaction of Allene with Iodobenzene

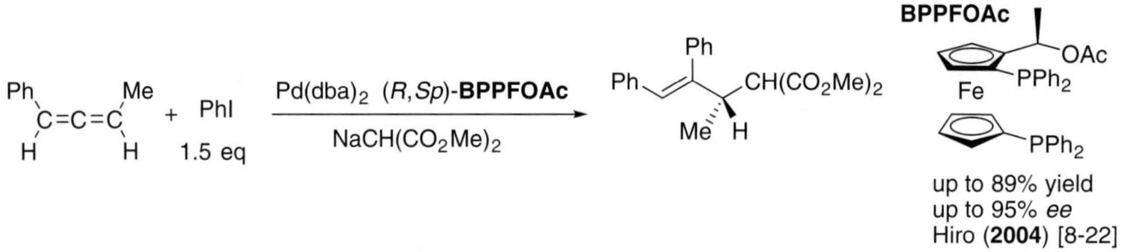

Ph\C=C=C/Me (H, H) + PhI (1.5 eq) — Pd(dba)$_2$ (R,Sp)-**BPPFOAc**, NaCH(CO$_2$Me)$_2$ → Ph-CH=C(Me)-CH(CO$_2$Me)$_2$ (Me, H)

BPPFOAc: Fc with -CH(OAc)-PPh$_2$ and -PPh$_2$ substituents

up to 89% yield
up to 95% ee
Hiro (**2004**) [8-22]

A.1.4
Regioselective Control Concerned with Unsymmetrically 1,3-Disubstituted 2-Propenyl Acetate

A.1.4.1 Regio- and Enantioselective Palladium-Catalyzed Allylic Substitution

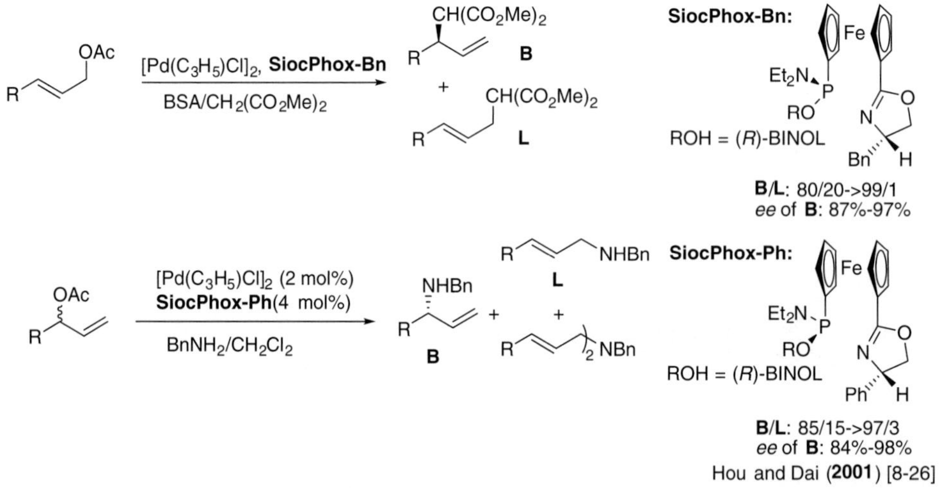

R-CH=CH-CH$_2$-OAc — [Pd(C$_3$H$_5$)Cl]$_2$, **SiocPhox-Bn**, BSA/CH$_2$(CO$_2$Me)$_2$ → R-CH(CH(CO$_2$Me)$_2$)-CH=CH$_2$ (**B**) + R-CH=CH-CH$_2$-CH(CO$_2$Me)$_2$ (**L**)

SiocPhox-Bn: Fc-oxazoline with Et$_2$N-P(OR)-N, Bn, H
ROH = (R)-BINOL

B/L: 80/20->99/1
ee of **B**: 87%-97%

R-CH(OAc)-CH=CH$_2$ — [Pd(C$_3$H$_5$)Cl]$_2$ (2 mol%), **SiocPhox-Ph** (4 mol%), BnNH$_2$/CH$_2$Cl$_2$ → R-CH(NHBn)-CH=CH$_2$ (**B**) + R-CH=CH-CH$_2$-NHBn (**L**) + R-CH=CH-CH$_2$-NBn

SiocPhox-Ph: Fc-oxazoline with Et$_2$N-P(OR)-N, Ph, H
ROH = (R)-BINOL

B/L: 85/15->97/3
ee of **B**: 84%-98%
Hou and Dai (**2001**) [8-26]

* **B**: Branched isomer, **L**: Linear isomer

A.1.4.2 Allylic Substitution of 2-Cyanopropionate

Up to 93 % yield
Up to 99% ee
Ito (**1996**) [10-22]

A.1.4.3 Allylic Substitution of the Conjugated Dienyl Acetates

SiocPhox-Bn:

ROH = (*R*)-BINOL
B/L: 94/6-98/2
ee of **B**: 56%-93%

SiocPhox-Ph:

ROH = (*R*)-BINOL
B/L: 94/6->98/2
ee of **B**: 88%-94%
Hou (**2005**) [8-28]

* **B**: Branched isomer, **L**: Linear isomer

A.1.4.4 Allylic Substitution with Acyclic Ketone Enolate as Nucleophile

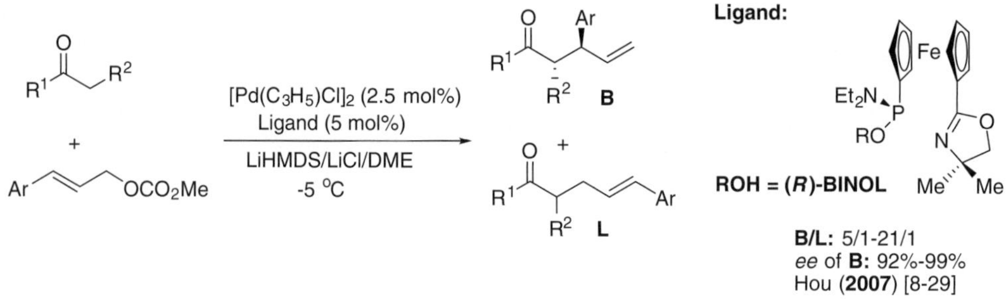

[Pd(C$_3$H$_5$)Cl]$_2$ (2.5 mol%)
Ligand (5 mol%)
LiHMDS/LiCl/DME
−5 °C

Ligand:

ROH = (R)-BINOL

B/L: 5/1–21/1
ee of **B**: 92%–99%
Hou (**2007**) [8-29]

* **B**: Branched isomer, **L**: Linear isomer

A.1.4.5 Allylic Alkylation with N,N-Disubstituted Amides as Nucleophile

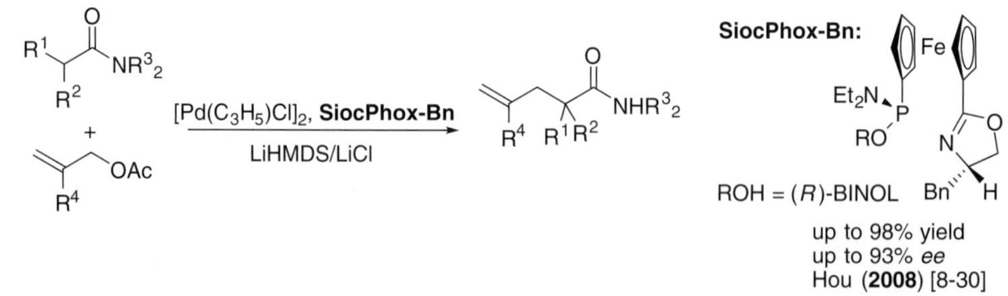

[Pd(C$_3$H$_5$)Cl]$_2$, **SiocPhox-Bn**
LiHMDS/LiCl

SiocPhox-Bn:

ROH = (R)-BINOL, Bn, H

up to 98% yield
up to 93% ee
Hou (**2008**) [8-30]

A.1.4.6 Allylic Alkylation with Activated Cyclic Ketone as Nucleophile

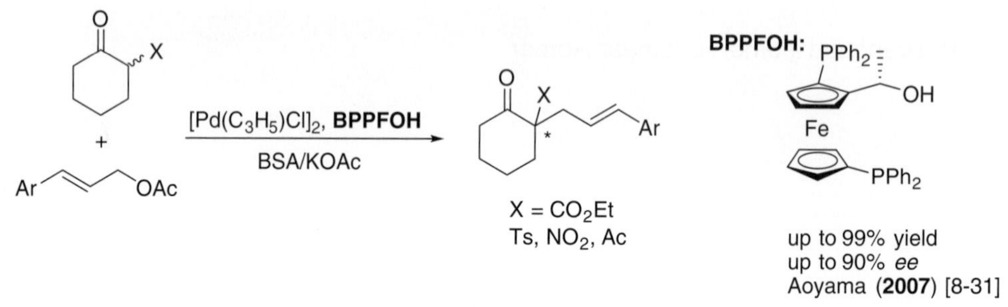

[Pd(C$_3$H$_5$)Cl]$_2$, **BPPFOH**
BSA/KOAc

X = CO$_2$Et
Ts, NO$_2$, Ac

BPPFOH:

up to 99% yield
up to 90% ee
Aoyama (**2007**) [8-31]

A.1.5
Applications of Palladium-Catalyzed Asymmetric Allylic Substitution Reaction

82% yield
90.3% ee
He and Bai (**2001**) [8-34]

A.2
Asymmetric Aldol Reactions

A.2.1
Gold-Catalyzed Asymmetric Aldol Reactions

A.2.1.1 Aldehyde with Isocyanoacetate

up to 98% yield
trans/cis up to 90/10
ee for trans (4S,5R) up to 96%
Hayashi (**1986**) [8-36]
Togni (**1989**) [8-37]

A.2.1.2 Imines with Isocyanoacetate

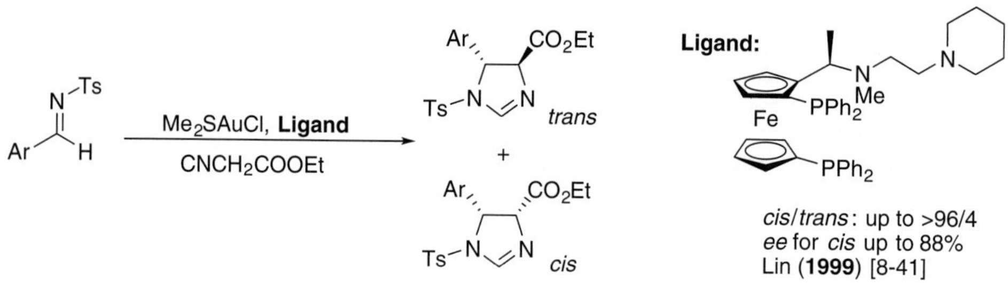

Reaction: Ar-CH=N-Ts + CNCH$_2$COOEt $\xrightarrow{\text{Me}_2\text{SAuCl, Ligand}}$ trans and cis oxazoline products (Ts-N, Ar, CO$_2$Et)

Ligand: Ferrocene-based with PPh$_2$, Me substituent, and N-CH(Me)-CH$_2$CH$_2$-N(piperidine)

cis/trans: up to >96/4
ee for cis up to 88%
Lin (**1999**) [8-41]

A.2.2
Silver-Catalyzed Asymmetric Aldol Reactions

R-CHO + CNCH$_2$SO$_2$Tol $\xrightarrow{\text{AgOTf, Ligand}}$ trans and cis oxazoline products (R, SO$_2$Tol)

Ligand: Ferrocene with PPh$_2$, NMeCH$_2$CH$_2$N(ring)$_n$, n = 1, 2

up to 97% yield
trans/cis: > 20/1
up to 86% ee
Hamashima (**1990**) [8-42]

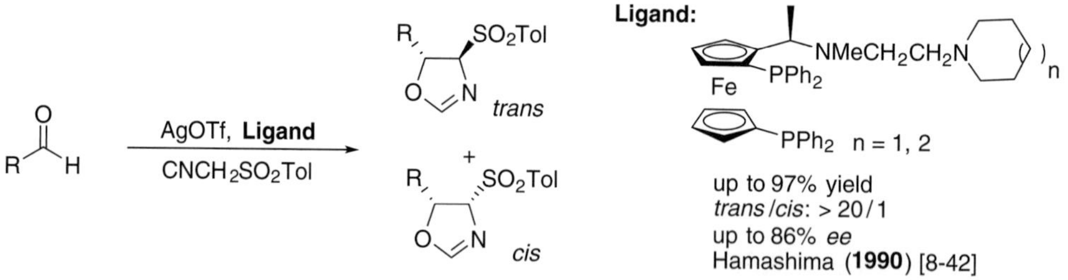

A.2.3
Applications of Gold or Silver-Catalyzed Asymmetric Aldol Reactions

n-C$_{13}$H$_{27}$-CH=CH-CHO + CNCH$_2$COOMe $\xrightarrow{[\text{Au}(c\text{-HexNC})_2]\text{BF}_4, \text{Ligand}}$ oxazoline intermediate → D-erythro-sphingosine (n-C$_{13}$H$_{27}$, NH$_2$, OH)

Ligand: Ferrocene with PPh$_2$, NMeCH$_2$CH$_2$N(morpholine)

trans: 80% yield
trans/cis: 89/11 and 93% ee
Hayashi (**1998**) [8-45]

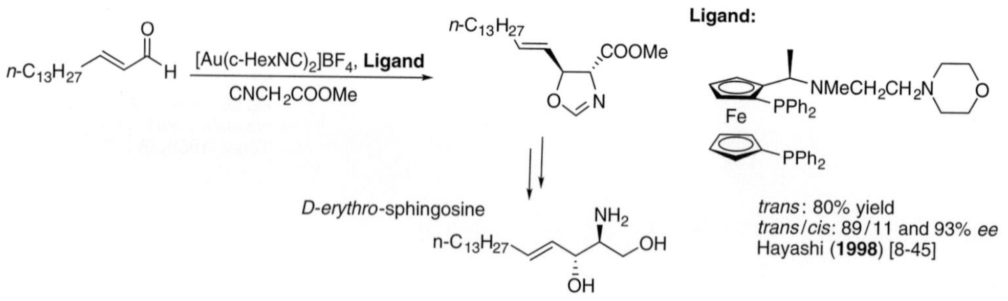

A.3
Asymmetric Cycloaddition Reactions

A.3.1
[3 + 2] Cycloaddition Reactions

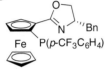

Ligand:

M = Ag
up to 99% yield
up to 98% ee
Zhou (**2005**) [5-86]

M = Ag
up to 98% yield
up to 92% ee
Zhou (**2007**) [5-87]

M = Cu
up to 85% yield
up to 92% de and 98% ee
Raghunath (**2005**) [5-88]

A.3.2
[4 + 2] Cycloaddition Reactions

Ligand:

up to 90% yield
up to 97% ee
Carretero (**2004**) [9-57]

A.3.3
[3 + 2] Cycloaddition Reactions

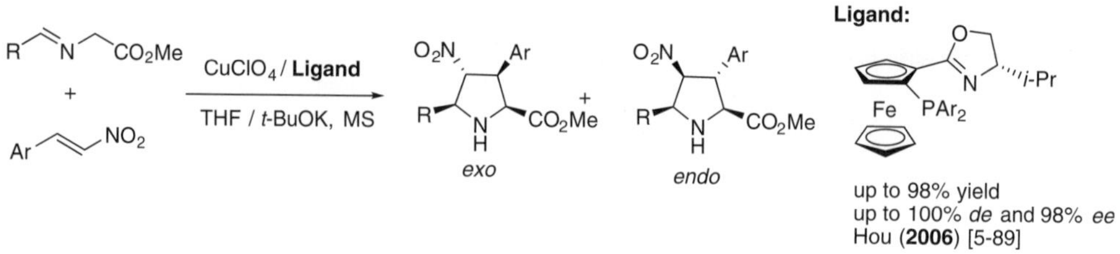

Ligand: oxazoline-ferrocene with PAr₂ and i-Pr

up to 98% yield
up to 100% de and 98% ee
Hou (**2006**) [5-89]

A.3.4
Asymmetric 1,3-Dipolar Cycloaddition of Azomethine Ylides with N-Phenyl Maleimide

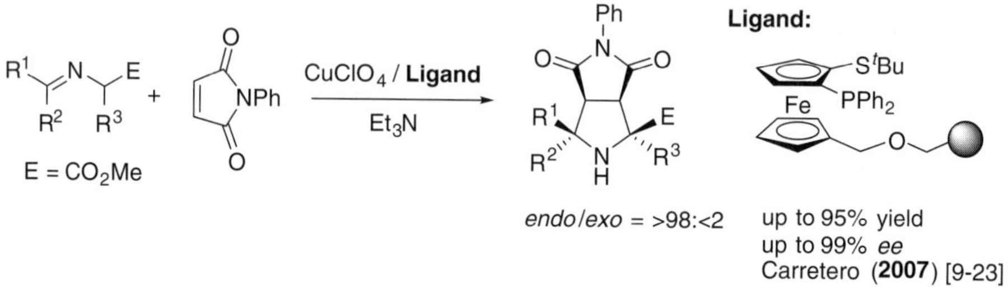

E = CO₂Me

endo/exo = >98:<2

Ligand: ferrocene with S^{t}Bu, PPh₂, polymer-supported

up to 95% yield
up to 99% ee
Carretero (**2007**) [9-23]

A.3.5
[3 + 2] Cycloaddition of Alkynes with Azomethine Imines

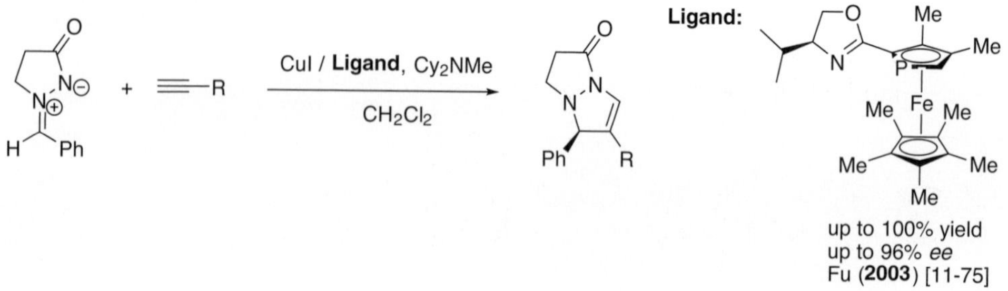

CuI / Ligand, Cy₂NMe
CH₂Cl₂

Ligand: i-Pr oxazoline, Me-substituted phosphaferrocene

up to 100% yield
up to 96% ee
Fu (**2003**) [11-75]

A.4 Asymmetric Hydrogenation

A.4.1 Asymmetric Hydrogenation of Alkenes (C=C)

A.4.1.1 Asymmetric Hydrogenation of Unfunctionalized Alkenes

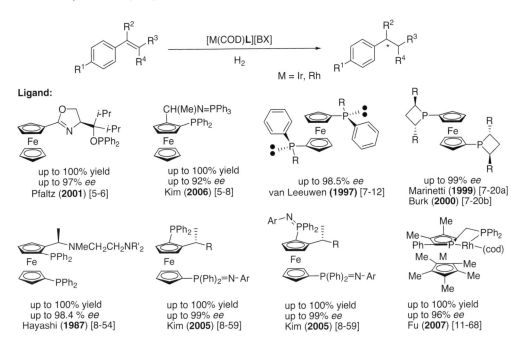

A.4.1.2 Asymmetric Hydrogenation of α,β-Unsaturated Esters

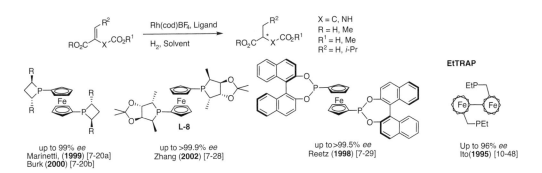

A.4.1.3 Hydrogenation of Dehydro α-Amino Acid

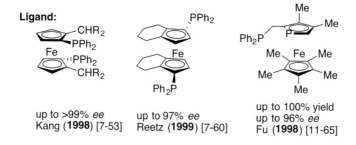

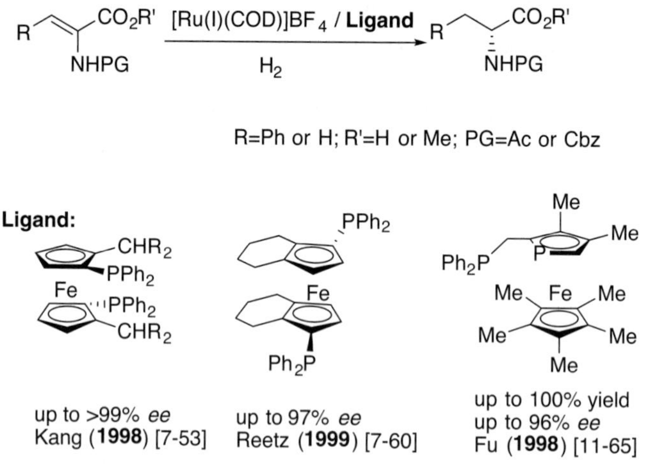

[Ru(I)(COD)]BF4 / Ligand
H2
CO2R'
NHPG
R=Ph or H; R'=H or Me; PG=Ac or Cbz
Ligand:
up to >99% ee
Kang (1998) [7-53]
up to 97% ee
Reetz (1999) [7-60]
up to 100% yield
up to 96% ee
Fu (1998) [11-65]

A.4.1.4 Asymmetric Hydrogenation of Activated Imine C=N Bond

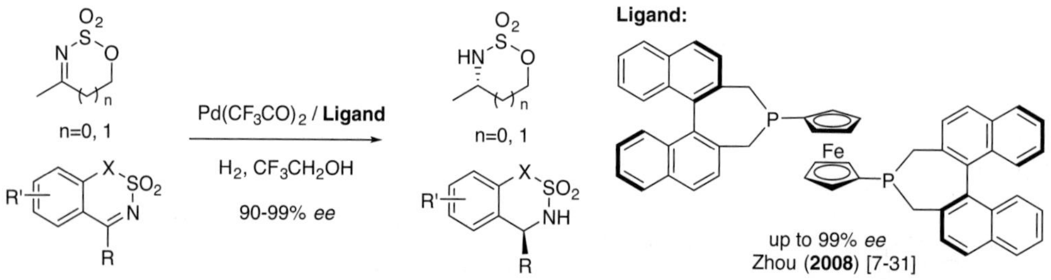

n=0, 1
Pd(CF3CO)2 / Ligand
H2, CF3CH2OH
90-99% ee
Ligand:
up to 99% ee
Zhou (2008) [7-31]

A.4.1.5 Applications: Hydrogenation of α,β-Unsaturated Acid

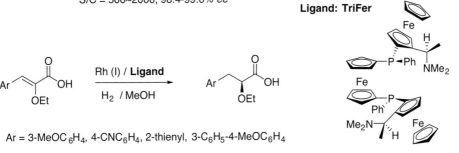

S/C = 500~2000, 98.4-99.6% ee

Intermediate of renin inhibitor **Aliskiren**

Ligand: TriFer

Ar = 3-MeOC$_6$H$_4$, 4-CNC$_6$H$_4$, 2-thienyl, 3-C$_6$H$_5$-4-MeOC$_6$H$_4$

S/C = 500~2000, 94.6-98.0% ee

Chen and McCormack (**2007**) [7-16e]

A.4.1.6 Asymmetric Hydrogenation of Alkyl α-Acetamido Acrylates

Ligand:

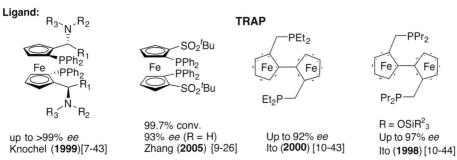

up to >99% ee
Knochel (**1999**)[7-43]

99.7% conv.
93% ee (R = H)
Zhang (**2005**) [9-26]

Up to 92% ee
Ito (**2000**) [10-43]

R = OSiR^2$_3$
Up to 97% ee
Ito (**1998**) [10-44]

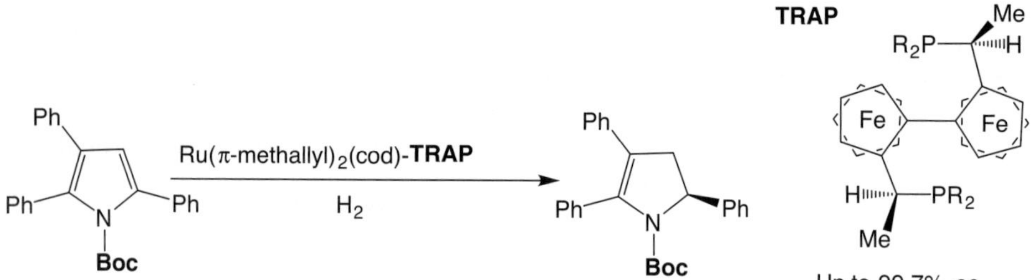

TRAP

Ru(π-methallyl)$_2$(cod)-**TRAP**, H$_2$

Up to 99.7% ee
Kuwano (**2008**) [10-53]

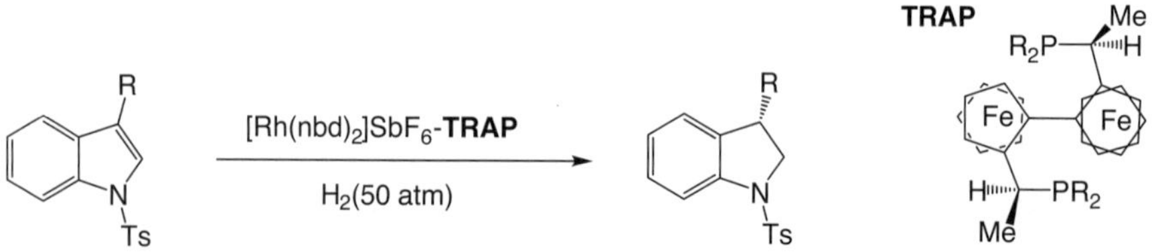

TRAP

[Rh(nbd)$_2$]SbF$_6$-**TRAP**, H$_2$(50 atm)

Up to 98% ee
Kuwano (**2000**) [10-50]
(**2006**) [10-51]

A.4.1.7 Industrial Applications

With Bophoz

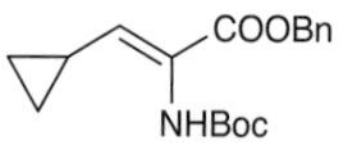

Rh / **BoPhoz**, ee 98%
s/c **2500**; tof ~2000 h^{-1}
feasibility study, Eastman

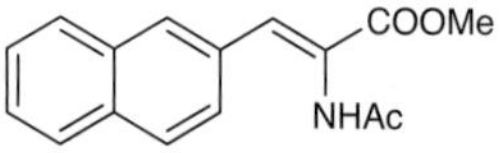

Rh / **BoPhoz**, ee 98%
s/c **2000**; tof n.a.
multi kg scale, Eastman

Ligand: BoPhoz [4-43]

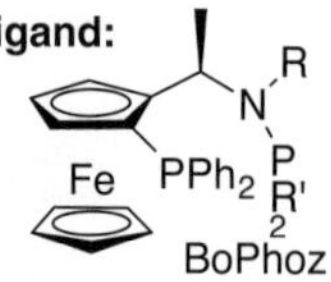

With Josiphos

For **Biotin** Production

Ligand: Josiphos-002 (Fe, Pt-Bu₂, PPh₂)

de 99%
ton **2,000**; tof n.a.
medium scale production
Lonza [4-11a]

Also effective in the hydrogenation of:

Rh / **J002**; ee 97%
ton **1,000**; tof 450 h⁻¹
pilot process, >200 kg
Lonza [4-11a]

Rh / **J002**, ee 94%
ton **350**; tof ~50h⁻¹
multi tons/year
Merck [4-11]

Ru / **J002**; ee 99%
ton **100**, tof ~4 h⁻¹
feasibility study
Merck [4-13]

Rh / **J002**; ee 92-97%
ton **100**, tof ~6 h⁻¹
feasibility study
Merck [4-14]

A.4.2
Asymmetric Hydrogenation of Alkenes (or C=N)

A.4.2.1 Ir-Catalyzed Asymmetric Hydrogenation of Quinolines

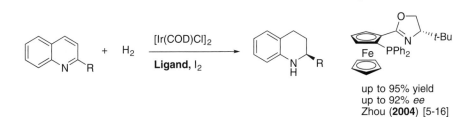

up to 95% yield
up to 92% ee
Zhou (**2004**) [5-16]

A.4.2.2 Industrial Applications

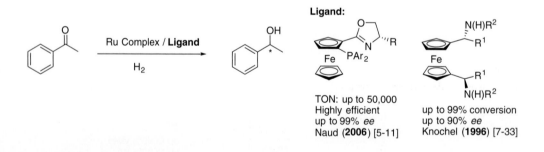

A.4.3 Asymmetric Hydrogenation of Ketones (C=O)

A.4.3.1 Asymmetric Hydrogenation of Acetophenone

A.4.3.2 Asymmetric Hydrogenation of β-Aminoketone

[Rh*]:
[Rh{(R,Sp)-**BPPFOH**}(NBD)]$^+$ClO$_4^-$

BPPFOH
up to 100% conv
up to 95% ee
Hayashi (**1979**)[8-53]

A.4.4 Asymmetric Hydroboration

Ligand:

up to 76% yield for **B**
up to 94% ee
Togni (**1996**) [5-47]

up to 97% yield for **B**
up to 92% ee
Knochel (**2003**) [5-49]

A.4.5 Asymmetric Hydrosilylation Reaction

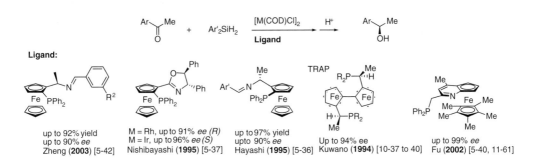

Ligand:

up to 92% yield
up to 90% ee
Zheng (**2003**) [5-42]

M = Rh, up to 91% ee (R)
M = Ir, up to 96% ee (S)
Nishibayashi (**1995**) [5-37]

up to 97% yield
upto 90% ee
Hayashi (**1995**) [5-36]

Up to 94% ee
Kuwano (**1994**) [10-37 to 40]

up to 99% ee
Fu (**2002**) [5-40, 11-61]

A.4.6
Asymmetric Transfer Hydrogenation

Reaction: Aryl methyl ketone + RuCl$_2$(PPh$_3$)$_3$ / **Ligand**, i-PrOH, base, rt → chiral aryl methyl carbinol

Ligand:

Ferrocene-oxazoline-PPh$_2$ ligand (with R substituent)
up to 99% yield
up to 99.7% ee
Uemura (**1999**) [5-22]

Ferrocene-PPh$_2$ with CH(Me)-N=CH-(2-HO-3,5-(NO$_2$)$_2$-C$_6$H$_2$) imine
up to 99% yield
up to 92% ee
Zheng (**2003**) [5-24]

Ferrocene-PCy$_2$/PPh$_2$ (Josiphos-002)
up to 99% Conv.
up to 99% ee
Baratta (**2007**) [4-15]

A.5
Pd-Catalyzed Asymmetric Heck Reaction

a) 2,3-dihydrofuran + PhOTf —[Pd(OAc)$_2$, **Ligand**, i-Pr$_2$NEt]→ 2-Ph-2,5-dihydrofuran (**U**) + 2-Ph-2,3-dihydrofuran (**I**)

U/I up to 99/1
up to 98% conv.
U: up to 98% ee
Dai and Hou (**2003**) [8-77]

b) N-CO$_2$Me-2,5-dihydropyrrole + ArOTf —[Pd(OAc)$_2$, **Ligand**, i-Pr$_2$NEt]→ N-CO$_2$Me-2-Ar-2,5-dihydropyrrole (**U**) + N-CO$_2$Me-2-Ar-2,3-dihydropyrrole (**I**)

Ar = Ph, 2-naphthyl, 4-MeO-C$_6$H$_4$, 4-F-C$_6$H$_4$

Ligand:

Ferrocene-oxazoline(^{t}Bu)-PPh$_2$

Ferrocene-oxazoline-PAr$_2$

U/I: up to 98/2
U: up to 99% ee
Dai and Hou (**2004**) [8-79]

* **U**: Unisomerized isomer, **I**: Isomerized isomer.

A.6
Addition of Organozinc Reagents

A.6.1
Addition of Dialkylzinc to Aldehydes

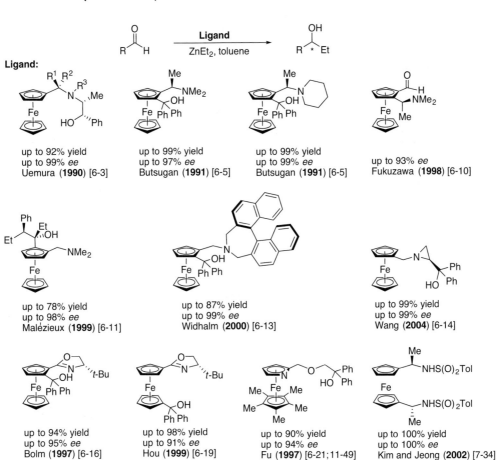

A.6.2
Addition of Diarylzinc to Aldehydes

A.6.3
Addition of Dialkylzinc to C=N Bond

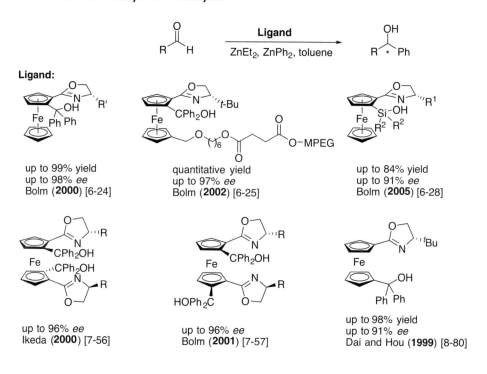

A.6.4
Addition of Phenylacetylene to Aldehydes

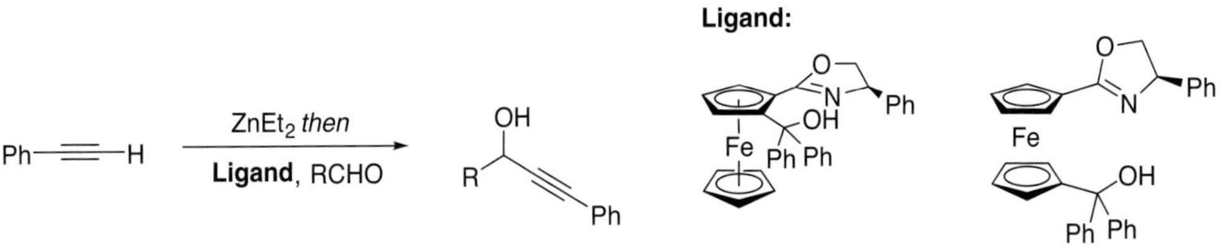

Ph—≡—H →(ZnEt₂ then Ligand, RCHO)→ R-CH(OH)-C≡C-Ph

Ligand:

up to 90% yield
up to 93% ee
Hou (**2004**) [6-29]

up to 88% yield
up to 93% ee
Hou (**2004**) [8-81]

A.6.5
Addition to Aldehydes with Boron Reagents

A.6.5.1 Aryl Transfer with Boronic Ester

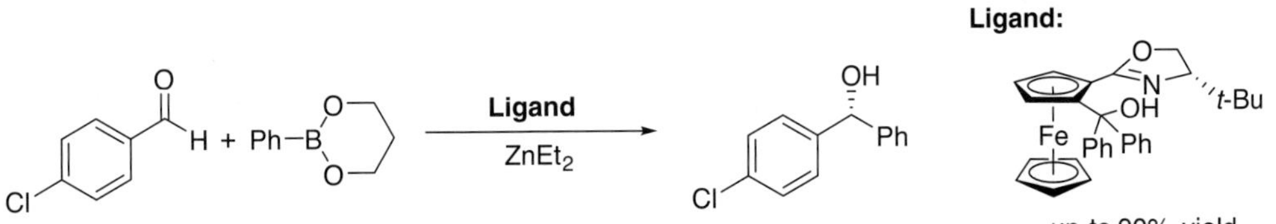

4-Cl-C₆H₄-CHO + Ph–B(OCH₂CH₂CH₂O) →(Ligand, ZnEt₂)→ 4-Cl-C₆H₄-CH(OH)-Ph

Ligand:

up to 90% yield
up to 97% ee
Bolm (**2002**) [6-30]

A.6.5.2 Aryl Transfer with Triphenylborane

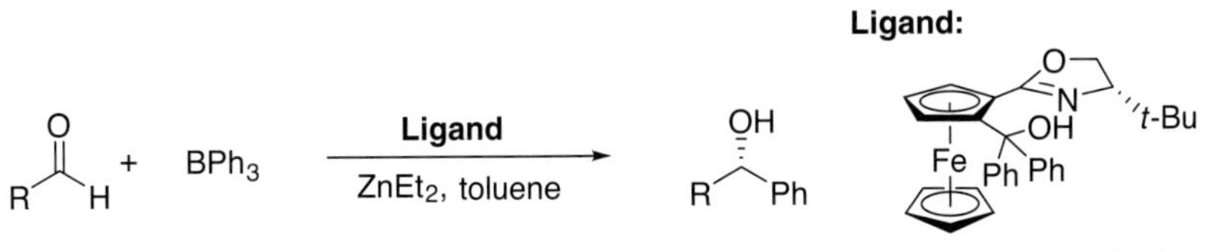

R-CHO + BPh₃ →(Ligand, ZnEt₂, toluene)→ R-CH(OH)-Ph

Ligand:

up to 99% yield
up to 98% ee
Bolm (**2004**) [6-31]

A.6.5.3 Aryl Transfer with Boronic Acids

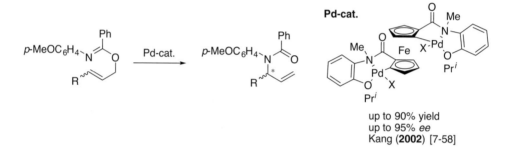

up to 93% yield
up to 98% ee
Bolm (**2002**) [6-30]

A.7
Asymmetric Rearrangement of Allylic Imidates

A.7.1
Rearrangement of N-Arylbenzimidates

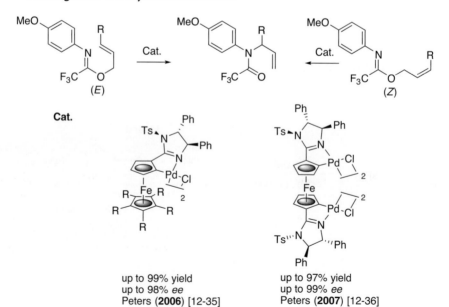

up to 90% yield
up to 95% ee
Kang (**2002**) [7-58]

A.7.2
Rearrangement of N-Aryl-trifluoracetimidate

up to 99% yield
up to 98% ee
Peters (**2006**) [12-35]

up to 97% yield
up to 99% ee
Peters (**2007**) [12-36]

A.7.3
Rearrangement with Added Nucleophiles

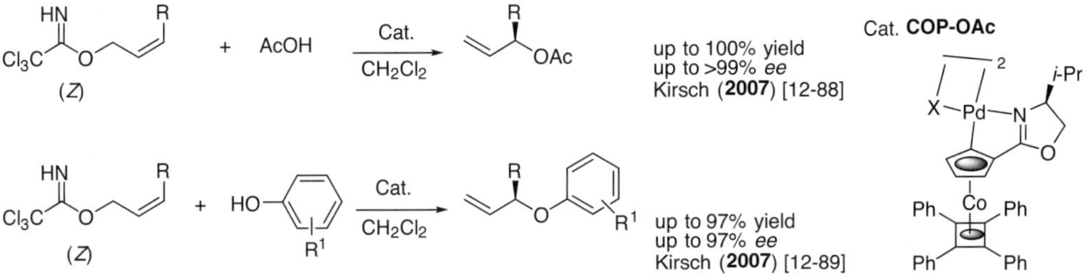

Cl_3C-C(=NH)-O-CH$_2$-CH=CH-R (Z) + AcOH $\xrightarrow{\text{Cat.}, CH_2Cl_2}$ CH$_2$=CH-CH(R)(OAc)

up to 100% yield
up to >99% ee
Kirsch (**2007**) [12-88]

Cat. **COP-OAc**

Cl_3C-C(=NH)-O-CH$_2$-CH=CH-R (Z) + HO-Ar(R^1) $\xrightarrow{\text{Cat.}, CH_2Cl_2}$ CH$_2$=CH-CH(R)(O-Ar-R^1)

up to 97% yield
up to 97% ee
Kirsch (**2007**) [12-89]

A.8
Cu-Catalyzed Cyclopropanation

A.8.1
With Diphenylethene

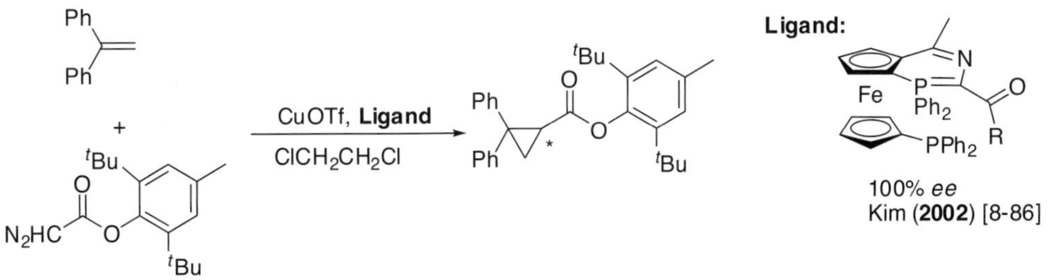

Ph$_2$C=CH$_2$ + N$_2$HC-C(=O)-O-Ar(^{t}Bu)$_2$(Me) $\xrightarrow{\text{CuOTf, Ligand}, ClCH_2CH_2Cl}$ cyclopropane product

Ligand: Fe-based ferrocenyl phosphine-oxazoline (Ph$_2$P, PPh$_2$, R)

100% ee
Kim (**2002**) [8-86]

A.8.2
With Aryl, Alkyl or Silylethene

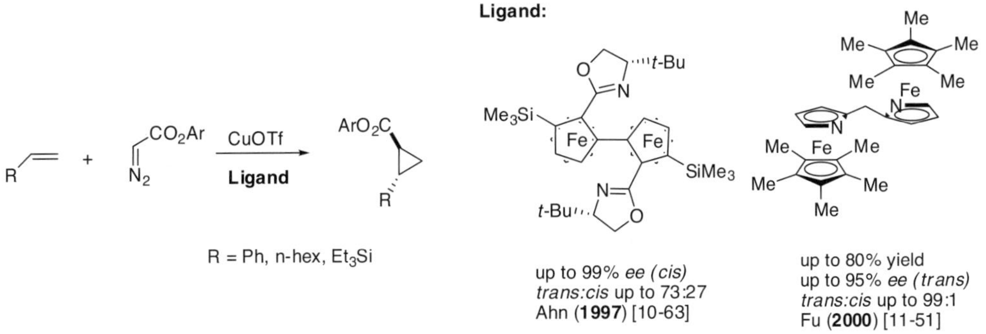

R-CH=CH$_2$ + N$_2$=CH-CO$_2$Ar $\xrightarrow{\text{CuOTf, Ligand}}$ cyclopropane with ArO$_2$C and R

R = Ph, n-hex, Et$_3$Si

Ligand: bis(oxazoline)-ferrocene with t-Bu, Me$_3$Si/SiMe$_3$

up to 99% ee (cis)
trans:cis up to 73:27
Ahn (**1997**) [10-63]

Pentamethylcyclopentadienyl Fe pyridine ligand

up to 80% yield
up to 95% ee (trans)
trans:cis up to 99:1
Fu (**2000**) [11-51]

A.9
Coupling Reaction of Vinyl Bromide and 1-Phenylethylzinc Chloride

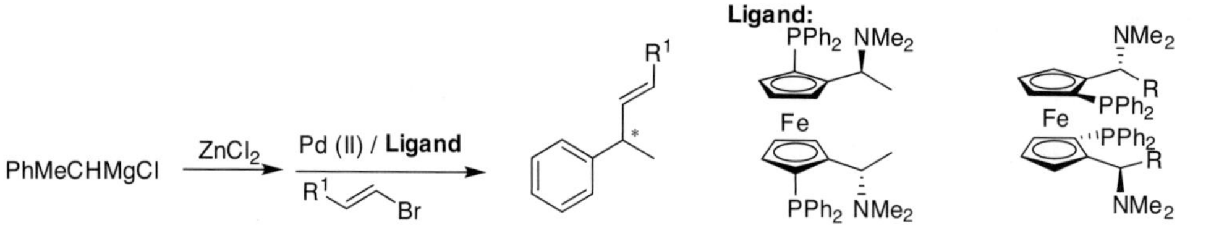

Ligand:
up to 93% ee
Hayashi (1989) [7-39]

up to 93% ee
Knochel (1998) [7-41]

A.10
Enantioselective Intramolecular Aminopalladation

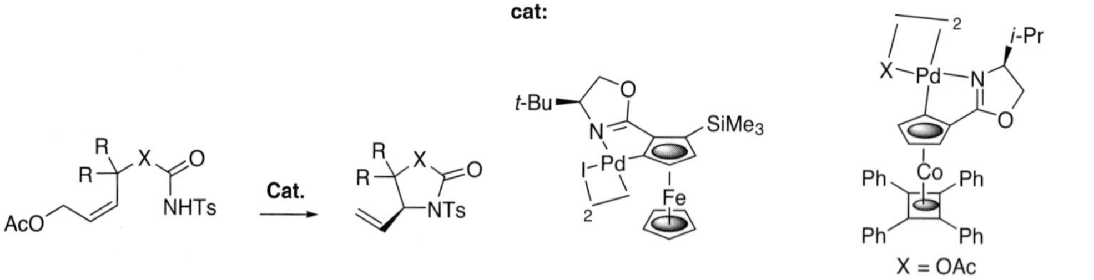

up to 96% yield
up to 96% ee
Overman (2002) [12-90]

up to 89% yield
up to 98% ee
Overman (2005) [12-91]

A.11
Nickel-Catalyzed Asymmetric Three-Component Coupling of Alkynes, Imines, and Organoboron Reagents

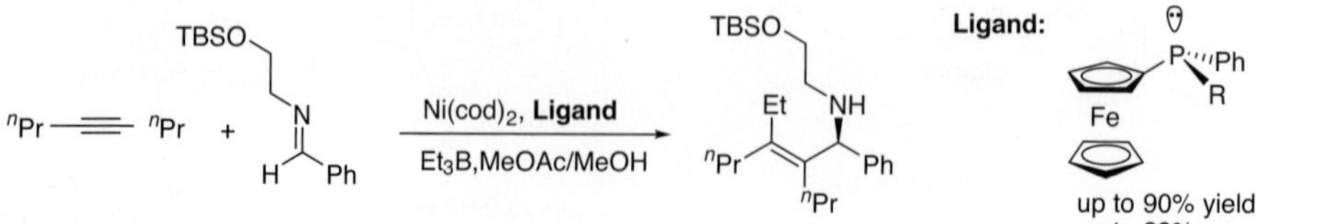

Ligand:

up to 90% yield
up to 89% ee
Jamison (2003) [3-10]

A.12
Reactions with Ketenes

A.12.1
O-Acylation with Ketenes

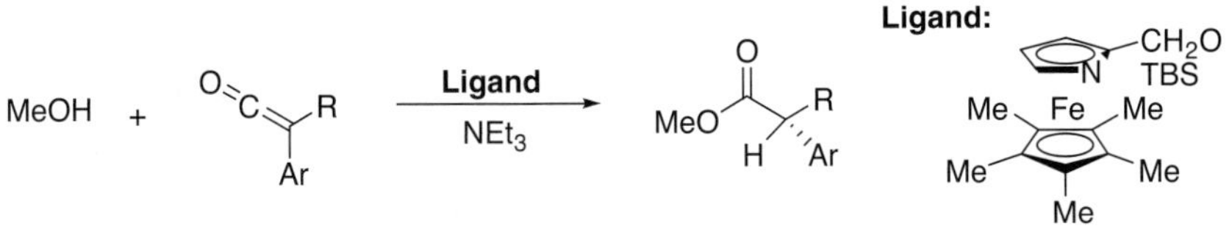

MeOH + ketene (R, Ar) → MeO-C(=O)-CH(R)(Ar)

Ligand: planar chiral DMAP-Fe (pyrrole with CH₂OTBS, Cp*)

up to 97% yield
up to 80% ee
Fu (**1999**) [11-30]

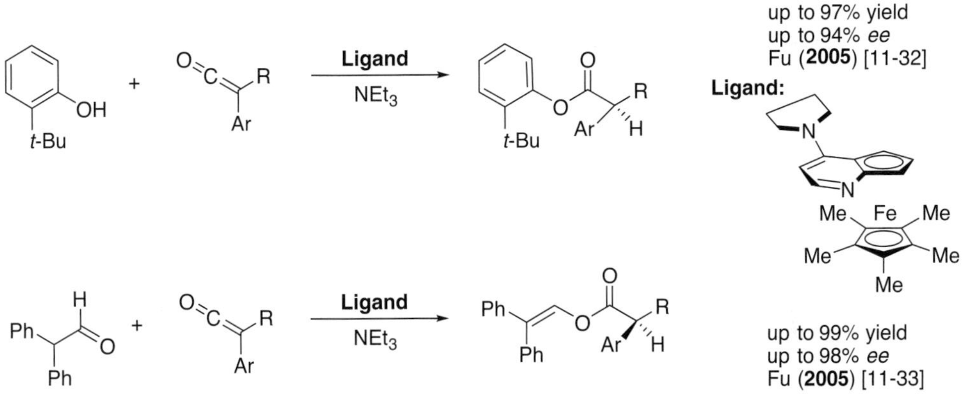

2-*t*-Bu-phenol + ketene → aryl ester

up to 97% yield
up to 94% ee
Fu (**2005**) [11-32]

Ligand: pyrrolidinyl-azaindole Fe-Cp*

Ph₂CH-CHO + ketene → Ph₂C=CH-O-C(=O)-CH(R)(Ar)

up to 99% yield
up to 98% ee
Fu (**2005**) [11-33]

A.12.2
Kinetic Resolution by O-Acylation

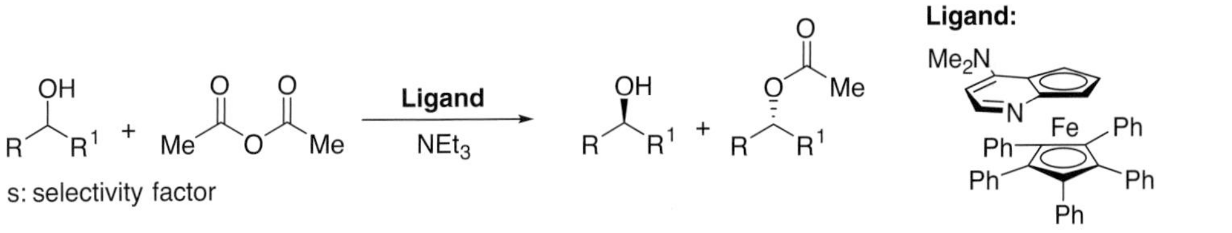

R-CH(OH)-R¹ + Ac₂O → R-CH(OH)-R¹ + R-CH(OAc)-R¹

s: selectivity factor

Ligand: Me₂N-azaindole Fe-(Ph)₄Cp

s up to 95
up to 99% ee @ 55% conv.
Fu (**1997**) [11-23 to 26]

A.12.3
N-Acylation with Ketenes

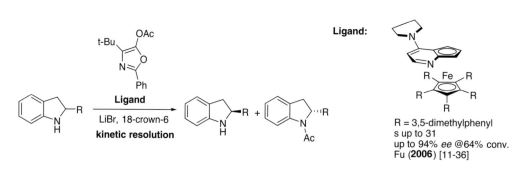

A.12.4
Kinetic Resolution by N-Acylation

A.12.5
C-Acylation via Rearrangement

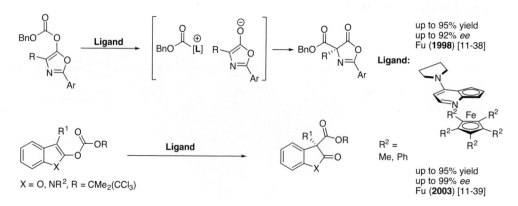

A.12.6
Cycloadditions

up to 99% yield
11:1 > cis: trans > 1:50
cis ee up to 98%
trans ee up to 99%
Fu (**2002**) [11-45]
Fu (**2005**) [11-46]

A.12.7
Copper-Catalyzed Insertions of Diazo Compounds

up to 98% yield
up to 98% ee
Fu (**2006**) [11-54]

A.12.8
Copper-Catalyzed Cycloadditions

up to 92% yield
up to 93% ee, dr > 6:1
Fu (**2007**) [11-56]

A.13
Ring Opening Reaction

A.13.1
Desymmetrization by Halogenations

up to 97% yield
up to 98% ee
Fu (**2001**) [11-43]

A.13.2
Ring Opening of Bicyclic Alkenes

up to >99% ee
Lautens (**2004**) [9-37]

Index

a
acetal   19, 43
– chiral   26
– diastereoselective DoM   31
– diastereoselective *ortho*-lithiation-
  functionalization   27
– planar chiral ligand   28
2-acetamido methyl acrylate   187
α-(acetamido)acrylate   292ff.
– β,β-disubstituted   293f.
– β-heterosubstituted   294
– β-substituted   293
α-(N-acetamido)acrylate   269ff.
(Z)-2-acetamidoacrylate   238
α-acetamidoacrylic acid   235
(Z)-2-acetamidocinnamate   238
α-acetamidocinnamic acid   270
2-(N-acetamino)acrylate   292
acetoacetate   77
1-acetonaphthone   103
acetophenone   106ff., 235, 291
acetylferrocene   291, 344ff.
(S)-N-acetylvaline   352
acrylic acid
– trisubstituted   236
α-acylamino cinnamic acid   178ff.
acylamino cinnamic ester   195
acylation   310ff.
– C   315
– N   313
– O   310ff.
N-acylhydrazone   69, 276
N-acylpyrrole   314
addition
– 1,4   287
– arylzinc   161
– asymmetric, *see* asymmetric addition
– conjugated   130
– copper-catalyzed   67, 130
– enantioselective   190f.
– enantioselective 1,6-addition   80
– nickel-catalyzed   66
– nucleophilic   299, 319
– organoaluminum to aldehydes and enones   65
– organometallic reagent to C=X bond   129
– organozinc   163, 190f.
– organozinc to aldehyde   149
– oxidative   348ff.
– titanium-catalyzed   190
– triethylaluminum to cyclohexenone   67
– trimethylaluminum to benzaldehyde   66
alcohol
– acylation   310ff.
– allylic   328f.
– homoallylic   311
– kinetic resolution   311
– secondary   139, 311
aldehydes
– addition of alkynylzinc   246
– addition of organoaluminum   65
– addition of organozinc   149ff.
– addition of zinc   244
– aromatic   190, 277
– aryl transfer   168
– asymmetric 1,2-addition   270
– asymmetric Aldol addition   269
– asymmetric alkylation   153f.
– asymmetric ethylation   155
– boron reagent   165
– coupling with ketene   313
– cyanosilylation   310
– enantiopure   348
– enantioselective alkynylation   166
– nucleophilic addition   299, 319
– phenyl addition   165

- phenylacetylene addition with ferrocene 170
- nickel-catalyzed asymmetric reductive coupling 56ff.
aldimine 131
- P 116
aldol condensation 91
aldol reaction, see asymmetric aldol reaction
aliskiren 180
alkene 98
- azabicyclic 267
- bicyclic 278
- heterobicyclic 267
alkenyl acetate
- cyclic 126
alkenyl transfer reaction 168
alkenylboronic acid 168f.
α-alkoxy-ketone 103
alkoxycyclization
- asymmetric 269
alkyl α-acetamido acrylate 194
alkyl aryl sulfoxide
- chiral 69, 276
alkyl chloride
- tertiary 317
5-alkyl-4-methoxycarbonyl-2-oxazoline 229
- optically active 229
alkylation
- asymmetric, see asymmetric alkylation
- enantioselective 160
- indole 263
alkylative ring opening
- asymmetric 267
alkylidene malonate 273
3-alkylphthalide 153
4-alkynal 90
alkynes
- nickel-catalyzed asymmetric reductive coupling 56ff.
- nucleophilic addition 299
alkynyl transfer
- asymmetric 166
alkynylation
- enantioselective 166
allene 223
allenylsilane 326
allyl acetate 226, 266
allyl alcohol 169
allyl carbonate 222f.
π-allyl palladium complex 223, 259ff., 287ff.
- M-type 259
- W-type 259ff.
allyl substrate
- chlorophenyl substituted 126

allyl trichlorosilane 276
allyl trifluoroborate 80
allylation
- asymmetric, see asymmetric allylation
- palladium-catalyzed 91, 222, 265
allylic acetate
- unsymmetrical 224
- (Z) 363
allylic alcohol 328f., 356
allylic alkylation
- asymmetric, see asymmetric allylic alkylation
- intramolecular 221
- iridium-catalyzed 126
- model substrate 75
- palladium-catalyzed 126, 217ff., 278
- regio- and enantioselective 226
- rhodium catalyzed 228
allylic amination
- enantioselective 224
- palladium-catalyzed 217ff.
allylic amine 356
allylic aryl ether 363
allylic N-arylbenzimidate 358ff.
allylic imidate rearrangement 356ff.
allylic oxidation
- asymmetric, see asymmetric allylic oxidation
allylic substitution 18, 201, 258ff.
- asymmetric, see asymmetric allylic substitution
- regio- and enantioselective 225ff.
(allyloxy) iminodiazapholidine 361
amide
- acyclic 227
- α-arylation 241f.
- γ,δ-unsaturated 227
amination
- intramolecular 249
- asymmetric reductive iridium-catalyzed 188
amine
- homoallylic 69
- kinetic resolution 313
- primary 313
- secondary 188
α-amino acetophenone 78
α-amino acid 292f.
- derivative 321
amino alcohol 158ff.
α-amino aldehyde 234
- optically active 234
2-amino-1-arylethanol 236
α-amino-β-keto ester 78
(S)-1-amino-2-methoxymethylpyrrolidine, see SAMP

(aminoalkyl)ferrocenylphosphine-gold(I) complex 233
aminomethyl aryl ketone 236
aminopalladation
– intramolecular 363
aminophosphine
– chiral 302
– ferrocene-based chiral 137
ancistroealanine A 129
anisaldehyde 164
AnisTRAP 290
antileishmanial alkaloid 129
π-arene chromium tricarbonyl 5
arm effect 217
aromatic electrophilic substitution 6
aryl alkyl ketone 314ff.
aryl alkyl secondary alcohol 69
α-aryl amino acid derivative 321
aryl azide 227
aryl bromide 80, 331
aryl ferrocenyl phosphine (aryl-MOPF) 65
aryl ferrocenyl sulfoxides 64
aryl glycine 321
aryl halide 266, 332
aryl ketone 79, 101
aryl methyl ketene 312
aryl transfer 165ff.
– boronic acid 171
– triphenylborane 166
aryl triflate 266
1-aryl-1,3-butadiene 111
3-aryl-2-ethoxyacrylic acid 180
arylaldehyde 192
arylation 139
– α-arylation 241f.
– asymmetric 164, 244
– palladium catalyzed 139
N-arylbenzimidate
– allylic 358ff.
– rearrangement 358
N-arylimine 77, 188
arylzinc addition 161
asymmetric addition 155, 314ff.
– diethylzinc 155ff., 244ff.
asymmetric aldol reaction 229ff., 288f.
– gold catalysis 229ff., 269
– silver catalysis 232
asymmetric alkylation 153ff., 192
– aldehyde 153f.
asymmetric alkylative ring opening 267
asymmetric alkynyl transfer 166
asymmetric allylation 69, 276
– palladium–rhodium dual catalysis 290
asymmetric allylic alkylation 199

– copper(I)-catalyzed 62
– palladium-catalyzed 196, 206, 326ff.
asymmetric allylic oxidation 138
– copper-catalyzed 138
asymmetric allylic substitution 199ff.
– application 228
– copper-catalyzed 64, 265
– palladium-catalyzed 121ff., 178, 197, 215ff., 258ff., 301
– substrate variant 221
asymmetric catalysis 307ff.
– scalemic planar chiral palladocyclic complexes 355f.
asymmetric cross-coupling 194ff., 239ff.
– nickel-catalyzed 239
– palladium-catalyzed 240ff.
asymmetric cyclopropanation 136, 248
– copper-catalyzed 302
– ruthenium catalyzed 136
asymmetric epoxidation 168ff., 277
– allylic alcohol 168
– vanadium-catalyzed 168ff.
asymmetric Heck reaction 126
– intermolecular 243, 267
– intramolecular 127, 242ff.
– palladium-catalyzed 127, 246, 266
asymmetric hydrogenation 97ff., 180ff., 194ff., 234ff.
– FerroTANE 184
– functionalized/unfunctionalized alkene 100
– Hectorite-supported 237
– heteroaromatics 103, 296ff.
– imine 104f.
– iridium-catalyzed 98ff., 188, 203
– olefins 292ff.
– rhodium-catalyzed 99, 178ff., 202, 269, 295
– ruthenium-catalyzed 297f.
asymmetric hydrosilylation (AHS) 109ff., 291f.
– carbon–carbon bond 270
– carbon–heteroatom bond 270
– palladium-catalyzed 110
– rhodium-catalyzed 113, 234f., 270, 291, 324
asymmetric ring-closing metathesis (ARCM) 42
– kinetic resolution of ferrocenylalkene 42
asymmetric transfer hydrogenation 106ff.
– ligand screening 108
– ruthenium-catalyzed 106ff.
aza Diels–Alder reaction 278
– asymmetric 273f.
aza-Baylis–Hillman reaction 139

aza-heterocyclic ferrocene
– planar chiral   149
aza-β-lactam   319
azabenzonorbornadiene   267
azaferrocene   160f., 307ff.
– catalysis   310
azlactone   68
– O-acylated   68, 314f.
azomethine imine   331
azomethine ylide   87, 132ff.
– asymmetric cycloaddition   132ff., 275
– 1,3-dipolar cycloaddition   87, 132, 275f.
– [3+2] cycloaddition   134

b
balanol   234
benzaldehyde   66, 152ff., 230, 319
– addition of diethylzinc with ferrocene   170
– epoxidation   277
benzimidazole P,S-ferrocene ligand   263
benzofuranone
– O-acylated   315
benzylamine   224ff.
N-benzyloxycarbonyl amine   321
bi-aryl ligand   4
BIFEP (2,2''-bis(diphenylphosphino)-1,1''-biferrocene)   283f., 299
– asymmetric synthesis   300
– P-chiral   85, 301
– synthesis   300f.
biferrocene ligand   283ff.
– chiral   302
– planar chirality   302
biferrocenoazepine   302f.
biferrocenyl bisphosphine   25
biferrocenyl diol   28
biferrocenylbisoxazoline   30
BINAP (2,2'-bis(diphenylphosphino)-1,1'-binaphthyl)   3, 283f.
– (R)   242
– (R)-(+)   354
– (S)-(–)   354
– C_2-symmetrical diphosphine ligand   175
(S,S)-f-binaphane   188
1,1'-binaphthalene derivative   128f.
binaphthalene skeleton   3
binaphthyl azepine   22
1,1-binaphthyl-phosphite   65
binaphthyl/phosphole   326
BINOL (binaphthol)   3, 224
– (S)   225
binuclear zinc catalytic mechanism   192
1,1-biphenyl-derived phosphite   66

bis(azaferrocene)   322ff.
– C_2 symmetric   319
bis(cyclohexyl isocynide) gold(I) tetrafluoroborate   230
1,8-bis(dimethylamino)naphthalene   359f.
(αR,α'R)- 2,2'-bis(α-N,N-dimethylaminopropyl)-(S,S)-1,1'-bis(diphenylphosphino)ferrocene   209
(–)-(S)-(S)-1,1'-bis(diphenylphosphino)-2,2'-bis(methoxycarbonyl)ferrocene   210
1,1'-bis(1-hydroxyalkyl)ferrocene   180
1,1'-bis[(S)-4-isopropyloxazolin-2-yl] ruthenocene   210
1,1'-bis(methylphenylphosphino)ferrocene   177
bis(oxazoline)
– ferrocene based   302
1,1'-bis(oxazolinyl)-1,1'-bis(diphenylphosphino)ferrocene   217
1,1'-bis(oxazolinyl)ferrocene   194ff.
bis-palladacycle   357
bis-palladation   344
bis(phosphaferrocene)   326
1,1'-bis(phosphino)ferrocene   186
bis(phosphino)ferrocenyloxazoline   220
bis(phospholane) ligand   183
bis-sulfone   269
bis-sulfone phosphine ligand   264
bis(sulfonyl)phosphino ferrocene ligand   265ff.
bis(tetrahydroindenyl)iron   202
1,1'-bis(tributylstannyl)ferrocene   217ff.
1,1'-bisoxazolines   30
1,1'-bisphosphanoferrocene   188
3,5-bistrifluoromethylacetophenone (BTMA)   102
BMPD   28
N-Boc-indole   90, 297
BoPhoz   21, 83ff.
boron reagent   165
boronic acid   171
boronic ester   165ff.
boronic reagent   241
BPE   183ff.
(R,R)-BPNF   178
BPPFA (N,N-dimethyl-1-[1',2-bis(diphenylphosphino)ferrocenyl] ethylamine)   21, 130, 215ff., 235, 248ff.
– (R,S_p)   216ff., 239ff.
– (S,R_p)   221ff., 234ff.
BPPFOAc
– (R,S_p)   230
– (S,R_p)   241
– (S,S_p)   223

BPPFOH (1-[1′,2-bis(diphenylphosphino)
  ferrocenyl]ethanol)   21
– (R,S$_p$)   235ff.
– rhodium complex   236
1,1′-bromo-ferrocenyloxazoline   245
1′-bromo-1-oxazolinyl ferrocene   220
β-bromostyrene   205
(E)-1-(2-bromovinyl)-4-methoxybenzene   241
Brønsted acid-base mechanism   314
2-butenyl acetate   224
2-butenylene dicarbamate   221
2-butenylene dicarbonate   221
BuTRAP   291
– (R,R)-(S$_p$,S$_p$)-i-BuTRAP   293ff.
tert-butyl thioether   263
(tert-butyldimethylsilyloxy) ethyl protecting
  group   61
tert-butyloxazoline diphosphine ligand   244
tert-butylsulfenylferrocene   277
(R,R$_p$)-1-(tert-butylsulfinyl)-2-
  (diphenylphosphino)-ferrocene
– synthesis   45f.
(R)-tert-butylsulfinylferrocene
– synthesis   45

c

C–C bond
– formation   121ff., 215
C=C bond
– asymmetric hydrosilylation   110
C=C hydrogenation   75
– model substrate   75
C–H activation
– diastereotopic   342
C=N bond
– asymmetric hydrosilylation   117
C=O bond
– asymmetric hydrosilylation   112
C=O hydrogenation   75
– model substrate   75
Cahn, Ingold, and Prelog (CIP) rule   15f.
Cahn–Ingold–Prelog (CIP) system   338
Candida antarctica   41ff.
Candida cylindracea lipase (CCL)   40f.
carbene
– N-heterocyclic   332
carbon–heteroatom bond formation   215
carbopalladation   249
carboxylic acid   99
catalysis
– asymmetric, see asymmetric catalysis
– nucleophilic   310
chalcone   130
chiral auxiliary   31

– diastereoselective directed ortho-metalation
  of ferrocenes   19
chiral base   36
chirality
– axial   218ff., 301
– central   159, 218ff., 230, 260
– planar   4, 15f., 40, 125, 159, 218f., 260f., 338
CHIRAPHOS   283f.
β-chloro alcohol   317
α-chloro ester   317
p-chlorobenzaldehyde   164ff.
chlorodicyclohexylphosphine   65
chlorohydrin   325
chlorotrimethylsilane   302
(Z)-cinnamic acid   90
cinnamyl acetate   224ff., 266
cinnamyl chloride   63
circular dichroism (CD)   338
cobalt catalysis   249
cobalt oxazoline palladacycle (COP)   360ff.
[2+2+2] cocyclization   249
complex induced proximate effect (CIPE)   5
complexation
– metal   197
conjugate addition
– asymmetric 247, 272f., 328f.
– rhodium-catalyzed   247
COP, see cobalt oxazoline palladacycle
copper-catalysis
– addition   66f.
– asymmetric allylic alkylation   62
– asymmetric allylic substitution   64, 265f.
– asymmetric conjugate addition   273, 328f.
– asymmetric intramolecular Kinugasa
  reaction   330
– asymmetric Kinugasa reaction   322
– conjugated addition   130
– Cu(I) complex   271ff.
– cycloaddition   322ff.
– [3+2] cycloaddition   134
– [4+1] cycloaddition   322
– cyclopropanation   248f., 302, 319
– enantioselective insertion   321
– FcPHOX   135
– Fesulphos   271ff.
– Josiphos catalyst   79
– nucleophilic addition   299
– reduction of activated C=C bond   78
– Taniaphos complex   86
coupling   80, 128
– enantioselective   315ff.
– palladium-catalyzed   80
– nickel-catalyzed asymmetric reductive
  55ff.

Cp ring
– ferrocene and ruthenocene   204
Cp–aryl ligand exchange   36
cross-coupling   126ff.
– asymmetric, see asymmetric cross-coupling
– palladium-catalyzed   331
crotonaldehyde   287
cumene hydroperoxide (CHP)   169
N-cumyl-N-ethylferrocenecarboxamide   36ff.
N-cumylcarboxamides   43
2-cyanocarboxylate   287
cyanoester   91
– α-   288
2-cyanopropionate   91, 287ff.
2-cyanopyrrole   314
cyanosilylation   310
cyclization   221
cycloaddition   317
– 1,3-dipolar   87, 132ff., 275
– [2+2]   317ff.
– [3+2]   132ff., 274
– [4+1]   322
– [4+2]   90, 272
– asymmetric   132, 272ff., 275
– asymmetric [4+1]   322
– copper-catalyzed   133f., 322ff.
cycloalkenyl acetate   223
tert-cyclobutanol   138f.
cyclohexanone
– α,α-disubstituted   227
cyclohexene derivative   67
cyclohexenone   67
2-cyclohexenyl acetate   26
cycloisomerization
– asymmetric   298
– ene-type   298
– 1,6-enyne   91
– palladium-catalyzed   298
1,5-cyclooctadiene   138
cyclopalladation   343
cyclopentadiene   111
cyclopentadienyl   308
cyclopropanation
– asymmetric   136, 248, 302
– copper-catalyzed   248f., 302, 319
– ruthenium-catalyzed   136, 248
cyclosporine   232
cyclotelluride   350
cyrhetrenyl-phox   115

d

N-decumylation   36
dehydroamino acid   180, 197
– α   200

– derivative   88, 99, 292
dehydroamino ester   197
desymmetrization   267
(DHQ)$_2$PYR   42
(DHQD)$_2$PYR   42
2,6-di-tert-butyl-4-methylphenyl (BHT)   320
1,1′-di(1-dimethylaminoethyl)ferrocene   342
dialkyl zinc   150, 273
diamine   350
α,β-diamino acid   231
– threo   294
diamino FerriPHOS   194f.
diaminoferrocene   189
1,3-diarylprop-2-enyl acetate   221
diastereoselectivity
– in DoM-substitution reaction   19
– ortho-functionalization of ferrocenes   18
diazo compound
– copper-catalyzed insertion   320
diazo ester   248, 322
– α-   321
diazoacetate   320
1,1′-dibromoferrocene   30, 217
dichlorobis(acetonitrile)palladium   195
dicyclohexylphosphino unit   267
Diels-Alder reaction
– asymmetric   136, 273
– iridium catalyzed   137
– rhodium catalyzed   137
diene
– 1,4   299
– 1,6   299
– Danishefsky-type   273f.
dienol
– chiral conjugated   58
dienyl acetate
– conjugated   226
diethylphosphonate   288
diethylzinc   149ff., 190f., 244ff., 319
– addition to benzaldehyde with ferrocene   170
– enantioselective addition to aldehyde   271
diferrocenyl dichalcogenide ligand   270
2,3-dihydro-4-(1H)-dihydropyridone   273
2,3-dihydrofuran   127, 266f., 322
dihydronaphthalene   79
cis-dihydronaphthol   267
dihydroxylation (AD)
– asymmetric   41
1,1′-dihydroxymethyl- 2,2′-diphenylphosphino ferrocene   199

(S_p)-N,N-diisopropyl-2-(diphenylphosphino)
  ferrocenecarboxamide 48
N,N-diisopropyl ferrocenecarboxamide 36
β-diketone 222
dilithiation 201
– o,o′ 194
1,1′-dilithioferrocene 7, 178
dimethoxy polyethylene glycol
  (DiMPEG) 163
1-(2-N,N-dimethyl amino methyl) ferrocenyl
  alcohol 155
dimethyl fumarate 275
dimethyl itaconate 187
dimethyl maleate 275
dimethyl malonate 221, 258ff., 326
(R,R)-N,N′-dimethyl-N,N′-di-(3,3-
  dimethylbutyl) cyclohexanediamine 38f.
2,2-dimethyl-2,3-dihydrofuran 127
N,N-dimethyl-1-(2-diphenylphosphino-
  ferrocenyl)ethylamine (PPFA) 97
– chiral 99
– (R,R_p)-PPFA 127
– (R,S_p)-PPFA 19, 44f., 110, 127f.
– (S,R_p)-PPFA 127
– (S_p)-PPFA 127
– synthesis 44f.
(S)-N,N-dimethyl-1-[(R)-2-
  (diphenylphosphino)ferrocenyl]
  ethylamine 141
N,N-dimethyl-1-ferrocenylethylamine 342
4,4-dimethyl-1-phenylpenta-1,2-diene 241
(dimethylaminomethyl)ferrocene 39, 352
– (R,R)-TMCDA-mediated enantioselective
  ortho-lithiation 39
2-(1-dimethylaminomethyl)-ferrocene
  carbaldehyde 155
N,N-dimethylaminopyridine, see DMAP
4-dimethylaminostyrene 111
N,N-dimethylbenzylamine 341
N,N′-dimethylcyclohexyldiamine 182
N,N-dimethylethylaminoferrocene (N,N-
  dimethylferrocenylethylamine, Ugi's
  amine) 6, 15ff., 42, 99, 342f.
cis-2,3-dimethylindoline 297
dineopentylzinc 63
DIOP (2,3-O-isopropylidene-2,3-dihydroxy-1,4-
  bis(diphenylphosphino)butane) 3, 180,
  283
– C_2-symmetrical 175
DIOP* 3
dioxolane
– chiral 26
DIPAMP (1,2-ethanediylbis(o-methoxyphenyl)
  phenylphosphine)

– C_2-symmetric P-chiral ligand 177
– C_2-symmetrical chelating diphosphine
  ligand 175
DiPFc 183ff.
N,N-diphenyl amide 227
(S)-1,3-diphenyl-1-butene 194
1,3-diphenyl-2-propenyl acetate 207, 258
diphenylacetylene 355
1,3-diphenylallyl acetate 326
(S,R_p)-2-(α-diphenylhydroxymethyl)ferrocenyl-
  5-tert-butyloxazoline 172
diphenylphosphine 182, 207
– trisubstituted 36
diphenylphosphino group 7
– diastereoselective introduction 23
trans-2-(diphenylphosphino)
  cyclohexanol 354
1′-(diphenylphosphino)-1-
  ferrocenecarboxaldehyde 220
(S,S_p)-2-[2-(diphenylphosphino)ferrocenyl]-4-
  (1-methylethyl)oxazoline 47, 142
diphenylphosphino naphthalene 283
diphenylphosphino sulfinylferrocene 220
N-diphenylphosphinoyl aromatic
  aldimine 130
1,3-diphenylprop-2-en-1-yl acetate 178
1,3-diphenylpropenyl acetate 121
1,1′-diphosphaferrocene 309
1,1′-diphosphetanylferrocene ligand 183
diphosphine
– achiral 202
– 1,2-diphosphine 25
– ferrocene-derived 202
diphosphine ligand
– C-centered chiral 180
– P-centered chiral 177
– chiral 199
– ferrocene-based C_2-symmetric 180
– rhutenocene-based 207
– C_2-symmetrical 182, 197
diphosphine-oxazoline ferrocene ligand
– planar chiral 243
diphosphinoferrocene-borane complex 182
directed metalation group (DMG) 5, 16ff.
– displacement 43
– α-methoxybenzyl 33
– P-chiral 43
directed ortho-metalation (DoM) 5, 15f.
– diastereoselective 17ff., 33ff.
– diastereoselectivity 19
– enantioselective 17, 36
DMAP (N,N-dimethylaminopyridine) 9, 66,
  310ff.
– ferrocene-based planar chiral 67

L-DOPA 175
DuPhos 183ff.

e

Ellman's reagent t-Bu-S(O)-St-Bu 24
enamide 77
– cyclic 295ff.
enamine 196
enantio-induction 218
enol ester
– α-stereocenter 313
enolate
– enantioface-selection 288
enone 322ff.
– addition of organoaluminum 65
– cyclic 130
enzymatic kinetic resolution 16f., 40
enyne
– conjugated 326
– nickel-catalyzed asymmetric reductive coupling 59f.
1,3-enyne 58
1,6-enyne 269
– chiral 58
– cycloisomerization 91
– intramolecular ene reaction 298
epoxidation 277
– asymmetric, see asymmetric epoxidation
ester amide
– ferrocene-based 209
esterification
– lipase-catalyzed 40
(R,S_p)-Et-BPPFA 242
ethyl 2-(benzenesulfonylmethyl)-2-propenyl carbonate 221
ethyl isocyanoacetate 231
ethyl transfer 152
ethylation 158
– asymmetric 155f.
– enantioselective 156
ethylene polymerization 323
(R,R)-(S_p,S_p)-EtTRAP 293
(S_p,S_p)-EtTRAP-H 286ff.

f

FcPHOX (ferrocenyloxazolinylphosphine) 121ff., 138
– asymmetric allylic substitution 125
– asymmetric hydrogenation 102
– aza-Baylis–Hillman reaction 139
– iridium complex 100
– planar chirality 125
– synthesis 46

FcSAMP hydrazone
– diastereoselective DoM 31
FERRIPHOS 176, 194ff.
Ferro-TANE 176ff.
– asymmetric hydrogenation 184
– synthesis 183
ferrocene 1ff.
– N,O-bidentate 168
– chiral ligand 9
– chiral monodentate 55
– Cp ring 204
– diastereoselective directed ortho-metalation 19
– diastereoselectivity in ortho-functionalization 18
– 1,1′-disubstituted 159f.
– 1,2-disubstituted 25, 36, 202
– electrophilic substitution mechanism 340
– enantioselective DoM 36
– MeO-PEG-supported 163
– norephedrine-derived 152
– 1,2-P,N 121ff.
– planar chiral 15ff., 155
– polymer supported 122, 163
– pyrrolidinyl-containing 157
– scalemic planar chiral 350
– structure 308
– C_1-symmetrical 202
– Ugi–Kagan type 26, 43
ferrocene carboxaldehyde 344ff.
– planar chiral 1,2-disubstituted 27
– 1,2-disubstituted 28
ferrocene carboxylic acid 29
ferrocene derivative
– planar chiral 322, 354
– C_2-symmetrical 1,1′,2,2′-tetrasubstituted 207
ferrocene dicarbaldehyde
– 1,2,1′-trisubstituted planar chiral 27
– 1,2,1′,2′-tetrasubstituted planar chiral 27
ferrocene diol
– chiral C_2-symmetrical 202
ferrocene ligand 98
– central and planar chirality 260ff.
– N,S 259f.
– N,S bidentate 259f.
– 1,2-P,N 98, 121ff.
– P,N 135, 268
– P,N-ligand containing N-heteroaromatic ring 121
– phosphinamine type 125
– P,S 261ff.
– P,S bidentate 260
– pyridine-containing 105

- 1,2-S,P 124
- S,S 265
- thio- and seleno-substituted oxazoline-containing 258
NS-ferrocene oxazoline-Zn catalysis 270
ferrocene phosphine oxazoline ligand 266
ferrocene scaffold
- bulkiness 9
- derivatization 6
- rigidity 9
- stability 9
ferrocenecarbaldehyde 155
1,1'-ferrocenedicarbonyl dichloride 194
N-ferrocenoyl aziridinyl-2-carboxylic ester 160
N-ferrocenoyl-2-[(diphenylphosphino)methyl]-pyrrolidine 130
ferrocenyl aldehyde
- 1,2-disubstituted 40
ferrocenyl alkene 42
ferrocenyl alkylamine
- Boc-protected 8
ferrocenyl amine 63
ferrocenyl amino alcohol 150ff.
- 1,2-disubstituted 153
- planar and axial chirality 157
ferrocenyl aminophosphine ligand 23
ferrocenyl aziridino alcohol 157
- C_2-symmetrical 1,1'-disubstituted 191
ferrocenyl binaphthyl azepine 23
ferrocenyl bis-palladacycle 344ff., 359
ferrocenyl boronic acid 26, 43
ferrocenyl carbaldehyde 8
ferrocenyl carbinol
- 1,2-disubstituted 40
α-ferrocenyl carbinol 33
- diastereoselective DoM 33
α-ferrocenyl carbocation effect 32
ferrocenyl carboxylic acid 6f.
ferrocenyl diamide
- tetrasubstituted planar chiral 200
1,1'-ferrocenyl dicarboxaldehyde 180ff.
(R,R)-ferrocenyl diol 208
1,2-ferrocenyl diphosphine ligand
- bidentate 73ff.
- classification 74
- both PR_2 groups attached to the Cp ring 74
- both PR_2 groups attached to side chains 87
- one PR_2 group attached to the Cp ring, one PR_2 group attached to the α-position of the side chain 75
- one PR_2 group attached to the Cp ring, one PR_2 group attached to the β-position of the side chain 83
- one PR_2 group attached to the Cp ring, one PR_2 group attached to other positions of the side chain 85
ferrocenyl diphosphine ligand 199
- tetrasubstituted 193
ferrocenyl diphosphonite ligand 186
ferrocenyl DMAP 25
ferrocenyl DMG 43
ferrocenyl imidazoline 34
- diastereoselective ortho-lithiation 34
ferrocenyl imidazoline palladacycle (FIP) 359
ferrocenyl ketones 62
ferrocenyl ligand 21
- anionic bidentate 338
- boron-containing 237
- complexation 338
- monodentate chiral 55
- multidentate chiral 30
- 1,1'-N,O 244
- 1,2-N,O 244
- N,O-bidentate 149ff.
- N,P 104ff.
- N,P-bidentate 138
- P,N-bidentate 118
- 1,2-P,N-bidentate 97
- planar chiral 31, 261
- planar chirality 4, 198
- rhodium complex 237
- selenium-sontaining 257ff.
- sulfur-containing 257ff.
- symmetrical 1,1'-bidentate 175ff.
- symmetrical 1,1'-disubstituted 177
- C_2-symmetrically chiral 193
- tetrasubstituted 198ff.
- 1,1',2,2'-tetrasubstituted 193
- unsymmetrical 1,1'-bidentate 215ff.
α-ferrocenyl lithium 32
ferrocenyl methylamines
- chiral α-substituted 20
N-ferrocenyl methylephedrine 151
ferrocenyl oxazoline 29, 343
- diastereoselective ortho-lithiation 29f., 286
- macrocyclic 30
- planar chiral ferrocene ligand 31
- synthesis 142
ferrocenyl oxazoline palladacycle (FOP) 358ff.
- planar chiral 358
1-ferrocenyl oxazoline-1'-phosphine 18
ferrocenyl oxazolinyl alcohol 158ff.
- planar chiral 161
ferrocenyl oxazolinyl hydroxamic acid 168
ferrocenyl phosphine
- chiral 235

– monodentate 56ff.
– P-chiral, monodentate 56f.
– planar chiral monodentate 64
– pyrazole containing 122
ferrocenyl phosphine amine ligand 119
ferrocenyl phosphine oxide
– diastereoselective DoM 35
ferrocenyl phosphine-imine ligand 112
– chiral 112
ferrocenyl phosphine-rhodium complex 236
ferrocenyl phosphonite ligand 65f.
ferrocenyl silanol 167ff.
ferrocenyl sulfide 24
ferrocenyl sulfoxide 23ff., 69
– diastereoselective ortho-lithiation 24
– 1,2-disubstituted planar chiral ligand 25
– planar chiral ligand 26
– (S) 104
ferrocenyl sulfoximines 34
– diastereoselective ortho-lithiation 35
ferrocenyl sulfur ylide 277
ferrocenyl thiolate ligand 266
(R,R)-ferrocenyl-1,1′-diamine 190
(S)-1-ferrocenyl-N,N-dimethylethylamine 285
ferrocenyl-1,1′-disulfonamide 190
(S)-2-ferrocenyl-4-(1-methylethyl)oxazoline 46
2-ferrocenyl-1,3-oxazoline 97, 343
4-ferrocenyl-1,3-oxazoline 343
ferrocenyl-oxazoline-carbinol 259
1,1-ferrocenyl-P,N-ligand 224
N-(1-ferrocenylalkyl) N-alkylephedrine 151
1,1′-ferrocenyldicarboxyl dichloride 192
ferrocenyldiphosphine 182, 208
ferrocenyldiketone 189
α-ferrocenylethyldimethylamine 44, 141
ferrocenylimine 344
ferrocenyloxazoline-derived P,N-ligand 132
ferrocenyloxazolinylphosphine (Fc-Phox) 101
– iridium complex 100
– synthesis 46
ferrocenylphosphine-gold(I) complex 230
ferrocenylphosphine-ketimine ligand 116
ferrocenylphosphine ligand 230
– sulfur-containing 230
ferrocenyltributyltin 68
(S,R_p)-FerroNPS 262
FerroPHOS 199
Fesulphos 25f., 257ff., 268ff.
– asymmetric ring opening 268
– Cu(I) complex 271ff.
– palladium 278

– (R_p) 273
Fesulphos dinaphthyl phosphine 271
FIP, see ferrocenyl imidazoline palladacycle
FOP, see ferrocenyl oxazoline palladacycle
Friedel–Crafts acylation reaction 6
fumarodinitrile 275

g
gold catalysis 268f.
– application 232
– asymmetric aldol reaction 229ff.
Grignard reagent 128ff., 266
Grubbs metathesis reaction 36

h
halogenation 316
Heck reaction
– asymmetric, see asymmetric Heck reaction
– intramolecular 79
– Pd-catalyzed 18
Hectorite 237
(–)-hennoxazole A 311
heterocyclopentadienyl 307
(E)-2-hexadecenal 233
homoallylsilane 299
homocoupling
– copper-catalyzed 22
(–)-huperzine A 229
hydrazoic acid 314
hydrazone 69, 347
hydroboration 202, 238
– asymmetric 118ff.
– rhodium catalyzed 118ff.
hydroesterification
– asymmetric 138, 249
– palladium-catalyzed 249
hydroformylation 178
– asymmetric 178f.
hydrogen bonding
– change of transition state 133
hydrogenation
– asymmetric, see asymmetric hydrogenation
– BoPhoz 84
– enantioselective 77, 88, 177ff., 200
– iridium-catalyzed 78
– palladium-catalyzed 188
– rhodium-catalyzed 78ff., 91, 177ff., 197, 235ff.
– ruthenium catalyzed 77
– TRAP-catalyzed 90f.
hydrosilylation
– asymmetric, see asymmetric hydrosilylation
– palladium-catalyzed 326
– rhodium-catalyzed 234f.

– titanium-catalyzed 202
α-hydroxy ester 321
β-hydroxy-α-amino acid 293
(2S,3R,4R,6E)-3-hydroxy-4-methyl-2-
  (methylamino)oct-6-enoic acid (MeBmt)
  232
(2S)-N-(1-hydroxy-3-methylbutyl)
  ferrocenamide 142
β-hydroxyamino acid
– optically active 230
1-hydroxymethyl-2-dimethyl
  aminomethylferrocene 154

i

Ikeda's ligand 176
2-imidazolidinone
– N-tosyl vinyl substituted 364
imidazoline 18
– 2-imidazoline 231
– pentaphenylferrocene-derived 344
– N-sulfonyl-substituted 344
imidazopyridine
– tricyclic 102
imine 104
– activated 188
– acyclic 188
– α-amino alcohol-derived 347
– asymmetric 1,2-addition 270
– asymmetric hydrogenation 104f., 203
– asymmetric hydrosilylation 116f., 270
– N-sulfonated 139
imine-first mechanism 318
iminoester 135
(iminophosphoranyl)ferrocene 99, 227
– chiral 238ff.
indole 263
– 2-substituted 296
– 3-substituted 297
indoline
– kinetic resolution 314
induction
– asymmetric 159
insertion
– copper-catalyzed 320f.
2-iodo ferrocenylphosphine oxide 22
1-iodo-2-substituted ferrocene 150
iodobenzene 223
iodoferrocene 286
– 2-iodoferrocene 35
– planar chiral 300
2-iodoferrocenecarboxaldehyde 348
iridium ferrocenyloxazolinylphosphine (Fc-
  Phox) complex 100
iridium-catalysis 114

– allylic alkylation 126
– asymmetric Diels Alder 137
– asymmetric hydrogenation 188, 203
– asymmetric hydrosilylation 117
– asymmetric reductive amination 188
– hydrogenation 78, 98ff.
isocyano ester 269
α-isocyano Weinreb amide 234
isocyanoacetate 229ff.
isomerization
– asymmetric 328f.
isopropyl 2-cyanopropionate 290
isopropylferrocene 36
isopropylsulfinylferrocene 70
itaconate 90, 296
itaconic acid derivative 77, 88
itaconic ester 237

j

Johnson–Corey epoxidation
– asymmetric 277
Josiphos 21, 76ff.
– analog 82
– copper-catalyst 79
– dendrimer-bound 82
– derivative 77
– functionalized 76ff.
– immobilized 80f.
– ionic liquid and water soluble 82
– naming of derivative 77
– polymer-bound 82
– SiO_2-bound 82
– structure 77ff.
– transformation 81

k

ketene 314ff.
– acylation 312
ketene-first mechanism 318
P-ketimine 116
β-keto ester 228
ketone
– allylic substitution 227
– asymmetric hydrogenation 101ff., 183, 236
– asymmetric hydrosilylation 112ff., 270,
  291, 324
– asymmetric reductive amination 188
– asymmetric transfer hydrogenation 107f.
– hydrosilylation 202
– Mannich-type addition 271
– nickel-catalyzed asymmetric reductive
  coupling 60
– prochiral aromatic 103
– reduction 324

– α,β-unsaturated   90
ketone enolate   226
α-ketonitrile   316
kinetic resolution   17, 40ff., 68f.
– enzymatic   16f., 40
– non-enzymatic   17, 41
– oxidative   139
Kinugasa reaction
– asymmetric   322
– asymmetric intramolecular   330
Kumada coupling   80
– nickel-catalyzed   239

l

β-lactam   322ff.
– cis   317
– trans   317
β-lactone   319
(+)-lasubines I and II   273f.
(S)-leucine   342
ligand effect   61
lipase   40ff.
lithiation
– asymmetric   39
– diastereoselective   33ff., 65, 343ff.
– matched pair   39
– ortho, see ortho-lithiation
– reductive   33
2-lithio-N,N-
    dimethylaminomethylferrocene   355
lithioferrocene   7, 16ff., 286, 302
lithium amide base
– chiral   37
lithium-bromide exchange   30

m

MAA   90
MAC   90
macrocycle   320
Mannich reaction   271
– asymmetric induction   159
– asymmetric Mannich-type reaction   272
matched pair lithiation   39
matched–mismatched pair interaction   39
menthyl p-toluenesulfinate   24
(–)-menthylferrocenesulfonate   39f.
mercuration   340
mercury-catalysis
– rearrangement of allylic imidate   361
metal
– complexation   197
metal complex
    – twist angle   207
metalation   339

– ortho   339
metalation reaction   7
metallacycles   337
– metallocene-based   339
– planar chiral   341
metallocene   1f.
metallocycles
– configuration   338
– planar chiral   338
metallocyclic complex
– asymmetric synthesis   339
– planar chiral   339
– scalemic planar chiral   354
metallocyclic ferrocenyl ligand   337ff.
methacrolein   287
N-methoxyl-N-methylamide   288
α-methoxybenzyl DMG   33f.
trans-4-(methoxycarbonyl)-5-(E)-1-
    pentadecenyl)-2-oxazoline   233
4-(methoxycarbonyl)-5-phenyl-2-oxazoline
    230
N-methoxycarbonyl-2-pyrroline   244
methoxycarbonylation
– palladium-catalyzed   80, 137ff.
methoxycyclization   268
(S)-(2-methoxymethylpyrrolidin-1-yl)
    ferrocene   22
N-(4-methoxyphenyl)
    trifluoroacetimidate   359
(E)-2-[3-(3-methoxypropoxy)-4-
    methoxybenzylidene]-3-methylbutanoic
    acid   180
methyl α-acetamidoacrylate derivative   238
methyl (Z)-α-acetamidocinnamate   237
methyl acetylacetate   221
methyl (Z)-2-(N-acetylamino)cinnamate   293
methyl acrolein   275
methyl acrylate   221, 275
methyl α-isocyanoacetate   233
methyl ketone   291
methyl phosphinite borane   56, 177
methyl 2-tributylstannylpropanoate   241
methyl vinyl ketone   221
(±)-methyl-2-(benzamidomethyl)-3-
    oxobutanoate   176
methylamine
– α-substituted   19
– γ-chiral   22
methylarylketone   109
p-methylbenzyladehyde   163
(E)-2-methylcinnamic acid   238
2-methylene-1,3-propanediol diacetate   228
methylenecyclopentane   221f.
O-methylephedrine

– derivative 23
– diastereoselective DoM 31
– diastereoselective ortho-lithiation 24
S-metolachlor 44
(R,R)-(S$_p$,S$_p$)-MeTRAP 295
Michael addition 79f., 91
– asymmetric 287ff.
– enantioselective 80
Mitsunobu cyclization 103
mnemonic rule 41
monolithioferrocene 300
monooxazoline ligand
– C_1-symmetrical 192
MOP 3
MOPF 25f.
Mucor miehei 41
(–)-cis-myrtanylamine 346

n
2-naphthyl triflate 267
(R)-1-naphthylethylamine 346
1-naphthylphosphine 273
neomenthyldiphenylphosphine 61
nickel-catalysis
– addition of trimethylaluminum to benzaldehyde 66
– asymmetric reductive coupling 55ff.
– [2+2+2]cocyclization 249
– ethylene polymerization 323
– iminoazaferrocene 323
nitroalkene 135
2-nitrocycloketone 222
nitrostyrene 275
norbornene 110ff.
NTs-indole 90
nucleophile
– prochiral 222ff.
– asymmetric induction 159
1,2- and 1,4-nucleophilic addition
– asymmetric 270
nucleophilic catalysis
– asymmetric 66, 276

o
octaethyldiphosphaferrocene 331
1-octene 178
olefin
– copper-catalyzed cyclopropanation 319
– cyclic tetrasubstituted 237
– functionalized 183
– polymerization 322
organoaluminum
– addition to aldehydes and enones 65
organometallic reagent

– addition to C=X bond 129
organosilanol 164
organotrifluoroborate 79
organozinc 328
– addition 163
– enantioselective 1,2-addition to aldehyde 149
ortho-diphenylphosphino group
– diastereoselective introduction 23
ortho-functionalization 36
ortho-lithiation 302
– chiral lithium amide base mediated 37
– diastereoselective 21ff., 35, 64, 286
– enantioselective 38f.
– 1,2-P,N-ferrocene ligand 98
– O-methylephedrine 24
– (–)Sparteine-mediated 37f., 200
– (R,R)-TMCDA-mediated 39
– Ugi's amine 21
ortho-lithiation-functionalization
– diastereoselective 27
ortho-magnesiation 34
Overman rearrangement
– asymmetric 8
oxabenzonorbornadiene 247f., 267
[2.2.1]-oxabicyclic alkene 131, 267
[3.2.1]-oxabicyclic alkene 131
meso oxabicyclic alkene 267
oxazaphospholidine borane 56
oxazaphospholidine-oxide 18
– diastereoselective ortho-lithiation 35
oxazolidinone 221
– 2-oxazolidinone 364
oxazoline 19, 43
– chiral 29
– 4,5-disubstituted 230
– DMG 30
– (S)-tert-leucine-derived 353
– trans-2-oxazoline 231
– trans-oxazoline 232
– (S)-valine-derived 353
1-oxazolinyl-1'-(phenylthio)ferrocene 259
oxazolinylferrocenyl phosphine 106f.
oxetane 320
oxime
– asymmetric hydrosilylation 118
β-oxo-α-(acetamido)acrylates 294
β-oxo-α-(acetamido)carboxylates 294

p
P–C cross coupling 184
P-Phos 4
1,1'-P,N ligands 32
palladacycle 337, 352ff.

- bridged 346
- bridged planar chiral 340
- halogen-bridged 362
- ketimine-based 342
- scalemic planar chiral 354
- stereoisomer 340
palladacyclic ligand 8
palladation 339ff.
- enantioselective 352
palladium
- π-allyl complex 223, 259ff., 287ff.
- [(η^3-C$_3$H$_5$)PdCl]$_2$ 326
- Pd(0) 289, 348ff.
- Pd(II) 196ff., 356
- [Pd(π-C$_3$H$_5$)(cod)]BF$_4$ 287
- Pd(π-C$_3$H$_5$)(Cp) 290
palladium catalysis 80, 111, 356
- allylation 91, 222, 290
- allylic alkylation 217ff., 278
- allylic amination 217ff.
- allylic substitution 225ff., 258ff.
- arylation 139
- α-arylation 241
- asymmetric alkylative ring opening 267
- asymmetric allylic alkylation 196, 206, 326ff.
- asymmetric allylic substitution 121ff., 178, 197ff., 215ff., 301
- asymmetric cross coupling 194, 205, 240f.
- asymmetric Heck reaction 243ff.
- asymmetric ring opening 268, 278
- cross-coupling 331
- cyclization 221
- ethylene polymerization 323
- Fc-PHOX 128
- Fesulphos 278
- hydrogenation 188
- hydrosilylation 326
- intramolecular allylic alkylation 221
- methoxycarbonylation 137ff.
- rearrangement of allylic imidate 361
- ring opening 79
- SiocPhox 225ff.
- Suzuki–Miyaura reaction 241
Pauson–Khand reaction 249
pentaphenylferrocene derivative 344ff.
pentaphenylferrocene palladacycle 359
pericyclic [3.3]-sigmatropic rearrangement 356
phenyl addition 165f.
phenyl boronic acid 331
phenyl boronic ester 166
N-phenyl maleimide 275f.

phenyl transfer 162ff.
- aldehyde with ferrocene 171
- aldehyde with triphenylborane 171
- asymmetric 161ff.
(S)-α-phenyl-η-ferrocenylbutyric acid 19
phenylacetylene 165
- addition to aldehyde with ferrocene 170
- diethylzinc-mediated enantioselective addition 165
α-phenylacrylic acid 236
- β-disubstituted 236
1-phenylethylmagnesium chloride 194ff.
phenyloxazoline i-Pr-Phox 133
phenyltriflate 267
phosphacyclopentadienyl ligand 309
phosphaferrocene 309ff.
- bidentate 327
- catalysis 324ff.
- methyl-substituted 328
- planar chiral 309
phosphaferrocene–oxazoline 328ff.
phosphaferrocene–pyridine 326
phosphapalladacycle 341ff.
- scalemic 353
phospharuthenocene 326
phosphetane
- 2,4-disubstituted 183
phosphinamide borane 56
phosphinamidite-thioether ligand 262
- bidentate 262
phosphine 57
- P-chiral 22
- prochiral 352
- recognition 354
phosphine ferrocenyl ligand
- 1,1'-disubstituted P-chiral bidentate 177
phosphine oxide 285f., 300
- DMG 43
- P-chiral 34
phosphine-gold complex 269
P,N,N-phosphine-heteroaryl imine ligand 136
- ferrocene-derived 136
phosphine-hydrazone ligand 220
phosphine-palladium-complex 110
phosphine-triazine ligand 122
phosphinite-boranes 300
phosphinite-oxazoline ligand 98ff.
phosphinite-oxazolinyl ferrocene 99
1-phosphino 2-sulfenyl ferrocene (Fesulphos ligand) 263
2-phosphino 1-sulfenyl ferrocene 257
- chiral planar 257
phosphino-ferrocene carboxylic acid

– planar chiral 223
phosphinocyrhetrenyloxazoline ligand 115
(phosphinoferrocenyl)oxazoline 101
phosphinooxazoline P,N ligand
– chiral 133
phosphinooxazoline pentamethylferrocene 125
phosphinylimine 77
phosphate
– 1,1-biphenyl-derived 66
phosphorodiamidites
– chiral ferrocenyl 35
– diastereoselective lithiation 35
Phox 3, 267
– ferrocene substituted 122
4-phthalimidobutanal 234
PhTRAP 287
– (R,R)-(S,S)- 284
Pigiphos 21
pinacol borane 332
PingFer 22, 86
planar chiral ferrocene 15
– enantioselective construction 18
– stereoselective synthesis 15ff.
planar chirality 15
– asymmetric allylic substitution 125
– asymmetric induction 159
– configuration notation 16
– double asymmetric induction 40
– FcPHOX 125
– pseudo-C_2-symmetric ligand 264
platinacycle 342ff.
platination 340ff.
polyethylene glycol monomethyl ether (MeO-PEG-OH, MPEG) 163
polymethylhydrosiloxane (PMHS) 78, 270
1,3-polyol 363
PPF-OMe 128
(R,S_p)-PPF-pyrrolidine 139
PPFA, see N,N-dimethyl-1-[2-(diphenylphosphino)ferrocenyl]ethylamine
proline 298
– (S) 341
2-propenyl acetate
– symmetrical 1,3-disubstituted 216
– unsymmetrical 1,3-disubstituted 224
propiophenone 292
prostaglandine 355
protein kinase C inhibitor 234
PrTRAP 291
– [Rh(cod)$_2$]BF$_4$–(R,R)-(S_p,S_p) 294
– [Rh(nbd)$_2$]PF$_6$–(S,S)-(R_p,R_p) 295
pyrazolidinone 331
pyrazolyl-P,N-ferrocenyl ligand 119f.

pyridine N-oxide
– planar chiral 316
1-pyridine-2-yl methanamine
– R-substituted (RPyme) 78
pyrrole 297f.
– 2,3,5-trisubstituted 298
pyrrolidine 275
pyrrolidinone 364
4-pyrrolidinopyridine nucleophilic catalysts
– ferrocene-based C_2-symmetric 68
(S)-1-(2-pyrrolidinylmethyl) pyrrolidine 28

q
quinoline 104

r
rearrangement
– O-acylated azlactone 68
– allylic imidate 356ff.
– pericyclic [3.3]-sigmatropic 356
– [1,2]-Wittig 33
reduction 325
– activated C=C bond 78
– asymmetric 269
– copper-catalyzed 90
– enantioselective 79, 90
– rhodium catalyzed 324
renin inhibitor 89, 180
resolution 17, 341
– kinetic 68f.
– lipase-catalyzed 40
– nonenzymatic kinetic 41
– stoichiometric chemical 41
reverse reagent combination 167
rhodium catalysis
– alkyl-TRAP complex 91
– allylic alkylation 228
– asymmetric alcoholysis 247
– asymmetric allylation 290
– asymmetric Diels Alder 137
– asymmetric hydroboration 118ff., 238
– asymmetric hydrogenation 178ff., 202, 235ff., 269, 295ff.
– asymmetric hydrosilylation 113f., 234f., 291, 324
– asymmetric isomerization 328f.
– asymmetric Michael addition 287
– BoPhoz 84
– enantioselective hydrogenation 177ff., 200
– hydrogenation 78ff., 99, 186, 197, 278
– phosphine complex 237
– PhTRAP 288
– reduction 324
– Rh(acac)(CO)$_2$ 288ff.

- [Rh(cod)$_2$]BF$_4$ 291
- [Rh(cod)$_2$]BF$_4$–(R,R)-(S$_p$,S$_p$)-BuTRAP   293
- RhH(CO)(PPh$_3$)$_3$ 287
- [Rh(nbd)$_2$]SbF$_6$ 296
- ring opening   248
- Walphos   88
ring closing metathesis
- asymmetric, see asymmetric ring closing metathesis
ring opening
- asymmetric   268ff.
- nucleophilic   79
- palladium-catalyzed   79, 131
- rhodium-catalyzed   247f.
Rphos   6
ruthenium
- asymmetric hydrosilylation   118
- Fc-phox complex   101
- RuCl(p-cymene)   287
- [RuCl(p-cymene){(S,S)-(R$_p$,R$_p$)-PhTRAP] Cl   297
- mer-trans-[RuCl$_2$[P(MeO)$_3$]$_2$ {K^2(P,N)-FcPN}]   108f.
- RuCl$_2$(PPh$_3$)$_3$ 106
- fac-[RuCl$_2$(PTA)$_2$ {K^2(P,N)-FcPN}]   108f.
- [RuI$_2$(p-cymene)]$_2$ 301
- Ru(π-methallyl)$_2$(cod)   297
ruthenium catalysis
- asymmetric cyclopropanation   136, 248
- asymmetric hydrogenation   102, 297f.
- asymmetric transfer hydrogenation   106f.
- enantioselective hydrogenation   77
- oxidative kinetic resolution   139
- Ru-phox   107
ruthenocene
- chiral   206
- Cp ring   204
    - C$_2$-symmetrical chiral   203
- tetrasubstituted   205
1,1′-ruthenocenyl dicarbonyl dichloride   206
ruthenocenyl diphosphine ligand
- C$_2$-symmetrical   206
ruthenocenyl ligand
- symmetrical 1,1′-bidentate   203
- C$_2$-symmetrical 1,1′-bidentate   205
ruthenocene P,N-chelating ligand
    - C$_2$-symmetrical tetrasubstituted   206
1,1′- ruthenocenyl ligand
- C$_2$-symmetrical chiral   206

s

Salen system   3
SAMP ((S)-1-amino-2-(methoxymethyl)-pyrrolidine)   31, 43, 220, 346

- FcSAMP   31
- hydrazone   32
Schlögl's rule   16
SDP   4
Segphos   4
selectivity factor   311
selenide ferrocene ligand   266
(3S,4R)-selenoalcohol   153
semicorrin   320
L-serine methyl ester hydrochloride   192
Sharpless asymmetric dihydroxylation (AD) reaction   41
silicon trick   19
- desilylation   30
silver catalysis
- asymmetric aldol reaction   232
- asymmetric cycloaddition   133f., 276
- 1,3-dipolar cycloaddition   132
silyl enol ether   271
silyl enolate   294
silyl ketene acetal   315f.
silyl ketene imine   316
silylcyanation
- yttrium-catalyzed   28
SiocPhox (2-{1-[1′-(2-oxazolinyl)ferrocenyl]}(diethylamino)phosphinooxy-1,1′-binaphthyl-2′-oL)   224ff.
- quaternary carbon center   226
SIPHOX   4
sodium enolate   222
(+)-sparteine   36
(–)-sparteine   36ff.
- enantioselective ortho-lithiation   38
- ortho-lithiation   37
- ortho-metalation   200
sphingosine
- D-erythro   233
- D-threo   233
Sphos   6
spirobiindane   4
SIPIROBOX   4
SPINOL   4
SPP100   89
Staudinger reaction   317f.
stereoselective synthesis
- planar chiral ferrocene   15ff.
trans-stilbene oxide   277
stoichiometric chemical resolution   41
styrene   80, 110ff., 137, 302
- asymmetric hydroformylation   178f.
- asymmetric hydroesterification   249
substitution
- allylic, see allylic substitution

– aromatic electrophilic 6
– asymmetric allylic, *see* asymmetric allylic substitution
– nucleophilic 300
sulfide ferrocenyl bis-phosphine ligand 269
sulfinyl ferrocene ligand 265, 276
1-sulfinyl 2-sulfonamido ferrocene ligand 271
N-sulfonyl aldimine
– ketone, ester and thioester 271
N-sulfonyl imine 231, 272ff.
N-sulfonylated-α-dehydroamino acid 77
sulfoxide 19, 269
– chiral 23
sulfoxide-lithium exchange 26
sulfoximines 18, 43
Suzuki coupling 129
Suzuki reaction 332
Suzuki–Miyaura cross coupling
– asymmetric 64f.
Suzuki–Miyaura reaction 241
– palladium-catalyzed 241

t
TADDOL (2,2-dimethyl-α,α,α′,α′-tetraaryl-1,3-dioxolane-4,5-dimethanol) 3, 272
Taniaphos 22ff., 85ff.
– copper complex 86
– structure 86
– synthetic application 87
(–)-terpestacin 58
2,2,6,6-tetrachlorocyclohexanone 317
tetrahydrofuran 320
(*S*)-tetrahydrofurfurylamine 346
1,4,5,6-tetrahydropyrazine 295
1,1′-*N,N,N′,N′*-tetraisopropyl ferrocene dicarboxamide 36
α-tetralone 107
N,N,N′,N′-tetramethyl-(1*R*,2*R*)-cyclohexanediamine [(*R*,*R*)-TMCDA] 38
– asymmetric lithiation of (dimethylaminomethyl)ferrocene 39
2-thienylsulfonyl imines 271
α-thio-aldehyde 154
(3*S*,4*R*)-thioalcohol 153
thiolate ferrocene ligand 266
2-thiophenecarbaldehyde 167
thrombin fibrinolysis
– inhibitor 183ff.
titanium-catalysis
– enantioselective addition 190
– hydrosilylation 202
(*R*,*R*)-TMCDA, see *N,N,N′,N′*-tetramethyl-(1*R*,2*R*)-cyclohexanediamine

(*S*,*S*)-TMCDA 38
tocopherol 229
Tol-BINAP 267ff.
o-TOL-Fesulphos 273
p-tolyl sulfoxide 24
N-tosyl 2,3-dihydropyridone 274
N-tosyl FIP 359
N-tosyl imine 274, 317
N-tosylindole 297
tosylmethyl isocyanide 232
transcyclometalation 352
– asymmetric 352
transfer hydrogenation 79
– aryl ketone 101
– asymmetric 106ff.
– osmium-catalyzed 79
– ruthenium-catalyzed 79
transition-metal catalysis 319
transmetalation 339
– inverse palladium to mercury 350
– oxidative 350
TRAP (*trans*-chelating phosphine) 21f., 90f., 283ff.
– *cis*-chelating TRAP–metal complex 287
– asymmetric hydrogenation 292
– metal complex 286
– *trans*-[PdBr$_2$(TRAP)] 286
– *trans*-[PtCl$_2$(*i*-PrTRAP)] 286
– rhodium-catalyzed 90
– (*R*,*R*)-(*S*$_p$,*S*$_p$) 285
– (*S*,*S*)-(*R*$_p$,*R*$_p$) 284ff.
– structure 284
– synthesis 285
TRAP-H 286
trialkylphosphine 285
tricarbonylchromium complex
– planar chiral 139
(*E*)-trichloroacetimidate 360ff.
trichloroacetimide 356
TriFer 180
trifluoroacetate salt 202
trifluoroacetimidate 360ff.
triethylaluminum 67
N-triflyl imine 317
trimethylaluminum 66
trimethylchlorosilane (TMSCl) 19
trimethylsilyl (TMS) 19
– trick 43
α-(trimethylsilyl)benzyl-Grignard reagent 241
2-trimethylsilylethanol 321
triphenylborane 166f.
– aryl transfer 166
– phenyl transfer 171

Tsuji–Trost reaction 303
Tunaphos 4
twist angle
– metal complex 207

u

Ugi's amine (*N,N*-dimethylethylamino-
   ferrocene) 6, 17ff., 150, 342f.
– diastereoselective *ortho*-lithiation 21
– methoxy analogues 43
– structure 74
– synthesis 99, 140
Ugi–Kagan ferrocenylmethylamine 43
Ullman coupling 286, 300

v

vanadium catalysis
– asymmetric epoxidation 168ff.
(*S*)-verapamil 316
vinyl bromide 80, 128, 205

vinyl tosylate 80
3-vinyl-cyclohexanone 221
vinyl-tetrahydro-2*H*-pyran derivative 223
vinylchroman 229
vinylcyclopropane 222
2-vinylcyclopropane-1,1-dicarboxylate 221
1-vinylferrocene
– 2-substituted 41
vinyloxazolidine 221

w

Walphos 21f., 88ff.
– derivative 88
– rhodium catalyst 88
– synthetic application 89
Weinreb amide 288
[1,2]-Wittig rearrangement 33

z

zinc addition 244